P9-CKY-449

Radio Propagation and Remote Sensing of the Environment

N. A. Armand

and

V. M. Polyakov

CRC PRESS

Boca Raton London New York Washington, D.C.

Library of Congress Cataloging-in-Publication Data

Armand, N. A.
Radio propagation and remote sensing of the environment / N.A. Armand, V.M. Polyakov.
p. cm.
Includes bibliographical references and index.
ISBN 0-415-31735-5 (alk. paper)
1. Radio wave propagation. 2. Earth sciences—Remote sensing. I. Poliakov, Valerii Mikhailovich. II. Title.

TK6553.A675 2004
621.36'78—dc22

2004047816

Direct all inquiries to CRC Press, 2000 N.W. Corporate Blvd., Boca Raton, Florida 33431.

Visit the CRC Press Web site at www.crcpress.com

International Standard Book Number 0-415-31735-5
Library of Congress Card Number 2004047816
Printed in the United States of America 1 2 3 4 5 6 7 8 9 0
Printed on acid-free paper

Contents

Introduction

Airborne instruments designed for remote sensing of the surface of the Earth and its atmosphere are important sources of information regarding the processes occurring on Earth. This information is used widely in the fields of meteorology, geography, geology, and oceanology, among other branches of the sciences. Also, the data gained by satellite observation have been applied to an increasing number of other areas, such as cartography, land surveying, agriculture, forestry, building construction, and protection of the environment, to name a few. Most developed countries have a space agency within their governmental organizations, a central task of which is development of remote sensing systems. Airborne and ground-based technologies for remote sensing are developed along with the space systems.

Recently, increased attention has been paid to development of microwave technology for remote sensing, particularly synthetic aperture radars and microwave radiometers. This interest is due primarily to two circumstances. The first is connected with the fact that spectral channels other than optical are considered, thus providing a new way to obtain additional information on the natural processes of the Earth and its atmosphere. The second circumstance is the transparency of clouds for radiowaves, which allows effective operation of radio systems regardless of weather. In addition, radiowave devices for remote sensing do not require illumination of the territory being observed so data can be collected at any moment of the day.

The information gained from data collected by microwave instruments depends on the medium being studied. Interpretation of these data is impossible without analyzing the various mechanisms of interaction that may be present. Such mechanisms are the primary scientific basis for designing any device of remote sensing, particularly with regard to choosing the frequency band, polarization, dynamic range, and sensitivity. Stating remote sensing problems requires addressing the principles of radio propagation and such processes as absorption, reflection, scattering, and so on. The interpretation algorithms for remote sensing data are properly based on these processes.

This book has generally been written in two parts based on the two circumstances just discussed. The first part describes the processes of radio propagation and the phenomena of absorption, refraction, reflection, and scattering. This discussion is intended to demonstrate determination of coupling between the radiowave parameters of amplitude, phase, frequency, and polarization and characteristics of the media (e.g., permittivity, shape). Solutions of well-posed problems provide a basis for estimation of the strength of various effects and demonstrate the importance of media parameters on the appearance of these effects, the possibility of detection of that or other effects against a background of noise and other masking phenomena, and so on.

It is necessary to point out that only rather simple models can be analyzed; therefore, the numerical relations between the observed effects and parameters of a medium itself and radiowaves are of primary importance. Only rarely can natural media be described by simple models, and we must rely on experimental data when determining quantitative relations. It is necessary to keep this in mind with regard to the problems and solutions provided here.

The second part of the book is dedicated to analysis of problems that used to be referred to as *inverse problems*. This analysis attempts to answer the question of how knowledge of radiowave properties allows us to estimate the parameters of the medium studied. Very often, inverse problems are ill posed; as a matter of fact, any measurement happens against a background of noise, and in all cases this concept of noise must be addressed sufficiently with regard to its additive interference with a signal. Inaccuracy of a model itself is also a factor to be considered. A typical peculiarity of ill-posed problems is their instability, manifested in the fact that a small error in the initial data (the data of measurement, in our case) can lead to a big error in the problem solution, in which case additional data must be inserted to remove the instability. These data are often referred to as *a priori*, and they bound possible solutions of the posed problem.

The approximation of models of many natural media requires developing empirical methods to interpret remote sensing data. Some of these methods are described in this book, together with brief descriptions of the operational principles of microwave devices used in remote sensing. Here, the authors do not intend to delve deeply into either the details of device construction or the algorithms of their data processing, as it is very difficult to do so within the limited framework of this book; therefore, only the principal fundamentals are presented.

The authors wish to thank the publisher for help in preparation of this book.

1 Electromagnetic Field Equations

1.1 MAXWELL'S EQUATIONS

It is well known that the electromagnetic field is generally described by Maxwell's equations as electric and magnetic fields **E** and **H**. Let us suppose the time dependence is $e^{-i\omega t}$. Such a notion of time dependence is especially convenient for narrowband oscillations — that is, for such oscillations that have a spectrum close to the assigned frequency ω and often referred to as the *carrier*. In this case, the conditions of complex oscillation propagation are practically the same as the propagation conditions for the time-harmonic oscillation of the carrier frequency. This accepted supposition is also advantageous for broad-bandwidth oscillations. The time-harmonic oscillation should only be considered as one of the harmonic components of oscillation (Fourier's theorem). Later on, magnetic media will not be involved, so permeability is equal to unity. On this basis, Maxwell's equations may be written as:

$$\nabla\times\mathbf{E}=ik\mathbf{H},\qquad \nabla\cdot\mathbf{H}=0,\qquad \nabla\times\mathbf{H}=-ik\mathbf{D}+\frac{4\pi}{c}\mathbf{j},\qquad \nabla\cdot\mathbf{D}=4\pi\rho \tag{1.1}$$

in the Gaussian system of units. Here, $k = \omega/c = 2\pi/\lambda$ is the wave number, where ω is the cyclic frequency and $c = 3\cdot 10^8$ cm/sec is the light velocity; **D** is the electric induction vector; and **j** is the external current density. The continuity equation resulted from Equation (1.1) is defined as:

$$i\omega\rho=\nabla\cdot\mathbf{j}\,. \tag{1.2}$$

Material equations connecting **E** and **D** vectors are now introduced. In the case of an isotropic medium, this relation is given as:

$$\mathbf{D}(\mathbf{r})=\varepsilon(\omega,\mathbf{r})\mathbf{E}(\mathbf{r}), \tag{1.3}$$

where $\varepsilon(\omega,\mathbf{r})$ is the permittivity of the medium, which, in general, is a function of frequency ω and coordinates defined by vector **r**. This local dependence on the coordinates of **r** means that spatial dispersion is not taken into account.

The permittivity is a complex value; that is,

$$\varepsilon(\omega)=\varepsilon'(\omega)+i\varepsilon''(\omega), \tag{1.4}$$

where the coordinate dependence is omitted. Here, $\varepsilon''(\omega)$ describes Joule losses in the medium. In particular, if for static conductivity σ, the corresponding component of the imaginary part is:

$$\varepsilon''_{\sigma} = \frac{4\pi\sigma}{\omega} . \tag{1.5}$$

In an anisotropic medium (for example, in the ionosphere), due to the magnetic field of the Earth, a connection such as Equation (1.3) is substituted for the tensor:

$$D_{\alpha} = \varepsilon_{\alpha\beta} E_{\beta} , \tag{1.6}$$

which is a summation over the repeating indexes. The tensor components $\varepsilon_{\alpha\beta}$ are complex functions of frequency (and coordinates in the general case).

It is necessary to input boundary conditions at media boundaries. Let **n** be the unitary normal to the surface (referred to as the *normal* in this text). The boundary conditions usually described are the continuities of the tangential field components:

$$\left[\mathbf{n}\times(\mathbf{E}_1 - \mathbf{E}_2)\right] = 0, \quad \left[\mathbf{n}\times(\mathbf{H}_1 - \mathbf{H}_2)\right] = 0 . \tag{1.7}$$

Here, $\mathbf{E}_1$, $\mathbf{E}_2$, $\mathbf{H}_1$, and $\mathbf{H}_2$ are field components on either side of the boundary. In some cases, we will come across problems when the tangential field components are broken due to the electric and magnetic surface currents of densities $\mathbf{K}_e$ and $\mathbf{K}_m$; that is,

$$\left[\mathbf{n}\times(\mathbf{E}_1 - \mathbf{E}_2)\right] = -\frac{4\pi}{c}\mathbf{K}_m, \quad \left[\mathbf{n}\times(\mathbf{H}_1 - \mathbf{H}_2)\right] = \frac{4\pi}{c}\mathbf{K}_e . \tag{1.8}$$

To the boundary conditions, as shown in Equation (1.7), we must add the conditions of radiation, and only divergent (going away) waves must equal infinity.

Sometimes, it is more convenient to use gap-type boundary conditions:

$$\left(\mathbf{n}\cdot(\varepsilon_1\mathbf{E}_1 - \varepsilon_2\mathbf{E}_2)\right) = 0, \quad \left(\mathbf{n}\cdot(\mathbf{H}_1 - \mathbf{H}_2)\right) = 0, \tag{1.9}$$

where ε_1 and ε_2 are the permittivities of media divided by the boundary concerned. They are equivalent to the conditions shown in Equation (1.7), and our use of them is only a matter of convenience. The maintenance conditions (Equations (1.7) and (1.9)) for field **H** are the equality magnetic fields on both sides of the boundary; that is,

$$\mathbf{H}_1 = \mathbf{H}_2 . \tag{1.10}$$

In the case of surface currents, it is necessary to insert surface charges δ_e and δ_m relative to the currents using the continuity equation:

$$i\omega\delta_e = \nabla\cdot\mathbf{K}_e, \quad i\omega\delta_m = \nabla\cdot\mathbf{K}_m . \tag{1.11}$$

According to this, we can then use the following boundary conditions:

$$\left(\mathbf{n}\cdot(\varepsilon_1\mathbf{E}_1 - \varepsilon_2\mathbf{E}_2)\right) = 4\pi\delta_e, \quad \left(\mathbf{n}\cdot(\mathbf{H}_1 - \mathbf{H}_2)\right) = 4\pi\delta_m . \tag{1.12}$$

Note that it is possible to formulate an independent equation for electric and magnetic fields. Here, we assumed permittivity coordinate independence and an absence of radiation sources in the area of space being considered. To achieve this aim, it is sufficient to use the ∇ operator on the first and third equations in Equation (1.1) and to assume that the field divergence in both cases is equal to zero. Then, it is easy to show that both fields satisfy the equations:

$$\nabla^2\mathbf{E} + \varepsilon k^2\mathbf{E} = 0, \quad \nabla^2\mathbf{H} + \varepsilon k^2\mathbf{H} = 0 . \tag{1.13}$$

All coordinate components of fields satisfy similar equations in the Cartesian system of coordinates. In this case, these equations are called *wave equations*.

1.2 ENERGETIC RELATIONSHIPS

Let us now input energetic relationships characterizing propagation and absorption of electromagnetic field. In this case, we will be dealing with squared values, so it is necessary to set up rules of calculation for the complex values and harmonic dependence on time. Assume that:

$$a(t) = a_0\cos(\omega t + \varphi_a) \Rightarrow a_0 e^{-i\omega t - i\varphi_a}, \quad b(t) = b_0\cos(\omega t + \varphi_b) \Rightarrow b_0 e^{-i\omega t - i\varphi_b} , \tag{1.14}$$

where the pointers indicate agreement between the representations of oscillations in the real and complex forms. From here on, as is customary in the theory of harmonic signals, let us consider squared values as an average in time (we are reminded about the notion of effective voltage in electrical engineering). Then it is easy to show that this means:

$$\overline{a(t)b(t)} = \frac{a_0 b_0}{2}\cos(\varphi_a - \varphi_b) = \frac{1}{2}\mathrm{Re}\, a(t)\, b^*(t). \tag{1.15}$$

The line above the first term indicates averaging for the time interval of the period.

Thus, the vector of the power flow density (Poynting's vector **S**) is determined as:

$$\mathbf{S} = \frac{c}{8\pi}\,\mathrm{Re}\left[\mathbf{E}\times\mathbf{H}^{*}\right]. \tag{1.16}$$

The divergence of this density is described by the transport equation:

$$\nabla\cdot\mathbf{S} = \frac{\omega}{8\pi}\,\mathrm{Im}(\mathbf{E}\cdot\mathbf{D}^{*}) - \frac{1}{2}\,\mathrm{Re}(\mathbf{E}\cdot\mathbf{j}^{*}). \tag{1.17}$$

The value:

$$Q = -\frac{\omega}{8\pi}\,\mathrm{Im}(\mathbf{E}\cdot\mathbf{D}^{*}) \tag{1.18}$$

describes the mean for the period losses of the electromagnetic power density, and

$$W = \frac{1}{2}\,\mathrm{Re}(\mathbf{E}\cdot\mathbf{j}^{*}) \tag{1.19}$$

is the work done by external currents (per unit volume).

In isotropic media, when Equation (1.3) is valid:

$$Q = \frac{\omega\varepsilon''(\omega)}{8\pi}\left|\mathbf{E}\right|^{2}. \tag{1.20}$$

In anisotropic media,

$$Q = \frac{\omega}{8\pi}\,\mathrm{Im}\left(\varepsilon^{*}_{\alpha\beta}E_{\alpha}E^{*}_{\beta}\right). \tag{1.21}$$

So, Equation (1.17) describes the balance of generation and absorption of electromagnetic energy in unit volume.

1.3 SOLVING MAXWELL'S EQUATIONS FOR FREE SPACE

Let us now deal with the case of electromagnetic wave generation in space with $\varepsilon = 1$ and not limited by any bodies. It this case, it is convenient to use Fourier's transform technique. We can represent the fields, currents, and charges as integrals:

$$\mathbf{E}(\mathbf{r}) = \int \tilde{\mathbf{E}}(\mathbf{q})\, e^{i\mathbf{q}\cdot\mathbf{r}} d^{3}\mathbf{q}, \quad \mathbf{H}(\mathbf{r}) = \int \tilde{\mathbf{H}}(\mathbf{r})\, e^{i\mathbf{q}\cdot\mathbf{r}} d^{3}\mathbf{q}. \tag{1.22}$$

Inserting these into Equation (1.1), we obtain this system of algebraic equations:

$$\left[\mathbf{q}\times\tilde{\mathbf{E}}\right]=k\tilde{\mathbf{H}},\qquad \left[\mathbf{q}\times\tilde{\mathbf{H}}\right]=-k\tilde{\mathbf{E}}-\frac{4\pi i}{c}\tilde{\mathbf{j}}$$
$$(\mathbf{q}\cdot\tilde{\mathbf{E}})=-4\pi i\rho,\quad (\mathbf{q}\cdot\tilde{\mathbf{H}})=0,\quad \tilde{\rho}=\frac{1}{\omega}(\mathbf{q}\cdot\tilde{\mathbf{j}}). \tag{1.23}$$

It is easy then to obtain:

$$\tilde{\mathbf{E}}=\frac{4\pi i}{\omega}\frac{\left[k^2\tilde{\mathbf{j}}-\mathbf{q}\left(\mathbf{q}\cdot\tilde{\mathbf{j}}\right)\right]}{q^2-k^2},\quad \tilde{\mathbf{H}}=\frac{4\pi i}{c}\frac{\left[\mathbf{q}\times\tilde{\mathbf{j}}\right]}{q^2-k^2}. \tag{1.24}$$

First, we will calculate field **H** because it is a shorter calculation. In accordance with Equation (1.22),

$$\mathbf{H}(\mathbf{r})=\frac{4\pi i}{c}\int\frac{\left[\mathbf{q}\times\tilde{\mathbf{j}}(\mathbf{q})\right]}{q^2-k^2}e^{i(\mathbf{q}\cdot\mathbf{r})}d^3\mathbf{q}. \tag{1.25}$$

As vector $\tilde{\mathbf{j}}(\mathbf{q})$ does not depend on coordinates, the last expression can be represented by:

$$\mathbf{H}(\mathbf{r})=[\nabla\times\mathbf{A}], \tag{1.26}$$

where the vector potential is:

$$\mathbf{A}(\mathbf{r})=\frac{4\pi}{c}\int\frac{\tilde{\mathbf{j}}(\mathbf{q})}{q^2-k^2}e^{i(\mathbf{q}\cdot\mathbf{r})}d^3\mathbf{q}. \tag{1.27}$$

Vector $\tilde{\mathbf{j}}(\mathbf{q})$ represents the spatial spectrum of the current density, and, according to the Fourier transform theory,

$$\tilde{\mathbf{j}}(\mathbf{q})=\frac{1}{8\pi^3}\int\mathbf{j}(\mathbf{r}')e^{-i(\mathbf{q}\cdot\mathbf{r}')}d^3\mathbf{r}'. \tag{1.28}$$

Therefore

$$\mathbf{A}(\mathbf{r})=\frac{1}{c}\int\mathbf{j}(\mathbf{r}')g(\mathbf{r}-\mathbf{r}')\,d^3\mathbf{r}'. \tag{1.29}$$

Here,

$$g(\mathbf{R}) = \frac{1}{2\pi}\int \frac{e^{i(\mathbf{q}\cdot\mathbf{R})}}{q^2 - k^2} d^3q = \frac{e^{ikR}}{R} \tag{1.30}$$

is the Green function. As a result,

$$\mathbf{H}(\mathbf{r}) = \frac{1}{c}[\nabla_r \times \int \mathbf{j}(\mathbf{r}')g(\mathbf{r}-\mathbf{r}')\, d^3\mathbf{r}'] = -\frac{1}{c}\int \left[\mathbf{j}(\mathbf{r}') \times \nabla_r g(\mathbf{r}-\mathbf{r}')\right] d^3\mathbf{r}'. \tag{1.31}$$

The subscript r indicates that the operation of differentiation is taken with respect to the observed point coordinates.

The corresponding expression for **E** has the form:

$$\mathbf{E}(\mathbf{r}) = \frac{i}{k}\left[k^2\mathbf{A} + \nabla(\nabla\cdot\mathbf{A})\right]. \tag{1.32}$$

If we apply Equation (1.29) here, we obtain:

$$\mathbf{E}(\mathbf{r}) = \frac{i}{\omega}\int \left[k^2\mathbf{j}(\mathbf{r}')g(\mathbf{r}-\mathbf{r}') + \left(\mathbf{j}(\mathbf{r}')\cdot\nabla_r\right)\nabla_r g(\mathbf{r}-\mathbf{r}')\right] d^3\mathbf{r}'. \tag{1.33}$$

Notice that, because the Green function satisfies the wave equation:

$$\nabla^2 g + k^2 g = -4\pi\delta(\mathbf{r}) \tag{1.34}$$

where $\delta(\mathbf{r})$ is the delta-function, then the vector potential satisfies a similar equation:

$$\nabla^2\mathbf{A} + k^2\mathbf{A} = -\frac{4\pi}{c}\mathbf{j}(\mathbf{r}). \tag{1.35}$$

The expression obtained can be considerably simplified in the case of sufficient distance from the region where the radiation currents are found. In this zone, which is called the *far zone*, $r \gg r'$. Then, $R = |\mathbf{r}-\mathbf{r}'| \cong r - (\mathbf{r}\cdot\mathbf{r}')/r$ and

$$\mathbf{A} = -ik\mathbf{P}\frac{e^{ikr}}{r} \tag{1.36}$$

where the vector

$$\mathbf{P} = -\frac{1}{i\omega}\int \mathbf{j}(\mathbf{r}')\exp\left(-ik\frac{(\mathbf{r}\cdot\mathbf{r}')}{r}\right) d^3\mathbf{r}'. \tag{1.37}$$

If $k\mathrm{r} >> 1$ (*wave zone*), then operation ∇ is equivalent to multiplication on $\nabla \Rightarrow ik\mathbf{r}/\mathrm{r}$ such that:

$$\mathbf{E} = -k^2\left[\frac{\mathbf{r}}{\mathrm{r}}\times\left[\frac{\mathbf{r}}{\mathrm{r}}\times\mathbf{P}\right]\right]\frac{e^{ik\mathrm{r}}}{\mathrm{r}}, \quad \mathbf{H} = \left[\frac{\mathbf{r}}{\mathrm{r}}\times\mathbf{E}\right]. \tag{1.38}$$

This expression shows that, in the wave zone, vectors **E** and **H** are orthogonal to each other and to the radiation direction **r**/r, which means, in this case, that the source field has transverse waves in the wave zone.

Poynting's vector in the wave zone is:

$$\mathbf{S} = \frac{\mathrm{c}}{8\pi}\frac{\mathbf{r}}{\mathrm{r}}\left|\mathbf{E}\right|^2 = \frac{k^4\mathrm{c}}{8\pi}\frac{\mathbf{r}}{\mathrm{r}^3}\left[\left|\mathbf{P}\right|^2 - \left|\frac{\mathbf{r}}{\mathrm{r}}\cdot\mathbf{P}\right|^2\right]. \tag{1.39}$$

1.4 DIPOLE RADIATION

With the constraint of a small volume being occupied by the currents, precisely subject to the inequality $k\mathrm{r}' << 1$, the exponent in Equation (1.37) can be expanded as a series and in the first approximation:

$$\mathbf{P} = -\frac{1}{i\omega}\int \mathrm{j}(\mathbf{r}')\,d^3\mathbf{r}' + \frac{1}{\mathrm{c}}\int \mathrm{j}(\mathbf{r}')\left(\frac{\mathbf{r}\cdot\mathbf{r}'}{\mathrm{r}}\right)d^3\mathbf{r}'. \tag{1.40}$$

First, we will consider the first term of this formula, which is indicated here as **p**. It is easy to show that:[2]

$$\mathbf{p} = \int \mathbf{r}'\rho(\mathbf{r}')\,d^3\mathbf{r}'. \tag{1.41}$$

The latter equation is the definition of the electrical dipole moment of the charges system in the investigated volume. So, the first term of Equation (1.40) describes the dipole radiation of the currents system.

The simplest model of electrical dipole is known to be a short compared to wavelength wire segment with length l and current J. Then,

$$\mathrm{p} = -\frac{Jl}{i\omega}. \tag{1.42}$$

Let us use χ to represent the angle between the dipole direction and the direction of radiation, which is defined by Poynting's vector direction. The angular power flow distribution at that point will be described by the formula:

$$S = \frac{k^4 c}{8\pi^2 r^2} |\mathbf{p}|^2 \sin^2 \chi . \tag{1.43}$$

The total power of radiation is given by integrating solid angle Ω:

$$W = \oint_{4\pi} \left(\mathbf{S} \cdot \frac{\mathbf{r}}{r} \right) r^2 d\Omega = \frac{k^4 c}{3} |\mathbf{p}|^2 . \tag{1.44}$$

By comparing this with the power flow of an isotropic radiator:

$$S_{is} = \frac{W}{4\pi r^2} \tag{1.45}$$

we obtain the directivity of the dipole:

$$G(\theta) = \frac{S}{S_{is}} = \frac{3}{2} \cos^2 \theta . \tag{1.46}$$

Here, $\theta = \pi/2 - \chi$. The value $\theta = 0$ corresponds to the radiation maximum, and it is easy to conclude that $G(0) = 1.5$ for the dipole.

The second term in Equation (1.40) becomes the main one for the system studied here, for which the electrical dipole moment is zero or is sufficiently small. First, we note that:

$$(\mathbf{r} \cdot \mathbf{r}')\mathbf{j} = -\frac{1}{2}\left[\mathbf{r} \times [\mathbf{r}' \times \mathbf{j}]\right] + \frac{1}{2}\left[(\mathbf{r} \cdot \mathbf{r}')\mathbf{j} + (\mathbf{r} \cdot \mathbf{j})\mathbf{r}'\right].$$

Now we introduce the magnetic dipole moment:

$$\mathbf{m} = \frac{1}{2c} \int \left[\mathbf{r}' \times \mathbf{j}(\mathbf{r}')\right] d^3\mathbf{r}' . \tag{1.47}$$

Then, the second item in Equation (1.40) can be rewritten as:

$$\mathbf{p}' = -\left[\frac{\mathbf{r} \times \mathbf{m}}{r}\right] + \mathbf{p}^{(2)} . \tag{1.48}$$

The first component in Equation (1.48) describes the field of the magnetic dipole. The simplest model of a magnetic dipole is a plain loop with current. If the loop square is Σ, then the absolute value of moment **m** is:

$$\text{m} = \frac{1}{\text{c}} J \Sigma \,, \tag{1.49}$$

and it is directed perpendicularly to the loop surface according to the right-hand screw rule or Ampère rule.

By analogy with the first item in Equation (1.40) which is equal to electrical moment, we can determine the density of magnetic current on the base of the ratio:

$$\mathbf{m} = -\frac{1}{i\omega} \int \mathbf{j}_m(\mathbf{r}')\, d^3\mathbf{r}' \,. \tag{1.50}$$

Then, a comparison with Equation (1.47) shows that:

$$\mathbf{j}_m = -\frac{ik}{2} \left[\mathbf{r} \times \mathbf{j} \right]. \tag{1.51}$$

By analogy, we may introduce magnetic charges on the basis of the continuity equation:

$$i\omega \rho_m = \nabla \cdot \mathbf{j}_m \,. \tag{1.52}$$

The formulas of radiated power are the same as in Equations (1.43) to (1.45), where a magnetic dipole should be substituted for an electrical one.

Densities of electrical and magnetic currents for dipole sources can be given by:

$$\mathbf{j} = -i\omega \mathbf{p} \delta(\mathbf{r} - \mathbf{r}_0) \,, \quad \mathbf{j}_\text{m} = -i\omega \mathbf{m} \delta(\mathbf{r} - \mathbf{r}_0). \tag{1.53}$$

where $\mathbf{r}_0$ is the radius vector of the point dipole location. In particular, we will obtain for the electrical dipole field at $\mathbf{r}_0 = 0$:

$$\mathbf{A} = -ik\mathbf{p} g(\mathbf{r}) \,, \quad \mathbf{E} = k^2 \mathbf{p} g + (\mathbf{p} \cdot \nabla) \nabla g \,, \quad \mathbf{H} = ik\mathbf{p} \times \nabla g \,. \tag{1.54}$$

Substituting Equation (1.30) for Green's function here, we obtain the following:

$$\begin{gathered}
\mathbf{E} = k^2 \left[\left[\frac{\mathbf{r} \times \mathbf{p}}{\text{r}} \right] \times \frac{\mathbf{r}}{\text{r}} \right] \frac{e^{ik\text{r}}}{\text{r}} + 2\left(\frac{1}{\text{r}} - ik \right) \left(\frac{\mathbf{r} \cdot \mathbf{p}}{\text{r}} \right) \frac{\mathbf{r}}{\text{r}} \frac{e^{ik\text{r}}}{\text{r}^2} \\
\mathbf{H} = k^2 \left[\frac{\mathbf{r} \times \mathbf{p}}{\text{r}} \right] \left(1 - \frac{1}{ik\text{r}} \right) \frac{e^{ik\text{r}}}{\text{r}} .
\end{gathered} \tag{1.55}$$

The expressions obtained are true not only in the wave area but also in the quasi-static area where $kr << 1$. It is important to observe the inequality $r >> l$ to see that the dipole field is not completely transversal in the quasi-static area.

For a magnetic dipole, the appropriate part of the vector potential is:

$$\mathbf{A} = ik\left[\frac{\mathbf{r}\times\mathbf{m}}{\mathbf{r}}\right]g = [\nabla\times(g\,\mathbf{m})]. \tag{1.56}$$

If the magnetic vector potential is considered by using the formula:

$$\mathbf{A}_{\mathrm{m}} = -ik\,\mathbf{m}\frac{e^{ikr}}{r} \tag{1.57}$$

with the generalization:

$$\mathbf{A}_{\mathrm{m}} = \frac{1}{c}\int \mathbf{j}_{\mathrm{m}}(\mathbf{r}')\frac{e^{ik|\mathbf{r}-\mathbf{r}'|}}{|\mathbf{r}-\mathbf{r}'|}d^3r', \tag{1.58}$$

then we will obtain the following for the corresponding components of the electromagnetic field:

$$\begin{aligned} &\mathbf{A} = -\frac{1}{ik}\nabla\times\mathbf{A}_{\mathrm{m}}, \quad \mathbf{E}_{\mathrm{m}} = -\nabla\times\mathbf{A}_{\mathrm{m}}, \\ &\mathbf{H}_{\mathrm{m}} = -\frac{1}{ik}\nabla\times\nabla\times\mathbf{A}_{\mathrm{m}} = \frac{i}{k}\left[\nabla(\nabla\cdot\mathbf{A}_{\mathrm{m}}) + k^2\mathbf{A}_{\mathrm{m}}\right]. \end{aligned} \tag{1.59}$$

In this case, we took into account the fact that the magnetic vector potential satisfies the wave equation:

$$\nabla^2\mathbf{A}_{\mathrm{m}} + k^2\mathbf{A}_{\mathrm{m}} = -\frac{4\pi}{c}\mathbf{j}_{\mathrm{m}}. \tag{1.60}$$

The fields $\mathbf{E}_{\mathrm{m}}$ and $\mathbf{H}_{\mathrm{m}}$ satisfy Maxwell's equations, which in this case must be written as:

$$\nabla\times\mathbf{E}_{\mathrm{m}} = ik\mathbf{H}_{\mathrm{m}} - \frac{4\pi}{c}\mathbf{j}_{\mathrm{m}}, \quad \nabla\times\mathbf{H}_{\mathrm{m}} = -ik\mathbf{E}_{\mathrm{m}}. \tag{1.61}$$

We leave to others consideration of Equation (1.48), which describes quadruple radiation and is outside our range of interest in this text.

1.5 LORENTZ'S LEMMA

Let us suppose that the fields $\mathbf{E}_1$, $\mathbf{H}_1$, $\mathbf{E}_2$, and $\mathbf{H}_2$ are created according to the currents $\mathbf{j}_1$ and $\mathbf{j}_2$. Both fields are described by Maxwell's equations, from which it is easy to derive the following:

$$\mathbf{H}_1 \cdot \nabla \times \mathbf{E}_2 - \mathbf{E}_2 \cdot \nabla \times \mathbf{H}_1 = ik\left(\mathbf{E}_1 \cdot \mathbf{E}_2 + \mathbf{H}_1 \cdot \mathbf{H}_2\right) - \frac{4\pi}{c}\mathbf{j}_1 \cdot \mathbf{E}_2.$$

Subtracting the second equation from the first one, we have:

$$\nabla \cdot \left\{\left[\mathbf{E}_1 \times \mathbf{H}_2\right] - \left[\mathbf{E}_2 \times \mathbf{H}_1\right]\right\} = \frac{4\pi}{c}\left(\mathbf{j}_1 \cdot \mathbf{E}_2 - \mathbf{j}_2 \cdot \mathbf{E}_1\right). \quad (1.62)$$

By using the identity $\nabla \cdot [\mathbf{a} \times \mathbf{b}] = \mathbf{b} \cdot \nabla \times \mathbf{a} - \mathbf{a} \cdot \nabla \times \mathbf{b}$, we obtain the equality known as Lorentz's lemma. If Equation (1.62) is integrated over volume V surrounded by surface S and both currents are included, then the corresponding integral form of Lorentz's lemma becomes:

$$\oint_S \left\{\left[\mathbf{E}_1 \times \mathbf{H}_2\right] - \left[\mathbf{E}_2 \times \mathbf{H}_1\right]\right\} \cdot \mathbf{n}\, d^2\mathbf{r} = \frac{4\pi}{c}\int_V \left(\mathbf{j}_1 \cdot \mathbf{E}_2 - \mathbf{j}_2 \cdot \mathbf{E}_1\right) d^3\mathbf{r}. \quad (1.63)$$

Here, $\mathbf{n}$ is the outward normal. If we direct surface S to infinity, then the source field on it has to satisfy the condition of radiation tending to transform the integral along S into zero. As a result, we obtain a formula for the mutuality theorem:

$$\int \left(\mathbf{j}_1 \cdot \mathbf{E}_2\right)\, d^3\mathbf{r} = \int \left(\mathbf{j}_2 \cdot \mathbf{E}_1\right)\, d^3\mathbf{r}\,. \quad (1.64)$$

Assume that the sources of fields $\mathbf{E}_1$ and $\mathbf{E}_2$ are dipoles with moments $\mathbf{p}_1$ and $\mathbf{p}_2$ and that they are located at points $\mathbf{r}_1$ and $\mathbf{r}_2$. Using Equation (1.53), we can formulate the mutuality theorem as:

$$\mathbf{p}_1 \cdot \mathbf{E}_2\left(\mathbf{r}_1\right) = \mathbf{p}_2 \cdot \mathbf{E}_1\left(\mathbf{r}_2\right). \quad (1.65)$$

Lorentz's lemma also applies in the case of magnetic currents and can be written as:

$$\oint_S \left\{\left[\mathbf{E}_1 \times \mathbf{H}_2\right] - \left[\mathbf{E}_2 \times \mathbf{H}_1\right]\right\} \cdot \mathbf{n}\, d^2\mathbf{r} = \frac{4\pi}{c}\int_V \left(\mathbf{j}_{m2} \cdot \mathbf{H}_1 - \mathbf{j}_{m1} \cdot \mathbf{H}_2\right) d^3\mathbf{r}\,. \quad (1.66)$$

In the case of magnetic dipoles, analogous to Equation (1.65), we have:

$$\mathbf{m}_1 \cdot \mathbf{H}_2(\mathbf{r}_1) = \mathbf{m}_2 \cdot \mathbf{H}_1(\mathbf{r}_2). \tag{1.67}$$

1.6 INTEGRAL FORMULAS

In this part, we obtain an expression for the field inside secluded volume V restricted by surface S (Figure 1.1). Let the components of the field, $\mathbf{E}(\mathbf{r}')$ and $\mathbf{H}(\mathbf{r}')$, be specified on surface S, where $\mathbf{r}'$ is a radius vector of the point on the surface S. It is necessary to find the electromagnetic field at point r inside volume V. Let us put the electric dipole, with moment p of any direction, at point r. The fields of this dipole are then indicated as $\mathbf{E}'$ and $\mathbf{H}'$. Then, according to Equation (1.63) of Lorentz's lemma, we may write:

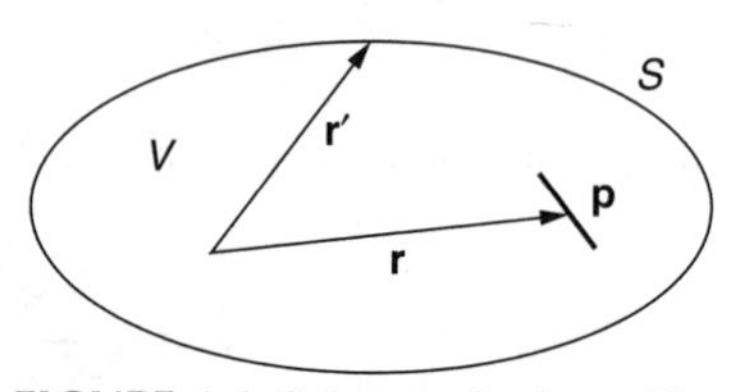

FIGURE 1.1 Scheme of volume V.

$$4\pi ik(\mathbf{p}\cdot\mathbf{E}) = \oint_S \left\{\mathbf{H}'(\mathbf{r}')\cdot\left[\mathbf{n}\times\mathbf{E}(\mathbf{r}')\right] + \mathbf{E}'(\mathbf{r}')\cdot\left[\mathbf{n}\times\mathbf{H}(\mathbf{r}')\right]\right\} d^2\mathbf{r}'. \tag{1.68}$$

The formula of rotation with scalar multiplication by the vector product of two other vectors was used in this case. It can be easily verified with the help of Equation (1.54) that:

$$\mathbf{H}'\cdot(\mathbf{n}\times\mathbf{E}) = ik(\mathbf{p}\times\nabla' g)\cdot(\mathbf{n}\times\mathbf{E}) = -ik\mathbf{p}\cdot[(\mathbf{n}\times\mathbf{E})\times\nabla' g].$$

In a similar manner,

$$\mathbf{E}'\cdot\left[\mathbf{n}\times\mathbf{H}\right] = \mathbf{p}\cdot\left\{k^2\left[\mathbf{n}\times\mathbf{H}\right]g + \left(\left[\mathbf{n}\times\mathbf{H}\right]\cdot\nabla'\right)\nabla' g\right\}.$$

The prime added to operator ∇ indicates differentiation by coordinates of the points of integration.

The performed transformation allows us to represent the integrand in Equation (1.68) in terms of vector:

$$-ik\left[\left[\mathbf{n}\times\mathbf{E}\right]\times\nabla' g\right] + k^2\left[\mathbf{n}\times\mathbf{H}\right] + \left(\left[\mathbf{n}\times\mathbf{H}\right]\nabla'\right)\nabla' g,$$

scalarly multiplied by moment **p** of the auxiliary dipole. As the value and direction of vector **p** are arbitrary, we can write:

$$\mathbf{E}(\mathbf{r}) = -\frac{1}{4\pi}\oint_S \left\{ik\left[\mathbf{n}\times\mathbf{H}\right]g + \left[\left[\mathbf{n}\times\mathbf{E}\right]\times\nabla' g\right] - \frac{1}{ik}\left(\left[\mathbf{n}\times\mathbf{H}\right]\cdot\nabla'\right)\nabla' g\right\} d^2\mathbf{r}'. \tag{1.69}$$

For the magnetic field, it is necessary to use Lorentz's lemma for magnetic dipoles. In this case,

$$-4\pi ik(\mathbf{m}\cdot\mathbf{H})=\oint_S\{\mathbf{H}'\cdot[\mathbf{n}\times\mathbf{E}]+\mathbf{E}'\cdot[\mathbf{n}\times\mathbf{H}]\}d^2\mathbf{r}'\ . \tag{1.70}$$

The same calculation results in:

$$\mathbf{H}(\mathbf{r})=-\frac{1}{4\pi}\oint_S\left\{-ik[\mathbf{n}\times\mathbf{E}]+[[\mathbf{n}\times\mathbf{H}]\times\nabla' g]+\frac{1}{ik}([\mathbf{n}\times\mathbf{E}]\cdot\nabla')\nabla' g\right\}d^2\mathbf{r}'. \tag{1.71}$$

The third components in Equations (1.69) and (1.71) may be transformed in the following way. It is well known that the nabla can be treated as a vector and that it must be placed before the differentiated function. Thus, we obtain the following:

$$([\mathbf{n}\times\mathbf{H}]\cdot\nabla)=(\mathbf{n}\cdot[\mathbf{H}\times\nabla])\Rightarrow(\mathbf{n}\cdot[\nabla\times\mathbf{H}])=-ik\,(\mathbf{n}\cdot\mathbf{E})$$
$$([\mathbf{n}\times\mathbf{E}]\cdot\nabla)=(\mathbf{n}\cdot[\mathbf{E}\times\nabla])\Rightarrow(\mathbf{n}\cdot[\nabla\times\mathbf{E}])=ik\,(\mathbf{n}\times\mathbf{H}).$$

Taking into consideration the above equations, Equations (1.69) and (1.71) can be rewritten in the more usual way of:

$$\mathbf{E}(\mathbf{r})=-\frac{1}{4\pi}\oint_S\{ik[\mathbf{n}\times\mathbf{H}]g+[[\mathbf{n}\times\mathbf{E}]\times\nabla' g]+(\mathbf{n}\cdot\mathbf{E})\cdot\nabla' g\}d^2\mathbf{r}', \tag{1.72}$$

$$\mathbf{H}(\mathbf{r})=-\frac{1}{4\pi}\oint_S\{-ik[\mathbf{n}\times\mathbf{E}]g+[[\mathbf{n}\times\mathbf{H}]\times\nabla' g]+(\mathbf{n}\cdot\mathbf{H})\cdot\nabla' g\}d^2\mathbf{r}'. \tag{1.73}$$

If some currents are inside volume V, then integrals (1.31) and (1.33) should be added to these expressions. The obtained formulas are referred to as the Stratton–Chu equations.[1–4] They give us the opportunity to calculate the field inside volume V according to the boundary values of the components. It is also recognized that these equations are an analytical formulation of the Huygens–Fresnel principle.

It should be noted that, in the obtained equalities, Green's function is not obliged to conform to Equation (1.30); it must only satisfy Equation (1.34), which means that solution of heterogeneous Equation (1.34), represented by Equation (1.30), may be added to the solution of the homogenous wave equation. The choice of the latter is determined by convenience. Note that the boundary values of fields **E** and **H** cannot be chosen independently because these vectors are related by Maxwell's equations; nevertheless, it is often done in the case of approximate calculations.

We have written Equations (1.72) and (1.73) for the case when the observation point is inside the closed surface S. These equations must be extended for the case when the point falls outside of the examined surface by using a synthetic method for a surface surrounded by a sphere of large radius R (surface S_R). Then, the formulas are valid for the total surface $S + S_R$, as the observation point is now inside the volume bounded by these surfaces. The radius of surface S_R tends to infinity, and by using the radiation conditions it may be shown that the electromagnetic field on this surface rather rapidly tends to zero to convert the corresponding integral also to zero.

Equations (1.72) to (1.73) are considerably simplified in the case when S is an infinite plane. Transformation to the infinite plane occurs in terms of the integral limit according to the limited element of the plane surface and the semisphere supported by it. As the semisphere radius increases toward infinity, the integral corresponding to it, as described above, approaches zero. As a result, only the integral along the infinite plane is left. In this case, it is convenient to choose Green's function as:

$$g = \frac{e^{ikR}}{R} - \frac{e^{ikR''}}{R''}, \tag{1.74}$$

where:

$$R = \sqrt{\left(x - x'\right)^2 + \left(y - y'\right)^2 + \left(z - z'\right)^2}, \quad R'' = \sqrt{\left(x - x'\right)^2 + \left(y - y'\right)^2 + \left(z + z'\right)^2}.$$

Here, x, y, and z are coordinates of the observation point and x′, y′, and z′ are coordinates of the point of integration. The value z′ = 0 corresponds to surface S. It is easy to show that Green's function is equal to zero for z′ = 0. Its derivatives along x′ and y′ also become zero, and

$$\frac{\partial g}{\partial z'} = -2\frac{\partial}{\partial z'}\left(\frac{e^{ikR}}{R}\right),$$

where we should consider $R = \sqrt{(x - x')^2 + (y - y')^2 + z^2}$ after performing the differentiation operation. Then, simple calculations give us the result:

$$\mathbf{E}\left(\mathbf{r}\right) = -\frac{1}{2\pi}\frac{\partial}{\partial z}\int_{-\infty}^{\infty}\int_{-\infty}^{\infty}\mathbf{E}\left(x', y'\right)\frac{e^{ikR}}{R}dx'dy'. \tag{1.75}$$

In the particular case, when field $\mathbf{E}$ is uniform and equal to $\mathbf{E}_0$ on surface S,

$$\mathbf{E}\left(\mathbf{r}\right) = -\mathbf{E}_0\frac{\partial}{\partial z}\int_0^{\infty}\frac{e^{ik\sqrt{\rho^2 + z^2}}}{\sqrt{\rho^2 + z^2}}\rho d\rho .$$

The integral is calculated using elementary mathematics, and, according to the Huygens–Fresnel principle, we obtain the expression for the plane wave propagating toward axis z:

$$\mathbf{E}(\mathbf{r}) = \mathbf{E}_0 e^{ikz}. \tag{1.76}$$

An expression similar to Equation (1.75) can also be obtained for the magnetic field.

Equation (1.75) permits further simplification for when field **E** changes slowly on surface *S*. In this case, the main area of integration in Equation (1.75) is concentrated close to the point $x' = x$, $y' = y$, which corresponds to the principles of stationary phase. We now have the opportunity to substitute, in the case of sufficient remoteness from the surface *S*,

$$R \cong z + \frac{(x-x')^2 + (y-y')^2}{2z}. \tag{1.77}$$

Then,

$$\mathbf{E}(x,y,z) \cong -\frac{ik}{2\pi}\frac{e^{ikz}}{z}\int_{-\infty}^{\infty}\int_{-\infty}^{\infty}\mathbf{E}(x',y')\exp\left[ik\frac{(x-x')^2+(y-y')^2}{2z}\right]dx'\,dy'. \tag{1.78}$$

The way in which the obtained approximation is referred to varies depending on the method of derivation (e.g., small angle, parabolic, diffusive). The considerable integration in Equation (1.78) is practically limited by the circle (Fresnel zone) described by the equation $(x-x')^2 + (y-y')^2 = \lambda z$ and having the radius:

$$\rho_F = \sqrt{\lambda z}\,. \tag{1.79}$$

The small angle approximation is the solution of a parabolic equation widely used in the theory of diffraction. In this approximation, the field may be represented as:

$$\mathbf{E}(x,y,z) = e^{ikz}\,\mathbf{U}(x,y,z)\,. \tag{1.80}$$

Combining this expression with the wave equation gives us:

$$2ik\frac{\partial \mathbf{U}}{\partial z} + \frac{\partial^2 \mathbf{U}}{\partial z^2} + \nabla_\perp^2 \mathbf{U} = 0\,.$$

Here, $\nabla_\perp^2$ is the Laplace operator (Laplacean) with respect to transversal to the z axis coordinate variables (i.e., x and y). Let us remember that we are dealing with small angle approximation, which means that the scale of field change with transversal variables is much more than the wavelength. After having determined the rapidly changed factor exp(ikz), the longitudinal scale of the field change is also becoming large. It is greater than the transversal scale. We now have the opportunity to ignore the second derivative along z in the last equation and obtain the parabolic equation for the field:

$$2ik\frac{\partial \mathbf{U}}{\partial z}+\nabla_{\perp}^{2}\mathbf{U}=0. \tag{1.81}$$

It is easy to prove that the solution of the parabolic equation leads to Equation (1.78) in the case when the radiation source is a hole in a plane screen.

Let us now consider the radiation of currents on a limited-size surface in the far zone. In this case, it is easy to obtain formulas similar to Equation (1.38):

$$\mathbf{E}=-k^{2}\left[\frac{\mathbf{r}}{\mathrm{r}}\times\hat{\mathbf{P}}_{e}\right]\frac{e^{ik\mathrm{r}}}{\mathrm{r}},\quad \mathbf{H}=-k^{2}\left[\frac{\mathbf{r}}{\mathrm{r}}\times\hat{\mathbf{P}}_{h}\right]\frac{e^{ik\mathrm{r}}}{\mathrm{r}}, \tag{1.82}$$

where:

$$\hat{\mathbf{P}}_{e}=-\frac{1}{4\pi ik}\oint_{S}\left\{\left[\mathbf{n}\times\mathbf{E}\right]-\left[\frac{\mathbf{r}}{\mathrm{r}}\times\left[\mathbf{n}\times\mathbf{H}\right]\right]\right\}\exp\left(-ik\frac{\mathbf{r}\cdot\mathbf{r}'}{\mathrm{r}}\right)d^{2}\mathbf{r}' \tag{1.83}$$

$$\hat{\mathbf{P}}_{h}=-\frac{1}{4\pi ik}\oint_{S}\left\{\left[\mathbf{n}\times\mathbf{H}\right]+\left[\frac{\mathbf{r}}{\mathrm{r}}\times\left[\mathbf{n}\times\mathbf{E}\right]\right]\right\}\exp\left(-ik\frac{\mathbf{r}\cdot\mathbf{r}'}{\mathrm{r}}\right)d^{2}\mathbf{r}'. \tag{1.84}$$

We can now compare the wave diffraction on so-called additional screens. Let us imagine a plane screen with a hole. We denote the screen area as A_{∞} and the square of the hole as A. A screen that fully closes this hole and has the same form is called an additional screen. We can now analyze the diffraction processes on the represented screens. It should be considered for simplicity that in both cases the question is one of plane wave diffraction of single amplitude falling perpendicularly to the screens. In the first case (the screen with the hole), the diffraction field is written as:

$$\mathbf{E}_{1}=-\frac{\mathbf{g}_{i}}{2\pi}\frac{\partial}{\partial z}\int_{A}\frac{e^{ikR}}{R}d^{2}\mathbf{r}', \tag{1.85}$$

where $\mathbf{g}_i$ is the polarization vector of the falling plane wave. In the second case of having an additional screen, the diffraction field will be set by the integral:

$$\mathbf{E}_2 = -\frac{\mathbf{g}_i}{2\pi}\frac{\partial}{\partial z}\int_{A_\infty - A}\frac{e^{ikR}}{R}d^2\mathbf{r}'. \tag{1.86}$$

The sum of these integrals is equal to the integral along the entire infinity plane, which gives a falling plane wave according to Equation (1.76). Also,

$$\mathbf{E}_1 + \mathbf{E}_2 = \mathbf{g}_i e^{ikz}. \tag{1.87}$$

The obtained result, which applies to other diffraction fields of additional screens, is referred to as the *theorem or principle of Babinet.* It is valid for any form of additional screen and any kinds of falling waves. In particular, this principle allows us to substitute the diffraction problem of the finite-size screen with the problem of diffraction on the hole.

1.7 APPROXIMATION OF KIRCHHOFF

The integral representation provided above is used especially often in the theory of radiowave diffraction. It means that surface fields excited by a field of incident waves are the result of a complex interaction between the electromagnetic field and the body (or bodies) where the diffraction occurs. In turn, it means, as was pointed out earlier, that it is impossible to arbitrarily set values **E** and **H** on the surface of a body, and it is necessary to solve the corresponding boundary problem. However, the number of such problems with analytical expressions is very limited and is obtained by bodies of simple shape. In the majority of cases, one should use approximate methods of calculation.

One of these methods is based on Kirchhoff's approximation, which was first formulated for solving problems of wave diffraction through holes in a screen. Let us consider a metallic screen with a cross section as shown in Figure 1.2. The hole in this screen has surface S_0. The rest of the screen is opaque and consists of surface S. Let electromagnetic wave $\mathbf{E}_i$, $\mathbf{H}_i$ generated by any source be incident on the screen from the left. It is often considered to be a plane wave. It is necessary to find the field of diffraction (or scattering) $\mathbf{E}_S$, $\mathbf{H}_S$ to the right of the screen. The problem can be reduced to integration along surface $S + S_0$ on the basis of integrals, as shown in Equations (1.72) and (1.73); however, as it was pointed out, the fields in the integrand themselves are, in principle, the object of solving the problem. The point is that Equations (1.72) and (1.73) are integral equations and rather often are the basis for numerical solution of diffraction problems.

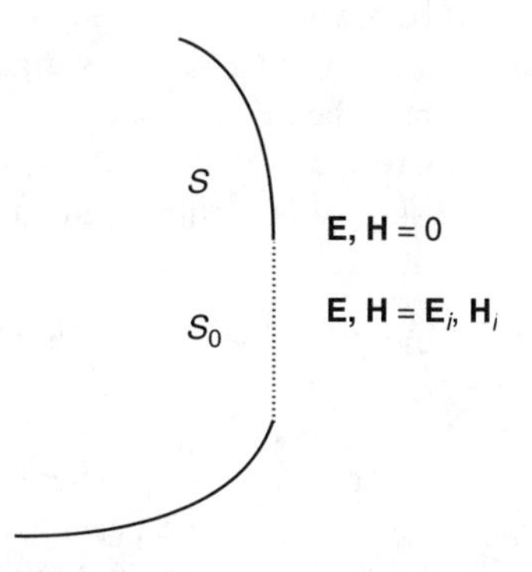

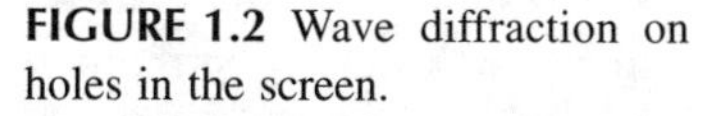

FIGURE 1.2 Wave diffraction on holes in the screen.

The essence of Kirchhoff's approximation is that the field is equal to zero on the reverse side of the screen (reverse according to the incident waves) and is equal to the incident wave field on the hole. Thus, for example, for the electrical field,

$$\mathbf{E}_S = -\frac{1}{4\pi}\int\!\!\int_{S_0} \left\{ ik\left[\mathbf{n}\times\mathbf{H}_i\right] + \left[\left[\mathbf{n}\times\mathbf{E}_i\right]\times\nabla' g\right] + \left(\mathbf{n}\cdot\mathbf{E}\right)\nabla' g \right\} d^2\mathbf{r}'. \quad (1.88)$$

The expression for $\mathbf{H}_S$ is obtained in the same way.

In the approximation here, the problem is reduced to the process of integration only, which considerably simplifies the problem. We will not go into the details regarding the basis of the procedure and will point out only that the application area of Kirchhoff's approximation is defined by two conditions. The first one demands that all the objects causing the diffraction (in the case described here, they are holes in the screen) must be much large compared to the wavelength. The second condition is connected with the fact that Kirchhoff's approximation correctly describes the radiation in the directions that are close to that following from the geometrical optics statements (in the given case, the direction of the incident wave). The small angle approximation, Equation (1.78), is rather convenient in these terms.

It is possible to further broaden the application of Kirchhoff's approximation. In some cases, the screen may be considered to be semitransparent, and tangential components of the field may be given as:

$$\left[\mathbf{n}\times\mathbf{E}\right] = T_{\mathrm{e}}\left[\mathbf{n}\times\mathbf{E}_i\right], \quad \left[\mathbf{n}\times\mathbf{H}\right] = T_{\mathrm{h}}\left[\mathbf{n}\times\mathbf{H}_i\right]. \quad (1.89)$$

Here, T_{e} and T_{h}, as functions of the surface point, are transmission coefficients. In particular, they may be random functions of coordinates.

If backward scattering is examined, Kirchhoff's approximation suggests that the field at every point of the surface is set according to the reflection laws from the plane boundary of two media. It means that the surface may be locally considered as a quasi-plane or faceted. This approximation is acceptable when the main radii of the surface curvature are much greater than the wavelength.

1.8 WAVE EQUATIONS FOR INHOMOGENEOUS MEDIA

We often encounter the problem of radiowave propagation in inhomogeneous media; that is, in media where permittivity is a function of the coordinates. In this case, it can be assumed that the field will be considered in that part of the space where no source is present. Furthermore, we shall consider isotropic media. Maxwell's equation may be written, according to these terms, as:

$$\nabla\times\mathbf{E} = ik\mathbf{H}, \quad \nabla\cdot\mathbf{H} = 0, \quad \nabla\times\mathbf{H} = -ik\varepsilon\mathbf{E}, \quad \nabla\cdot\mathbf{E} = -\mathbf{E}\cdot\nabla\ln\varepsilon. \quad (1.90)$$

We will now consider cases of plane layered media where the permittivity depends on one coordinate (say, z). Let us first derive the equation for the electrical field vector. For this purpose, we will act on Equation (1.90) with the nabla operator. In this case, the relation $\nabla \times (\nabla \times \mathbf{a}) = \nabla(\nabla \cdot \mathbf{a}) - \nabla^2 \mathbf{a}$, which is valid for any vector **a**, should be taken into account. As a result, we obtain:

$$\nabla^2 \mathbf{E} + k^2 \varepsilon(\mathbf{r})\mathbf{E} + \nabla(\mathbf{E} \cdot \nabla \ln \varepsilon) = 0. \tag{1.91}$$

The same operations with the magnetic field vector lead to the equation:

$$\nabla^2 \mathbf{H} + k^2 \varepsilon(\mathbf{r})\mathbf{H} + \nabla \ln \varepsilon \times (\nabla \times \mathbf{H}) = 0. \tag{1.92}$$

The equations obtained will be found useful for researching the problem of wave propagation in stratified media.

1.9 THE FIELD EXCITED BY SURFACE CURRENTS

In this part, we will solve the problem of a field of surface currents situated at the boundary of two media separated by a plane surface (Figure 1.3). This problem is more significant than it may appear at first. In reality, any fields incident on the plane of two separated media generate surface currents, which are sources of a reflected or scattered field.

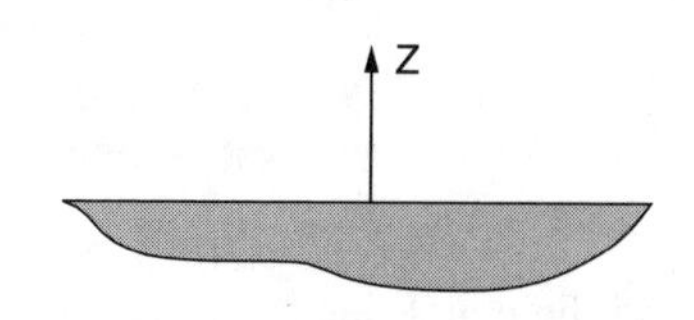

FIGURE 1.3 Geometry of two media separated by plane surface.

According to the geometry of Figure 1.3, let us assume that the lower media has permittivity ε 1, and the permittivity of the upper media is equal to one. The border of separation is described by the equation z = 0. The boundary conditions, due to Equations (1.8) and (1.12), are written as:

$$[e_z \times (\mathbf{E}_1 - \mathbf{E}_2)] = -\frac{4\pi}{c}\mathbf{K}_m, \quad [e_z \times (\mathbf{H}_1 - \mathbf{H}_2)] = \frac{4\pi}{c}\mathbf{K}_e \tag{1.93}$$

$$e_z \cdot (\mathbf{E}_1 - \varepsilon \mathbf{E}_2) = 4\pi\delta_m, \quad e_z \cdot (\mathbf{H}_1 - \mathbf{H}_2) = 4\pi\delta_e . \tag{1.94}$$

Here, $\mathbf{E}_1$ and $\mathbf{H}_1$ are the fields in the upper semispaces, $\mathbf{E}_2$ and $\mathbf{H}_2$ are the fields in the lower semispaces, and $\mathbf{e}_z$ is the unitary vector for the **z** direction.

Let us begin by considering vector **E**. We want to solve the appropriate wave equation, Equation (1.13), in the form of the two-dimensional Fourier's integral:

$$\mathbf{E}(\mathbf{r}) = \int \mathbf{G}(z,\mathbf{w}) e^{i\mathbf{w}\cdot\mathbf{s}} d^2\mathbf{s} . \tag{1.95}$$

Here, vector **s** is in the plane x, y. It is easy to obtain the equation for vector **G**:

$$\frac{d^2\mathbf{G}}{dz^2}+\left(\varepsilon k^2-\mathrm{w}^2\right)=0, \tag{1.96}$$

which has two solutions: $\exp(iz\sqrt{\varepsilon k^2-s^2})$ and $\exp(-iz\sqrt{\varepsilon k^2-s^2})$. It is necessary to choose the first solution for the upper semispace (z > 0) in accordance with the principle of radiation, letting ε = 1. For the lower semispace (z < 0), the second solution is preferable. So,

$$\mathbf{E}(\mathbf{r})=\int\tilde{\mathbf{E}}_1(\mathbf{w})\exp(i\mathbf{q}_1\cdot\mathbf{r})d^2\mathbf{w},\quad \mathbf{E}_2(\mathbf{r})=\int\tilde{\mathbf{E}}_2(\mathbf{w})\exp(i\mathbf{q}_2\cdot\mathbf{r})d^2\mathbf{w}, \tag{1.97}$$

where:

$$\mathbf{q}_1=\mathbf{w}+\mathbf{e}_z\gamma_1,\quad \mathbf{q}_2=\mathbf{w}-\mathbf{e}_z\gamma_2\,. \tag{1.98}$$

We have noted that:

$$\gamma_1=\sqrt{k^2-\mathrm{w}^2},\qquad \gamma_2=\sqrt{\varepsilon k^2-\mathrm{w}^2}=\sqrt{\gamma_1^2+k^2(\varepsilon-1)}\,, \tag{1.99}$$

and the equalities

$$q_1^2=k^2,\quad q_2^2=\varepsilon k^2,\quad \mathbf{q}_2=\mathbf{q}_1-(\gamma_1+\gamma_2)\mathbf{e}_z,\quad \gamma_2^2-\gamma_1^2=k^2(\varepsilon-1) \tag{1.100}$$

will be useful for the future. Expressions similar to Equation (1.97) also apply to field **H**.

Maxwell's equations for both media are written as:

$$\left[\mathbf{q}_1\times\tilde{\mathbf{E}}_1\right]=k\tilde{\mathbf{H}}_1,\quad \left[\mathbf{q}_1\times\tilde{\mathbf{H}}_1\right]=-k\tilde{\mathbf{E}}_1,\quad \left(\mathbf{q}_1\cdot\tilde{\mathbf{E}}_1\right)=\left(\mathbf{q}_1\cdot\tilde{\mathbf{H}}_1\right)=0, \tag{1.101}$$

$$\left[\mathbf{q}_2\times\tilde{\mathbf{E}}_2\right]=k\tilde{\mathbf{H}}_2,\quad \left[\mathbf{q}_2\times\tilde{\mathbf{H}}_2\right]=-\varepsilon k\tilde{\mathbf{E}}_2,\quad \left(\mathbf{q}_2\cdot\tilde{\mathbf{E}}_2\right)=\left(\mathbf{q}_2\cdot\tilde{\mathbf{H}}_2\right)=0. \tag{1.102}$$

The last two equalities in Equations (1.101) and (1.102) reflect the conditions of transversality, defined by the electrical and magnetic field divergence being equal to zero in both media considered. The boundary conditions are reduced to the following equalities:

$$\left[\mathbf{e}_z\times(\tilde{\mathbf{E}}_1-\tilde{\mathbf{E}}_2)\right]=-\frac{4\pi}{c}\tilde{\mathbf{K}}_m,\qquad \left[\mathbf{e}_z\times(\tilde{\mathbf{H}}_1-\tilde{\mathbf{H}}_2)\right]=\frac{4\pi}{c}\tilde{\mathbf{K}}_e, \tag{1.103}$$

$$\left(\mathbf{e}_z \cdot (\tilde{\mathbf{E}}_1 - \varepsilon\tilde{\mathbf{E}}_2)\right) = 4\pi\tilde{\delta}_e, \qquad \left(\mathbf{e}_z \cdot (\tilde{\mathbf{H}}_1 - \tilde{\mathbf{H}}_2)\right) = 4\pi\tilde{\delta}_m. \tag{1.104}$$

Here, $\tilde{\mathbf{K}}_e$, $\tilde{\mathbf{K}}_m$ and $\tilde{\delta}_e$, $\tilde{\delta}_m$ are spatial Fourier images of the surface currents and charges, respectively. It follows from the equations of continuity, Equation (1.11), that:

$$\begin{aligned} \omega\tilde{\delta}_e &= \left(\mathbf{w} \cdot \tilde{\mathbf{K}}_e\right) = \left(\mathbf{q}_1 \cdot \tilde{\mathbf{K}}_e\right) = \left(\mathbf{q}_2 \cdot \tilde{\mathbf{K}}_e\right), \\ \omega\tilde{\delta}_m &= \left(\mathbf{w} \cdot \tilde{\mathbf{K}}_m\right) = \left(\mathbf{q}_1 \cdot \tilde{\mathbf{K}}_m\right) = \left(\mathbf{q}_2 \cdot \tilde{\mathbf{K}}_m\right). \end{aligned} \tag{1.105}$$

Now, we have to work through some not very complicated calculations to obtain the obvious expressions for Fourier images of fields. First, note that it is evident from the boundary conditions, Equation (1.104), and from the equation of continuity that:

$$\left(\mathbf{e}_z \cdot \tilde{\mathbf{E}}_2\right) = \frac{1}{\varepsilon}\left(\mathbf{e}_z \cdot \tilde{\mathbf{E}}_1\right) - \frac{4\pi}{\varepsilon\omega}\left(\mathbf{q}_1 \cdot \tilde{\mathbf{K}}_e\right). \tag{1.106}$$

If we multiply Equation (1.103) vectorially by $\mathbf{e}_z$ and use the result, Equation (1.106), we obtain:

$$\tilde{\mathbf{E}}_2 = \tilde{\mathbf{E}}_1 - \frac{\varepsilon - 1}{\varepsilon}\mathbf{e}_z\left(\mathbf{e}_z \cdot \tilde{\mathbf{E}}_1\right) - \frac{4\pi}{c}\left[\frac{\mathbf{e}_z}{\varepsilon k}\left(\mathbf{q}_1 \cdot \tilde{\mathbf{K}}_e\right) + \left[\mathbf{e}_z \times \tilde{\mathbf{K}}_m\right]\right]. \tag{1.107}$$

Let us multiply this equation scalarly by $\mathbf{q}_2$ and use the equation of transversity and relation, Equation (1.100), to obtain the equality:

$$\left(\mathbf{e}_z \cdot \tilde{\mathbf{E}}_1\right) = \frac{4\pi}{c\left(\varepsilon\gamma_1 + \gamma_2\right)}\left\{\frac{\gamma_2}{k}\left(\mathbf{q}_1 \cdot \tilde{\mathbf{K}}_e\right) - \varepsilon\left(\mathbf{q}_1 \cdot \left[\mathbf{e}_z \times \tilde{\mathbf{K}}_m\right]\right)\right\}. \tag{1.108}$$

If the magnetic fields in the second boundary conditions, Equation (1.104), are expressed in terms of the electrical fields with the help of Maxwell's equations, we get the intermediate result:

$$\begin{aligned} &\left(\gamma_1 + \gamma_2\right)\tilde{\mathbf{E}}_1 - \frac{1}{\varepsilon}\left[\left(\varepsilon - 1\right)\mathbf{q}_1 + \left(\gamma_1 + \varepsilon\gamma_2\right)\mathbf{e}_z\right]\left(\mathbf{e}_z \cdot \tilde{\mathbf{E}}_1\right) = \\ &= -\frac{4\pi}{c}\left\{k\tilde{\mathbf{K}}_e - \frac{\mathbf{q}_2 + \gamma_2\mathbf{e}_z}{\varepsilon k}\left(\mathbf{q}_1 \cdot \tilde{\mathbf{K}}_e\right) - \gamma_2\left[\mathbf{e}_z \times \tilde{\mathbf{K}}_m\right]\right\} \end{aligned} \tag{1.109}$$

after some complicated calculations. Let us substitute Equation (1.108) here to obtain:

$$(\gamma_1-\gamma_2)\tilde{\mathbf{E}}_2=\frac{4\pi}{c}\left\{-k\tilde{\mathbf{K}}_e+\frac{\gamma_1+\gamma_2}{k(\varepsilon\gamma_1+\gamma_2)}\left[\mathbf{q}_1+(\gamma_2-\gamma_1)\mathbf{e}_z\right]\left(\mathbf{q}_1\cdot\tilde{\mathbf{K}}_e\right)+\right.$$

$$\left.+\gamma_2\left[\mathbf{e}_z\times\tilde{\mathbf{K}}_m\right]-\frac{(\varepsilon-1)\mathbf{q}_1+(\gamma_1+\varepsilon\gamma_2)\mathbf{e}_z}{\varepsilon\gamma_1+\gamma_2}\left(\mathbf{q}_1\cdot\left[\mathbf{e}_z\times\tilde{\mathbf{K}}_m\right]\right)\right\}. \qquad (1.110)$$

Having replaced the scalar product according to the rule

$$\left[\mathbf{a}\times(\mathbf{b}\times\mathbf{c})\right]=\mathbf{b}(\mathbf{a}\cdot\mathbf{c})-\mathbf{c}(\mathbf{a}\cdot\mathbf{b}),$$

we obtain:

$$\tilde{\mathbf{E}}_1=\frac{4\pi}{c(\varepsilon\gamma_1+\gamma_2)}\left\{\varepsilon\left[\mathbf{q}_1\times\tilde{\mathbf{K}}_m\right]+\frac{1}{k}\left[\mathbf{q}_1\times\left[\mathbf{q}_1\times\tilde{\mathbf{K}}_e\right]\right]+\right.$$

$$\left.+\frac{\varepsilon-1}{\gamma_1+\gamma_2}\left\{k\left[\mathbf{q}_1\left[\mathbf{e}_z\times\tilde{\mathbf{K}}_e\right]\right]+\left[\mathbf{e}_z\times\mathbf{q}_1\right]\left(\mathbf{q}_1\cdot\tilde{\mathbf{K}}_m\right)\right\}\right\}. \qquad (1.111)$$

Similar expressions may be obtained for the other components of the field in both media. We will not do this here, as the method for obtaining them is obvious. Knowing $\tilde{\mathbf{E}}_1$ gives us the opportunity to determine $\tilde{\mathbf{E}}_2$ in accordance with Equation (1.107) and to get the expression for the magnetic fields via Maxwell's equations.

1.10 ELEMENTS OF MICROWAVE ANTENNAE THEORY

The theory of microwave antennae whose size, as a rule, is much larger than the wavelength is usually based on the theory of radiowave diffraction in a hole in a plane screen. The hole has the form of an antenna aperture, and the field distribution inside (aperture field) is specified according to another supposition (characteristics of a radiator, mirror, lens, etc.). In this case, the basis for further calculations is Equation (1.75). Taking into account that the radiated field in the wave zone is of particular interest, when $(kl^2/\mathrm{r}) << 1$, where l is any maximal antenna size, we can use the approximation $R \cong \mathrm{r} - (\mathbf{r}\cdot\mathbf{r}')/\mathrm{r}$, which means that the diffraction in the Fraunhofer zone is under consideration. In this approach, the density of the radiated field power flow is described by the formula:

$$S(\theta,\varphi)=\frac{W_t G(\theta,\varphi)}{4\pi \mathrm{r}^2}, \qquad (1.112)$$

where W_t is the total radiated power. The spherical angles of observation θ and φ are introduced here, and the function $G(\theta,\varphi)$ represents the *directivity*. We came across this definition when we considered the dipole field of radiation in Equation (1.46). Directivity shows how much more the power flow density is radiated by a real antenna in a given direction compared to an isotropic radiator flow when the transmitted power of radiation is the same in both cases. The directivity is related to the antenna effective area (effective aperture), $A_e(\theta,\varphi)$, as:

$$G = \frac{4\pi A_e}{\lambda^2}, \tag{1.113}$$

where:

$$A_e = \frac{\left(\frac{\mathbf{e}_z \cdot \mathbf{r}}{r}\right)^2 \left| \int\limits_A \mathbf{E}_i(\mathbf{r}') \exp\left[-ik\frac{\mathbf{r}' \cdot \mathbf{r}}{r}\right] d^2\mathbf{r}' \right|^2}{\int\limits_A \left| \mathbf{E}_i(\mathbf{r}') \right|^2 d^2\mathbf{r}'}. \tag{1.114}$$

Here, $\mathbf{E}_i$ is the aperture field. Recall that:

$$\frac{\mathbf{r}' \cdot \mathbf{r}}{\mathrm{r}} = (\mathrm{x}' \cos\varphi + \mathrm{y}' \sin\varphi)\sin\theta.$$

The effective area of the antenna characterizes its quality of reception. If the plane radiowave with density power flow S is incident from direction (θ,φ), the power received by the antenna is:

$$W_r = SA_e(\theta,\varphi). \tag{1.115}$$

Let us suppose that the maximum radiation is in direction $\theta = 0$. The function:

$$\Psi(\theta,\varphi) = \frac{G(\theta,\varphi)}{G(0,\varphi)} \tag{1.116}$$

describes the angular density flow distribution of the radiated power and is considered the *normalized radiation pattern* (directional diagram). It is easy to see from Equation (1.113) that the pattern for radiation and for reception coincide for a common antenna.

For many estimating calculations, the aperture field can be considered uniform, and $A_e(0,\varphi) = A$; that is, the effective area is equal to the aperture square in the direction of the pattern maximum. The normalized pattern itself is expressed by the simple integral:

$$\Psi(\theta,\varphi)=\left|\frac{\cos\theta}{A}\int_A \exp\left(-ik\frac{\mathbf{r}\cdot\mathbf{r}'}{r}\right)d^2\mathbf{r}'\right|^2 . \tag{1.117}$$

Let us consider, for example, a circular aperture of radius a. In this case,

$$\Psi(\theta)=\left[\frac{2J_1(ka\sin\theta)}{ka\sin\theta}\right]^2\cos^2\theta . \tag{1.118}$$

Here, $J_1(x)$ is a Bessel function of the first order. This relation is sometimes called the *Airy formula*. The half-power beamwidth is estimated by the value:

$$\Delta\theta = 0.512\frac{\lambda}{a} = 1.02\frac{\lambda}{2a} . \tag{1.119}$$

The result obtained corroborates the rule that the beamwidth of an antenna of size D has a value of the order λ/D. Essentially, this rule is the result of the wave diffraction laws and follows from the principle of uncertainty typical of a Fourier transform. This rule means, in particular, that the cross section of the pattern (footprint) has the scale:

$$\rho=\frac{\lambda}{D}L \tag{1.120}$$

at distance L. This scale defines the spatial resolution of radio devices (in particular, radar and radiometers) for remote sensing. For example, two sources of radiation or two targets may be resolved if the distance between them is more than ρ. Such a statement was known in optics long ago and was called the *criteria of Rayleigh*.

Let us now examine the common characteristics of the antenna effective area and its directivity. The integral of Poynting along a spherical surface of radius r is equal to the total radiated power. Thus, it follows from Equation (1.112) that:

$$\int_{4\pi} G(\theta,\varphi)\,d\Omega=\int_0^{\pi}\int_0^{2\pi} G(\theta,\varphi)\sin\theta\,d\theta\,d\varphi=4\pi . \tag{1.121}$$

The result is the equality

$$\int_{4\pi} A_e(\theta,\varphi)\,d\Omega=\lambda^2 . \tag{1.122}$$

In some cases, to simplify the calculations and some estimations, it is convenient to use a modeled normalized pattern and corresponding antenna parameters of a circular aperture:

$$\Psi(\theta)=\exp\left(-\frac{\theta^2}{\theta_0^2}\right),\quad G(0)=\frac{4}{\theta_0^2},\quad A_e(0)=\frac{\lambda^2}{\pi\theta_0^2}. \tag{1.123}$$

1.11 SPATIAL COHERENCE

The spatial coherence problem appears in at least two cases. The first one is connected with the radiation of sources randomly arranged in space. The second case is connected with the problem of wave propagation in a medium with random scattering (rain, for example). Every scatterer is the source of secondary radiation and, as a result, waves with statistical characteristics appear.

Let us consider a system of a number of generators arranged in any area of space. In doing so, we will compare the characteristics of total radiation for two spaced points. The comparison of wave characteristics in the transversal plane relative to the direction of wave propagation is of interest. The comparison in the longitudinal direction equals the comparison in different time delay. For simplicity, let us consider the scalar case. We will describe the field characteristics by parabolic equation. The function of spatial coherence can be defined as:

$$\Gamma_E(\mathbf{s}_1,\mathbf{s}_2,z)=\frac{1}{2}\langle E(\mathbf{s}_1,z)E^*(\mathbf{s}_2,z)\rangle=\frac{1}{2}\langle U(\mathbf{s}_1,z)U^*(\mathbf{s}_2,z)\rangle=\Gamma_U. \tag{1.124}$$

We can derive an equation to describe the spatial behavior of a coherence function defined in such a way. For this purpose, we use the following equations for the corresponding fields:

$$2ik\frac{\partial U(\mathbf{s}_1,z)}{\partial z}+\nabla^2_{\mathbf{s}_1}U(\mathbf{s}_1,z)=0,\quad -2ik\frac{\partial U(\mathbf{s}_2,z)}{\partial z}+\nabla^2_{\mathbf{s}_2}U(\mathbf{s}_2,z)=0.$$

Multiply the first equation by $U^*(\mathbf{s}_2,z)$, and the second one by $U(\mathbf{s}_1,z)$, then subtract the second result from the first and average over the statistical ensemble to obtain:

$$2ik\frac{\partial\Gamma_U}{\partial z}+\left(\nabla^2_{\mathbf{s}_1}-\nabla^2_{\mathbf{s}_2}\right)\Gamma_U=0. \tag{1.125}$$

This is a well-known equation of parabolic type, and its solution requires knowledge of boundary conditions (for example, at z = 0). We will suppose from here on that:

$$\Gamma_U(\mathbf{s}_1,\mathbf{s}_2,0)=\Gamma_0(\mathbf{s}_1,\mathbf{s}_2)=\frac{1}{2}\langle U_0(\mathbf{s}_1)U_0^*(\mathbf{s}_2)\rangle. \tag{1.126}$$

Conceptually, we should have to calculate the real part of the defined coherence function and apply the corresponding equation for $\mathrm{Re}\Gamma_U$; however, it is easier to employ the equation for the complex coherent function and to seek its real part in the final solution.

Let us introduce the coordinates of the gravity center, $\mathbf{s}_1 + \mathbf{s}_2 = 2\mathbf{S}$, and the differential coordinates $\mathbf{s}_1 - \mathbf{s}_2 = \mathbf{s}$. Equation (1.125) is written using these coordinates in the form:

$$ik\frac{\partial \Gamma_U}{\partial z} + \frac{\partial^2 \Gamma_U}{\partial \mathbf{S}\,\partial \mathbf{s}} = 0. \tag{1.127}$$

Here, derivatives by vectors represent gradients in corresponding directions. Let us now write the solution in the form of the Fourier integral:

$$\Gamma_U(\mathbf{S},\mathbf{s},z) = \int \tilde{\Gamma}_U(\mathbf{W},\mathbf{s},z)e^{i\mathbf{W}\cdot\mathbf{S}}d^2\mathbf{W},$$

which leads particularly to the equation for the Fourier transform image:

$$k\frac{\partial \tilde{\Gamma}_U}{\partial z} + \left(\mathbf{W}\frac{\partial \tilde{\Gamma}_U}{\partial \mathbf{s}}\right) = 0. \tag{1.128}$$

The standard way to solve such an equation involves, first of all, writing a system of characteristic equations which, in the given case, has the form:

$$\frac{dz}{k} = \frac{dx}{W_x} = \frac{dy}{W_y},$$

where x and y are coordinates of vector **s**. Solution of the characteristic equations is obvious and can be written as $\mathbf{s} - \mathbf{W}z/k = \mathbf{c}$, where **c** is a constant vector,

$$\tilde{\Gamma}_U(\mathbf{W},\mathbf{s},z) = \tilde{\Gamma}_U\left(\mathbf{W},\ \mathbf{s} - \frac{\mathbf{W}}{k}z\right), \tag{1.129}$$

and

$$\Gamma_U(\mathbf{S},\mathbf{s},z) = \int \tilde{\Gamma}_U\left(\mathbf{W},\mathbf{s} - \frac{\mathbf{W}}{k}z\right)e^{i\mathbf{W}\cdot\mathbf{S}}d^2\mathbf{W}. \tag{1.130}$$

We will now use the boundary condition represented by Equation (1.126) to define $\tilde{\Gamma}_U$. It follows that:

$$\tilde{\Gamma}_U(\mathbf{W},\mathbf{s}) = \frac{1}{4\pi^2}\int \Gamma_0\left(\mathbf{S}'+\frac{\mathbf{s}}{2},\ \mathbf{S}'-\frac{\mathbf{s}}{2}\right)e^{-i\mathbf{W}\cdot\mathbf{S}'}d^2\mathbf{S}'. \tag{1.131}$$

and we obtain:

$$\Gamma_U = \frac{1}{4\pi^2}\int\int d^2\mathbf{S}'d^2\mathbf{W}\Gamma_0\left(\mathbf{S}'+\frac{\mathbf{s}}{2}-\frac{\mathbf{W}z}{2k},\ \mathbf{S}'-\frac{\mathbf{s}}{2}+\frac{\mathbf{W}z}{2k}\right)e^{i\mathbf{W}\cdot(\mathbf{S}-\mathbf{S}')}. \tag{1.132}$$

A further simplification is connected with the supposition that:

$$\Gamma_0(\mathbf{s}_1,\mathbf{s}_2) = I(\mathbf{S})\gamma_0(\mathbf{s}). \tag{1.133}$$

In this formula, $I(\mathbf{S})$ describes the intensity distribution on the aperture of a radiator, and $\gamma_0(\mathbf{s})$ is the spatial correlation coefficient of radiating sources. Such a statement reasonably reflects many real situations. It turns out that

$$\Gamma_U = \frac{k^2}{4\pi^2 z^2}\int\int I(\mathbf{S}')\gamma_0(\mathbf{s}')\exp\left(i\frac{k}{z}(\mathbf{S}-\mathbf{S}')\cdot(\mathbf{s}-\mathbf{s}')\right)d^2\mathbf{S}'d^2\mathbf{s}'. \tag{1.134}$$

Formally, the integration is carried out to infinite limits; however, in practice, the integration over $\mathbf{s}'$ is conducted within the scale limits of correlation length l_0 of the primary source and with respect to $\mathbf{S}'$ in the size limits of the radiating area A. At sufficient distances from the radiating system, $2\pi a l_0/\lambda z \ll 1$; therefore,

$$\Gamma_U = \frac{k^2}{z^2}\tilde{\gamma}_0\left(\frac{k\mathbf{S}}{z}\right)\exp\left(i\frac{k\mathbf{S}\cdot\mathbf{s}}{z}\right)\int_A I(\mathbf{S}')\exp\left(-i\frac{k\mathbf{s}\cdot\mathbf{S}'}{z}\right)d^2\mathbf{S}'. \tag{1.135}$$

Here, the function:

$$\tilde{\gamma}_0(\mathbf{q}) = \frac{1}{4\pi^2}\int\gamma_0(\mathbf{s})e^{-i\mathbf{q}\cdot\mathbf{s}}d^2\mathbf{s} \tag{1.136}$$

is the normalized spatial spectrum of the sources field. Index A under the integral in Equation (1.135) indicates that the integration is carried out within the frame of the limited area (aperture).

The expression obtained is the mathematical expression of the van Cittert–Zernike theorem,[11] confirming that the coherence function of random sources of radiation is proportional to the Fourier transform of their spatial distribution intensity. Equation (1.135) is sometimes referred to as the generalized theorem of van Cittert–Zernike. The theorem itself was proven primarily for the case $l_0 = 0$, when the spatial correlation function of the sources of radiation is expressed in terms of the δ-function. It should be assumed that $S<\left(2\lambda z^3/\pi\right)^{1/4}$ because our arguments are relevant to small-angle approximations (where the parabolic equation is in the base). Therefore, the maximum of kS/z has a value of order $\left(2^5\pi^3/\lambda^3 z\right)^{1/4}$. On the other hand, the spatial spectrum differs substantially from zero within the wave number interval $0 < q < 2\pi/l_0$. So, if the scale of correlation is much less than $\left(\pi\lambda^3 z/2\right)^{1/4}$, which is often the case, then the argument in the expression for the spatial spectrum in (1.135) can be replaced by zero; however, $\int \gamma_0(\mathbf{s})d^2\mathbf{s} = \pi l_0^2$ by definition of the correlation scale. Therefore, $\tilde{\gamma}_0(0) = l_0^2/4\pi$.

As a result of these calculations, the real part can be introduced as:

$$\operatorname{Re}\Gamma_U = \frac{k^2 l_0^2}{4\pi z^2}\int_A I(\mathbf{S}')\cos\left(\frac{k\mathbf{s}}{z}\cdot(\mathbf{S}-\mathbf{S}')\right)d^2\mathbf{S}'.$$

Further, it is more convenient for our purposes to use the correlation coefficient of the radiated field; that is, the value:

$$\gamma_U(\mathbf{S},\mathbf{s},z) = \frac{1}{P}\int_A I(\mathbf{S}')\cos\left(\frac{k\mathbf{s}}{z}\cdot(\mathbf{S}-\mathbf{S}')\right)d^2\mathbf{S}'. \tag{1.137}$$

Here, $P = \int_A I(\mathbf{S}')d^2\mathbf{S}'$ represents the power radiated through the aperture.

Let us now introduce the transversal correlation radius of the radiation field using the definition:

$$\pi s_\perp^2(\mathbf{S},z) = \int \gamma_U(\mathbf{S},\mathbf{s},z)\,d^2\mathbf{s}.$$

It is characteristic that the value of the correlation radius depends on the point of space where this radius is being defined. Integration by **s** gives the δ-function and we have:

$$s_\perp^2 = \frac{4\pi}{P}\int_A I(\mathbf{S}')\,\delta\left[\frac{k}{z}(\mathbf{S}-\mathbf{S}')\right]d^2\mathbf{S}'.$$

The result of the integration depends on whether or not the vector **S** point falls into the area of the square of aperture A. If it does not, then the integral to the right is zero and the value of correlation radius is also zero. In the opposite case, according to the rules of integration with the δ-function,

$$s_{\perp}^{2}(\mathbf{S},z)=\frac{\lambda^{2}z^{2}}{\pi P}I(\mathbf{S}). \tag{1.138}$$

In particular, if the intensity inside of the aperture is uniform, then:

$$s_{\perp}=\frac{\lambda z}{\sqrt{\pi A}}. \tag{1.139}$$

So, the transverse correlation radius of the radiated field is the ratio of the area of the Fresnel zone squared to the aperture dimension; it increases, naturally, with the distance between the point of observation and the radiating aperture.

2 Plane Wave Propagation

2.1 PLANE WAVE DEFINITION

In the previous chapter, we defined a plane wave as a wave whose characteristics depend on only one Cartesian coordinate. We also noted that the plane wave is excited by a system of sources distributed uniformly on an infinite plane. Because a source of infinite size is an abstraction, the notion of a plane wave is also abstract.

We also established in the previous chapter that, far from real sources (sources of limited sizes), radiated waves can be considered to be spherical and that the phase of radiated waves close to the pattern maximum is constant on a spherical surface of radius R, where R is the distance from the source. Locally, a spherical surface of large radius differs little from a plane and may be supposed to be a plane in the defined frames of space. Let us now refine the bounds of these frames.

Ignoring unimportant details, the spherical wave may be described by the following expression ($\varepsilon = 1$):

$$\mathrm{E} = \mathbf{T}\frac{e^{ikR}}{R},$$

where the vector **T** does not depend on R and describes the wave polarization, amplitude, and distribution angle. Here, $R = \sqrt{\mathrm{x}^2 + \mathrm{y}^2 + \mathrm{z}^2}$ where x, y, and z are coordinates of the observation point. Accordingly, the start of the coordinate system is situated at the point where the radiator is located.

Let us imagine that the pattern maximum is close to the direction defined by the condition $r = \sqrt{\mathrm{x}^2 + \mathrm{y}^2} = 0$. If we consider that, in some of the space, area $r^2 << z^2$, then approximately:

$$\mathbf{E} = \mathbf{T}\frac{\exp\left(ikz + ikr^2/2z\right)}{z}. \tag{2.1}$$

It is now easy to set up the conditions such that the field depends on only coordinate z. In this case, the front of the spherical wave may be considered to be locally plane; thus, the wave is also thought to be locally plane. The wave phase of Equation (2.1) changes in plane z = *const* due to the law:

$$\Phi = kz + \frac{kr^2}{2z}. \tag{2.2}$$

By convention, the wave phase is considered to be constant in the stated plane if coordinates x and y change within such limits so that the value of $kr^2/2z < \pi$, or:

$$r < \sqrt{\lambda z} = \rho_F. \tag{2.3}$$

So, the wave radiated by the real sources can be considered to be a plane within the Fresnel zone.

Let us point out that the size of the fixed area increases according to the distance from the radiating source and can be rather large, which gives us the opportunity to analyze various wave phenomena within the frame of the plane wave approximation and to approach essential results with acceptable accuracy. Also, a plane wave is one of the simplest types of waves.

2.2 PLANE WAVES IN ISOTROPIC HOMOGENEOUS MEDIA

The explanation of the plane wave concept provided above does not include all types of waves, as the form of a plane wave depends on the propagation media characteristics. The simplest is the case of homogeneous and isotropic media. In this case, as was shown in the first chapter, the electromagnetic field vectors satisfy the wave equation, Equation (1.13). It follows, then, that every component of electrical and magnetic fields satisfies a scalar wave equation, thus we can examine the propagation of any one and extend the results to others.

Let us denote the chosen component field component as u. Because all of its parameters depend on one coordinate (for example, z), it must satisfy the equation:

$$\frac{d^2u}{dz^2} + \varepsilon k^2 u = 0. \tag{2.4}$$

The common solution of this equation is:

$$u = u_1 e^{ik\sqrt{\varepsilon}z} + u_2 e^{-ik\sqrt{\varepsilon}z}, \tag{2.5}$$

where constants u_1 and u_2 are defined from the exiting and boundary conditions. The first term in Equation (2.5) represents a wave propagating in the direction of positive values of z. These waves are usually referred to as *direct*. The second term in Equation (2.5) describes a back wave propagating in the direction of negative values of z. If all the sources are to the left along the z-axis and no obstacles are causing wave reflection, then the back wave has to be absent and we suppose that $u_2 = 0$.

For simplification and convenience of further calculations, let us introduce the value:

$$n = n' + in'' = \sqrt{\varepsilon}. \tag{2.6}$$

The value:

$$n' = \sqrt{\frac{1}{2}\left(\sqrt{\varepsilon'^2 + \varepsilon''^2} + \varepsilon'\right)} = \sqrt{\frac{1}{2}\left(|\varepsilon| + \varepsilon'\right)} \tag{2.7}$$

is called the *refractive index* (the term is borrowed from optics), and

$$n'' = \sqrt{\frac{1}{2}\left(\sqrt{\varepsilon'^2 + \varepsilon''^2} - \varepsilon'\right)} = \sqrt{\frac{1}{2}\left(|\varepsilon| - \varepsilon'\right)} \tag{2.8}$$

is called the *index of absorption.*

Taking into consideration the time dependence, the expression for the plane wave can be represented as:

$$u = u_0 e^{i\Phi - \gamma z}. \tag{2.9}$$

The value u_0 is called the *initial wave amplitude*, where

$$\Phi = kn'z - \omega t \tag{2.10}$$

is the phase, and

$$\gamma = kn'' \tag{2.11}$$

is the *coefficient of attenuation* (absorption) of the wave. Because of the absorption dielectric, wave amplitude decreases with distance, according to the exponential law:

$$|u| = |u_0| e^{-\gamma z}. \tag{2.12}$$

By convention, it is supposed that at depth $z = d_s$, such that $\gamma d_s = 1$, the field is practically faded. The value:

$$d_s = \frac{1}{\gamma} = \frac{1}{kn''} \tag{2.13}$$

is called the *depth of penetration* or *skin depth.*

It is a simple matter to determine the phase velocity from Equation (2.10):

$$v_\Phi = \frac{c}{n'}. \tag{2.14}$$

At $n' > 1$, which usually occurs, the phase velocity is less than the speed of light. In the case of plasma, when $n' < 1$, the phase velocity is more the velocity of light.

The wave number in dielectric k' equals kn', and the wavelength:

$$\lambda' = \frac{2\pi}{k'} = \frac{\lambda}{n'} \tag{2.15}$$

differs from the same in vacuum.

Let us consider specific cases. Very often $\varepsilon'' << \varepsilon'$, which corresponds to the case of low absorbed media. Than $n' \cong \sqrt{\varepsilon'}$ and $n'' \cong \varepsilon'' / (2\sqrt{\varepsilon'})$. In this case,

$$v_\Phi = \frac{c}{\sqrt{\varepsilon'}}, \quad \lambda' = \frac{\lambda}{\sqrt{\varepsilon'}}, \quad \gamma = \frac{\omega\varepsilon''}{2c\sqrt{\varepsilon'}}, \quad d_s = \frac{2c\sqrt{\varepsilon'}}{\omega\varepsilon''}. \tag{2.16}$$

In the opposite case of high absorbed media, $\varepsilon'' >> \varepsilon'$, and $n' \cong n'' \cong \sqrt{\varepsilon''/2}$. For this case,

$$v_\Phi = \sqrt{\frac{2}{\varepsilon''}}c, \quad \lambda' = \sqrt{\frac{2}{\varepsilon''}}\lambda, \quad \gamma = \frac{\omega}{c}\sqrt{\frac{\varepsilon''}{2}}, \quad d_s = \frac{c}{\omega}\sqrt{\frac{2}{\varepsilon''}}. \tag{2.17}$$

The plane wave may be introduced as follows:

$$\mathbf{E}, \mathbf{H} = \mathbf{E}_0, \mathbf{H}_0 e^{i\mathbf{q}\cdot\mathbf{r}}$$

in the case of an arbitrary direction of propagation. Vector $\mathbf{q}$ is the *wave vector* and defines the direction of the wave propagation. Substitution in Maxwell's equations gives:

$$\left[\mathbf{q}\times\mathbf{E}\right] = k\mathbf{H}, \quad \left(\mathbf{q}\cdot\mathbf{H}\right) = 0, \quad \left[\mathbf{q}\times\mathbf{H}\right] = -\varepsilon k\mathbf{E}, \quad \left(\mathbf{q}\cdot\mathbf{E}\right) = 0, \tag{2.18}$$

assuming current density $\mathbf{j} = 0$ and density of charge $\rho = 0$. It follows from these equations that vectors $\mathbf{E}$ and $\mathbf{H}$ are orthogonal to each other and to the direction of wave propagation at real vector $\mathbf{q}$. It is supposed in this case that $\varepsilon = 0$, which can occur in the case of plasma. Except, for instance, vector $\mathbf{H}$ from Equation (2.18), we may easily obtain the equation of dispersion:

$$\mathrm{q}^2 = k^2\varepsilon \,. \tag{2.19}$$

The fact that the modulus of vector $\mathbf{q}$ equals a complex number in the general case means that it is a complex vector itself; that is, $\mathbf{q} = \mathbf{q}' + i\mathbf{q}''$, where vectors $\mathbf{q}'$ and

$\mathbf{q}''$ are real because their projections on the coordinates axes are real numbers. It follows from Equation (2.19) that:

$$q'^2 - q''^2 = k^2\varepsilon', \qquad 2(\mathbf{q}'\cdot\mathbf{q}'') = k^2\varepsilon''. \tag{2.20}$$

Vectors $\mathbf{q}'$ and $\mathbf{q}''$ do not have to be parallel to each other, which means that for some plane waves (inhomogeneous plane waves) the planes of equal phase and equal amplitude do not coincide. Such waves do not correspond to the plane wave definition given at the beginning of this chapter because their different characteristics (the amplitude and the phase) depend on one coordinate. In homogeneous media, inhomogeneous plane waves are not excited; however, they appear upon electromagnetic wave propagation in inhomogeneous media.

Equations (2.18–19) permit us to state that the complex amplitudes of vectors **E** and **H** are connected with the equality:

$$\mathbf{H}=\sqrt{\varepsilon}\,\mathbf{E}\,. \tag{2.21}$$

Poynting's vector of a plane wave in the general case is defined by the formula:

$$\mathbf{S} = \frac{c|\mathbf{E}|^2}{8\pi k}\mathbf{q}', \tag{2.22}$$

from which we may conclude that the plane wave is directed along a ray orthogonal to the plane of uniform phase.

In conclusion, we have shown that investigating fields with respect to spatial Fourier integrals, which we have already used in Chapter 1, involves nothing more than expansion of the fields with respect to plane waves.

2.3 PLANE WAVES IN ANISOTROPIC MEDIA

As we have already pointed out, the connection between vectors **D** and **E** in anisotropic media is a tensor one. So, Equation (1.6) should be employed, and Equation (2.18) becomes:

$$[\mathbf{q}\times\mathbf{E}] = k\mathbf{H}, \quad (\mathbf{q}\cdot\mathbf{H}) = 0, \quad [\mathbf{q}\times\mathbf{H}] = -k\mathbf{D}, \quad (\mathbf{q}\cdot\mathbf{D}) = 0. \tag{2.23}$$

Excluding vector **H** from these equations, we obtain:

$$\mathbf{q}(\mathbf{q}\cdot\mathbf{E}) - q^2\mathbf{E} + k^2\mathbf{D} = 0. \tag{2.24}$$

Let us suppose that the wave propagates along the z-axis, in which case $q_x = q_y = 0$, and we substitute $q_z = kn$ for the z-component of vector $\mathbf{q}$. Then, the following system of equations can be obtained from Equation (2.24):

$$\left(\varepsilon_{xx} - n^2\right)E_x + \varepsilon_{xy}E_y + E_z = 0,$$

$$\varepsilon_{yx}E_x + \left(\varepsilon_{xx} - n^2\right)E_y + \varepsilon_{yz}E_z = 0, \tag{2.25}$$

$$\varepsilon_{zx}E_x + \varepsilon_{zy}E_y + \varepsilon_{zz}E_z = 0$$

Furthermore, it is useful to exclude the z-component of electric field from this system so we have:

$$E_z = -\frac{\varepsilon_{zx}E_x + \varepsilon_{zy}E_y}{\varepsilon_{zz}}, \tag{2.26}$$

which, after substitution of this expression in the other two expressions of Equation (2.25), leads to a simpler system of equations:

$$\left(A - n^2\right)E_x + C_{xy}E_y = 0, \qquad C_{yx}E_x + \left(B - n^2\right)E_y = 0. \tag{2.27}$$

Here,

$$A = \varepsilon_{xx} - \frac{\varepsilon_{xz}\varepsilon_{zx}}{\varepsilon_{zz}}, \; B = \varepsilon_{yy} - \frac{\varepsilon_{yz}\varepsilon_{zy}}{\varepsilon_{zz}}, \; C_{xy} = \varepsilon_{xy} - \frac{\varepsilon_{xz}\varepsilon_{zy}}{\varepsilon_{zz}}, \; C_{yx} = \varepsilon_{yx} - \frac{\varepsilon_{zx}\varepsilon_{yz}}{\varepsilon_{zz}}. \tag{2.28}$$

Because Equation (2.27) is a system of linear uniform algebraic equations relative to the components of an electric field, the conditions of the nontrivial solution require the determinant of the system to approach zero. The dispersion equation can be stated by the following expression, which is reduced to a biquadrate equation relative to refractive index n:

$$\left(A - n^2\right)\left(B - n^2\right) - C_{xy}C_{yx} = 0 \tag{2.29}$$

with the obvious solution:

$$n_{1,2}^2 = \frac{1}{2}\left[A + B \pm \sqrt{\left(A - B\right)^2 + 4C_{xy}C_{yx}}\right]. \tag{2.30}$$

Let us now apply the common expressions obtained for the case of plasma, which is of special interest because of radiowave propagation in the ionosphere. We will not develop the expression for tensor components of magnetic active plasma, as it may be found elsewhere (for example, in Ginsburg[12] from which we have taken some necessary expressions). Moreover, let us point out that waves are weakly absorbed in the ionosphere because microwaves are discussed throughout this book; therefore, we neglect the absorption to avoid complicating the problem.

Let us introduce some definitions. The value:

$$\omega_{\mathrm{p}} = \sqrt{\frac{4\pi e^2 N}{m}}, \qquad f_{\mathrm{p}} = \frac{\omega_{\mathrm{p}}}{2\pi} \cong 9 \cdot 10^3 \sqrt{N} \tag{2.31}$$

is called the plasma frequency. Here, N is the concentration of electrons in plasma, $e = 4.8 \cdot 10^{-10}$ CGS electrostatic system (CGSE) is the electron charge, and $m = 9.1 \cdot 10^{-28}$ g is its mass. For the ionosphere of Earth, where the maximal value of the electron concentration is N_{m} $2 \cdot 10^6$ cm^{-3}, the maximal value of the plasma frequency is about 10 MHz. So, in microwaves the ratio:

$$v = \frac{\omega_{\mathrm{p}}^2}{\omega^2} << 1 \tag{2.32}$$

always occurs.

Now let us introduce the cyclotron frequency defined by the equality:

$$\omega_{\mathrm{H}} = \frac{eH_0}{mc}, \qquad f_{\mathrm{H}} = \frac{\omega_{\mathrm{H}}}{2\pi} \cong 2.8 \cdot 10^6 H_0, \tag{2.33}$$

where H_0 is the strength of the magnetic field of Earth and has a value about 0.5 Oersted (Gauss). Therefore, the cyclotron frequency is equal to about 1.5 MHz, and the ratio:

$$u = \frac{\omega_{\mathrm{H}}^2}{\omega^2} << 1 \tag{2.34}$$

exists in the microwave region.

We shall suppose that the magnetic field of Earth lies in the z0y plane at angle β to the z-axis, which, we recall, coincides with the direction of wave propagation. The components of the permittivity tensor are described by:[12]

$$\varepsilon_{\mathrm{xx}} = 1 - \frac{v}{1-u}, \qquad \varepsilon_{\mathrm{xy}} = -\varepsilon_{\mathrm{yx}} = \frac{iv\sqrt{u}\cos\beta}{1-u},$$

$$\varepsilon_{xz} = -\varepsilon_{zx} = -\frac{iv\sqrt{u}\sin\beta}{1-u}, \quad \varepsilon_{yy} = 1 - \frac{v\left(1-u\sin^2\beta\right)}{1-u}, \tag{2.35}$$

$$\varepsilon_{yz} = \varepsilon_{zy} = \frac{uv\sin\beta\cos\beta}{1-u}, \quad \varepsilon_{zz} = 1 - \frac{v\left(1-u\cos^2\beta\right)}{1-u}$$

in the chosen coordinate system. The substitution of these expressions in Equation (2.28) permits us to calculate the values A, B, C_{xy}, and C_{yx} and then to obtain an expression for the refractive index:

$$n_{1,2}^2 = 1 - \frac{2v(1-v)}{2(1-v) - u\sin^2\beta \pm \sqrt{u^2\sin^4\beta + 4u(1-v)^2\cos^2\beta}}. \tag{2.36}$$

Having two solutions for the refraction index means that two types of waves occur in magnetic active plasma: the ordinary one, to which the (+) sign corresponds in Equation (2.36), and the extraordinary one, for which the (–) sign would be chosen. Equations (2.32) and (2.34) can be used to represent Equation (2.36) more simply as:

$$n_{1,2}^2 = 1 - v\left[1 + \frac{u}{2}\sin^2\beta \mp \frac{1}{2}\sqrt{u^2\sin^2\beta + 4u\cos^2\beta}\right]. \tag{2.37}$$

It is often supposed that $u = 0$ for ultra-high-frequency (UHF) and microwave regions, in which case the ordinary and extraordinary waves do not differ, and only one wave exists in the plasma and has the index of refraction:

$$n = \sqrt{1-v} = \sqrt{1 - \frac{\omega_p^2}{\omega^2}} \cong 1 - \frac{\omega_p^2}{2\omega^2}. \tag{2.38}$$

Later, we will define more precisely when it is sufficient to use the approximation for wave propagation in the ionosphere, but for now we will say only that refraction index $n < 1$ in this approximation, which means that the phase velocity of waves in plasma is greater than the velocity of light.

Equation (2.26) allows us to express the longitudinal component of field E_z via the transversal components according to the equality:

$$E_z \cong iv\sqrt{u}\sin\beta[E_x - \sqrt{u}E_y\cos\beta]. \tag{2.39}$$

The expression presented here shows that the longitudinal components of waves are smaller than the transversal ones; therefore, waves in the ionosphere at high enough frequencies can be considered transversal in all events.

Finally, the polarization coefficient is an important characteristic which is described by the relation:

$$K_{1,2} = \frac{E_x}{E_y} = i\frac{2\cos\beta}{\sqrt{u\sin^4\beta + 4\cos^2\beta} \pm \sqrt{u}\sin^2\beta}. \tag{2.40}$$

at high enough frequencies. Because u is small, the condition $2\cos\beta >> \sqrt{u}\sin^2\beta$ is true over a wide range of angles and leads to quasi-longitudinal propagation at UHF and microwave ranges. Thus, the approximations:

$$n_1 = n_o \cong 1 - \frac{v}{2}\left(1 - \sqrt{u}\cos\beta\right), \quad n_2 = n_e \cong 1 - \frac{v}{2}\left(1 + \sqrt{u}\cos\beta\right). \tag{2.41}$$

are correct for both wave types. The polarization coefficients are defined as follows:

$$K_1 = K_o \cong i, \qquad K_2 = K_e \cong -i. \tag{2.42}$$

Hence, the ordinary and extraordinary waves are circularly polarized with directions of rotation opposite those of the polarization planes.

2.4 ROTATION OF POLARIZATION PLANE (FARADAY EFFECT)

The possibility of the existence of two types of waves in magnetized plasma results in some specific effects, one of them being rotation of the plane of polarization, known as the *Faraday effect*. Let us imagine that a linearly polarized plane wave is incident on a layer of magnetized plasma. A plane with invariable linear polarization is not able to propagate in the plasma considered here, and, as we have just established, only the existence of circular polarized waves is possible, both ordinary and special waves. They are excited at the plasma input, adding in such a way that their sum is equal to the linearly polarized incident wave (taking into account, of course, the processes of reflection and penetration at the plasma boundary). If the phase velocities of ordinary and extraordinary waves are the same, then a wave with invariable linear polarization would propagate; however, in this case, the velocities are different, which means, for instance, that the electrical vectors of ordinary and extraordinary waves turn in opposite directions at different angles. This difference in angle rotation leads to rotation of the summary polarization vector, the electrical one, at an angle, and is known as the Faraday effect. The described rotation differs in essence from the rotation of electrical (and, of course, magnetic) vectors of circular polarized waves in that it rotates with the field frequency at each point of space. In this case, the polarization direction is left unchangeable at each point of the space and changes only during transition from point to point in the wave propagation direction.

Elementary calculations show that the value of the summary wave rotation angle is:

$$\Psi_{\mathrm{F}} = \frac{\omega}{2c}\left(n_o - n_{\mathrm{e}}\right)L \cong \frac{\omega v\sqrt{u}L\cos\beta}{2c}, \tag{2.43}$$

where L is the distance passed by the wave in plasma. On the basis of this formula, it is easy to establish that the Faraday angle of rotation is proportional to half of the phase difference of ordinary and extraordinary waves when they pass distance L. By using our expressions for u and v in Equations (2.32) and (2.34), we obtain:

$$\Psi_{\mathrm{F}} = \frac{\omega_{\mathrm{p}}^2\omega_{\mathrm{H}}L\cos\beta}{2\omega^2 c} = \frac{e^3 NH_0 L\cos\beta}{2\pi m^2 c^2 f^2} = 2.36\cdot 10^4\,\frac{NH_0 L\cos\beta}{f^2}. \tag{2.44}$$

The estimations carried out for the ionosphere of the Earth show that the angle of Faraday rotation is sizeable even at frequencies of hundreds of megahertz, and it should be taken into consideration when designing radio systems of this range. Measurement of the plane polarization angle rotation can be used for estimating the electron content, as the magnetic field strength of the Earth is known.

The reduced formulas help to answer the question of when we should take into consideration the terms with $\sqrt{u}\cos\beta$ in Equation (2.41). If the Faraday angle is small, then the difference between ordinary and extraordinary waves is insignificant; otherwise, it is necessary to take this difference into account, at least, while analyzing polarization phenomenon.

2.5 GENERAL CHARACTERISTICS OF POLARIZATION AND STOKES PARAMETERS

Linear and circular polarization, as discussed previously, are particular cases. In this section, we will consider general characteristics of polarization and interpolate parameters that describe these characteristics with sufficient complexity. Let us choose the z-axis as the wave propagation direction. We shall assume that the waves are completely transversal with the components of the electrical field:

$$\mathrm{E}_{\mathrm{x}} = \hat{\mathrm{E}}_{\mathrm{x}}\exp\left(iqz\text{-}i\omega t + i\Phi_{\mathrm{x}}\right), \qquad \mathrm{E}_{\mathrm{y}} = \hat{\mathrm{E}}_{\mathrm{y}}\exp\left(iqz\text{-}i\omega t + i\Phi_{\mathrm{y}}\right). \tag{2.45}$$

Here, Φ_{x} and Φ_{y} are initial phases of the x- and y-components of the field. So, amplitudes E_{x} and E_{y} can be considered as real values. If we write Equation (2.45) in the view of real expressions and take the real part of the right part and exclude the phase $qz - \omega t$, then it is easy to establish that the summary field vector $\mathbf{E} = E_x\mathbf{e}_x + E_y\mathbf{e}_y$ is elliptically polarized. The ellipse of polarization is described by the equation:

$$\left(\frac{E_x}{\hat{E}_x}\right)^2 + \left(\frac{E_y}{\hat{E}_y}\right)^2 - 2\frac{E_x E_y}{\hat{E}_x \hat{E}_y}\cos\Phi = \sin^2\Phi, \tag{2.46}$$

where $\Phi = \Phi_x - \Phi_y$ is the phase difference of the vector $\mathbf{E}$ components. The large axis of ellipse is inclined by the angle ψ to the x-axis. It is easy to define this angle with the equality:

$$tg2\psi = \frac{S_2}{S_1}, \tag{2.47}$$

where:

$$S_2 = 2\hat{E}_x\hat{E}_y\cos\Phi = 2\,\mathrm{Re}\left(E_x E_y^*\right), \quad S_1 = \hat{E}_x^2 - \hat{E}_y^2 = \left|E_x\right|^2 - \left|E_y\right|^2. \tag{2.48}$$

The ellipse radii are defined as follows:

$$a^2 = S_0 + \sqrt{S_1{}^2 + S_2{}^2}, \quad b^2 = S_0 - \sqrt{S_1{}^2 + S_2{}^2}. \tag{2.49}$$

where

$$S_0 = \hat{E}_x^2 + \hat{E}_y^2 = \left|E_x\right|^2 + \left|E_y\right|^2 \tag{2.50}$$

is the wave intensity.

The values S_0, S_1, S_2, and

$$S_3 = 2\hat{E}_x\hat{E}_y\sin\Phi = 2\,\mathrm{Im}\left(E_x E_y^*\right) \tag{2.51}$$

are called *Stokes parameters* and characterize the polarization property of transversal plane waves. It is easy to be convinced of the truth of the relation:

$$S_0^2 = S_1^2 + S_2^2 + S_3^2, \tag{2.52}$$

which takes place for coherent waves. The problem of combined coherent waves and noise radiation will be examined later.

The ellipse radii a and b and the angle of inclination, ψ, are defined at polarization measurements. Stokes parameters are calculated according to the formulas:

$$S_0 = \frac{a^2 + b^2}{2}, \quad S_1 = \frac{\left(a^2 - b^2\right)\cos 2\psi}{2}, \quad S_2 = \frac{\left(a^2 - b^2\right)\sin 2\psi}{2}, \quad S_3 = ab, \tag{2.53}$$

and correspondingly the field components are calculated as:

$$\hat{E}_x = \sqrt{\frac{S_0 + S_1}{2}}, \quad \hat{E}_y = \sqrt{\frac{S_0 - S_1}{2}}, \quad \tan\Phi = \frac{S_3}{S_2}. \tag{2.54}$$

Equation (2.52) allows us to consider the normalized Stokes parameters:

$$s_1 = \frac{S_1}{S_0}, \quad s_2 = \frac{S_2}{S_0}, \quad s_3 = \frac{S_3}{S_0} \tag{2.55}$$

as the components of a three-dimensional vector (*Stokes vector*) of unitary length. Its point lies on a sphere of unitary radius, which is called the Poincare sphere. The position of the Stokes vector point on the Poincare sphere determines the polarization property of the wave.

Equation (2.52), as has already been pointed out, is valid only for coherent waves. It is not valid in the presence of causal waves (thermal radiation waves, waves of random scattering, etc.). We will now analyze this problem in more detail. Let the wave be a mixture of coherent and completely casual waves. Then, the field components may be represent as:

$$E_x = \hat{E}_x \exp\left(iqz - i\omega t + i\Phi_x\right) + N_x, \quad E_y = \hat{E}_y \exp\left(iqz - i\omega t + i\Phi_y\right) + N_y. \tag{2.56}$$

Here, N_x and N_y describe the casual components of the field, and it is supposed that $\langle N_x \rangle = \langle N_x \rangle = 0$. Because of the statistical character of the problem, the Stokes parameters should be considered as statistically averaged values. Then,

$$S_0 = \left\langle \left|E_x\right|^2 \right\rangle + \left\langle \left|E_y\right|^2 \right\rangle = \hat{E}_x^2 + \hat{E}_y^2 + \left\langle \left|N_x\right|^2 \right\rangle + \left\langle \left|N_y\right|^2 \right\rangle, \tag{2.57}$$

$$S_1 = \left\langle \left|E_x\right|^2 \right\rangle - \left\langle \left|E_y\right|^2 \right\rangle = \hat{E}_x^2 - \hat{E}_y^2 + \left\langle \left|N_x\right|^2 \right\rangle - \left\langle \left|N_y\right|^2 \right\rangle, \tag{2.58}$$

$$S_2 = 2\,\mathrm{Re}\left\langle E_x E_y^* \right\rangle = 2\hat{E}_x \hat{E}_y \cos\Phi + 2\,\mathrm{Re}\left\langle N_x N_y^* \right\rangle, \tag{2.59}$$

$$S_3 = 2\,\mathrm{Im}\left\langle E_x E_y^* \right\rangle = 2\hat{E}_x \hat{E}_y \sin\Phi + 2\,\mathrm{Im}\left\langle N_x N_y^* \right\rangle. \tag{2.60}$$

Generally, the power difference of the noise orthogonal components and the existence of correlation between them may be assumed. Further, let us represent the noise components in the form:

$$N_x = |N_x| e^{i\gamma_x}, \quad N_y = |N_y| e^{i\gamma_y}, \qquad \gamma = \gamma_x - \gamma_y.$$

Taking into account the introduced designations, Equation (2.52) should be replaced by:

$$S_0^2 - S_1^2 - S_2^2 - S_3^2 = 4\Big\{\hat{E}_x^2\left\langle |N_y|^2 \right\rangle + \hat{E}_y^2\left\langle |N_x|^2 \right\rangle + \left\langle |N_x| \right\rangle \left\langle |N_y|^2 \right\rangle -$$

$$-\left[\left\langle |N_x||N_y| \cos\gamma \right\rangle\right]^2 - \left[\left\langle |N_x||N_y| \sin\gamma \right\rangle\right]^2 - 2\hat{E}_x\hat{E}_y\left\langle |N_x||N_y| \cos(\Phi - \gamma) \right\rangle\Big\}. \quad (2.61)$$

It is not difficult to establish that, in the general case,

$$S_0^2 \geq S_1^2 + S_2^2 + S_3^2. \quad (2.62)$$

Therefore, the coefficient

$$m = \frac{S_1^2 + S_2^2 + S_3^2}{S_0^2} \quad (2.63)$$

defines the degree of wave coherence, where $m = 1$ for fully coherent waves.

The usefulness of applying the Stokes parameters is determined by two circumstances. The first one is connected with the fact that the Stokes parameters of a combined wave are equal to the sums of corresponding parameters of the partial waves (additive property); thus, incoherence of the partial waves is assumed. The second circumstance is connected with changes of the Stokes parameters in the processes of linear wave transformation. The change of polarization generally takes place in the processes of wave propagation or transformations in devices. The linear transformation will be considered later. For now, let E_x^i and E_y^i be the orthogonal field components at the input of the device, and let E_x^t and E_y^t be the same components at its output. The linear relations between them are:

$$E_x^t = d_{xx}E_x^i + d_{xy}E_y^i, \qquad E_y^t = d_{yx}E_x^i + d_{yy}E_y^i. \quad (2.64)$$

The second expression permits us to connect the Stokes matrix at the output of the device with the Stokes matrix at the input.[65] This connection appears as:

$$\mathbf{S}_t = \mathbf{MS}_i. \tag{2.65}$$

Here, matrix **M** is the *Mueller matrix*. We do not provide the components of the Mueller matrix here, as they are overall rather cumbersome (for details, see, for example, O'Neill[65]).

2.6 SIGNAL PROPAGATION IN DISPERSION MEDIA

Previously, we considered sine-wave propagation in general, pointing out that propagation laws of complex in spectral structure waves (signals) will not differ essentially from the same ones for sine waves. In general terms, these propagation laws are valid only for media for which characteristics vary only a little bit within the spectral band of signals (i.e., media with a weakly expressed frequency dispersion), as occurs very often for natural media; however, we also noted the strong frequency dependence of the refraction coefficient when we discussed the problems of wave propagation in plasma. The frequency dispersion is well defined in this case. It means, in particular, that the phase velocity of waves depends on the frequency. Because the complex signal is the sum of the spectral components, the phase relations between them change during the propagation process due to phase velocity difference, and, correspondingly, the sum itself changes also, which produces distortion of the signal form.

The aim of this section is to consider this phenomenon. We will address scalar waves but will confine our discussion to the case of absorption absence, thus our examples will refer to waves in isotropic plasma. Let us consider the case when a wave propagates along the positive direction of the z-axis. Let $E(0,t)$ be the primary signal form (at $z = 0$). Its spectrum can be written as:

$$\tilde{E}(\omega) = \frac{1}{2\pi}\int_{-\infty}^{\infty} E(0,t)e^{i\omega t}dt. \tag{2.66}$$

The function $E(0,t)$ is the real function; therefore, the spectrum has the property $\tilde{E}(-\omega) = \tilde{E}^*(\omega)$. Hence, it follows that, if $\mu(\omega) = \arg E(\omega)$, then $\mu(-\omega) = -\mu(\omega)$. In the propagation process, every spectral component satisfies $\tilde{E}(\omega)\exp[iq(\omega)z\text{-}i\omega t]$, where $q(\omega) = (\omega/c)\sqrt{\varepsilon(\omega)} = (\omega/c)n(\omega)$ is the wave number in the media. So, at distance z, the signal will be described, as a whole, by the Fourier integral:

$$E(z,t) = \int_{-\infty}^{\infty} \tilde{E}(\omega)e^{iq(\omega)z-i\omega t}d\omega. \tag{2.67}$$

We can show that the very first oscillations reach point z with velocity *c*. The corresponding calculations show that weak oscillation arises just after time moment $t = z/c$. These oscillations are usually referred to as *precursors*.

Here, we will confine ourselves to the case when $t > z/c$ so the stationary phase method may be used for asymptotic calculation of the integral shown in Equation (2.67). The points of the stationary phase are found from:

$$z q'(\omega_s) = t \tag{2.68}$$

in accordance with the procedure of this method. Before moving on to analysis of the roots of this last equation, we will recap the basic analytic properties of $q(\omega)$, which are evident from dielectric theory. We will restrict ourselves to the case of weak wave absorption and neglect the imaginary part of permittivity so we may consider $\varepsilon(\omega)$ to be an even function of frequency. Correspondingly, the wave number is an odd function of frequency, and its first derivative is an even one. So, Equation (2.68) has at least two real roots differing in symbol (under the condition that the derivative of the wave number is not constant). The second derivative of the wave number has to be an odd function of frequency. In terms of the analytical properties of the spectra, we have:

$$E(z,t) = 2\left|\tilde{E}(\omega_s)\right| \sqrt{\frac{\pi}{z\left|q''(\omega_s)\right|}} \cos\left[z q(\omega_s) - \omega_s t + \mu(\omega_s) + \operatorname{sgn} q''(\omega_s)\frac{\pi}{4}\right]. \tag{2.69}$$

The traits mean differentiation with respect to argument. In the case of plasma, $q(\omega) = 1/c\,\sqrt{\omega^2 - \omega_p^2}$, and Equation (2.68) has the obvious solutions:

$$\omega_s = \pm\frac{\omega_p t}{\sqrt{t^2 - z^2/c^2}}.$$

Note that as $t \to z/c$ and $\omega_s \to \infty$ precursors generally are formed at the expense of the high-frequency part of the signal spectrum.

Let us now address the propagation problem in a dispersion medium of narrow-band signals by examining signals for which the spectra are concentrated close to the carrier frequency ω_0 and satisfy:

$$\tilde{E}(\omega) = g(\omega - \omega_0). \tag{2.70}$$

The maximum of function *g* corresponds to the frequency $\omega = \omega_0$; therefore, if the stationary phase point ω_s differs considerably from the carrier frequency ω_0, then the spectral density of the precursor is small and does not carry the main signal power. The main signal power occurs when $\omega_s = \omega_0$ at the moment:

$$t_g = \mathrm{z}\,\mathrm{q}'(\omega_0). \tag{2.71}$$

The time moment introduced in such a way is referred to as the *group time*, and the main signal energy propagates with group velocity:

$$v_g = \frac{\mathrm{z}}{t_g} = \frac{1}{\mathrm{q}'(\omega_0)}. \tag{2.72}$$

Generally speaking, the stationary phase method cannot be used close to the maximum of the narrow-band signal spectral density. Slow change by all integrands (except for the fast-oscillating exponent sought by the stationary phase method) is usually assumed in the stationary phase method; however, the signal spectra are not slowly changing functions because of the narrow band. This is the reason why more detailed analyses of processes close to the maximum of the signal spectral density are required. Precursors should not be mentioned in this case, as we are dealing with the main part of the signal. For our purposes here, let us make some transformations to give us a more convenient representation of the narrow-band signal. Its spectral expansion now has the form:

$$E(\mathrm{z},t) = \int_{-\infty}^{\infty} g(\omega - \omega_0)\, e^{i\mathrm{q}(\omega)\mathrm{z} - i\omega t}\, d\omega.$$

It is logical to introduce a new variable, $\Omega = \omega - \omega_0$, and the integration is made within the narrow spectral band of the signal, so the expansion $\mathrm{q}(\omega) \cong \mathrm{q}(\omega_0) + \mathrm{q}'(\omega_0)\Omega + \mathrm{q}''(\omega_0)\Omega^2/2$ can be used. As a result, we obtain the next form of signal presentation:

$$E(\mathrm{z},t) = e^{i\mathrm{q}(\omega_0)\mathrm{z} - i\omega_0 t}\, \hat{\mathrm{E}}(\mathrm{z},t), \tag{2.73}$$

where:

$$\hat{\mathrm{E}}(\mathrm{z},t) = \int_{-\infty}^{\infty} g(\Omega) \exp\left[-i\Omega\left(t - \frac{\mathrm{z}}{v_g} \right) + i\,\mathrm{sgn}\,\mathrm{q}''(\omega_0) \frac{\tilde{\tau}^2 \Omega^2}{4} \right] d\Omega\,. \tag{2.74}$$

where the time constant is:

$$\tilde{\tau} = \sqrt{2\left|\mathrm{q}''(\omega_0)\right| \mathrm{z}}\,. \tag{2.75}$$

It is easy to conclude that:

$$g(\Omega)=\frac{1}{2\pi}\int_{-\infty}^{\infty}\hat{\mathrm{E}}(0,t)e^{i\Omega t}dt. \tag{2.76}$$

Sometimes, it is convenient to express the envelope via the time domain form of the primary signal. By combining Equations (2.74) and (2.76), the frequency integration can be easily realized:

$$\hat{\mathrm{E}}(\mathrm{z},t)=\frac{e^{\operatorname{sgn} q''(\omega_0)i\pi/4}}{\sqrt{\pi}}\int_{-\infty}^{\infty}\hat{\mathrm{E}}\left(0,\ t-\mathrm{z}/v_{\mathrm{g}}+\tilde{\tau}\xi\right)e^{-\operatorname{sgn} q''(\omega_0)i\,\xi^2}d\xi. \tag{2.77}$$

Note the effective spectral width of the signal in terms of $\Delta\Omega$.

We will now consider the case when $\Delta\Omega^2\tilde{\tau}^2 << 1$. As the integration in Equation (2.74) is carried out within the effective spectral band of the signal, we may neglect the quadratic by Ω item and write:

$$\hat{\mathrm{E}}(\mathrm{z},t)=\int_{-\infty}^{\infty}g(\Omega)\exp\left[-i\Omega\left(t-\mathrm{z}/v_{\mathrm{g}}\right)\right]d\Omega=\hat{\mathrm{E}}\left(0,t-\mathrm{z}/v_{\mathrm{g}}\right).$$

Thus, under the conditions assumed here, the signal envelope propagates at group velocity without form distortion. These conditions occur when the dispersion is weak, when the wave passes a small distance, or when the spectral band of the signal is sufficiently narrow. These arguments suggest the main role of time τ at the signal form distortion with the propagation in dispersion medium.

The introduced value τ is the conceptual time of medium relaxation; this time depends not only on the characteristics of the medium but also on the distance passing by the wave; hence, we can say that signals with spectra wider than $1/\tau$ will exhibit form distortion. The degree of this distortion can be estimated by the coefficient of similarity (convolution):

$$K_{\mathrm{d}}=\operatorname{Re}\frac{1}{W}\int_{-\infty}^{\infty}\hat{\mathrm{E}}(\mathrm{z},t)\mathrm{E}^{*}\left(0,t-\mathrm{z}/v_{\mathrm{g}}\right)dt, \tag{2.78}$$

where the value:

$$W=\int_{-\infty}^{\infty}\left|\hat{\mathrm{E}}(\mathrm{z},t)\right|^2dt=\int_{-\infty}^{\infty}\left|\hat{\mathrm{E}}(0,t)\right|^2dt=2\pi\int_{-\infty}^{\infty}\left|g(\Omega)\right|^2d\Omega \tag{2.79}$$

is proportional to the total signal energy. The corresponding calculation gives the result:

$$K_{\mathrm{d}} = \frac{\int_{-\infty}^{\infty} \left|g(\Omega)\right|^2 \cos\left(\Omega^2\tilde{\tau}^2/4\right) d\Omega}{\int_{-\infty}^{\infty} \left|g(\Omega)\right|^2 d\Omega}. \tag{2.80}$$

A comparison of the shape of the signal passing at a definite distance in the medium with its undistorted sample is provided by this convolution coefficient.

It is not difficult to approximate the two extreme cases by ignoring the concrete spectrum view of the envelope. The integration in Equation (2.80) is made within the spectral width of signal $\Delta\Omega$. In those cases when $\Delta\Omega\tilde{\tau} << 1$, $\cos(\Omega^2\tilde{\tau}^2/4) \cong 1$ and $K_{\mathrm{d}} \cong 1$, and the lack of distortion in this case should be noted. In the opposite case, when $\Delta\Omega\tilde{\tau} >> 1$, the main area of integration in the upper integral of Equation (2.80) is concentrated within the interval $\pm 1/\tilde{\tau}$. Therefore,

$$\int_{\infty}^{\infty} \left|g(\Omega)\right|^2 \cos\left(\Omega^2\tilde{\tau}^2/4\right) d\Omega \cong \left|g(0)\right|^2 \int_{-\infty}^{\infty} \cos\left(\Omega^2\tilde{\tau}^2/4\right) d\Omega = \frac{\sqrt{2\pi}}{\tilde{\tau}} \left|g(0)\right|^2.$$

The estimation:

$$\int_{\infty}^{\infty} \left|g(\Omega)\right|^2 d\Omega \cong \left|g(0)\right|^2 \Delta\Omega$$

is applicable for the lower integral. As a result:

$$K_{\mathrm{d}} \cong \frac{\sqrt{2\pi}}{\Delta\Omega\tilde{\tau}} << 1, \tag{2.81}$$

and the similarity is practically absent in this case because of strong distortion of the signal form. For these reasons, we must define the medium pass band (i.e., the spectral interval within which the signal passes a given distance without noticeable distortion). Such a bandwidth can be estimated on the base of the condition $\Delta\Omega^2\tilde{\tau}^2/4 = \pi/2$, or:

$$\Delta F = \frac{\Delta\Omega}{2\pi} = \frac{1}{\sqrt{2\pi}\tilde{\tau}}. \tag{2.82}$$

Equation (2.74) can be rewritten as:

$$\hat{E}(z,\tau_g) = \int_{-\infty}^{\infty} g(\Omega)\exp\left[-i\Omega\tau_g + i\frac{q''(\omega)}{2}z\Omega^2\right]d\Omega. \tag{2.83}$$

If we now consider the envelope as a function of independent variables z and τ_g, then it is easy to obtain for it the parabolic equation:

$$\frac{\partial \hat{E}}{\partial z} + \frac{iq''(\omega)}{2}\frac{\partial^2 \hat{E}}{\partial \tau_g^2} = 0. \tag{2.84}$$

This equation has the imaginary coefficient of diffusion:

$$D(\omega) = \frac{iq''(\omega)}{2} = \frac{i}{2}\frac{d}{d\omega}\left[\frac{1}{v_g(\omega)}\right] = -\frac{i}{2v_g^2}\frac{dv_g}{d\omega}. \tag{2.85}$$

The imaginary quantity of the diffusion coefficient permits us to compare the frequency dispersion processes with the diffraction phenomenon, which is also often described by a parabolic equation. In particular, the impulse diffusion due to frequency dispersion may be compared with the wave scattering in the diffraction process.

Let us, finally, apply the obtained results for the plasma. Here,

$$q(\omega) = \frac{1}{c}\sqrt{\omega^2 - \omega_p^2}, \quad v_g(\omega) = c\sqrt{1 - \frac{\omega_p^2}{\omega^2}}, \quad \tilde{\tau} = \sqrt{\frac{2\omega_p^2 z}{c\left(\omega^2 - \omega_p^2\right)^{3/2}}}. \tag{2.86}$$

By assuming that $\omega >> \omega_p$ for the ionosphere, then:

$$\tilde{\tau} = \frac{\omega_p}{\omega^{3/2}}\sqrt{\frac{2z}{c}} = \frac{1}{f^{3/2}}\sqrt{\frac{e^2 Nz}{\pi^2 mc}} = \frac{2.92\cdot 10^{-2}\sqrt{Nz}}{f^{3/2}}. \tag{2.87}$$

And the pass band is:

$$\Delta F = \frac{13.6}{\sqrt{Nz}} f^{3/2}. \tag{2.88}$$

2.7 DOPPLER EFFECT

In all the cases considered above, the sources and receivers were assumed to be motionless relative to each other. Now, we will analyze the situation when mutual movement of a source and a receiver occurs with constant velocity, which leads to new circumstances — in particular, the frequency change of electromagnetic waves, or the phenomenon known as the *Doppler effect.* To consider this problem, let us introduce two Cartesian coordinate systems (x,y,z) and (x′,y′,z′). Let us suppose that the coordinate axes of both systems are directed similarly and that the touched system moves evenly and rectilinearly with velocity **v** relative to the untouched system. These systems of coordinates are inert with respect to each other. The coordinates and time are transformed using Lorentz's transform formulae in accordance with the theory of relativity[15,16] and which are written in the general case as:

$$\mathbf{r} = \mathbf{r}' + \mathbf{v}\left[\frac{(\mathbf{r}'\cdot\mathbf{v})}{\mathrm{v}^2}\left(\frac{1}{\sqrt{1-\beta^2}} - 1\right) + \frac{t'}{\sqrt{1-\beta^2}}\right], \tag{2.89}$$

$$t = \frac{t' + \dfrac{(\mathbf{r}'\cdot\mathbf{v})}{c^2}}{\sqrt{1-\beta^2}}. \tag{2.90}$$

Here, **r** and t are the radius vector of the point and the time in the non-dotted coordinate system, respectively, and **r**′ and t' are the same values in the dotted system; $\beta = \mathrm{v}/c$.

To be more certain, we will assume that source is at the beginning of the dotted coordinate system and moves together with it. Because the electromagnetic wave velocity is constant in both systems (permittivity is equal to unity everywhere), the radiowave phase is the invariant; that is, it does not change due to transition from one coordinate system to another. In other words, the equality $\mathbf{qr} - \omega t = \mathbf{q}'\mathbf{r}' - \omega' t'$, where $\mathbf{q}$, $\mathbf{q}'$, ω, and ω' are the wave vectors and the frequencies in the dotted and nondotted systems, respectively, is correct. If we instead use $\mathbf{q} = \omega/c\mathbf{e}$ and $\mathbf{q}' = \omega'/c\mathbf{e}'$, where **e** and **e**′ are unit vectors of the wave direction in the coordinate systems being considered, then the previous equality can be rewritten as:

$$\omega\left[t - \tfrac{1}{c}(\mathbf{e}\cdot\mathbf{r})\right] = \omega'\left[t' - \tfrac{1}{c}(\mathbf{e}'\cdot\mathbf{r}')\right].$$

Let us now substitute the untouched coordinates and time for dotted ones using the Lorentz transform. The relation established in this way can be written out only in dotted variables:

$$\omega\left\{\frac{t'+\frac{(\mathbf{r}'\cdot\mathbf{v})}{c^2}}{\sqrt{1-\beta^2}}-\frac{1}{c}\left[(\mathbf{e}\cdot\mathbf{r}')+(\mathbf{e}\cdot\mathbf{v})\left[\frac{(\mathbf{r}'\cdot\mathbf{v})}{\mathrm{v}^2}\left(\frac{1}{\sqrt{1-\beta^2}}-1\right)+\frac{t'}{\sqrt{1-\beta^2}}\right]\right]\right\}=$$

$$=\omega'\left[t'-\frac{1}{c}(\mathbf{e}'\cdot\mathbf{r}')\right].$$

Equating all the components with time t', we obtain the following formula for the frequency transformation:

$$\omega=\frac{\omega'\sqrt{1-\beta^2}}{1-\frac{(\mathbf{e}\cdot\mathbf{v})}{c}}=\frac{\omega'\sqrt{1-\beta^2}}{1-\beta\cos\vartheta}, \tag{2.91}$$

where ϑ is the angle between the wave propagation direction in the nondotted system of coordinates and the radiator movement velocity. Equating the components with the radius vector of the observation point in the dotted coordinate system, we have:

$$\frac{\mathbf{v}}{c}-\mathbf{e}\sqrt{1-\beta^2}-\frac{\mathbf{v}(\mathbf{e}\cdot\mathbf{v})}{\mathrm{v}^2}\left(1-\sqrt{1-\beta^2}\right)=-\mathbf{e}'\left(1-\frac{\mathbf{e}\cdot\mathbf{v}}{c}\right).$$

We multiply this equality by $\mathbf{v}$ to obtain:

$$\mathbf{e}\cdot\mathbf{v}=\frac{\mathbf{e}'\cdot\mathbf{v}+\frac{\mathrm{v}^2}{c}}{1+\frac{\mathbf{e}'\cdot\mathbf{v}}{c}}$$

or, in more common form:

$$\cos\vartheta=\frac{\cos\vartheta'+\beta}{1+\beta\cos\vartheta'}. \tag{2.92}$$

The last formula describes the aberration phenomenon, or the change of wave propagation direction due to relative movement of the source and the receiver.

As noted earlier, a frequency change due to mutual movement of a source and a receiver is known as the Doppler effect, which we will now discuss in more detail. We will first consider the case when $\vartheta = 0$, which occurs when the source and the receiver of radiation are moving toward each other. Then,

$$\omega = \omega' \sqrt{\frac{1+\beta}{1-\beta}}, \tag{2.93}$$

which occurs when the frequency of the received wave is more than the frequency of the radiation.

In the opposite case of mutual moving away, when $\vartheta = \pi$:

$$\omega = \omega' \sqrt{\frac{1-\beta}{1+\beta}}, \tag{2.94}$$

and it is easy to come to the conclusion that in this case the received wave frequency is less than the frequency of the transmitter.

When $\vartheta = \pi/2$:

$$\omega = \omega' \sqrt{1-\beta^2} < \omega'. \tag{2.95}$$

We have a right to suppose that $\beta = v/c << 1$ in all cases; therefore, with the limited accuracy of the second order, we obtain:

$$\frac{\omega = \omega'}{\omega'} = \frac{\Delta f_d}{f} \beta \cos\vartheta + \frac{\beta^2}{2} \cos 2\vartheta . \tag{2.96}$$

In particular, the very relative transversal Doppler effect ($\vartheta = \pi/2$) has value order $\beta^2/2$ and is particularly noticeable at space velocities. For example, at $v = 7$ km/sec, $\beta^2/2 = 2.72 \cdot 10^{-10}$, which means the Doppler shift value is 0.27 Hz at a radiation frequency of 1 GHz, which can be registered by modern devices. Further, as a rule, we will restrict ourselves to a linear velocity approximation of the Doppler effect when, as it is easy to see, the frequency shift is proportional to the radial by the observation beam component of the mutual velocity. It is convenient, in this case, to establish a formula to describe this Doppler effect. Let us consider the wave phase as a function of time and the reception point coordinates relative to the point of radiation. The coordinates are also functions of time because of mutual movement; therefore, we have to consider $\Phi = \Phi[t, \mathbf{r}(t)]$, and the frequency can be determined as:

$$\omega = -\frac{d\Phi}{dt} = -\frac{\partial\Phi}{\partial t} = \mathbf{v} \cdot \nabla\Phi . \tag{2.97}$$

The (–) sign is used here because the time dependence is expressed in the form $e^{-i\omega t}$. When $\Phi = kr - \omega t$, the Doppler frequency shift is:

$$\Delta f = \frac{\Delta\omega}{2\pi} = \frac{\omega - \omega_0}{2\pi} = -\frac{\mathbf{v} \cdot \nabla \mathbf{r}}{\lambda} = -\frac{\mathbf{v} \cdot \mathbf{e}}{\lambda} . \tag{2.98}$$

3 Wave Propagation in Plane-Layered Media

3.1 REFLECTION AND REFRACTION OF PLANE WAVES AT THE BORDER OF TWO MEDIA

Natural media — the atmosphere, earth, and others — can be supposed to be homogeneous only within bounded area of space. In reality, the permittivity of these media is a function of the coordinates and, in general terms, time, which is ignored, as a rule, because of the comparative sluggishness of natural processes. As usual, the time required for electromagnetic wave propagation in a natural medium is much less than typical periods of medium property changes.

In the first approximation, natural media can be considered to be plane layered; that is, their permittivity changes in only one direction. If a Cartesian coordinates system is chosen in such a way that one of the coordinates (for example, z) coincides with this direction, then we may say that permittivity depends on this coordinate only. For the atmosphere of Earth, a concept of spherical-layered media would be more correct, and this idea is considered in the following chapters; however, we should point out that the curvature radius of media layers of the Earth is so large that the concept of plane stratification is sufficient in many cases.

In this chapter, we will study only plane wave propagation in plane-layered media. We will assume waves as the plane only conventional because surfaces with constant phase and constant amplitude are not planes in the cases of wave propagation direction inclined to the layers. On the whole, wave parameters cannot depend on only one particular Cartesian coordinate in all cases. In most of the cases that we will consider here, media parameters change little compared to wavelength; therefore, surfaces of equal phase or amplitude have sufficiently large curvature radii to be considered locally plane. It is quite acceptable to talk about plane waves in these terms.

In some situations, medium properties can be changed sharply and on a wavelength scale; however, regions of such great change are usually rather thin layers and the waves are plane, outside of the layers just discussed. These layers correspond, in the extreme, to the areas jumpy changes of permittivity and occur at the boundary of two media. The air–ground boundary is an example of this idea.

We turn now to consideration of plane wave reflection and transmission processes at a plane interface. We shall assume for the sake of simplicity that the permittivity of the medium from where the wave originates is equal to unity. Air is an example of such a medium. The problem of a wave incident on the ground from the air is investigated in such a way. The permittivity in this case is indicated by ε. The problem of wave reflection and transmission in media separated by a plane interface is well known and has been analyzed in many texts regarding electromagnetic waves; therefore, we will not investigate this problem in detail here and will examine only essential formulae.

The electrical and magnetic fields of the incident wave are described by the vectors:

$$\mathbf{E}_i = \mathbf{E}_i^0 e^{i\mathbf{q}_i \cdot \mathbf{r}},\ \mathbf{H}_i = \mathbf{H}_i^0 e^{i\mathbf{q}_i \cdot \mathbf{r}},\ \mathbf{H}_i^0 = \frac{1}{k}\left[\mathbf{q}_i \times \mathbf{E}_i^0\right],\ \mathbf{E}_i^0 = -\frac{1}{k}\left[\mathbf{q}_i \times \mathbf{H}_i^0\right],\ \mathrm{q}_i = k. \tag{3.1}$$

The boundary of the two media will be assumed to be a plane that is perpendicular to the z-axis. Let θ_i represent the incidence angle (see Figure 3.1) in such a way that:

$$\cos\theta_i = -\frac{1}{k}\left(\mathbf{e}_z \cdot \mathbf{q}_i\right). \tag{3.2}$$

Note that vector $\mathbf{q}_i$ can be defined as:

$$\mathbf{q}_\mathrm{i} = \mathbf{w}_\mathrm{i} - \sqrt{k^2 - \mathrm{w}_i^2}\,\mathbf{e}_z,\quad \mathbf{w}_\mathrm{i} \cdot \mathbf{e}_z = 0, \tag{3.3}$$

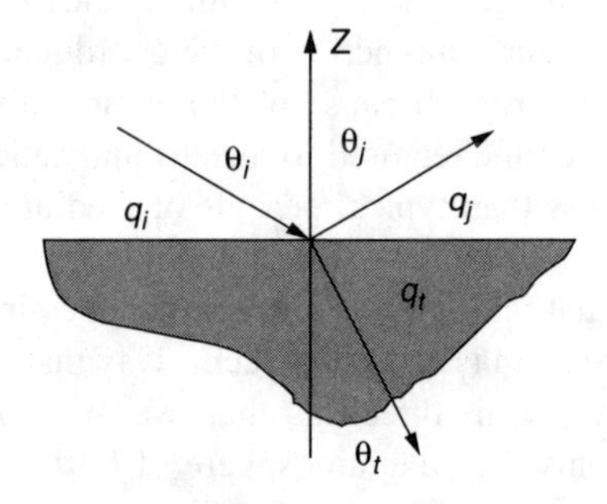

FIGURE 3.1 Plane wave incidence at the plane boundary.

from which it follows that $\mathrm{w}_i = k\sin\theta_i$.

Let us call the field excited by the incident wave in the air the *reflected wave* and represent its fields by $\mathbf{E}_\mathrm{r}$ and $\mathbf{H}_\mathrm{r}$. The wave in the ground is the *refracted wave* or *transmitted wave*, and the fields of this wave are represented by $\mathbf{E}_\mathrm{t}$ and $\mathbf{H}_\mathrm{t}$. It is a simple matter to establish that the reflected and refracted waves are also found in the plane and that their wave vectors lie in the same plane as the vector of the incident wave. The wave vector of the reflected wave is:

$$\mathbf{q}_\mathrm{r} = \mathbf{w}_i + \sqrt{k^2 - \mathrm{w}^2}\,\mathbf{e}_z, \tag{3.4}$$

and the wave vector of the refracted wave is:

$$\mathbf{q}_\mathrm{t} = \mathbf{w}_i - \sqrt{\varepsilon k^2 - \mathrm{w}^2}\,\mathbf{e}_z. \tag{3.5}$$

The angle between vector $\mathbf{q}_\mathrm{r}$ and the z-axis is the *angle of reflection*. It is determined by the relation:

$$\cos\theta_\mathrm{r} = \frac{1}{k}\left(\mathbf{e}_z \cdot \mathbf{q}_\mathrm{r}\right), \tag{3.6}$$

from which it follows that $\theta_r = \theta_i$; that is, the angle of reflection equals the angle of incidence.

In the case of complex permittivity ε, vector $\mathbf{q}_r$ is complex and the transmitted wave is, in general, the inhomogeneous plane wave. The refraction angle determined from the equation:

$$\cos\theta_t = -\frac{1}{k\sqrt{\varepsilon}}\left(\mathbf{e}_z \cdot \mathbf{q}_t\right) = \sqrt{\varepsilon k^2 - w_i^2} \tag{3.7}$$

is also complex in the general case. It is simple, however, to derive Snell's law through the formula:

$$\sqrt{\varepsilon}\sin\theta_t = \sin\theta_i. \tag{3.8}$$

In the case of a weak absorptive medium, when we can neglect the imaginary part of ε, angle θ_i is real, and we can fix the propagation direction of the refracted wave.

Let us point out for later that:

$$\mathbf{q}_r = \mathbf{q}_i - 2\mathbf{e}_z\left(\mathbf{e}_z \cdot \mathbf{q}_i\right), \quad \mathbf{q}_t = \mathbf{q}_i - \left[\left(\mathbf{e}_z \cdot \mathbf{q}_i\right) + \sqrt{k^2(\varepsilon - 1) + \left(\mathbf{e}_z \cdot \mathbf{q}_i\right)^2}\right]\mathbf{e}_z. \tag{3.9}$$

The problem discussed here can be divided into the two cases of horizontal and vertical polarization. In the case of horizontal polarization, the amplitudes of the reflected and refracted waves are connected linearly with the amplitude of the incident wave:

$$\mathbf{E}_r^0 = F_h \mathbf{E}_i^0, \qquad \mathbf{E}_t^0 = T_h \mathbf{E}_i^0, \tag{3.10}$$

where

$$F_h = \frac{\cos\theta_i - \sqrt{\varepsilon - \sin^2\theta_i}}{\cos\theta_i + \sqrt{\varepsilon - \sin^2\theta_i}} \tag{3.11}$$

is the *coefficient of reflection* for the horizontally polarized waves, and

$$T_h = 1 + F_h = \frac{2\cos\theta_i}{\cos\theta_i + \sqrt{\varepsilon - \sin^2\theta_i}} \tag{3.12}$$

is the coefficient of transmission.

The magnetic and electrical wave components exchange places, in some sense, for the case of vertical polarization. In this case, the equations

$$\mathbf{H}_r^0 = F_v \mathbf{H}_i^0, \qquad \mathbf{H}_t^0 = T_v \mathbf{H}_i^0, \tag{3.13}$$

are valid, where the reflection coefficient of the vertically polarized waves is:

$$F_v = \frac{\varepsilon \cos\theta_i - \sqrt{\varepsilon - \sin^2\theta_i}}{\varepsilon \cos\theta_i + \sqrt{\varepsilon - \sin^2\theta_i}}, \tag{3.14}$$

and, correspondingly, the coefficient of transmission:

$$T_v = 1 + F_v = \frac{2\varepsilon \cos\theta_i}{\varepsilon \cos\theta_i + \sqrt{\varepsilon - \sin^2\theta_i}}. \tag{3.15}$$

In these equations, the coefficients F_h and F_v are referred to as the *Fresnel coefficients of reflection*. They are complex values in the general case:

$$F_h = |F_h| e^{i \arg F_h}, \qquad F_v = |F_v| e^{i \arg F_v}, \tag{3.16}$$

and, therefore, wave reflection and refraction are accompanied not only by their amplitude change but also by phase rotation.

Let us now express the fields of the reflected wave for the case of incidence on the interface the plane wave of any linear polarization. For convenience, we will set the vector of direction of the incident, reflected, and transmitted waves using the formulae:

$$\mathbf{q}_i = k\mathbf{e}_i, \quad \mathbf{q}_r = k\mathbf{e}_r, \quad \mathbf{q}_t = k\mathbf{e}_t, \tag{3.17}$$

and we can obtain the expressions:

$$\mathbf{E}_r^0\left[1-(\mathbf{e}_z\cdot\mathbf{e}_i)^2\right] = F_v(\mathbf{e}_z\cdot\mathbf{E}_i^0)\left\{\left[1-2(\mathbf{e}_z\cdot\mathbf{e}_i)^2\right]\mathbf{e}_z + (\mathbf{e}_z\cdot\mathbf{e}_i)\mathbf{e}_i\right\} + \\ + F_h\mathbf{e}_z\times\mathbf{e}_i(\mathbf{e}_z\cdot\mathbf{H}_i), \tag{3.18}$$

$$\mathbf{H}_r^0\left[1-(\mathbf{e}_z\cdot\mathbf{e}_i)^2\right] = F_h(\mathbf{e}_z\cdot\mathbf{H}_i^0)\left\{\left[1-2(\mathbf{e}_z\cdot\mathbf{e}_i)^2\right]\mathbf{e}_z + (\mathbf{e}_z\cdot\mathbf{e}_i)\mathbf{e}_i\right\} - \\ - F_v\mathbf{e}_z\times\mathbf{e}_i(\mathbf{e}_z\cdot\mathbf{E}_i) \tag{3.19}$$

for the reflected wave. A simple expression for the field amplitudes:

$$\left(\mathrm{E}_{\mathrm{r}}^{0}\right)^{2}=\left(\mathrm{H}_{\mathrm{r}}^{0}\right)^{2}=\frac{F_{\mathrm{v}}^{2}\left(\mathbf{e}_{\mathrm{z}}\cdot\mathbf{E}_{i}^{0}\right)^{2}+F_{\mathrm{h}}^{2}\left(\mathbf{e}_{\mathrm{z}}\cdot\mathbf{H}_{i}^{0}\right)}{1-\left(\mathbf{e}_{\mathrm{z}}\cdot\mathbf{e}_{i}\right)^{2}} \tag{3.20}$$

can be established from the formulae provided above. Equations (3.18) and (3.19) can be reduced to:

$$\mathbf{E}_{\mathrm{r}}^{0}=F_{\mathrm{h}}\mathbf{E}_{\mathrm{i}}^{0}+\frac{\left(\mathbf{e}_{\mathrm{z}}\cdot\mathbf{E}_{\mathrm{i}}^{0}\right)}{1-\left(\mathbf{e}_{\mathrm{z}}\cdot\mathbf{e}_{\mathrm{i}}\right)^{2}}\left\{\mathbf{e}_{\mathrm{z}}\left[F_{\mathrm{v}}\left[1-2\left(\mathbf{e}_{\mathrm{z}}\cdot\mathbf{e}_{\mathrm{i}}\right)^{2}\right]-F_{\mathrm{h}}\right]+\left(F_{\mathrm{h}}+F_{\mathrm{v}}\right)\left(\mathbf{e}_{\mathrm{z}}\cdot\mathbf{e}_{\mathrm{i}}\right)\mathbf{e}_{\mathrm{i}}\right\}, \tag{3.21}$$

$$\mathbf{H}_{\mathrm{r}}^{0}=F_{\mathrm{v}}\mathbf{H}_{\mathrm{i}}^{0}+\frac{\left(\mathbf{e}_{\mathrm{z}}\cdot\mathbf{H}_{\mathrm{i}}^{0}\right)}{1-\left(\mathbf{e}_{\mathrm{z}}\cdot\mathbf{e}_{\mathrm{i}}\right)^{2}}\left\{\mathbf{e}_{\mathrm{z}}\left[F_{\mathrm{h}}\left[1-2\left(\mathbf{e}_{\mathrm{z}}\cdot\mathbf{e}_{\mathrm{i}}\right)^{2}\right]-F_{\mathrm{v}}\right]+\left(F_{\mathrm{h}}+F_{\mathrm{v}}\right)\left(\mathbf{e}_{\mathrm{z}}\cdot\mathbf{e}_{\mathrm{i}}\right)\mathbf{e}_{\mathrm{i}}\right\}. \tag{3.22}$$

We can obtain similar formulae for the transmitted wave field.

As was mentioned above, the processes of reflection and refraction are due to changes in the radiowave amplitude and phase. In particular, the tendency is for the reflected and refracted waves to become elliptically polarized by incidence on the interface the plane wave of arbitrary linear polarization. In other words, a change of the wave polarization takes place. As we already know, the polarization character can be described by a Stokes matrix. The processes of reflection and refraction can be considered as linear transforms; however, calculation of a Mueller matrix is rather a complicated procedure in this case. It is easier to do direct calculation of Stokes parameters. Let us represent the incident wave in the form $\mathbf{E}_{\mathrm{i}}=E_{\mathrm{h}}^{(i)}\mathbf{e}_{\mathrm{h}}^{(i)}+E_{\mathrm{v}}^{(i)}\mathbf{e}_{\mathrm{v}}^{(i)}$, where the unitary vectors $\mathbf{e}_{\mathrm{h}}^{(i)}$ and $\mathbf{e}_{\mathrm{v}}^{(i)}$ are directed toward the vectors of horizontal and vertical polarization and form the coordinate basis for the coordinate system relevant to the incident wave. We may use a the similar coordinate basis for the reflected wave and present its field in the form $\mathbf{E}_{\mathrm{r}}=E_{\mathrm{h}}^{(\mathrm{r})}\mathbf{e}_{h}^{(\mathrm{r})}+E_{\mathrm{v}}^{(\mathrm{r})}\mathbf{e}_{\mathrm{v}}^{(\mathrm{r})}$.

Now we will establish the relation between the orthogonal amplitude components of the reflected and incident waves. It is easy to do this for the horizontal components, where the required relation has the form $E_{\mathrm{h}}^{(\mathrm{r})}=F_{\mathrm{h}}E_{\mathrm{h}}^{(i)}$. As for the vertical components, it is necessary to note that in this case $E_{\mathrm{v}}^{(\mathrm{r})}=-F_{\mathrm{v}}E_{\mathrm{v}}^{(i)}$. Let us represent the Stokes parameters of the incident wave by $S_{n}^{(i)}\left(n=0\div 3\right)$ and the Stokes parameters of the reflected wave by $S_{n}^{(\mathrm{r})}$. Simple calculations allow us to set the relations between two systems of parameters and to establish the Stokes matrix transformation law for reflection of the wave. These relations have the form:

$$S_0^{(r)} = \frac{1}{2}\left[S_0^{(i)}\left(|F_h|^2 + |F_v|^2\right) + S_1^{(i)}\left(|F_h|^2 - |F_v|^2\right)\right], \tag{3.23a}$$

$$S_1^{(r)} = \frac{1}{2}\left[S_0^{(i)}\left(|F_h|^2 - |F_v|^2\right) + S_1^{(i)}\left(|F_h|^2 + |F_v|^2\right)\right], \tag{3.23b}$$

$$S_2^{(r)} = |F_h F_v|\left(S_3^{(i)} \sin\Theta_r - S_2^{(i)} \cos\Theta_r\right), \tag{3.23c}$$

$$S_3^{(r)} = -|F_h F_v|\left(S_3^{(i)} \cos\Theta_r + S_2^{(i)} \sin\Theta_r\right). \tag{3.23d}$$

Here, $\Theta_r = \arg F_h - \arg F_v$. Similar relations can be obtained for the refracted wave.

Partial polarization appears at reflection of the noise radiation from the interface. In particular, the coefficient of polarization has the form:

$$m = \frac{|F_h|^2 - |F_v|^2}{|F_h|^2 + |F_v|^2}. \tag{3.24}$$

Let us point out, in conclusion, that if it is a question of a wave incident on a medium with permittivity ε_1 at the border of a medium whose permittivity equals ε_2, then in all previous formulae ε is equal to the relative permittivity ($\varepsilon = \varepsilon_2/\varepsilon_1$), and for wave number k it is necessary to substitute $\sqrt{\varepsilon_1}k$.

Let us also mention the specific relation connecting the reflection coefficients of vertically and horizontally polarized waves. The following relations are obtained from Equation (3.11):

$$\sqrt{\varepsilon - \sin^2\theta_i} = \frac{1 - F_h}{1 + F_h}\cos\theta_i, \quad \varepsilon = \frac{1 + F_h^2 - 2F_h \cos 2\theta_i}{\left(1 + F_h\right)^2}.$$

By inserting these expressions into the formula for the reflection coefficient of the vertically polarized waves, we obtain the unknown relation:

$$F_v = \frac{F_h\left(F_h - \cos 2\theta_i\right)}{1 - F_h \cos 2\theta_i}. \tag{3.25}$$

3.2 RADIOWAVE PROPAGATION IN PLANE-LAYERED MEDIA

Now, we will consider radiowave propagation in a medium whose permittivity is, in the general case, an arbitrary complex function of Cartesian coordinates; in this case, we choose z. As has been pointed out, such media are referred to as plane layered (stratified). It was noted, too, that Maxwell equations written in the form of Equations (1.91) to (1.92) are convenient to use. These equations are simplified essentially because $\varepsilon = \varepsilon(z)$. Hence, it should be taken into account that a plane wave of any polarization propagates in one plane in this case and can be represented as the sum of two wave types. Let the y0z plane be the wave propagation plane, which means that the field does not depend on the x-coordinate and the operator $\partial/\partial x = 0$. Then, the basic waves are E-waves (or waves of horizontal polarization), for which the electrical field vector is directed perpendicularly to the plane of propagation (i.e., it is represented in the form $\mathbf{E} = \Pi_e \mathbf{e}_x$), and H-waves (or vertical polarized waves), for which $\mathbf{H} = \Pi_m \mathbf{e}_x$. For E-waves, the field is described by the equation:

$$\frac{\partial^2 \Pi}{\partial y^2} + \frac{\partial^2 \Pi}{\partial z^2} + k^2 \varepsilon(z) \Pi = 0. \tag{3.26}$$

The equation for H-waves is more complicated:

$$\frac{\partial^2 \Pi_m}{\partial x^2} + \frac{\partial^2 \Pi_m}{\partial z^2} - \frac{1}{\varepsilon} \frac{d\varepsilon}{dz} \frac{\partial \Pi_m}{\partial z} + k^2 \varepsilon(z) \Pi_m = 0. \tag{3.27}$$

Equation (3.27) differs from the wave equation but is easily reduced to it by the substitution of $\Pi_m = \sqrt{\varepsilon} \hat{\Pi}_m$:

$$\frac{\partial^2 \hat{\Pi}_m}{\partial x^2} + \frac{\partial^2 \hat{\Pi}_m}{\partial z^2} + k^2 \varepsilon_e(z) \Pi_m = 0 , \tag{3.28}$$

where the effective permittivity is determined by the equality:

$$\varepsilon_e = \varepsilon - \frac{\sqrt{\varepsilon}}{k^2} \frac{d^2}{dz^2} \left(\frac{1}{\sqrt{\varepsilon}} \right). \tag{3.29}$$

The solutions of Equations (3.26) and (3.28) we are seeking as:

$$\Pi(y,z) = Q(z) e^{ik\eta y}. \tag{3.30}$$

Then, the problem is reduced to solution of the common differential equation:

$$\frac{d^2Q}{dz^2}+k^2\left[\varepsilon(z)-\eta^2\right]Q=0. \tag{3.31}$$

The constant of separation (η) is determined as follows. Let us suppose that ε changes with the z-coordinate beginning from some distance z_0, and, at z z_0, $\varepsilon(z) = \varepsilon_0 = const.$ Let a plane wave described by $\exp[ik\,(y \sin\theta_i + z\cos\theta_i)]\sqrt{\varepsilon_0}$ be incident on the described medium from the area $z < z_0$. Thus,

$$\eta=\sqrt{\varepsilon_0}\sin\theta_i \tag{3.32}$$

because the incident and exited waves are both matched dependent on y.

Equation (3.31), in the general case, has no solution in the analytical form. It is expressed through known functions only in some cases with a particular view of the function $\varepsilon(z)$. We will find one such partial solution in the next section.

3.3 WAVE REFLECTION FROM A HOMOGENEOUS LAYER

To solve the problem of waves in plane-layered media, let us first study the case of two media with permittivities ε_1 and ε_3, separated by a homogeneous layer of thickness d, and with permittivity ε_2 (Figure 3.2). A layer of ice floating on water is an example of such a natural object.

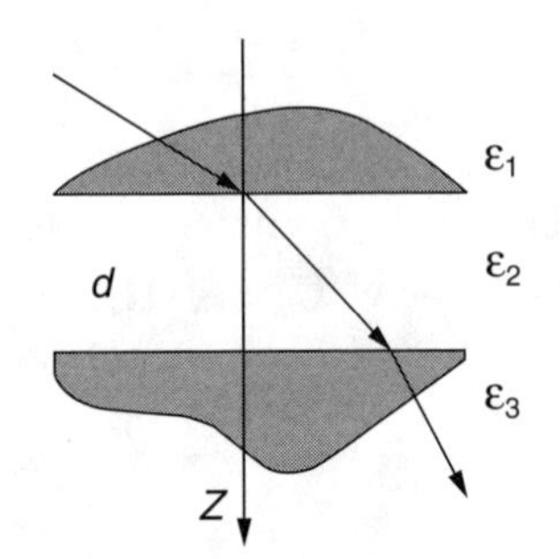

FIGURE 3.2 Plane wave propagation in a homogeneous layer.

It is necessary to analyze two individual problems for E- and H-waves. Let us begin with the E-wave by assuming that the plane wave occurs in the semispace $-\infty < z < 0$ and the reflected wave appears as a result of interaction with this layer. Let us represent potential Π_e in the form:

$$\Pi_e(y,z)=Q_e(z)e^{ik\sqrt{\varepsilon_1}\,y\sin\theta_i}, \tag{3.33}$$

where θ_i is the incident angle. We can now describe the function $Q_e(z)$ in the area $-\infty < z < 0$ as the sum:

$$Q_e(z)=e^{ik\sqrt{\varepsilon_1}\,z\cos\theta_i}+F_e e^{-ik\sqrt{\varepsilon_1}\,z\cos\theta_i}.$$

The first term corresponds to the incident wave, for which the amplitude is assumed to be equal to unity. The second item corresponds to the reflected wave, and F_e is the coefficient of reflection.

The function $Q_e(z)$ satisfies the equation:

$$\frac{d^2 Q_e}{dz^2} + k^2\left(\varepsilon_2 - \varepsilon_1 \sin^2\theta_i\right) Q_e = 0$$

inside the layer $0 < z < d$ and has the general solution:

$$Q_e(z) = \alpha \exp\left(ik\sqrt{\varepsilon_2 - \varepsilon_1 \sin^2\theta_i z}\right) + \beta \exp\left(-ik\sqrt{\varepsilon_2 - \varepsilon_1 \sin^2\theta_i z}\right).$$

Finally, in the third medium ($z > d$):

$$Q_e(z) = T_e \exp\left(ik\sqrt{\varepsilon_3 - \varepsilon_1 \sin^2\theta_i z}\right),$$

where T_e is the coefficient of transmission. The values F_e, T_e, α, and β can be defined as solutions of equations derived from the boundary conditions.

It is useful next to employ Snell's law by introducing the angles θ_2 and θ_3:

$$\sqrt{\varepsilon_1}\sin\theta_i = \sqrt{\varepsilon_2}\sin\theta_2 = \sqrt{\varepsilon_3}\sin\theta_3. \tag{3.34}$$

If ε_1, ε_2, and ε_3 are real numbers and if $\varepsilon_2 - \varepsilon_1\sin^2\theta_i > 0$ and $\varepsilon_3 - \varepsilon_1\sin^2\theta_i > 0$ (i.e., no total inner reflection), then these angles characterize the directions of the wave propagation in the media considered here.

The boundary conditions, Equation (1.7), lead to continuity of function Π_e and its first derivative over z at $z = 0, d$. Four algebraic equations are obtained as a result:

$$1 + F_e = \alpha + \beta, \quad \sqrt{\varepsilon_1}\left(1 - F_e\right)\cos\theta_i = \sqrt{\varepsilon_2}\left(\alpha - \beta\right)\cos\theta_2,$$

$$\alpha e^{i\varphi_2} + \beta e^{-i\varphi_2} = T_e e^{i\varphi_3}, \quad \sqrt{\varepsilon_2}\left(\alpha e^{i\varphi_2} - \beta e^{-i\varphi_2}\right)\cos\theta_2 = \sqrt{\varepsilon_3} T_e \cos\theta_3 e^{i\varphi_3}.$$

Here,

$$\varphi_2 = kd\sqrt{\varepsilon_2}\cos\theta_2, \qquad \varphi_3 = kd\sqrt{\varepsilon_3}\cos\theta_3. \tag{3.35}$$

Solutions to these equations have the following forms:

- For the reflection coefficient from the layer:

$$F_e = \frac{F_e^{12}(\theta_i) + F_e^{23}(\theta_2)e^{2i\varphi_2}}{1 + F_e^{12}(\theta_i)F_e^{23}(\theta_2)e^{2i\varphi_2}}, \tag{3.36a}$$

- For the amplitudes of the directed and reflected waves inside the layer:

$$\alpha = \frac{1+F_e}{1+F_e^{23}(\theta_2)e^{2i\varphi_2}}, \quad \beta = \alpha F_e^{23}(\theta_2)e^{2i\varphi_2}, \tag{3.36b}$$

- For the coefficient of transmission:

$$T_e = \frac{(1+F_e)\left[1+F_{ee}^{23}(\theta_2)\right]e^{i(\varphi_2-\varphi_3)}}{1+F_e^{23}(\theta_2)\,e^{2i\varphi_2}}. \tag{3.36c}$$

Here,

$$F_e^{12}(\theta_i) = \frac{\sqrt{\varepsilon_1}\cos\theta_i - \sqrt{\varepsilon_2 - \varepsilon_1 \sin^2\theta_i}}{\sqrt{\varepsilon_1}\cos\theta_i + \sqrt{\varepsilon_2 - \varepsilon_1 \sin^2\theta_i}} \tag{3.37}$$

is the reflection coefficient of the horizontally polarized waves from the interface of media 1 and 2, and

$$F_e^{23}(\theta_2) = \frac{\sqrt{\varepsilon_2}\cos\theta_2 - \sqrt{\varepsilon_3 - \varepsilon_2 \sin^2\theta_2}}{\sqrt{\varepsilon_2}\cos\theta_2 + \sqrt{\varepsilon_3 - \varepsilon_2 \sin^2\theta_2}} \tag{3.38}$$

is the corresponding coefficient of the reflection from the interface of media 2 and 3.

The same results are obtained for H-waves. It is necessary, in the previous formulae, to substitute the reflection coefficients of horizontally polarized waves for the equivalent ones of vertically polarized waves.

We will not analyze the general formulae, as doing so can be rather complicated. Instead, we will confine ourselves to the particular case of a vertical wave incident on the layer when the reflection coefficients for the E- and H-waves are similar. We will suppose for simplicity that $\varepsilon_1 = 1$; that is, assume, for example, that we have a wave incident on the layered ground from the air. Then,

$$F^{12} = \frac{1-\sqrt{\varepsilon_2}}{1+\sqrt{\varepsilon_2}}, \qquad F^{23} = \frac{\sqrt{\varepsilon_2}-\sqrt{\varepsilon_3}}{\sqrt{\varepsilon_2}+\sqrt{\varepsilon_3}}. \tag{3.39}$$

Let us calculate the reflection coefficient module using the designations:

$$F^{12} = \left|F^{12}\right| e^{i\psi_{12}}, \quad F^{23} = \left|F^{23}\right| e^{i\psi_{23}}, \quad \varphi_2 = kd\sqrt{\varepsilon_2} = \psi + i\frac{\tau}{2}, \quad \psi = kdn_2', \quad \tau = 2kdn_2'' \cdot \tag{3.40}$$

Thus, we have:

$$\left|F\right|^2 = \frac{\left|F^{12}\right|^2 + \left|F^{23}\right|^2 e^{-2\tau} + 2\left|F^{12}F^{23}\right| e^{-\tau}\cos\left(2\psi + \psi_{23} - \psi_{12}\right)}{1 + \left|F^{12}F^{23}\right|^2 e^{-2\tau} + 2\left|F^{12}F^{23}\right| e^{-\tau}\cos\left(2\psi + \psi_{23} + \psi_{12}\right)}. \tag{3.41}$$

In the case of sufficiently strong absorption inside the layer (when $\tau >> 1$), the reflection coefficient is $|F|^2 = |F^{12}|^2$, which suggests that a sufficiently thick layer — a layer with a thickness that is many times greater then the skin depth — is equivalent to the semispace from the point of view of the wave reflection processes. Such layers are called *absorptive*.

We will refer to layers with moderate absorption inside as *half-absorptive*. In these cases, the reflective coefficient module oscillates with the frequency change due to interference of waves reflected from the layer boards. The phase shift of the wave reflected from interfaces 2 and 3 varies with the frequency change, and, as a result, the waves reflected from the layer boards are by turns summed in phase or antiphase, which is why the oscillations are dependent on frequency. The amplitude of these oscillations and their quasi-period depend on the layer thickness and its complex permittivity. In particular, the amplitude of oscillations decreases with increased absorption and tends to zero for the absolute absorptive layer.

Let us simplify the problem by considering the case of the dielectric layer. The imaginary part of ε_2 is small, and $\psi_{12} = \pi$ in this case. Then,

$$\left|F\right|^2 = 1 - \frac{\left(1 - \left|F^{12}\right|^2\right)\left(1 - \left|F^{23}\right|^2 e^{-2\tau}\right)}{1 + \left|F^{12}F^{23}\right|^2 e^{-2\tau} - 2\left|F^{12}F^{23}\right| e^{-\tau}\cos\left(2\psi + \psi_{23}\right)}. \tag{3.42}$$

Although we assume the value ε_2'' is small, the absorption coefficient in the layer may be large:

$$\tau \cong \frac{kd\varepsilon_2''}{\sqrt{\varepsilon_2'}} \tag{3.43}$$

The frequency dependence of the reflection coefficient is basically determined by the phase shift value:

$$\psi = \frac{\omega}{c}\sqrt{\varepsilon_2'}\,d. \tag{3.44}$$

The reflection coefficient minima are achieved at frequencies:

$$\omega_n = \frac{c}{d\sqrt{\varepsilon_2'}}\left(n\pi - \frac{\psi_{23}}{2}\right), \quad (n = 1, 2, \ldots). \tag{3.45}$$

By this,

$$|F|^2_{\min} = \frac{\left(|F^{12}| - |F^{23}|e^{-\tau}\right)^2}{\left(1 - |F^{12}F^{23}|e^{-\tau}\right)^2}. \tag{3.46}$$

Maxima take place at the frequencies:

$$\omega_n = \frac{c}{d\sqrt{\varepsilon_2'}}\left[\left(n - \frac{1}{2}\right)\pi - \frac{\psi_{23}}{2}\right], \quad (n = 1, 2, \ldots). \tag{3.47}$$

The maximum value of the reflection coefficient is equal to:

$$|F|^2_{\max} = \frac{\left(|F^{12}| + |F^{23}|e^{-\tau}\right)^2}{\left(1 + |F^{12}F^{23}|e^{-\tau}\right)^2}. \tag{3.48}$$

The interference waves reflected from both interfaces is especially clear if, by using the geometrical progression formula, we replace Equation (3.36a) by the sum:

$$F = \left(F^{12} + F^{23}e^{2i\varphi_2}\right)\sum_{s=0}^{\infty}(-1)^s\left(F^{12}F^{23}\right)^s e^{2is\varphi_2}. \tag{3.49}$$

The expression reported actually represents the sum of waves reflected sequentially from the layer interfaces. For example, the first item, F_{12}, corresponds to the wave reflected from the interface between the first and the second media;

$[1-(F^{12})^2]F^{23}e^{2i\varphi_2}$ represents the wave transmitted inside the layer through the first interface and then reflected from the second interface, finally coming out after the second interface to cross the first interface. This factor describes the wave decrease related to the reflection from the second border, while the first one represents the decrease caused by the dual wave passing through the first border, and, finally, the third factor describes the phase shift appearance and corresponding wave attenuation due to absorption with dual passing of the layer. Note that the items in Equation (3.49) describe the waves reflected from the layer interfaces many times.

Such representation is especially convenient for describing the reflection of pulse oscillations, for which the spectra have limited bandwidths. Let such a spectrum be described by the function $\tilde{E}(\omega)$. In this instance, we will assume that we are dealing with radar sounding the ground above, illuminating it into the nadir (in our calculations, the z-axis is directed downward). The reflected signal form is described by a function of the form:

$$E(z,t)=\int_{-\infty}^{\infty}\tilde{E}(\omega)F(\omega)\,e^{-i\omega t-ikz}d\omega\ . \tag{3.50}$$

If we substitute Equation (3.49) here and assume that the role of the permittivity frequency dispersion is weak in the frame of the signal bandwidth and that the absorption in the layer does not depend on the frequency, then the result of calculations will be:

$$E(t,z)=F^{12}E_i(t,-z)+\left[1-\left(F^{12}\right)^2\right]F^{23}E_i\left(t,-z+2\sqrt{\varepsilon_2'}d\right)e^{-2\tau}+\cdots. \tag{3.51}$$

Here,

$$E_i(t,z)=\int_{-\infty}^{\infty}\tilde{E}(\omega)e^{-i\omega t+ikz}d\omega$$

is the incident field. Equation (3.51) vividly describes the processes of impulse reflections from the layer interfaces, which we have already discussed above.

In this simple case, the reflected impulse form reiterates the form of the incident (sounding) impulse due to the absence of the dispersion and frequency dependence of wave attenuation due to absorption. In the opposite case, it is necessary to take into account the dispersion signal distortion. If it is a question of natural media, then these distortions are, as a rule, insignificant.

It is logical to establish the definition of an energetic reflective coefficient in the case of waves with a finite-frequency bandwidth. No amplitudes of incident and reflected waves are compared, which is senseless in the given case, but it is more logical compared to their energy flows. If the following is the power flow density of the incident wave:

$$S_i = \frac{c}{8\pi}\left|E_i\right|^2$$

then the flow density of its energy is:

$$\Pi_i = \frac{c}{8\pi}\int_{-\infty}^{\infty}\left|E_i(t)\right|^2 dt = \frac{c}{4}\int_{-\infty}^{\infty}\left|\tilde{E}(\omega)\right|^2 d\omega\,. \tag{3.52}$$

The Parseval equality[18] can be used for conversion from integration over time to integration with respect to frequency. The convenience of introducing the energy flow density is that it does not depend on time. For the reflected wave,

$$\Pi_r = \frac{c}{4}\int_{-\infty}^{\infty}\left|\tilde{E}(\omega)F(\omega)\right|^2 d\omega\,. \tag{3.53}$$

The energetic reflective coefficient may be determined as:

$$\overline{F}^2 = \frac{\Pi_r}{\Pi_i} = \frac{\int_{-\infty}^{\infty}\left|\tilde{E}_i(\omega)F(\omega)\right|^2 d\omega}{\int_{-\infty}^{\infty}\left|E_i(\omega)\right|^2 d\omega}\,. \tag{3.54}$$

These formulae easily allow us to express the reflective coefficient for waves of thermal radiation. In this case, as is known, the electromagnetic oscillations can be represented as white noise, for which the energetic spectrum is uniform in the frequency band (for example, ω_1 and ω_2). Let us assume:

$$\left|\tilde{E}(\omega)\right|^2 = E_0^2 \quad \text{where } \omega_1 < \omega < \omega_2\,, \tag{3.55}$$

and $\left|\tilde{E}(\omega)\right|^2 = 0$ outside the considered frequency band. Then,

$$\overline{F}^2 = \frac{1}{\omega_2 - \omega_1}\int_{\omega_1}^{\omega_2}\left|F(\omega)\right|^2 d\omega\,. \tag{3.56}$$

So, the energetic reflective coefficient for the noise radiation is the squared module of the common coefficient of the sine waves averaged in the frequency band.

Let us now consider the case of the dielectric layer. The result of integration in Equation (3.56) gives:

$$\overline{F}^2 = 1 - \frac{A}{(\beta_2 - \beta_1)\sqrt{1-B^2}} \arctan \frac{\sqrt{1-B^2}\sin(\beta_2 - \beta_1)}{\cos(\beta_2 - \beta_1) - B\cos(\beta_2 + \beta_1)}. \quad (3.57)$$

Here,

$$\beta_{1,2} = \frac{\omega_{1,2}}{c}\sqrt{\varepsilon_2'}d + \frac{\psi_{23}}{2}, \quad A = \frac{\left(1 - \left|F^{12}\right|^2\right)\left(1 - \left|F^{23}\right|^2 e^{-2\tau}\right)}{1 + \left|F^{12}F^{23}\right|^2 e^{-2\tau}}, \quad B = \frac{2\left|F^{12}F^{23}\right| e^{-\tau}}{1 + \left|F^{12}F^{23}\right|^2 e^{-2\tau}}.$$

At a given frequency range (the difference $\beta_2 - \beta_1$ is constant), the energetic reflective coefficient remains the oscillating function of the central frequency determined by the sum $\omega_2 + \omega_1$; however, enlarging the frequency range (the difference $\beta_2 - \beta_1$ increases) results in a decrease in the amplitude of these oscillations due to the averaging effect of summation over frequencies. This averaging becomes practically full at $\beta_2 - \beta_1 = \pi$. Then,

$$\overline{F}^2 = F_0^2 = 1 - \frac{A}{\sqrt{1-B^2}} = \frac{\left|F^{12}\right|^2 + \left|F^{23}\right|^2 e^{-2\tau} - 2\left|F^{12}F^{23}\right|^2 e^{-2\tau}}{1 - \left|F^{12}F^{23}\right|^2 e^{-2\tau}}. \quad (3.58)$$

3.4 WENTZEL–KRAMERS–BRILLOUIN METHOD

Analyzing the problems of wave propagation in stratified media, we realize that we must solve an equation of the following type:

$$\frac{d^2u}{dz^2} + k^2 p(z)u = 0. \quad (3.59)$$

In the general case, as was pointed out, this equation has no solution open to analysis. The corresponding solutions in known functions take place in a very limited number of analytical dependencies, $p(z)$; therefore, it is necessary to use approximate methods of solution of equations such as Equation (3.59), based on same peculiarities of the function $p(z)$.

One of these methods is the Wentzel–Kramers–Brillouin (WKB) method.[18,19] It can be employed for the case when the scale of function $p(z)$ changes little compared to wavelength λ. The analytical properties of the indicated function are determined by the analytical characteristics of the permittivity of the medium; therefore, we are dealing conceptually with a slow change of $\varepsilon(z)$ in the terms formulated above. Initially, we will focus on the reasoning behind this method. For this purpose, in the case of permanent coefficient p, the solution for the direct wave has the form $u = \exp[i\varphi(z)]$, where phase $\varphi(z) = k\sqrt{p}z$. Naturally, at slow behavior $p(z)$, as was mentioned above, the phase change at small segment dz is $d\varphi(z) = k\sqrt{p(z)}dz$. The differential equation for the phase has this obvious solution:

$$\varphi(z) = k\int_0^z \sqrt{p(\varsigma)}\,d\varsigma. \tag{3.60}$$

The wave amplitude in the medium with variable $\varepsilon(z)$ cannot be constant now and must change, but only slowly. The solution to Equation (3.59), then, is:

$$u(z) = Q(z)e^{i\varphi(z)}, \tag{3.61}$$

where $\varphi(z)$ is given by Equation (3.60). By combining Equations (3.59) and (3.61), we obtain a new equation:

$$\frac{d^2Q}{dz^2} + 2ik\sqrt{p}\frac{dQ}{dz} + \frac{ik}{2\sqrt{p}}\frac{dp}{dz}Q = 0. \tag{3.62}$$

No simplifications have been performed yet. We have just substituted one equation for another. And, although the obtained equation is not particularly simpler, it does allow further simplification due to the slowness concept. For this, our approach is the same used to obtain a parabolic equation, but only the wave propagating in one direction (in the given case, the direct one) is considered. The fast oscillated multiplier related to the phase change on the wavelength scale is selected in the same manner, and corresponding equations for slow changed components are simplified. Now, the next step of our study is to simplify Equation (3.62).

Let the scale of the permittivity change and correspondingly the function p be equal to Λ; thus, $dp/dz = d\varepsilon/dz \propto \varepsilon/\Lambda$. We may assume that the scale of change of amplitude Q is the same. Then, it is simple enough to establish that the first item in Equation (3.62) has a value of Q/Λ^2, while the other terms have values of the order $Q/\Lambda\lambda$. It follows that the first summand may be eliminated if $\lambda \ll \Lambda$, so

$$Q \cong \frac{Q_0}{p^{1/4}}, \tag{3.63}$$

where Q_0 is the constant. So, the solution for the direct wave is:

$$u \cong \frac{Q_0}{p^{1/4}} \exp\left[ik \int_0^z \sqrt{p(\varsigma)}\, d\varsigma \right]. \tag{3.64}$$

The solution for the backward wave is obtained from Equation (3.64) by replacing i by $-i$.

If we insert the variable $\sigma = z/\Lambda$ into Equation (3.62), we obtain:

$$\frac{1}{k\Lambda}\frac{d^2Q}{d\sigma^2} + 2i\sqrt{p}\frac{dQ}{d\sigma} + \frac{i}{2\sqrt{p}}\frac{dp}{d\sigma}Q = 0.$$

It is clear from this that eliminating the item with the second derivative is equivalent to ignoring values of the order $(k\Lambda)^{-1}$. This, in turn, leads to the idea that writing the solution to Equation (3.59) in view of the Debye series:

$$u = e^{ik\psi}\sum_{s=0}^{\infty}\frac{u_s}{(ik)^s} \tag{3.65}$$

is the logical deduction of the WKB approximation. Series of this kind are asymptotic. Inserting this series into Equation (3.59) and making equal the items with a similar degree of wave number k give us:

$$\left(\frac{d\psi}{dz}\right)^2 = p(z), \quad 2\frac{d\psi}{dz}\frac{du_s}{dz} + \frac{d^2\psi}{dz^2}u_s = -\frac{d^2u_{s-1}}{dz}. \tag{3.66}$$

The solution of the first equation in this formula gives us the expression for the phase. In the process, both signs, either (+) or (–), are possible upon square root extraction from the right side of this equation, so the independent solution of a second-order differential equation, which Equation (3.59) is, has been taken into account. Equation (3.66) gives us the solution to Equation (3.63), and the first items of the Debye series correspond to the WKB approximation. Solving Equation (3.66) allows us to determine a more precise definition of this approximation. It is easy, however, to establish that the WKB approximation is valid by using the condition:

$$\frac{1}{k\Lambda} \propto \left|\frac{1}{kp}\frac{dp}{dz}\right| \cong \left|\frac{1}{k\varepsilon}\frac{d\varepsilon}{dz}\right| << 1, \tag{3.67}$$

which agrees with the concept of slowness. A second condition relates to the fact that the function $p(z)$ has not been converted to zero, which may occur in the case described by Equation (3.31) if somewhere $\varepsilon(z) < \varepsilon_0$ and angle θ_i is such that in the same point z_0:

$$\varepsilon_0 \sin\theta_i = \varepsilon(z_0). \tag{3.68}$$

The point z_0 is the *turning point*, and the total internal wave reflection occurs on the plane $z = z_0$. It is necessary to look for a solution more precise than the WKB approximation for the area near the turning point. We may obtain such a solution by using the method of reference equations.

3.5 EQUATION FOR THE REFLECTIVE COEFFICIENT

An equation for the reflective coefficient of any plane-layered medium is developed in this part. For simplicity, let us examine the case of radio propagation along the direction of the permittivity change. Let the z-axis coincide with this direction so that $\varepsilon = \varepsilon(z)$ and Maxwell's equations can be reduced to the following:

$$\frac{dE_x}{dz} = ikH_y, \quad \frac{dH_y}{dz} = ik\varepsilon E_x. \tag{3.69}$$

Let us represent the fields of directed and reflected waves in the form:

$$E_x = E_i(z)\left[1 + V(z)\right], \quad H_y = \sqrt{\varepsilon}E_i(z)\left[1 - V(z)\right]. \tag{3.70}$$

Here, $E_i(z)$ is the field of the direct wave. The function $V(z)$ looks like the coefficient of reflection, although it is not literally the same. Not only is it the ratio of the complex amplitudes of the reflected and the direct waves but it also has variable phase correlation of type shown in Equation (3.60) for both waves. However, at point $z = 0$, which is the assumed beginning of the permittivity variable, it is transverse in the reflective coefficient. From Equations (3.69) and (3.70), it follows that:

$$\frac{1}{E_i}\frac{dE_i}{dz}(1+V) + \frac{dV}{dz} = ik\sqrt{\varepsilon}(1-V), \quad \frac{1}{E_i}\frac{dE_i}{dz}(1-V) - \frac{dV}{dz} = ik\sqrt{\varepsilon}(1+V) - \frac{\varepsilon'}{2\varepsilon}(1-V).$$

It follows that the function $V(z)$ must satisfy the Riccati equation:

$$\frac{dV}{dz} = -2ik\sqrt{\varepsilon}V + \frac{\varepsilon'}{4\varepsilon}(1-V^2). \tag{3.71}$$

The reader can find the corresponding equations for inclined incidents of E and H polarized waves in Breshovskish.[21]

Equation (3.71) generally has no solutions in known functions; however, it is simple to solve numerically. Let us assume that permittivity is given by dependence of the form:

$$\varepsilon(z)=\begin{cases}\varepsilon_0=const & \text{for}<0,\\ \varepsilon(z) & \text{for } 0<z<d,\\ \varepsilon_\infty=const & \text{for}>d.\end{cases} \tag{3.72}$$

Such dependence is depicted in Figure 3.3. We have to suppose, because of the assumed dependence, that if z > d $d\varepsilon/dz$ = 0. No reflected waves are present; therefore, V(d) = 0, which may be formulated as the initial condition for Equation (3.71).

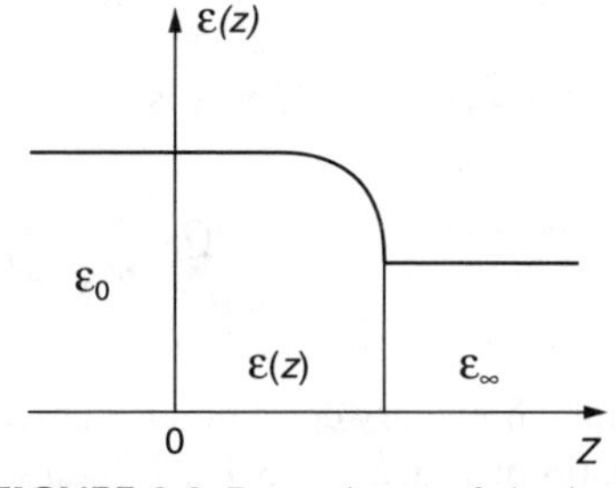

FIGURE 3.3 Dependence of the layer permittivity.

In the permittivity breaking points, the function V(z) also has a break, which makes the numerical calculation rather inconvenient. The function:

$$W(z)=\frac{H_y}{E_x}=\sqrt{\varepsilon}\frac{1-V(z)}{1+V(z)}, \tag{3.73}$$

is more suitable at this point. It make sense of the medium admittance at point z and is continuous everywhere. By substituting:

$$V(z)=\frac{\sqrt{\varepsilon}-W}{\sqrt{\varepsilon}+W} \tag{3.74}$$

in Equation (3.71), we can be easily convinced that this function satisfies the equation:

$$\frac{dW}{dz}=ik\left(\varepsilon-W^2\right), \tag{3.75}$$

which is simpler than Equation (3.71) from many aspects. The initial condition for Equation (3.75) is:

$$W(d)=\sqrt{\varepsilon_\infty}, \tag{3.76}$$

and

$$W = \frac{1}{ik} \frac{Q'}{Q} \tag{3.77}$$

replaces nonlinear Equation (3.75) with a linear one of the second order (wave equation):

$$\frac{d^2 Q}{dz^2} + k^2 \varepsilon Q = 0. \tag{3.78}$$

The boundary condition for it is:

$$\frac{dQ}{dz} - ik\sqrt{\varepsilon_\infty} Q = 0 \quad \text{for } z = d. \tag{3.79}$$

The second boundary condition is not required because, in the end, we are interested in the ratio of Equation (3.77), which is why one of the indefinite constants of the solution to Equation (3.78) is missing. In fact, the solution generally has the form:

$$Q(z) = AQ_1(z) + BQ_2(z), \tag{3.80}$$

where Q_1 and Q_2 are any partial linear and independent solutions of Equation (3.78), and A and B are constants to be determined. The function:

$$W(z) = \frac{1}{ik} \frac{Q_1'(z) + \frac{B}{A} Q_2'(z)}{Q_1(z) + \frac{B}{A} Q_2(z)} \tag{3.81}$$

depends only on the ratio B/A, which can be obtained from Equation (3.78). The result is:

$$\frac{B}{A} = -\frac{Q_1'(d) - ik\sqrt{\varepsilon_\infty} Q_1(d)}{Q_2'(d) - ik\sqrt{\varepsilon_\infty} Q_2(d)}. \tag{3.82}$$

If we insert this ratio into Equation (3.80) and define the reflective coefficient as:

$$F = V(-0) = \frac{\sqrt{\varepsilon_0} - W(0)}{\sqrt{\varepsilon_0} + W(0)}, \tag{3.83}$$

then we obtain:

$$F = \frac{Y_2(0)Y_1(d) - Y_1(0)Y_2(d)}{Z_1(0)Y_2(d) - Z_2(0)Y_1(d)} \tag{3.84}$$

after some not very complicated calculations. Here,

$$Y_l(z) = Q_l'(z) - ik\sqrt{\varepsilon(z)}Q_l(z), \quad Z_l(z) = Q_l'(z) + ik\sqrt{\varepsilon(z)}Q_l(z).$$

Let us point out one detail. If a permittivity break occurs at the point z = 0 (i.e., ε(z) changes at interface), then the value of function $V(0)$ in Equation (3.83) should be taken at z = –0 (i.e., left of the break), and the permittivity value in Equation (3.83) is also taken left of the break.

Finally, we will consider the case of slow permittivity variation in the layer. The WKB method may be used for finding the functions Q_1 and Q_2, written as:

$$Q_1 = \frac{e^{i\varphi}}{\varepsilon^{1/4}}, \quad \frac{dQ_1}{dz} = ik\varepsilon^{1/4}e^{i\varphi}, \quad Q_2 = \frac{e^{-i\varphi}}{\varepsilon^{1/4}}, \quad \frac{dQ_2}{dz} = -ik\varepsilon^{1/4}e^{-i\varphi}.$$

We introduced into the WKB approximation expressions for the functions as well as their derivatives to point out the accuracy required for their calculation (the derivative of the permittivity is neglected in the approximation considered here). Equation (3.82) thus can be written as:

$$\frac{B}{A} = \frac{\sqrt{\varepsilon(d)} - \sqrt{\varepsilon_\infty}}{\sqrt{\varepsilon(d)} + \sqrt{\varepsilon_\infty}} e^{2i\varphi(d)} = F_{23}(d)\, e^{2i\varphi(d)}.$$

Assumed here is the presence of a permittivity bound at the layer interface z = d. The Fresnel coefficient of reflection is calculated according to this bound. The function W is described by the formula:

$$W(z) = \sqrt{\varepsilon(z)} \frac{e^{2i\varphi(z) - 2i\varphi(d)} - F_{23}(d)}{e^{2i\varphi(z) - 2i\varphi(d)} + F_{23}(d)}. \tag{3.85}$$

It is a simple matter to discover that the main changes compared with the homogeneous layer are related to the substitutions:

$$\sqrt{\varepsilon}z \Rightarrow \int_0^z \sqrt{\varepsilon(\varsigma)}\, d\varsigma, \quad F_{23} = \frac{\sqrt{\varepsilon_2} - \sqrt{\varepsilon_3}}{\sqrt{\varepsilon_2} + \sqrt{\varepsilon_3}} \Rightarrow F_{23}(d) = \frac{\sqrt{\varepsilon(d)} - \sqrt{\varepsilon_\infty}}{\sqrt{\varepsilon(d)} + \sqrt{\varepsilon_\infty}}.$$

This result is intuitive but should be regarded as a first approximation. The problem of weak reflections will be analyzed in more detail later.

3.6 EPSTEIN'S LAYER

In this section, we will examine the case when the permittivity analytical dependence allows the wave equation solution to be expressed in known functions, although this is not our primary intent. The principal objective behind the problem discussed here is to reveal the continuous transformation of the coordinate dependence of permittivity from smooth to sharp and to formulate, for example, criteria for the validity of the Fresnel coefficient application in those cases when the permittivity does not change sharply.

The permittivity as a function of the z-coordinate is represented in the form:

$$\varepsilon(z)=\varepsilon_0+\left(\varepsilon_\infty-\varepsilon_0\right)\frac{e^{z/d}}{1+e^{z/d}}. \tag{3.86}$$

Such dependence was considered by Epstein which is why the layer of permittivity described here is often referred to as the *Epstein layer.* At $z\to-\infty$, $\varepsilon(z)\to\varepsilon_0$; at $z\to\infty$, $\varepsilon(z)\to\varepsilon_\infty$. If we substitute:

$$\xi=-e^{z/d}, \tag{3.87}$$

then Equation (3.78) transforms into:

$$\frac{d^2Q}{d\xi^2}+\frac{1}{\xi}\frac{dQ}{d\xi}+\frac{k^2d^2}{\xi^2\left(1-\xi\right)}\left(\varepsilon_0-\varepsilon_\infty\xi\right)Q=0\,.$$

The solutions of this equation are expressed through hypergeometric functions:

$$F_1\left(\alpha,\beta,\gamma,\xi\right)=1+\frac{\alpha\beta}{\gamma}\xi+\frac{\alpha\left(\alpha+1\right)\beta\left(\beta+1\right)}{\gamma\left(\gamma+1\right)1\cdot 2}\xi^2+\cdots, \tag{3.88}$$

which has been well studied.[26] The parameters of the described functions are determined through the equalities:

$$\begin{gathered}\alpha+\beta=\gamma-1,\qquad \alpha=ikd\left(\sqrt{\varepsilon_\infty}+\sqrt{\varepsilon_0}\right),\\ \beta=-ikd\left(\sqrt{\varepsilon_\infty}-\sqrt{\varepsilon_0}\right),\quad \gamma=1+2ikd\sqrt{\varepsilon_0}.\end{gathered} \tag{3.89}$$

We will not go into detail here regarding the method for solving this problem; the interested reader can find a relevant discussion in, for example, Reference 22.

Let us now introduce an expression for the coefficient of reflection:

$$F = \frac{\Gamma\left(1+2ikd\sqrt{\varepsilon_0}\right)\Gamma[-ikd\left(\sqrt{\varepsilon_\infty}+\sqrt{\varepsilon_0}\right)]\Gamma[1-ikd\left(\sqrt{\varepsilon_\infty}+\sqrt{\varepsilon_0}\right)]}{\Gamma\left(1-2ikd\sqrt{\varepsilon_0}\right)\Gamma[-ikd\left(\sqrt{\varepsilon_\infty}-\sqrt{\varepsilon_0}\right)]\Gamma[1-ikd\left(\sqrt{\varepsilon_\infty}-\sqrt{\varepsilon_0}\right)]}. \quad (3.90)$$

Here, Γ(x) is the gamma-function. The obtained formula is not simple for general analysis because of the complex behavior of the gamma-function of complex variable; however, we can obtain a rather simple expression for the reflective coefficient module, the analysis of which is not difficult. It is necessary to take into account that:

$$\left|\Gamma(ix)\right|^2 = \frac{\pi}{x\sinh \pi x}, \quad \Gamma(1+ix)\Gamma(1-ix) = \frac{\pi x}{\sinh \pi x}, \quad (3.91)$$

with the following result:

$$|F| = \left|\frac{\sinh\left[\pi kd\left(\sqrt{\varepsilon_\infty}-\sqrt{\varepsilon_0}\right)\right]}{\sinh\left[\pi kd\left(\sqrt{\varepsilon_\infty}+\sqrt{\varepsilon_0}\right)\right]}\right|. \quad (3.92)$$

If $\pi kd << 1$ (thin layer), then Equation (3.92) transforms into the Fresnel formula for the reflective coefficient. At $\left|\pi kd\left(\sqrt{\varepsilon_\infty}-\sqrt{\varepsilon_0}\right)\right| >> 1$ (the layer with a slow change of permittivity), we have:

$$|F| = \exp\left(-2\pi kd\sqrt{\varepsilon_0}\right) \text{ if } \varepsilon_\infty > \varepsilon_0, \quad |F| = \exp\left(-2\pi kd\sqrt{\varepsilon_\infty}\right) \text{ if } \varepsilon_\infty < \varepsilon_0. \quad (3.93)$$

It is clear that the reflective coefficient is exponentially small in the last case.

3.7 WEAK REFLECTIONS

Let us now consider the problem of a plane wave incident from the left on the layer with borders $z = 0, d$ (Figure 3.4). The permittivity changes by bound from the ε = 1 to ε_0 at the first border of z = 0. A similar bound from ε_d to ε_∞ takes place at the second border of z = *d*.

The permittivity varies arbitrarily but without a sharp bound inside the area between both interfaces. The problem of the first approximation was already analyzed in Section 3.5; here, we will discuss reflections inside the layer. The solution to Equation (3.71) takes the form:

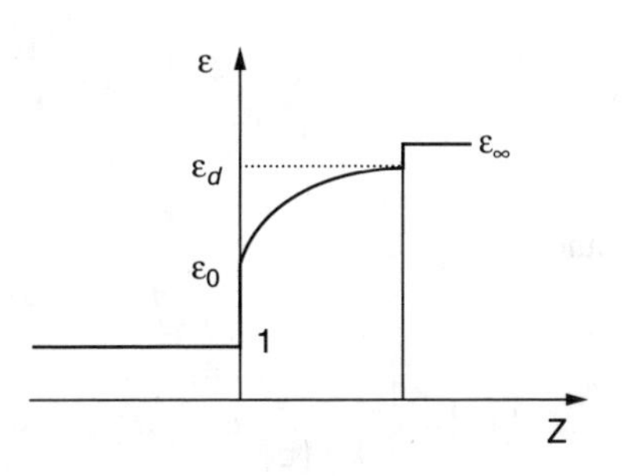

FIGURE 3.4 Permittivity changes in the layer with weak reflection.

$$V(z)=\left[F_{23}(d)+\Re(z)\right]e^{2i[\varphi(d)-\varphi(z)]},\quad \varphi(z)=k\int_0^z\sqrt{\varepsilon(\zeta)}\,d\zeta\,, \tag{3.94}$$

where, as before, $F_{23}(d)$ is the Fresnel coefficient of reflection from the back wall of the layer. The selection of the solution in such a form is based on physical arguments, according to which the local reflections inside the layer are added to the reflection from the back wall. Equation (3.71) can be reduced to the following:

$$\frac{d\Re}{dz}=\frac{\varepsilon'}{4\varepsilon}\left[e^{2i[\varphi(z)-\varphi(d)]}-\left(F_{23}+\Re\right)^2e^{2i[\varphi(d)-\varphi(z)]}\right]. \tag{3.95}$$

We assume that the reflection inside the layer has to be small because of the slowness of the permittivity changes. This assumption allows us to also assume that $|\Re|\ll 1$ and to neglect the square of this value in Equation (3.95). As a result, we obtain the equation:

$$\frac{d\Re}{dz}+\frac{\varepsilon'}{2\varepsilon}F_{23}\Re e^{2i[\varphi(d)-\varphi(z)]}=\frac{\varepsilon'}{4\varepsilon}\left[e^{2i[\varphi(z)-\varphi(d)]}-F_{23}^2e^{2i[\varphi(d)-\varphi(z)]}\right]. \tag{3.96}$$

Now, we are dealing with a linear equation, which greatly simplifies the problem. The solution of this equation, taking into account the condition that $\Re(d)=0$, is:

$$\Re(z)=\Im e^{-M(z)}+e^{-M(z)-2i\varphi(d)}\int_0^z\frac{dL(\varsigma)}{d\varsigma}e^{M(\varsigma)}d\varsigma-\frac{F_{23}(d)}{2}\left[1-e^{-M(z)}\right]. \tag{3.97}$$

Here,

$$\Im=-e^{-2i\varphi(d)}\int_0^d\frac{dL(\varsigma)}{d\varsigma}e^{M(\varsigma)}d\varsigma+\frac{F_{23}(d)}{2}\left[e^{M(d)}-1\right] \tag{3.98}$$

and

$$L(z)=\frac{1}{4}\int_0^z\frac{\varepsilon'(\varsigma)}{\varepsilon(\varsigma)}e^{2i\varphi(\varsigma)}d\varsigma,\quad M(z)=\frac{F_{23}(d)e^{2i\varphi(d)}}{2}\int_0^z\frac{\varepsilon'(\varsigma)}{\varepsilon(\varsigma)}e^{-2i\varphi(\varsigma)}d\varsigma. \tag{3.99}$$

Now, it is a simple matter to determine the value of the reflection coefficient from the layer:

$$F = \frac{1 - W(0)}{1 + W(0)} = \frac{F_{12}(0) + V(0)}{1 + F_{12}(0)V(0)}. \tag{3.100}$$

Here, $F_{12}(0)$ is the Fresnel coefficient of reflection at z = 0, and we obtain:

$$F = \frac{F_{12}(0) + \left[F_{23}(d) + \Im\right]e^{2i\varphi(d)}}{1 + F_{12}(0)\left[F_{23}(d) + \Im\right]e^{2i\varphi(d)}}. \tag{3.101}$$

A comparison with the results provided in Section 3.5 shows that the summed reflection from the layer elements is added to the reflections from the borders when inhomogeneous permittivity distribution occurs inside of the layer. This reflection is described by the value $\Im$ according to Equation (3.101). We assume this quantity is small, as we have neglected the square of the function $\Re(z)$. We also assume that $|M(d)| << 1$ under these conditions, so:

$$\Im = -\frac{e^{-2i\varphi(d)}}{4}\int_0^d \frac{\varepsilon'(\varsigma)}{\varepsilon(\varsigma)} e^{2i\varphi(\varsigma)} d\varsigma + \frac{F_{23}^2(d)e^{2i\varphi(d)}}{4}\int_0^d \frac{\varepsilon'(\varsigma)}{\varepsilon(\varsigma)} e^{-2i\varphi(\varsigma)} d\varsigma. \tag{3.102}$$

The sum obtained has a clear physical sense. The first item describes the process of direct wave local reflection. The second one describes waves generated in the process of local reflection of the backward wave occurring due to reflection at the second layer interface. These waves become direct, propagate to the second interface, reflect from it, and propagate backward. This explanation allows us to understand the proportionality of the second item to the square of the reflection coefficient $F_{23}(d)$.

We will now turn our attention to the slowness concept of permittivity variation and the small local reflection related to it. Usually, we consider the scale of the permittivity change to be greater than that of the wavelength; however, this requirement does not appear in our arguments here (based on the WKB approximation). Our approximation is based only on the assumption of the function $\Re(z)$ being small; correspondingly, the constant $\Im$ is also small. However, the last may be small and by the bound, more accurately fast on the wavelength scale change of the permittivity. By letting such a change take place from the value ε_1 to ε_3 near the point $z = z$ on a scale much smaller than the wavelength scale, the exponential factors in Equation (3.102) are not changed during integration near this point, so we have:

$$\Im = -\frac{1}{4}\ln\frac{\varepsilon_2}{\varepsilon_1}\left\{e^{2i[\varphi(z)-\varphi(d)]} - F_{23}^2(d)\,e^{2i[\varphi(d)-\varphi(z)]}\right\}. \tag{3.103}$$

The requirement for the value obtained to be small is easily reached when a small permittivity jump occurs in the considered point, or observance of the inequality $|\varepsilon_2 - \varepsilon_1| << \varepsilon_1$. Then,

$$-\frac{1}{4}\ln\frac{\varepsilon_2}{\varepsilon_1} = -\frac{1}{4}\ln\left(1+\frac{\varepsilon_2-\varepsilon_1}{\varepsilon_1}\right) \cong -\frac{\varepsilon_2-\varepsilon_1}{4\varepsilon_1},$$

which corresponds to the Fresnel reflective coefficient; otherwise, at the same conditions:

$$\frac{\sqrt{\varepsilon_1}-\sqrt{\varepsilon_2}}{\sqrt{\varepsilon_1}+\sqrt{\varepsilon_2}} \cong \frac{1}{2}\left(1-\sqrt{1+\frac{\varepsilon_2-\varepsilon_1}{\varepsilon_1}}\right) \cong -\frac{\varepsilon_2-\varepsilon_1}{4\varepsilon_1}.$$

So, it is also necessary to include a little by bound permittivity change in the concept of the small reflection.

In the case of really slow variation of permittivity, it is possible to calculate the integrals in Equation (3.102) using integration by parts:

$$\Im = \frac{1}{8ik}\left\{\frac{\varepsilon'(0)}{\left[\varepsilon(0)\right]^{3/2}}\left[e^{-2i\varphi(d)} + F_{23}^2(d)e^{2i\varphi(d)}\right] - \frac{\varepsilon'(d)}{\left[\varepsilon(d)\right]^{3/2}}\left[1+F_{23}^2(d)\right]\right\}. \qquad (3.104)$$

As an example, we can analyze the case of the linear layer when the permittivity is described by the following function:

$$\varepsilon(z) = \varepsilon_0 + (\varepsilon_d - \varepsilon_0)\frac{z}{d}. \qquad (3.105)$$

In this case

$$\varphi(d) = \frac{2kd}{3(\varepsilon_d - \varepsilon_0)}\left(\varepsilon_d^{3/2} - \varepsilon_0^{3/2}\right) \qquad (3.106)$$

and

$$\Im = \frac{\varepsilon_d - \varepsilon_0}{8ikd}\left\{\frac{1}{\varepsilon_0^{3/2}}\left[e^{-2i\varphi(d)} + F_{23}^2(d)\,e^{2i\varphi(d)}\right] - \frac{1}{\varepsilon_d^{3/2}}\left[1+F_{23}^2(d)\right]\right\}. \qquad (3.107)$$

If the reflection coefficients $F_{12}(0)$ and $F_{23}(d)$ are equal to zero (i.e., permittivity bounds at the layer interfaces are absent) or essentially small, then inner reflections are the dominant factors, and the coefficient of the reflection from the layer is:

$$F = -\frac{1}{4}\int_0^d \frac{\varepsilon'(\varsigma)}{\varepsilon(\varsigma)} e^{2i\varphi(\varsigma)} d\varsigma. \tag{3.108}$$

The formulae obtained here, in spite of their approximate character, describe the processes of radiowave reflection from natural media with sufficient accuracy.

3.8 STRONG REFLECTIONS

In this section, we discuss sharp permittivity changes inside a layer. Strictly speaking, in this case, the concept of strong reflection is restricted. For the convenience of further analysis, let us make the following substitution in Equation (3.71):

$$V(z) = \tanh J(z). \tag{3.109}$$

The reduced equation has the form:

$$\frac{dJ}{dz} = -ik\sqrt{\varepsilon}\sinh 2J + \frac{\varepsilon'}{4\varepsilon}. \tag{3.110}$$

We will suppose the permittivity continuity including the layer borders for the task simplification. It means that $V(d) = 0$ and correspondingly $J(d) = 0$. The equation is not solved, as before, in a general way. We should assume that the derivative of the permittivity is large and that the second item in the right-hand side of Equation (3.110) is dominant. This allows us to use the method of perturbations[22] to solve the equation. Let us represent the solution as a series:

$$J = J_0 + ikJ_1 + (ik)^2 J_2 + \cdots. \tag{3.111}$$

One should not confuse the function introduced here with Bessel functions. Although outwardly the wave number appears to be the small parameter in this series, in essence the expansion is made according to the small parameter kd. Combining Equations (3.111) and (3.110) and equating the members at the same degree ik, we obtain this system of "catching" equations:

$$\frac{dJ_0}{dz} = \frac{\varepsilon'}{4\varepsilon}, \quad \frac{dJ_1}{dz} = -\sqrt{\varepsilon}\sinh 2J_0, \quad \frac{dJ_2}{dz} = -2J_1\sqrt{\varepsilon}\cosh 2J_0. \tag{3.112}$$

The boundary condition for these equations is conversion of the solutions into zero at $z = d$. Then, we obtain:

$$J_0(z) = -\frac{1}{4}\int_z^d \frac{\varepsilon'(\varsigma)}{\varepsilon(\varsigma)} d\varsigma = \frac{1}{4}\ln\frac{\varepsilon(z)}{\varepsilon_d}, \tag{3.113}$$

$$J_1(z) = \int_z^d \sqrt{\varepsilon(\varsigma)}\ \sinh 2J_0(\varsigma) d\varsigma = -\frac{1}{2\sqrt{\varepsilon_d}}\int_z^d \left[\varepsilon_d - \varepsilon(\varsigma)\right] d\varsigma, \tag{3.114}$$

$$J_2(z) = 2\int_z^d \sqrt{\varepsilon(\varsigma)}\ J_1(\varsigma) \cosh 2J_0(\varsigma) d\varsigma = \frac{1}{\sqrt{\varepsilon_d}}\int_z^d \left[\varepsilon_d + \varepsilon(\varsigma)\right] J_1(\varsigma)\, d\varsigma. \tag{3.115}$$

The last expression can be reduced to the following:

$$J_2(z) = J_1^2(z) - 2z\sqrt{\varepsilon_d}\, J_1(z) - \int_z^d \varsigma\left[\varepsilon_d - \varepsilon(\varsigma)\right] d\varsigma \tag{3.116}$$

after simple transformations. Further, by expanding tanh$J$ into the corresponding Taylor series, we obtain the following expression for the coefficient of reflection from the layer:

$$F = V(0) = F_f - k^2\left(1 - F_f^2\right)\left[J_1^2(0)\left(1 - F_f\right) - I\right] + ik\left(1 - F_f^2\right)J_1(0). \tag{3.117}$$

Here,

$$F_f = \tanh J_0(0) = \frac{\sqrt{\varepsilon_0} - \sqrt{\varepsilon_d}}{\sqrt{\varepsilon_0} + \sqrt{\varepsilon_d}}, \tag{3.118}$$

and

$$I = \int_0^d \varsigma\left[\varepsilon_d - \varepsilon(\varsigma)\right] d\varsigma. \tag{3.119}$$

One might question the necessity of taking into account the second term of approximation in our calculations, but we have done so because the reflective coefficient module might be necessary in further calculations. It will be necessary to square the items of the perturbation series, which means it will also be necessary to calculate the values squared by the parameter of expansion; therefore, we must calculate these values from the very beginning to correctly take them into account. Thus, the formula for the module of the reflective coefficient can be written as:

$$|F|^2 = F_f^2 + k^2\left(1 - F_f^2\right)\left[J_1^2(0)\left(1 - F_f\right)^2 + I\right], \tag{3.120}$$

where the imaginary part of the permittivity is assumed to be zero. We point out, in conclusion, that the reflection will be small by $\varepsilon_d = \varepsilon_0$, and, for our expansion to be valid, it is also necessary to assume that $kd\sqrt{|\varepsilon|_{max}} >> 1$, where $|\varepsilon|_{max}$ is the maximum of the permittivity module inside the layer.

3.9 INTEGRAL EQUATION FOR DETERMINING THE PERMITTIVITY DEPTH DEPENDENCE

Determination of permittivity as a function of depth z is a remote sensing problem. The task can be formulated in the following manner. Measurement of the reflective coefficient from the interface of two media as a function of the frequency is performed by a radar system. It is necessary to define the permittivity depth distribution on the basis of this reflective coefficient frequency dependence.

Let us suppose that permittivity does not depend on the frequency inside the sounding frequency band. The reflective coefficient frequency dependence, in this case, is caused only by the depth distribution of the sounded medium permittivity. Further, we can analyze this dependence not on frequency ω but on wave number $k = \omega/c$, which is similar. Function W, determined by Equation (3.75), also depends on the wave number, and we will consider it to be the function $W(z,k)$. The coefficient of reflection:

$$F(k) = V(0,k)$$

is obtained by experiment. From here, referring to Equation (3.73), the following function can be calculated:

$$W(0,k) = \frac{1 - F(k)}{1 + F(k)}. \tag{3.121}$$

We have assumed that $\varepsilon(0) = 1$.

The next task is to establish the relation between $W(0,k)$ as a function of the wave number and $\varepsilon(z)$ as a function of depth z. Let us imagine that we have a model of the layer based on prior information. We will represent the model parameters by the letter m; that is, we will introduce permittivity $\varepsilon_m(z)$ and the corresponding admittance $W_m(z,k)$. By Equation (3.78), the function $Q_m(z,k)$ satisfies the equation:

$$\frac{d^2 Q_m}{dz^2} + k^2 \varepsilon_m Q_m = 0. \tag{3.122}$$

We can then multiply Equation (3.78) by Q_m, multiply Equation (3.122) by Q, and subtract one result from the other to obtain:

$$\frac{d}{dz}\left(Q_m \frac{dQ}{dz} - Q\frac{dQ_m}{dz}\right) = -k^2\left(\varepsilon - \varepsilon_m\right)QQ_m.$$

Using Equations (3.77) and (3.76), we obtain:

$$W(z,k) - W_m(z,k) = \left(\sqrt{\varepsilon_\infty} - \sqrt{\varepsilon_\infty^{(m)}}\right)\frac{Q(d,k)Q_m(d,k)}{Q(z,k)Q_m(d,k)} - \\ -ik\int_z^d\left[\varepsilon(\varsigma) - \varepsilon_m(\varsigma)\right]\frac{Q(\varsigma,k)Q_m(\varsigma,k)}{Q(z,k)Q_m(z,k)}d\varsigma. \tag{3.123}$$

In the process of obtaining this equation, it was assumed that the thickness of the studied layer is known with sufficient accuracy from existing data. Assuming that $z = 0$ in the last equation, we finally obtain:

$$W(0,k) - W_m(0,k) = \left(\sqrt{\varepsilon_\infty} - \sqrt{\varepsilon_\infty^{(m)}}\right)\frac{Q(d,k)Q_m(d,k)}{Q(0,k)Q_m(0,k)} - \\ -ik\int_0^d\left[\varepsilon(\varsigma) - \varepsilon_m(\varsigma)\right]\frac{Q(\varsigma,k)Q_m(\varsigma,k)}{Q(0,k)Q_m(0,k)}d\varsigma. \tag{3.124}$$

This integral equation is based on a comparison of experimental data with data from the model. Essentially, the equation concerns the function:

$$\left[\varepsilon(z) - \varepsilon_m(z)\right]\frac{Q(z,k)}{Q(0,k)}.$$

Only in the case of small deviation from the model of real spatial distribution of permittivity can we assume that $Q \cong Q_m$ and write:

$$W(0,k) - W_m(0,k) = \left(\sqrt{\varepsilon_\infty} - \sqrt{\varepsilon_\infty^{(m)}}\right)\left[\frac{Q_m(d,k)}{Q_m(0,k)}\right]^2 - \\ -ik\int_0^d\left[\varepsilon(\varsigma) - \varepsilon_m(\varsigma)\right]\left[\frac{Q_m(\varsigma,k)}{Q_m(0,k)}\right]^2 d\varsigma. \tag{3.125}$$

Now, the equation concerning the unknown function $\varepsilon(z) - \varepsilon_m(z)$ can be determined.

Because the difference between model and real inputs is supposed to be small, Equation (3.125) can be easily reformulated directly for the reflective coefficient:

$$F = \frac{1-W}{1+W} = \frac{1-W_m - W + W_m}{1+W_m + W - W_m} \cong F_m - \frac{2(W-W_m)}{(1+W_m)^2}.$$

It is easy to establish from this that:

$$F = F_m + \left(\sqrt{\varepsilon_\infty} - \sqrt{\varepsilon_\infty^{(m)}}\right) H(d,k) - ik \int_0^d \left[\varepsilon(\varsigma) - \varepsilon_m(\varsigma)\right] H(\varsigma,k)\,d\varsigma, \tag{3.126}$$

where

$$H(z,k) = 2k^2 \left[\frac{Q_m(z,k)}{Q'_m(0,k) + ikQ_m(0,k)}\right]^2. \tag{3.127}$$

This equation is especially convenient for conversion to the equation of the reflective coefficient module, which is important because the object of measurement is often not the reflective coefficient itself but its absolute magnitude (the power of the signal is measured). The form of the equation for the refractive coefficient module is obvious, which is why we do not write it out separately here.

4 Geometrical Optics Approximation

4.1 EQUATIONS OF GEOMETRICAL OPTICS APPROXIMATION

In this chapter, we will discuss wave propagation problems in a medium with an arbitrary law of permittivity coordinate dependence; that is, we will assume that the permittivity has the form $\varepsilon = \varepsilon(\mathbf{r})$ in the common case. The spatial variation slowness of the permittivity is assumed to be similar to that for the Wentzel–Kramers–Brillouin (WKB) approximation carried out in Chapter 3. We assume again a small change of permittivity at the wavelength scale. This property can be expressed as the inequality:

$$\left|\nabla \ln \varepsilon\right| << k. \tag{4.1}$$

As in the WKB method, we will utilize a solution to Maxwell's equations in the form of the asymptotic Debye series:

$$\mathbf{E} = e^{ik\psi} \sum_{s=0}^{\infty} \frac{\mathbf{E}_s(\mathbf{r})}{(ik)^s}, \qquad \mathbf{H} = e^{ik\psi} \sum_{s=0}^{\infty} \frac{\mathbf{H}_s(\mathbf{r})}{(ik)^s}. \tag{4.2}$$

The value ψ is referred to as the *eikonal value*. We arrive at the system of connected equations:

$$\begin{aligned} &\left[\nabla\psi \times \mathbf{H}_0\right] + \varepsilon \mathbf{E}_0 = 0, && \left[\nabla\psi \times \mathbf{E}_0\right] - \mathbf{H}_0 = 0, \\ &\left[\nabla\psi \times \mathbf{H}_1\right] + \varepsilon \mathbf{E}_1 = -[\nabla \times \mathbf{H}_0], && \left[\nabla\psi \times \mathbf{E}_1\right] - \mathbf{H}_1 = -[\nabla \times \mathbf{E}_0], \end{aligned} \tag{4.3}$$

after substitution of the Debye series in Maxwell's equations; members of the same degree of k are equal to each other.

Zero-order equations are a system of homogeneous linear algebraic equations. For the purpose of their nontrivial solution, it is necessary to reduce the determinant to zero. This requirement leads to the equation for ψ (the eikonal equation). We can obtain this equation fairly simply if $\mathbf{H}_0$ is expressed through $\mathbf{E}_0$; also, we must take into account the mutual orthogonality of vectors $\mathbf{E}_0$, $\mathbf{H}_0$, and $\nabla\psi$ that follows from the equations of the zero-order approximation. Now we can easily show that the eikonal equation may be written down in the form:

$$(\nabla\psi)^2 = \varepsilon. \tag{4.4}$$

The value:

$$\varphi = k\psi \tag{4.5}$$

represents the radiowave phase in the zero approximation, and eikonal ψ (in engineering terminology) represents the electrical length passed by the wave. We assume that the phases of components $\mathbf{E}_0$ and $\mathbf{H}_0$ do not depend on coordinates in the approximation. Furthermore, in this approximation, these vectors are believed to be real, including the initial wave phase at once in the wavelength. Certainly, small additions to this phase may be made by calculation of the following items of expansion.

In the zero-order approximation, the power flow density:

$$\mathbf{S}_0 = \frac{c}{8\pi}\left[\mathbf{E}_0 \times \mathbf{H}_0\right] = \frac{c}{8\pi}\mathrm{E}_0^2\nabla\psi \tag{4.6}$$

is directed along lines of the eikonal gradient. This fact allows us to refer to the zero approximation as the *geometrical optics approximation*, which corresponds to the small wavelength conversion (hence the term *optics*) and allows the wave propagation laws to be formulated in the language of geometry.

The validity of the geometrical optics approximation is defined by Equation (4.1). If, as before, the scale of permittivity change is designated Λ, then the Debye series is essentially expansion according to the inverse degree of large parameter $k\Lambda$. Other conditions will be formulated later.

Let us assume, in the beginning, that the permittivity is a real value. We can define the vector of wave propagation by the formula:

$$\nabla\psi = \sqrt{\varepsilon}\,\mathbf{s}, \tag{4.7}$$

where $\mathbf{s}$ is the unitary vector, which is orthogonal to the equiphase surfaces. The lines orthogonal to the surfaces of the eikonal constant value (to equiphase surfaces) are called *rays*. Vector $\mathbf{s}$ is tangential to the rays and describes the wave energy propagation direction.

If τ is the length along the ray, then the ray equation has the form $d\mathbf{r}/d\tau = \mathbf{s}$. Then, $\nabla\psi = \mathbf{s}d/\psi/d\tau$, and the eikonal equation becomes the common differential equation:

$$\frac{d\psi}{d\tau} = \sqrt{\varepsilon} \tag{4.8}$$

the solution of which is written in the form:

$$\psi(\tau) = \psi_0 + \int_0^{\tau} \sqrt{\varepsilon(\tau')}\, d\tau', \tag{4.9}$$

where ψ_0 is the initial value of the eikonal equation. Let us point out that use of the plus sign was determined by extraction of the square root in Equation (4.9). It is important to note that we are dealing with a direct wave. In the case of a backward wave, the minus sign should be used. It is clear that the eikonal form as a function of coordinates depends on the ray along which the integration is provided.

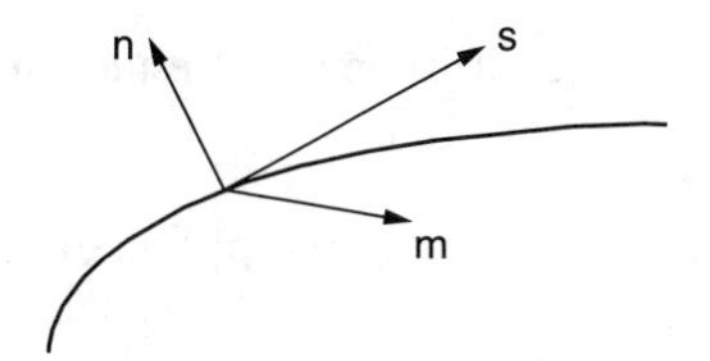

FIGURE 4.1 The orthogonal unitary vector system.

We may use the orthogonal unitary vector system of the normal **n** and the binormal **m** (Figure 4.1). Their changes along the ray characterize its bending and torsion. The Frenet–Serre formulae:

$$\frac{d\mathbf{s}}{d\tau} = \frac{\mathbf{n}}{\rho}, \qquad \frac{d\mathbf{n}}{d\tau} = -\frac{\mathbf{s}}{\rho} - \chi\,\mathbf{m}, \qquad \frac{d\mathbf{m}}{d\tau} = \chi\,\mathbf{n} \tag{4.10}$$

are known from differential geometry.[29] The value ρ is the ray curvature radius, and χ is its torsion. The vectors **s**, **n**, and **m** are the basis of the curved-line coordinate system formed by the ray ensemble and equiphase surfaces. This system is often referred to as the *ray coordinates*. Equation (4.9) is the eikonal equation solution in the ray coordinates system.

Let us use the eikonal equation in Equation (4.8) to calculate ρ and χ. We must take the gradient of both parts to obtain:

$$\frac{d\left(\mathbf{s}\sqrt{\varepsilon}\right)}{d\tau} = \nabla\sqrt{\varepsilon}\,.$$

Thus, it follows that:

$$\frac{d\mathbf{s}}{d\tau} = \nabla \ln\sqrt{\varepsilon} - \mathbf{s}\cdot\left(\mathbf{s}\nabla \ln\sqrt{\varepsilon}\right). \tag{4.11}$$

Equation (4.11), together with the first Frenet–Serre equation, allows us to determine the radius of ray curvature:

$$\frac{1}{\rho^2} = \left(\nabla \ln\sqrt{\varepsilon}\right)^2 - \left(\mathbf{s}\cdot\nabla \ln\sqrt{\varepsilon}\right)^2. \tag{4.12}$$

If angle α between the direction of the ray and the direction of the permittivity is introduced, then:

$$\frac{1}{\rho} = \left|\nabla \ln \sqrt{\varepsilon} \sin \alpha\right|. \tag{4.13}$$

Further, it is simple to establish that:

$$\begin{aligned} \mathbf{n} &= \rho\left[\nabla \ln \sqrt{\varepsilon} - \mathbf{s}\left(\mathbf{s} \cdot \nabla \ln \sqrt{\varepsilon}\right)\right] = \rho\left[\mathbf{s}\left(\mathbf{s} \cdot \nabla \ln \sqrt{\varepsilon}\right)\right], \\ \mathbf{m} &= [\mathbf{s} \cdot \mathbf{n}] = \rho\left[\mathbf{s} \cdot \nabla \ln \sqrt{\varepsilon}\right], \\ \chi &= \mathbf{n} \cdot \frac{d\mathbf{m}}{d\tau} = \left[\mathbf{n} \cdot \left(\mathbf{s} \cdot \nabla\right)\mathbf{m}\right]. \end{aligned} \tag{4.14}$$

We must now derive equations for fields $\mathbf{E}_0$ and $\mathbf{H}_0$. First of all, let us point out that the wave is transversal in zero-order approximation, so its components may be represented in the form:

$$\mathbf{E}_0 = E_n \mathbf{n} + E_m \mathbf{m}, \qquad \mathbf{H}_0 = \sqrt{\varepsilon}\left[\mathbf{s} \times \mathbf{E}_0\right] = \sqrt{\varepsilon}\left(E_n \mathbf{n} - E_m \mathbf{m}\right). \tag{4.15}$$

On the basis of Equation (4.15) and after some not very complicated calculations,[25] we can define the conditions that connect the electrical field components directed along the normal and along the binormal:

$$\begin{aligned} &\left(\left[2\sqrt{\varepsilon}\nabla E_n + E_n \nabla(\sqrt{\varepsilon})\right] \cdot \mathbf{s}\right) + \sqrt{\varepsilon} E_n \nabla \cdot \mathbf{s} + 2\chi\sqrt{\varepsilon} E_m = 0, \\ &\left(\left[2\sqrt{\varepsilon}\nabla E_m + E_m \nabla(\sqrt{\varepsilon})\right] \cdot \mathbf{s}\right) + \sqrt{\varepsilon} E_m \nabla \cdot \mathbf{s} - 2\chi\sqrt{\varepsilon} E_n = 0. \end{aligned} \tag{4.16}$$

Returning to the local cylindrical coordinate system:

$$E_n = \mathrm{E}_0 \cos\vartheta, \quad E_m = \mathrm{E}_0 \sin\vartheta\,, \tag{4.17}$$

we can easily obtain from Equation (4.16) the transfer equation:

$$\nabla \cdot \left(\sqrt{\varepsilon} \mathrm{E}_0^2 \mathbf{s}\right) = \nabla \cdot \mathbf{S} = 0 \tag{4.18}$$

and the equation of torsion:

$$(\mathbf{s}\cdot\nabla\vartheta)-\chi=0. \tag{4.19}$$

Equation (4.18) conveys the energy conservation law. The solution to Equation (4.19) may be written as:

$$\vartheta=\vartheta_0+\int_0^{\tau}\chi(\tau')\,d\tau', \tag{4.20}$$

which describes the law of wave polarization elliptical rotation without changing its form (Rytov's law).

We may rewrite Equation (4.18) in another form by using Equation (4.7). Then, in the ray coordinates,

$$\frac{dE_0}{d\tau}+E_0\frac{\nabla^2\psi}{2\sqrt{\varepsilon}}=0\,,$$

and the solution is obvious:

$$E_0(\tau)=E_0(0)\exp\left(-\frac{1}{2}\int_0^{\tau}\frac{\nabla^2\psi(\tau')}{\sqrt{\varepsilon(\tau')}}d\tau'\right). \tag{4.21}$$

This expression can be written down in another form by using the ray divergence. Let us insert the ray coordinates ξ, η, and τ. Coordinate τ is directed along the rays, while the other two coordinates are orthogonal to it and are directed, for example, along the vectors of normal and binormal. The Jacobian of the transition from Cartesian coordinates (x,y,z) to ray coordinates is given by the formula:

$$D(\xi,\eta,\tau)=\begin{vmatrix}\partial x/\partial\xi & \partial x/\partial\eta & \partial x/\partial\tau\\ \partial y/\partial\xi & \partial y/\partial\eta & \partial y/\partial\tau\\ \partial z/\partial\xi & \partial z/\partial\eta & \partial z/\partial\tau\end{vmatrix}. \tag{4.22}$$

The ray tube is defined as a ray family passing through the area $d\xi d\eta$ near a point with coordinates (ξ,η,τ). The square of the surface element perpendicular to the direction of the rays (equiphase surface) equals $Dd\xi d\eta$. The volume of the ray tube element equals $Dd\xi d\eta d\tau$. Applying vector analysis to Equation (4.8), it is a simple matter to obtain:

$$\int_V \nabla^2\psi d^3\mathbf{r} = \int_V \nabla\cdot(\nabla\psi)\,d^3\mathbf{r} = \oint_S (\mathbf{n}\cdot\nabla\psi)\,d^2\mathbf{r} = \oint_S \sqrt{\varepsilon}\,(\mathbf{n}\cdot\mathbf{s})\,d^2\mathbf{r}. \quad (4.23)$$

We will use this formula for the volume bounded by the ray coordinates and will reduce the volume to zero; moreover, we take into account that $(\mathbf{n}\cdot\mathbf{s}) = 0$ at the sides of the tube. We then have:

$$\nabla^2\psi D(\tau)\,d\xi\,d\eta\,d\tau = [D(\tau+d\tau)\sqrt{\varepsilon(\tau+d\tau)} - D(\tau)\sqrt{\varepsilon(\tau)}]\,d\xi\,d\eta$$

or

$$\frac{\nabla^2\psi}{\sqrt{\varepsilon}} = \frac{d}{d\tau}\ln\left[\sqrt{\varepsilon(\tau)}D(\tau)\right]. \quad (4.24)$$

Instead of Equation (4.21), we obtain:

$$\mathrm{E}_0(\tau) = \sqrt{\frac{\varepsilon(0)D(0)}{\varepsilon(\tau)D(\tau)}}\,\mathrm{E}_0(0). \quad (4.25)$$

This result is transparent from the physical point of view: Due to the law of energy conservation, the field amplitude changes together with changes in the cross section of the ray tube. The second condition of the geometrical optics approximation validity follows from Equation (4.25). It is not valid where $D(\tau) = 0$. Areas where the Jacobian is reduced to zero are called *caustics* and round out the ray family. In particular, the turning point, which we mentioned in Chapter 3 with regard to the WKB approximation, corresponds to the caustics plane. For more details about caustics, refer to Kravtsow and Orlov.[27]

Up to this point, it was assumed that permittivity is the real value. Let us now turn to the more realistic case of weak absorption in the media and, related to this, complex permittivity. We can still use Equation (4.4), but the eikonal itself must now be complex (i.e., $\psi = \psi' + i\psi''$). It is apparent that the eikonal imaginary part describes wave attenuation due to absorption and, being multiplied by the wave number, is equal to the coefficient of extinction. The separation of real and imaginary parts in Equation (4.4) leads to a pair of equations concerning $\nabla\psi'$ and $\nabla\psi''$. It is difficult to find the solution of these equations, particularly because it is necessary to know the angle between $\nabla\psi'$ and $\nabla\psi''$.

We will now consider a simple but common case of small absorption in the sense that $\varepsilon'' << \varepsilon'$ and $(\nabla\psi'') << (\nabla\psi')^2$. The pair of equations then acquires the form:

$$\left(\nabla\psi'\right)^2 = \varepsilon', \quad 2\left(\nabla\psi'\nabla\psi''\right) = \varepsilon''. \tag{4.26}$$

The first equation is solved as before and the second one is transformed to the form:

$$\frac{d\psi''}{d\tau} = \frac{\varepsilon''}{2\sqrt{\varepsilon'}}.$$

Hence, it follows that:

$$\psi''\left(\tau\right) = \frac{1}{2}\int_0^{\tau} \frac{\varepsilon''\left(\tau'\right)}{\sqrt{\varepsilon'\left(\tau'\right)}} d\tau'. \tag{4.27}$$

We will now briefly address the case of anisotropic media, including the ionosphere. In this case, the eikonal equation is broken down into two equations — one for ordinary and one for extraordinary waves:[28]

$$\left(\nabla\psi_{\text{o}}\right)^2 = n_{\text{o}}^2, \quad \left(\nabla\psi_e\right)^2 = n_e^2. \tag{4.28}$$

These waves, generally speaking, can be considered to be independent if the length of the beating between them is much smaller than the scale of the medium inhomogeneities. The beating length is estimated by the value:

$$l = \frac{2\pi c}{\omega\left(n_{\text{o}} - n_e\right)} \cong \frac{c f^2}{f_{\text{p}}^2 f_{\text{H}} \cos\beta} = \frac{1.3\cdot 10^{-4} f^2}{N H_0}. \tag{4.29}$$

The substitution of specific values ($f = 10^8$ Hz, $N = 2 \cdot 10^6$ cm^3, $H_0 = 0.5$ Oersted) gives the estimation $l \approx 10$ km. It would seem that the independence of ordinary and nonordinary waves can be broken with increasing frequency and, correspondingly, with increasing beating length. This is not so, however, because in this case the wave relation coefficient decreases with increases in frequency;[28] therefore, ordinary and extraordinary waves are practically always independent for the ultra-high-frequency and microwave bands for the ionosphere of Earth.

The ray trajectories of ordinary and extraordinary waves practically coincide, because their refractive indexes differ little in the range of waves being considered here, which allows us to develop a formula to calculate the polarization angle rotation value due to the Faraday effect:

$$\Psi_F = \frac{\omega}{2c}\int_0^{\tau}\left(n_o - n_e\right)d\tau' \cong \frac{e^3}{2\pi m^2 c^2 f^2}\int_0^{\tau} N\left(\tau'\right)H_0\left(\tau'\right)\cos\beta\left(\tau'\right)\,d\tau'. \tag{4.30}$$

Finally, we will calculate the Doppler frequency shift for wave propagation in an inhomogeneous medium. For this purpose, let us refer back to Equation (2.97) and rewrite it as follows:

$$\omega_d = -k\left(\mathbf{v}\cdot\nabla\psi\right) = -k\sqrt{\varepsilon}\left(\mathbf{s}\cdot\mathbf{v}\right). \tag{4.31}$$

It is easy to see that the Doppler shift value is proportional to the velocity component directed along the ray (ray velocity).

4.2 RADIOWAVE PROPAGATION IN THE ATMOSPHERE OF EARTH

The atmosphere of Earth can be considered, in the first approximation, as a spherically layered medium where the permittivity is a function of the radius beginning at the center of the Earth. We do not include in our consideration here the changes in atmospheric parameters along the surface of the Earth that take place at the transition from day to night (light to shadow), along frontal zones with significant changes of air temperature, and so on. It should be supposed that $\varepsilon = \varepsilon(R)$, $\nabla\varepsilon = \mathbf{R}/R(d\varepsilon/dR)$, etc. It is believed, that, on average, $g_\varepsilon = d\varepsilon/dR \cong -8\cdot 10^{-8}$ 1/m in the troposphere near the surface of Earth; therefore, the geometrical optics approximation is highly accurate for the given wave range. The vertical gradient in the ionosphere is even smaller, so applying the geometrical optics approximation is still appropriate. Although the permittivity of air does differ from unity, the difference is insignificant and we can assume it to be equal to unity without causing problems.

We can prove rather easily the permanency of vector $\sqrt{\varepsilon}[\mathbf{R}\times\mathbf{s}]$ along the ray trajectory; hence, we can make the statement that in the case of a spherically layered medium the ray trajectories are plane curves. The product:

$$\sqrt{\varepsilon\left(R\right)}R\sin\alpha\left(R\right) = \eta, \tag{4.32}$$

is invariant along the ray, where the constant η is determined from the initial conditions. If, for example, a ray left Earth at angle α_0, then $\eta = \sqrt{\varepsilon_0}\,a\sin\alpha_0$, where the radius of the Earth $a \cong 6.4\cdot 10^3$ km (see Figure 4.2). The starting point may be established on the ray prolongation until the surface of the Earth. Equation (4.32) is often referred to as Snell's law for spherically layered media.

We will now consider the situation when a ray passes by the surface of Earth (Figure 4.3). Far from the atmosphere of Earth, $\varepsilon(R) \to 1$, and $R\sin\alpha \to p$, where p is the *aimed distance* (a term borrowed from the theory of particle scattering). In this case, a ray turning point occurs at $R = R_m$, where $\alpha(R_m) = \pi/2$.

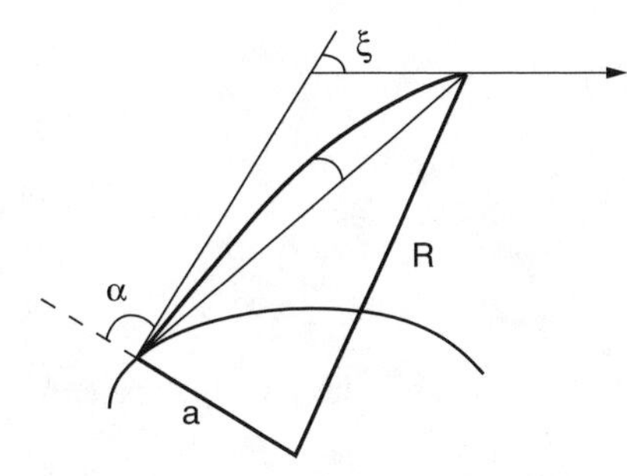

FIGURE 4.2 Radio propagation from the surface of the Earth.

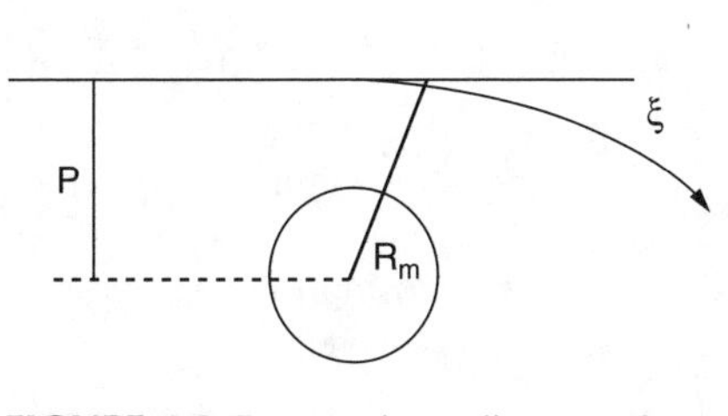

FIGURE 4.3 Propagation radio wave along the surface of the Earth.

It is convenient, in our case, to write the eikonal equation for the spherical coordinate system with the center coinciding with the center of the Earth. This system can be chosen in such a way as to take into account the plane character of the ray trajectories so that the eikonal will not depend on the azimuthal angle φ. The eikonal equation can be written as:

$$\left(\frac{\partial\psi}{\partial R}\right)^2+\frac{1}{R^2}\left(\frac{\partial\psi}{\partial\theta}\right)^2=\varepsilon(R) \tag{4.33}$$

in our chosen system of coordinates. Proceeding according to the variable separation method, we will seek a solution as the sum:

$$\psi(R,\theta)=\psi_R(R)+\psi_\theta(\theta). \tag{4.34}$$

Combining Equations (4.33) and (4.34) and taking into account the initial conditions, we obtain:

$$\psi(R,\theta)=\eta(\theta-\theta_0)\pm\int_{R_0}^{R}\sqrt{\varepsilon(R')-\frac{\eta^2}{R'^2}}dR', \tag{4.35}$$

where R_0 and θ_0 are the coordinates of the initial ray point. The sign before the integral is chosen depending on the type of ray branch: the plus sign for the ascendant branch and the minus sign for the descendent branch. For geometrical reasons,

$$\frac{Rd\theta}{dR}=\tan\alpha=\frac{\eta}{\pm R\sqrt{\varepsilon-\eta^2/R^2}} \tag{4.36}$$

and

$$\theta - \theta_0 = \pm\eta \int_{R_0}^{R} \frac{dR'}{R'^2\sqrt{\varepsilon - \eta^2 / R'^2}}. \tag{4.37}$$

The rule for choosing the appropriate sign is the same as for the previous case. As a result, we now have:

$$\psi(R,\theta) = \pm \int_{R_0}^{R} \frac{R'\varepsilon(R')\,dR'}{\sqrt{\varepsilon(R')R'^2 - \eta^2}}. \tag{4.38}$$

It follows from Equation (4.38) that:

$$\nabla^2\psi = \frac{1}{2}\frac{d/dR\left(\varepsilon - \eta^2/R^2\right)}{\sqrt{\varepsilon - \eta^2/R^2}}.$$

Taking into account Equation (4.21), the equality can be derived as:

$$E_0(R) = \left[\frac{\varepsilon(R_0) - \eta^2/R_0^2}{\varepsilon(R) - \eta^2/R^2}\right]^{1/4} E_0(0) = \sqrt{\frac{\sqrt{\varepsilon_0}\cos\alpha_0}{\sqrt{\varepsilon(R)}\cos\alpha}}\,E_0(0). \tag{4.39}$$

As before, the result obtained corresponds to the WKB approximation.

When the ray passes through point $R = R_m$ (Figure 4.3) it comes into contact with the caustic surface and additional caustic phase shift occurs.[12,19] Without going into detail regarding the calculation, we should point out that, in this case,

$$\psi(R,\theta) = \int_{R_m}^{R_0} \frac{R'\varepsilon(R')\,dR'}{\sqrt{\varepsilon(R')R'^2 - p^2}} + \int_{R_m}^{R} \frac{R'\varepsilon(R')\,dR'}{\sqrt{\varepsilon(R')R'^2 - p^2}} - \frac{\pi}{2}, \tag{4.40}$$

and cos α approaches zero close to the turning point; the amplitude, formally calculated in the geometrical optics approximation, tends to infinity, which emphasizes once again that the geometrical optics approximation is inapplicable in areas close to caustics.

The important ray parameter is the angle of refraction characterizing the degree of its bending. The differential of this angle is defined as the angle between the ray direction in infinity nearby points τ and $\tau + d\tau$ (Figure 4.4). The differential value (let us represent it as ξ) is determined from the equality $[\mathbf{s}(\tau)\times\mathbf{s}(\tau+d\tau)] = -\mathbf{m}\sin(d\xi)$. Let us now use the expansion:

$$\mathbf{s}(\tau+d\tau)=\mathbf{s}(\tau)+\frac{d\mathbf{s}}{d\tau}d\tau\,,$$

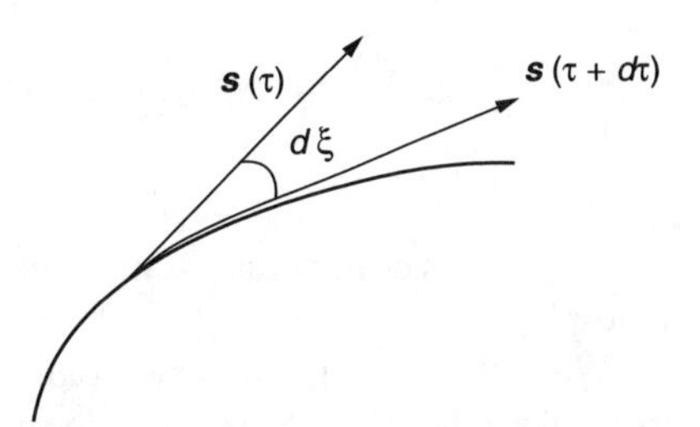

FIGURE 4.4 Refraction angle differentials and changing optical density.

Equation (4.11), and the relation $\left[\mathbf{s}\times\mathbf{R}/\mathrm{R}\right]=\mathbf{m}\sin\alpha$ allow us to obtain:

$$\frac{d\xi}{d\tau}=-\frac{1}{2\varepsilon}\frac{d\varepsilon}{dR}\sin\alpha\left(R\right). \tag{4.41}$$

Hence, it follows that:

$$\xi=-\frac{1}{2}\int_{\tau_0}^{\tau}\frac{1}{\varepsilon}\frac{d\varepsilon}{dR'}\sin\alpha d\tau'=-\frac{1}{2}\int_{R_0}^{R}\frac{1}{\varepsilon}\frac{d\varepsilon}{dR'}\tan\alpha dR'$$

and

$$\xi\left(R,\eta\right)=-\frac{\eta}{2}\int_{R_0}^{R}\frac{\frac{1}{\varepsilon\left(R'\right)}\frac{d\varepsilon\left(R'\right)}{dR'}dR'}{\sqrt{\varepsilon\left(R'\right)R'^2-\eta^2}}. \tag{4.42}$$

This expression will have the form:

$$\xi=-\frac{p}{2}\int_{R_m}^{R_0}\frac{\frac{1}{\varepsilon}\frac{d\varepsilon}{dR'}dR'}{\sqrt{\varepsilon R'^2-p^2}}-\frac{p}{2}\int_{R_m}^{R}\frac{\frac{1}{\varepsilon}\frac{d\varepsilon}{dR'}dR'}{\sqrt{\varepsilon R'^2-p^2}}. \tag{4.43}$$

for the rays shown in Figure 4.3. In particular, if R_0 and R are sufficiently large, the corresponding points of the ray trajectory are practically outside the atmosphere. The upper limit in the last integral can be set at infinity, and we can write:

$$\xi(p) = -p\int_{R_m}^{\infty} \frac{\frac{1}{\varepsilon}\frac{d\varepsilon}{dR}dR}{\sqrt{\varepsilon R^2 - p^2}}. \tag{4.44}$$

Angle ξ, in this critical case, is the angle of wave scattering in the spherically layered atmosphere.

When the reception of waves radiated by highly disposed sources (for example, artificial satellites) is realized on the surface of the Earth, angle ξ is equal to an error in determination of the angle position of these sources. This error in determination of the zenith angle can be written as:

$$\xi(\alpha_0) = -\frac{a\sqrt{\varepsilon_0}\sin\alpha_0}{2}\int_{a}^{\infty} \frac{\frac{1}{\varepsilon}\frac{d\varepsilon}{dR}dR}{\sqrt{\varepsilon R^2 - \varepsilon_0 a^2\sin^2\alpha_0}}. \tag{4.45}$$

By determining this error from data measured at different values of angle α_0, we can define the altitude profile of the atmospheric permittivity using the inverse problems technique.

Let us point out that Equation (4.45) can be rewritten as:

$$\xi(\alpha_0) = -\frac{a\sqrt{\varepsilon_0}\sin\alpha_0}{2}\int_{0}^{\infty} \frac{\frac{1}{\varepsilon(\varsigma)}\frac{d\varepsilon(\varsigma)}{d\varsigma}}{\sqrt{\varepsilon(\varsigma)(a+\varsigma)^2 - \varepsilon_0 a^2\sin^2\alpha_0}}d\varsigma. \tag{4.46}$$

Let us now extend the radius of the Earth to infinity, thereby transforming the problem to one involving a plane-layered atmosphere. As a result we have:

$$\xi(\alpha_0) = -\sqrt{\varepsilon_0}\sin\alpha_0\int_{0}^{\infty} \frac{\frac{1}{\varepsilon(\varsigma)}\frac{d\varepsilon(\varsigma)}{d\varsigma}}{\sqrt{\varepsilon(\varsigma) - \varepsilon_0\sin^2\alpha_0}}d\varsigma. \tag{4.47}$$

The last integral is calculated at any function $\varepsilon(\varsigma)$. For our purposes here, we can substitute $\sin\alpha(\varsigma) = \sqrt{\varepsilon_0/\varepsilon(\varsigma)}\sin\alpha_0$. Then,

$$\xi(\alpha_0) = \arcsin\left(\sqrt{\varepsilon_0}\sin\alpha_0\right) - \alpha_0. \tag{4.48}$$

So, the value of the refraction angle for a plane-layered medium depends only on the permittivity values at the terminal points of the ray trajectory (Laplace theorem).

Let us point out that this result is easily obtained using Snell's law for plane-layered media. The small deviation of the permittivity from unity allows the use of the Taylor expansion to obtain:

$$\xi\left(\alpha_0\right)=\frac{\varepsilon_0-1}{2}\tan\,\alpha_0. \tag{4.49}$$

As previously mentioned, the Doppler frequency shift must be considered in the case of moving sources:

$$\omega_{\mathrm{d}}=-k\sqrt{\varepsilon}\left[v_{\mathrm{R}}\cos\alpha(R)+R\Omega_{\theta}\sin\alpha(R)\right] \tag{4.50}$$

according to Equation (4.31). Here, $v_{\mathrm{R}} = dR/dt$ is the radial velocity and $\Omega_\theta = d\theta/dt$ is the angular velocity. The Doppler frequency shift depends not only on movement parameters but also on the atmospheric characteristics, because angle α is determined by the refraction phenomenon.

4.3 NUMERICAL ESTIMATIONS OF ATMOSPHERIC EFFECTS

Let us now turn to numerical estimation of atmospheric effects that occur during radio propagation of ultra-high-frequency and microwave bands. Let us begin with the troposphere. The altitude dependence of atmospheric air permittivity can be generally described by:[31]

$$\varepsilon\left(\varsigma\right)=1+\left(\varepsilon_0-1\right)\exp\left(-\frac{\varsigma}{H_{\mathrm{T}}}\right). \tag{4.51}$$

Here, as before, ς is the altitude above land, $\varepsilon_0 - 1 \cong 6 \cdot 10^{-4}$ defines the near-land value of the air permittivity, and $H_{\mathrm{T}} \cong 8$ km is the frequently used troposphere height. The parameters of this exponential model of the troposphere depend on meteorological conditions and, particularly, on geographical location. So, the parameters used here should be considered only as reference ones that are close to average, although the real values are not very different from these values. Estimations show that the integral of refraction for this model can be represented in the form:[32]

$$\xi\left(\alpha_0\right)=\frac{\varepsilon_0-1}{2}\tan\,\alpha_0\,Z\left(s_0\right), \tag{4.52}$$

where

$$Z(s_0) = 2s_0 e^{s_0^2} \int_{s_0}^{\infty} e^{-s^2} ds, \qquad s_0 = \sqrt{\frac{a_e}{2H_T}} \cos\alpha_0, \tag{4.53}$$

and a_e 8500 km is the so-called equivalent radius of the Earth. At $s_0 >> 1$, $Z(s_0) \cong 1$, and

$$\xi(\alpha_0) \cong \frac{\varepsilon_0 - 1}{2} \tan\alpha_0, \tag{4.54}$$

which corresponds to the plane-layered troposphere approximation. At $s_0 << 1$, which corresponds to the angles being close to $\pi/2$, $Z(s_0) \cong \sqrt{\pi}\, s_0$ and the refraction angle is defined as follows:

$$\xi \cong \frac{\varepsilon_0 - 1}{2} \sqrt{\frac{a_e \pi}{2H_T}}. \tag{4.55}$$

It follows from Equation (4.55) that the maximum value of the refraction angle is $\xi \cong 1.2 \cdot 10^{-2}$.

Let us now consider the influence of the troposphere effect on distance measurement — for example, for spacecraft using radio equipment. The time of the signal traveling from the object to the reception point is directly measured in this case. This time is easily converted to the electrical length of the propagation track (eikonal). The eikonal value depends on the properties of the troposphere, so we will now address formal calculation of the distance using a known eikonal value, as well as the error involved. This error value is equal to the difference $\Delta\psi = \psi - \psi_0'$, where ψ_0' is the eikonal value in the absence of the atmosphere. This difference is approximately equal to:[32]

$$\Delta\psi = \frac{(\varepsilon_0 - 1) H_T}{2\cos\alpha_0} Z(s_0). \tag{4.56}$$

If $s_0 >> 1$,

$$\Delta\psi \cong \frac{(\varepsilon_0 - 1) H_T}{2\cos\alpha_0}, \tag{4.57}$$

which corresponds to the plane-layered atmosphere approximation. When $s_0 << 1$,

$$\Delta\psi \cong \frac{\varepsilon_0 - 1}{2}\sqrt{\frac{\pi a_e H_T}{2}}. \tag{4.58}$$

The maximum value is $\Delta\psi \cong 10^2$ m for the parameters used here.

Now, we turn our attention to calculation of the frequency Doppler shift. In doing so, we will be interested in that part of the frequency change that is determined by the troposphere properties. Let us first discuss the bypass variant (Figure 4.3) using a case typical for planet occultation observations;[32] the method presented here has been used for measurement of troposphere parameters.[35] Let the transmitter be at point T, the receiver at infinity, and the ray come to the receiver at sighting distance *p*. Angle α in Equation (4.50) is measured, in this case, at point R_T. At this point, $\alpha(R_T) = \theta - \xi$, and $\varepsilon(R_T) = 1$ if radius R_T is large enough; $\xi = 0$ in the absence of atmosphere; and the corresponding frequency shift is:

$$\hat{\omega}_d = -\frac{v_= \omega}{c}, \tag{4.59}$$

where

$$v_= = v_R \cos\theta + R_T \Omega_\theta \sin\theta \tag{4.60}$$

is the platform velocity along the undisturbed (by atmosphere) ray (sighting line). Having done the expansion for small-angle ξ, we now have:

$$\frac{\Delta\omega_T}{\omega} = \frac{\omega_d - \hat{\omega}_d}{\omega} = -\xi\frac{v_\perp}{c}. \tag{4.61}$$

Here,

$$v_\perp = v_R \sin\theta - R_T \Omega_\theta \cos\theta \tag{4.62}$$

is the velocity component perpendicular to the sighting line. To estimate the effect, let us assume that $\xi = 2.4 \cdot 10^{-2}$ and $v_\perp = 8$ km/sec (for an artificial satellite of Earth). Thus, $\Delta\omega_T/\omega \cong 6.4 \cdot 10^{-7}$.

When the radiation reception occurs on Earth, we must refer to Equation (4.50), which we can rewrite to take into account the trajectory equation, Equation (4.32):

$$\omega_d = -k\left[\frac{v_R}{R}\sqrt{\varepsilon(R)R^2 - \varepsilon_0 a^2 \sin^2\alpha_0} + a\varepsilon_0\Omega_\theta \sin\alpha_0\right].$$

Expansion when parameters $\varepsilon_0 - 1$ and ξ are small leads to the formula:

$$\frac{\Delta\omega_T}{\omega} = -\frac{a}{c}\left[\frac{\varepsilon_0 - 1}{2}\sin\alpha_0 - \xi\cos\alpha_0\right]\left[\frac{v_R a \sin\alpha_0}{R\sqrt{R^2 - a^2\sin^2\alpha_0}} + \Omega_\theta\right]. \tag{4.63}$$

We have supposed so high altitude of the radiation source that we can assume $\varepsilon(R) = 1$. If now we use formula (4.52) for medium refraction the result will be

$$\frac{\Delta\omega_T}{\omega} = -\frac{(\varepsilon_0 - 1)a\sin\alpha_0}{2c}\left[1 - Z(s_0)\right]\left[\frac{v_R a\sin\alpha_0}{R\sqrt{R^2 - a^2\sin^2\alpha_0}} + \Omega_\theta\right]. \tag{4.64}$$

For the common example of an artificial satellite on a circular orbit around Earth, the radial velocity may be expected to be equal to zero. Further, we can assume that $a\Omega_\theta = 8$ km/sec, $a_0 = \pi/2$, and $\Delta\omega_T/\omega \cong 8 \cdot 10^{-9}$ at these parameters.

For the analysis of ionospheric effects, we will work primarily with spacecraft orbits that are above the ionospheric electron concentration maximum. Again, let us first estimate the refraction angle value. In this case, knowledge of the angle between the ray direction at the point of reception on Earth and the sight line to the radiation source is more important than knowledge of the refraction angle itself. Let us indicate this angle as δ. We can show that, approximately,[32]

$$\delta = \frac{e^2}{2\pi m f^2}\frac{a R_m \sin\alpha_0 A(R) N_t(R)}{A^3(R_m)\left[A(R) - a\cos\alpha_0\right]}, \tag{4.65}$$

where R is the spacecraft altitude (distance from the center of Earth), R_m is the altitude of the electron concentration maximum, $A(R) = \sqrt{R^2 - a^2\sin^2\alpha_0}$, and the function:

$$N_t(R) = \int_a^R N(R')dR' \tag{4.66}$$

is the electron content at the altitude of the radiation source. Let us suppose for numerous estimations that $R >> a$, $R_m = a + z_m$, and $z_m = 300$ km. We can propose the average estimation of $N_t(\infty) \cong 3 \cdot 10^{13}\,\text{cm}^{-2}$ for the total electron content. Then,

$$\delta_I = \frac{1.73 \cdot 10^{12} \sin\alpha_0}{f^2\left(1 - 0.91\sin^2\alpha_0\right)^{3/2}}. \quad (4.67)$$

From here, for instance, $\delta = 8.4 \cdot 10^{-4}$ (3′) for $f = 108$ Hz and $\alpha_0 = 60°$.

Let now estimate the dependence of the eikonal value (electrical length) on the ionosphere. Here, for the same reasons:

$$\Delta\psi = -\frac{e^2}{2\pi \mathrm{m} f^2}\frac{R_\mathrm{m} N_t(R)}{A\left(R_\mathrm{m}\right)}. \quad (4.68)$$

where $\Delta\psi$ has a negative value in the ionosphere due to the fact that the radiowave phase velocity in plasma is greater then the speed of light. The maximum effect takes place at the distance from Earth where the ray penetrates through the entire ionosphere thickness. In this case, we must substitute the total electron content value in Equation (4.68) to obtain:

$$\Delta\psi = -\frac{1.21 \cdot 10^{21}}{f^2\sqrt{1 - 0.91\sin^2\alpha_0}}. \quad (4.69)$$

For $\alpha_0 = 60°$ and $f = 10^8$ Hz, $\Delta\psi = -2.15 \cdot 10^3$ m. Note that the effect would be positive and $\Delta L_g = -\Delta\psi$ if we dealt not with the phase correction for the distance (i.e., for the eikonal) but with the so-called group correction ΔL_g (i.e., refractive correction on the distance defined by the signal group velocity).

The Doppler effect can be written as:

$$\frac{\Delta\omega_\mathrm{I}}{\omega} = \frac{a\delta\cos\alpha_0}{\mathrm{c}}\left(\frac{a\sin\alpha_0}{R\sqrt{R^2 - a^2\sin^2\alpha_0}}v_\mathrm{R} + \Omega_\theta\right). \quad (4.70)$$

The main dependence here, again, is connected with the electron content. Having already became standard, our estimations give, with $\alpha_0 = 60°$ and $f = 10^8$ Hz, a value of the order of 10^{-8} for the relative frequency change.

Finally, we now consider the Faraday rotation of the polarization plane. According to Equation (4.30),

$$\Psi_\mathrm{F} = \frac{e^3}{2\pi \mathrm{m}^2 c^2 f^2}\int_0^\tau N(\tau')\left(\mathbf{H}_0 \cdot d\mathbf{s}\right). \quad (4.71)$$

The magnetic field of Earth coincides with the field of the magnetic dipole, having magnetic moment M (M ≅ $8.1 \cdot 10^{25}$ CGSE);[33] therefore,

$$\mathbf{H}_0 \cong -\nabla\left(\frac{\mathbf{MR}}{R^3}\right) = -\nabla\left(\frac{\mathbf{M}\cos\mu}{R^2}\right), \tag{4.72}$$

where μ is the angle between $\mathbf{M}$ and $\mathbf{R}$. At these conditions,

$$\Psi_{\mathrm{F}} = -\frac{e^3\mathrm{M}}{2\pi\,\mathrm{m}^2c^2f^2}\int_0^{\tau} N(\tau')\frac{\partial}{\partial\tau'}\left(\frac{\cos\mu}{R^2}\right)d\tau' . \tag{4.73}$$

With these assumptions, we obtain the following approximate calculation:

$$\Psi_{\mathrm{F}} = -\frac{e^3\mathrm{M}}{2\pi\,\mathrm{m}^2c^2f^2}\left[\frac{R'}{\sqrt{R'^2 - a^2\sin^2\alpha_0}}\frac{\partial}{\partial\tau'}\left(\frac{\cos\mu}{R'^2}\right)\right]_{R'=\dot{R}} N_{\mathrm{t}}(R). \tag{4.74}$$

Here $\dot{R} = R_{\mathrm{m}}$ when $R > R_{\mathrm{m}}$ and $\dot{R} = R$ when $R < R_{\mathrm{m}}$. Kolosov et al.[31] provides an example of the calculation of $\partial/\partial\tau'(\cos\mu/R'^2)$. Here, we provide an estimation, having assumed on average that:

$$\frac{\partial}{\partial\tau'}\left(\frac{\cos\mu}{R'^2}\right) \cong -\frac{2\sin\mu_0}{R'^3},$$

where μ_0 is the magnetic latitude of the ground point, and we obtain:

$$\Psi_{\mathrm{F}} = \frac{2.12\cdot 10^{17}}{f^2\sqrt{1-0.91\sin^2\alpha_0}}, \tag{4.75}$$

assuming that $\langle\sin\mu_0\rangle = 0.5$ and substituting the previous values of the parameters for high-altitude sources (e.g., artificial satellites). We obtain $\Psi_{\mathrm{F}} \cong 38$ radians for the traditional $\alpha_0 = 60°$ and $f = 10^8$ Hz.

4.4 FLUCTUATION PROCESSES ON RADIOWAVE PROPAGATION IN A TURBULENT ATMOSPHERE

The atmosphere model adopted in the previous sections is assumed to be homogeneous on a large scale with a regional character and parameters dependent on climate conditions in general. These parameters (for example, temperature) change slowly throughout the day and vary according to season, so they have both diurnal and seasonal trends. Because it is large scale, this model allows us to use the spherically layered atmosphere approximation and gives us the opportunity, as we have shown, to describe radiowave propagation processes in general. The described effects change only slowly, with typical periods of at least 1 hour; however, along with these rather slow processes much faster ones also occur with periods on the order of seconds. They have a random character and are generated by turbulent pulsation in the atmosphere. As experience shows, these turbulent pulsations initiate small fluctuations of medium permittivity. Because of the accumulation phenomenon, fluctuations in the radiowave parameters (amplitude, phase, frequency) may not be small, and we must study these fluctuations, because, in some cases, they generate noticeable effects with regard to radiowave propagation in the troposphere and ionosphere of Earth.

We will now provide a brief description of the statistical properties of turbulent pulsation in the atmosphere. We must first consider permittivity as a stochastic function of coordinates and time. The last dependence is neglected in the first approximation; due to the great velocity of radiowave propagation, the signal manages to pass from the transmitter to the receiver in a time that is much less than typical periods of turbulent fluctuations. This allows the medium to be considered as unchangeable and we may take into account only spatial changes in the permittivity; hence, the approximation should be assumed to be a random function of coordinates.

Let us assume that the medium is statistically homogeneous and suppose for simplicity that $\langle \varepsilon(\mathbf{r}) \rangle = 1$. Then,

$$\varepsilon(\mathbf{r}) = 1 + \mu(\mathbf{r}), \tag{4.76}$$

where $\mu(\mathbf{r})$ is a random coordinate function with a zero average. Further, it is supposed that the intensity of the permittivity fluctuations is small; that is, $\langle \mu^2 \rangle \ll 1$. This gives us the opportunity to consider μ to be a small value on the whole, based on the small probability of its having large values; therefore, the probability of significant local fluctuations of the radiowave parameters is also small. We also need the spatial correlation function of the permittivity:

$$\mathrm{K}_{\varepsilon}(\mathbf{r}_1, \mathbf{r}_2) = \langle \mu(\mathbf{r}_1) \mu(\mathbf{r}_2) \rangle. \tag{4.77}$$

The introduced function depends only on the differential vector $\mathbf{r} = \mathbf{r}_2 - \mathbf{r}_1$ due to the statistic homogeneity of the turbulence field. In addition, we will assume statistical isotropy of the turbulence. On this basis, the correlation function depends only on the distance between the points of correlation and does not depend on the direction of vector $\mathbf{r}$ connecting these points. Let us introduce the spatial spectrum of turbulence:

$$\tilde{\mathrm{K}}_{\varepsilon}(\mathbf{q}) = \frac{1}{8\pi^3} \int \mathrm{K}(\mathbf{r})\, e^{-i\mathbf{q}\cdot\mathbf{r}} d^3\mathbf{r}. \tag{4.78}$$

The correlation function is expressed through its spatial spectrum by the relation

$$\mathrm{K}_{\varepsilon}(\mathbf{r}) = \int \tilde{\mathrm{K}}_{\varepsilon}(\mathbf{q})\, e^{i\mathbf{q}\cdot\mathbf{r}} d^3\mathbf{q}. \tag{4.79}$$

The function $\tilde{\mathrm{K}}_{\varepsilon}(\mathbf{q})$ represents the spectral density of the power of the permittivity fluctuations inside the wave vector interval ($\mathbf{q}$, $\mathbf{q}+d\mathbf{q}$). We can easily show that the isotropy of the correlation function leads to isotropy of the spectral density.

We will often use the following formula for the spectrum:

$$\tilde{\mathrm{K}}_{\varepsilon}(\mathrm{q}) = \frac{C_{\varepsilon}^{2} e^{-\mathrm{q}^2/\mathrm{q}_m^2}}{\left(1+\mathrm{q}^2/\mathrm{q}_0^2\right)^{\nu}}. \tag{4.80}$$

Here, by convention, the wave number q_0 corresponds to the maximum (outer) turbulence scale and q_m is the minimum (inner) scale. Let us call value ν the spectral index (although this term is often given to the double value 2ν). It is often assumed that $\nu = 11/6$ for the case of tropospheric turbulence. It corresponds to the model of homogeneous and isotropic turbulence, the properties of which were established in the works of Kolmogorov and Obukhov. The constant C_{ε}^{2} is proportional to the fluctuations intensity $\langle \mu^2 \rangle$. This relation is easily determined from the equality:

$$\left\langle \mu^2 \right\rangle = \mathrm{K}_{\varepsilon}(0) = \int \tilde{\mathrm{K}}_{\varepsilon}(\mathbf{q})\, d^3\mathbf{q}. \tag{4.81}$$

By applying Equation (4.80) and performing the integration in a spherical coordinate system, we obtain:

$$\left\langle \mu^2 \right\rangle = 4\pi C_{\varepsilon}^{2} \int_0^{\infty} \frac{e^{-\mathrm{q}^2/\mathrm{q}_m^2} \mathrm{q}^2 d\mathrm{q}}{\left(1+\mathrm{q}^2/\mathrm{q}_0^2\right)^{\nu}}.$$

The integral can be expressed via a confluent hypergeometric function; however, it may be proved more easily using the fact that usually $\mathrm{q}_m \gg \mathrm{q}_0$. The radius of the integral convergence has a value of about q_0. Then, the exponent in the integral can be substituted for unity, and we obtain:

$$\left\langle \mu^2 \right\rangle = \frac{\pi^{3/2}\Gamma\left(\nu - 3/2\right)}{\Gamma(\nu)} \mathrm{q}_0^3 C_{\varepsilon}^{2}. \tag{4.82}$$

Let us now determine the outer scale of permittivity fluctuations for this model spectrum as the sphere radius of volume:

$$\frac{4\pi}{3} l^3 = \frac{1}{\langle \mu^2 \rangle} \int K\varepsilon(\mathbf{r}) d^3\mathbf{r} = \frac{8\pi^3}{\langle \mu^2 \rangle} \tilde{K}_\varepsilon(0). \tag{4.83}$$

The use of the model spectrum gives:

$$l^3 = \frac{6\pi^2 C_\varepsilon^2}{\langle \mu^2 \rangle} = \frac{6\sqrt{\pi}\Gamma(\nu)}{\Gamma\left(\nu - \frac{3}{2}\right)} \frac{1}{q_0^3}. \tag{4.84}$$

Particularly, at $\nu = 11/6$:

$$l = \frac{1.55}{q_0}. \tag{4.85}$$

The outer scale of the atmosphere has a value of at least 100 m. So it should be assumed that:

$$kl >> \sqrt{\langle \mu^2 \rangle} .$$

This relation corresponds to Equation (4.1) and confirms in the first approach the validity of the geometrical optics conception by wave propagation in a turbulent atmosphere. The small intensity of permittivity fluctuation allows us to solve the geometrical optics equations by the method of disturbance. Let us address, first of all, the trajectory equation, Equation (4.11), which will now have the form:

$$\frac{d\mathbf{s}}{d\tau} = \frac{1}{2} \nabla\mu - \frac{1}{2} \mathbf{s}\left(\mathbf{s}\nabla\mu\right). \tag{4.86}$$

where $\mathbf{s} = \mathbf{s}_0 = const$ in the zero-order approximation. Thus, the trajectory is a straight line in the zero-order approach and is described by the equation $\mathbf{r} = \mathbf{s}_0\tau$.

Let us choose the coordinate system in such a way that vector $\mathbf{s}_0$ would be on the z-axis (i.e., $\mathbf{s}_0 = \mathbf{e}_z$). Then, the expression for the eikonal value will have the form:

$$\psi\left(\mathbf{r}\right) = z + \frac{1}{2} \int_0^z \mu\left(\mathbf{r}'\right) dz'. \tag{4.87}$$

The initial eikonal value is assumed to be zero. Furthermore, let L be the distance z traveled by the wave in the medium. In these symbols, the mean eikonal value is:

$$\langle \psi(\mathbf{r}) \rangle = L, \tag{4.88}$$

or the distance traveled by the wave through the turbulent medium. The fluctuation part of the eikonal value is:

$$\psi_1(\mathbf{r}) = \frac{1}{2} \int_0^L \mu(\mathbf{r}')\, dz'. \tag{4.89}$$

If $L >> l$, then this integral is the sum of a great number of statistically independent items. Thus, according to the central limit theorem, the eikonal fluctuations are distributed due to the normal law and it follows that their dispersion (or fluctuation power) fully characterizes the distribution. It is easy to establish that:

$$\sigma_\psi^2 = \langle \psi_1^2 \rangle = \frac{1}{4} \int_0^L \int_0^L K_\varepsilon(0, 0, |z'' - z'|)\, dz'\, dz''. \tag{4.90}$$

The double integral is easily transformed to a one-dimensional one by changing the variables ($z' + z'' = 2Z$ and $z'' - z' = z$):

$$\langle \psi_1^2 \rangle = \frac{1}{2} \int_0^L K_\varepsilon(0, 0, z)(L - z)\, dz. \tag{4.91}$$

Sufficient integration is found in the interval $0 < z < 1$. As distance L is supposed to be much more than the correlation radius, then, approximately:

$$\langle \psi_1^2 \rangle = \frac{L}{2} \int_0^\infty K_\varepsilon(z)\, dz = \frac{L}{4} \int_{-\infty}^\infty K_\varepsilon(z)\, dz = \pi^2 L \int_0^\infty \tilde{K}_\varepsilon(q_\perp, 0)\, q_\perp dq_\perp. \tag{4.92}$$

Here, $q_\perp = \sqrt{q_x^2 + q_y^2}$. Referring back to Equation (4.80) gives us:

$$\langle \psi_1^2 \rangle = \frac{\pi^2 C_\varepsilon^2 q_0^2 L}{2(\nu - 1)} = \left[\sqrt{\frac{\pi}{6}} \frac{\Gamma(\nu)}{\Gamma\left(\nu - \frac{3}{2}\right)} \right]^{2/3} \frac{\langle \mu^2 \rangle L l}{2(\nu - 1)}. \tag{4.93}$$

Here, we neglected again the difference between the exponential factor and unity. Equation (4.93) can take the form:

$$\left\langle \psi_1^2 \right\rangle = 0.25\left\langle \mu^2 \right\rangle L l. \tag{4.94}$$

It is characteristic that the intensity of the eikonal fluctuations and the phase are proportional to the distance traveled by the wave. The eikonal fluctuations are independent at the parts pass of separated by correlation interval l. The number of such parts is approximately $N \cong L/l$ at a line of length L. The intensity of the local eikonal fluctuations in each of these parts is about $\langle \mu^2 \rangle l^2$. The combined fluctuation intensity is of the order of $N\langle \mu^2 \rangle l^2$, from which Equation (4.94) follows.

Let us now consider the amplitude fluctuations. We can use Equation (4.21) to analyze the logarithm of amplitude fluctuations:

$$\kappa = \ln \frac{E_0(\tau)}{E_0(0)} = -\frac{1}{2}\int_0^{\tau} \frac{\nabla^2 \psi}{\sqrt{\varepsilon}}\, d\tau'. \tag{4.95}$$

Then we can substitute the expression for the eikonal value in Equation (4.87) to achieve the first approximation:

$$\kappa = -\frac{1}{4}\left[\mu(x,y,z) - \mu(x,y,0)\right] - \frac{1}{4}\int_0^L dz' \int_0^{z'} \nabla_\perp^2 \mu \; dz''.$$

Here, the transversal operator $\nabla_\perp^2 = \partial^2/\partial x^2 + \partial^2/\partial y^2$. It is easy to establish that the average value $\langle \kappa \rangle = 0$, and it follows that the logarithm of the wave amplitude is invariable, on average. The first item in this formula may be in the calculations of the amplitude logarithm fluctuation power. The estimations, which we have omitted, show that inclusion of the first member gives items on the order of $\langle \mu^2 \rangle$ and $\langle \mu^2 \rangle L/l$. They are tiny compared to the second item, which is on the order of $\langle \mu^2 \rangle L^3/l^3$. Before performing the necessary averaging, we can point out that integration by parts allows us to turn the double integral into a single one. So, the main item becomes:

$$\kappa = -\frac{1}{4}\int_0^L (L - z)\nabla_\perp^2 \mu \; dz, \tag{4.96}$$

and the average square is:

$$\left\langle \kappa^2 \right\rangle = \frac{\nabla_\perp^4}{16}\int_0^L \int_0^L (L - z')(L - z'') K_\varepsilon\left(0,0,\left|z'' - z'\right|\right) dz' dz''. \tag{4.97}$$

Here, $\nabla_{\perp}^2 = \left(\partial^2/\partial x^2 + \partial^2/\partial y^2\right)$. Of course, differentiation of the correlation function is derived first, and the corresponding variables tend to zero. We must now perform a transformation of variables similar to what we did to obtain Equation (4.91):

$$\left\langle \kappa^2 \right\rangle = \frac{\nabla_{\perp}^4}{8} \int_0^L \left[\frac{1}{3}\left(L^3 - z^3\right) - \frac{z}{2}\left(L^2 - z^2\right) \right] K_{\varepsilon}\left(0,0,z\right) dz \, .$$

Further, the reasoning used earlier with regard to the main integration interval at the condition $l << L$ allows us to derive the equalities:

$$\left\langle \kappa^2 \right\rangle = \frac{L^3 \nabla_{\perp}^4}{24} \int_0^{\infty} K_{\varepsilon}\left(0,0,z\right) dz = \frac{\pi^2 L^3}{12} \int_0^{\infty} q_{\perp}^5 \tilde{K}_{\varepsilon}\left(q_{\perp},0\right) dq_{\perp} . \tag{4.98}$$

The fact that the fluctuation intensity of the amplitude logarithm, in the geometrical optics approximation, is proportional to the cube of the distance requires some attention. We can show that the allowance for radiowave diffraction on the turbulent inhomogeneities establishes this normality within limited distances.[34] At rather large distances, value $\langle \kappa \rangle$ asymptotically becomes proportional to the first degree of the distance.

By substituting the expression for the model spectrum, we obtain:

$$\left\langle \kappa^2 \right\rangle = \frac{\pi^2 L^3 C_{\varepsilon}^2}{12} \int_0^{\infty} \frac{e^{-q_{\perp}^2/q_m^2} q_{\perp}^5 \, dq_{\perp}}{\left(1 + q_{\perp}^2 / q_0^2\right)^{\nu}} . \tag{4.99}$$

Here, it is necessary to take into account the exponent in the integral, as it will be divergent on infinity. As was already pointed out, such integrals are expressed through confluent hypergeometric functions; however, the smallness of the ratio q_0/q_m allows us, as more detailed calculations show, to substitute Equation (4.99) for an approximate one (because of good convergence at zero):

$$\left\langle \kappa^2 \right\rangle = \frac{\pi^2 L^3 C_{?}^2 q_0^{2\nu}}{12} \int_0^{\infty} e^{-q_{\perp}^2/q_m^2} q_{\perp}^{5-2\nu} dq_{\perp} . \tag{4.100}$$

The last integral is easily expressed via a gamma-function with the result:

$$\left\langle \kappa^2 \right\rangle = \frac{\pi^2}{24} C_{\varepsilon}^2 \Gamma\left(3 - \nu\right) q_0^{2\nu} q_m^{6-2\nu} L^3 , \tag{4.101}$$

and we have:

$$\left\langle \kappa^2 \right\rangle = \frac{\pi^2 \Gamma(7/6)}{24} C_\varepsilon^2 q_0^{11/3} q_m^{7/3} = 0.38 C_\varepsilon^2 q_0^{11/3} q_m^{7/3} \tag{4.102}$$

for $\nu = 11/6$.

5 Radiowave Scattering

5.1 CROSS SECTION OF SCATTERING

We have already considered radiowave propagation in inhomogeneous media; however, the rate of change of media parameters was so small in our examples that the geometrical optics technique was rather workable in the first approximation. In this chapter, we will consider situations when the spatial variation of the medium can be sharp on the wavelength scale including the jump changes. Examples include drop formation of clouds and rain, areas of vegetation, hailstones, etc. The primary phenomena with regard to radiowave interaction with such inhomogeneities are diffraction and the associated wave scattering. Thus, certainly, wave absorption in the inhomogeneities does take place. Both processes — scattering and absorption — lead to attenuation of the incident wave power flow, and a phenomenon known as *extinction* occurs.

The waves field can be represented, in the cases discussed here, as the sum of the incident and scattered waves; that is, $\mathbf{E} = \mathbf{E}_i + \mathbf{E}_s$ and $\mathbf{H} = \mathbf{H}_i + \mathbf{H}_s$. The field of scattered waves is described by:

$$\nabla \times \mathbf{E}_s = ik\mathbf{H}_s, \quad \nabla \times \mathbf{H}_s = -ik\mathbf{E}_s - ik\left[\varepsilon(\mathbf{r}) - 1\right]\mathbf{E}. \tag{5.1}$$

Here, $\varepsilon(\mathbf{r})$ is the permittivity of the scattering waves of the homogeneities. It is assumed, then, that the permittivity of the medium is equal to unity in the absence of inhomogeneities. We may consider the second term in Equation (5.1) to be a current with density:

$$\mathbf{j}_e(\mathbf{r}) = -\frac{ikc}{4\pi}\left[\varepsilon(\mathbf{r}) - 1\right]\mathbf{E} = -\frac{i\omega}{4\pi}\left[\varepsilon(\mathbf{r}) - 1\right]\mathbf{E}. \tag{5.2}$$

Then, the scattered field can be expressed in the form of an integral from the inserted equivalent current. The integral must be extended, in this case, to the volume taken up by all of the scattering particles (inhomogeneities). We shall first pay attention to the simple case of one particle and will consider the field at the wave zone. We must use Equations (1.37) and (1.38) to obtain:

$$\mathbf{E}_s = \frac{k^2 e^{ikr}}{4\pi r}\int_V \left[\frac{\mathbf{r}}{r} \times \left[\mathbf{E} \times \frac{\mathbf{r}}{r}\right]\right](\varepsilon - 1)\exp\left(-ik\frac{\mathbf{r}\cdot\mathbf{r}'}{r}\right)d^3\mathbf{r}', \quad \mathbf{H}_s = \left[\frac{\mathbf{r}}{r} \times \mathbf{E}_s\right]. \tag{5.3}$$

Here, integration is performed over the volume of the scattering inhomogeneities.

From this point forward, we will assume as a rule that the sources of radiation are so far away that the incident field within the scattering volume (or the particle size) does not differ significantly from the plane wave. As we have already seen, it is necessary for the size of the Fresnel zone to exceed the scale of the scattering inhomogeneities. Let us assume that the incident wave has the form $\mathbf{E}_i = \mathbf{g}_i \exp\left[ik\left(\mathbf{e}_i \cdot \mathbf{r}\right)\right]$ and $\mathbf{H}_i = \left[\mathbf{e}_i \times \mathbf{E}_i\right] = \left[\mathbf{e}_i \times \mathbf{g}_i\right]\exp\left[ik\left(\mathbf{e}_i \cdot \mathbf{r}\right)\right]$. Here, single vector $\mathbf{g}_i$ is the vector of linear polarization, and $\mathbf{e}_i$ is the single vector of the propagation direction of the incident wave. The incident wave amplitude is assumed to be equal to unity. The vector $\mathbf{e}_s = \mathbf{r}/r$ determines the direction of the wave scattering.

The scattering characteristics of a particle (inhomogeneity) are defined by the vector:

$$\mathbf{f}\left(\mathbf{e}_i, \mathbf{e}_s, \mathbf{g}_i\right) = \frac{k^2}{4\pi} \int_V \left[\mathbf{e}_s \times \left(\mathbf{E} \times \mathbf{e}_s\right)\right]\left(\varepsilon - 1\right)\exp\left[-ik\left(\mathbf{e}_s \cdot \mathbf{r}'\right)\right] d^3\mathbf{r}', \qquad (5.4)$$

called the scattering amplitude. The scattered wave is described in these notations by the expression:

$$\mathbf{E}_s = \mathbf{f}\left(\mathbf{e}_i, \mathbf{e}_s, \mathbf{g}_i\right)\frac{e^{ikr}}{r}, \quad \mathbf{H}_s = \left[\mathbf{e}_s \times \mathbf{f}\right]\frac{e^{ikr}}{r}. \qquad (5.5)$$

Equation (5.5) has a general nature and is not automatically a consequence of Equation (5.3); in particular, a value of $\varepsilon = \infty$ (a metallic particle) in Equation (5.3) makes the current definition, Equation (5.2), meaningless. Equation (5.5), however, keeps its meaning in such a case. Equation (5.3) itself is, in essence, an integral form of the initial Maxwell equations. Field E under the integral is not generally known beforehand and should be found by solving the problem of radiowave diffraction on the inhomogeneity being considered. It can be determined only in very few cases from any previous findings, usually with the same approach.

The amplitude of scattering is a function of the incident wave direction and its polarization, as well as the direction of scattering. The scattering is, in the general case, accompanied by a change of polarization. The dependence of the scattering amplitude on the incident wave propagation direction and its polarization is obvious; therefore, as a rule we will omit the corresponding vectors from our list of scattering amplitudes and arguments and, for brevity, will keep only the dependence on vector $\mathbf{e}_s$.

The power flow density of the incident wave is $S_i = c/8\pi$. Poynting's vector of scattered wave is:

$$\mathbf{S}_s = S_i \left|\mathbf{f}\left(\mathbf{e}_s\right)\right|^2 \frac{\mathbf{e}_s}{r^2}. \qquad (5.6)$$

The following value is the differential cross section of scattering:

$$\sigma_d(\mathbf{e}_i, \mathbf{e}_s, \mathbf{g}_i) = \left| f(\mathbf{e}_i, \mathbf{e}_s, \mathbf{g}_i) \right|^2. \tag{5.7}$$

It has the dimensions of a square and corresponds to the power flow density scattered in direction $\mathbf{e}_s$ within a solid angular unit. It should be pointed out that in radar science a rather different definition of the scattering differential cross section is used and is connected with the one introduced here by the simple equality $\sigma_d^{(r)} = 4\pi\sigma_d$.

The total flow of the power scattered in all directions is characterized by the cross section of scattering:

$$\sigma_s(\mathbf{e}_i, \mathbf{g}_i) = \int_{4\pi} \sigma_d(\mathbf{e}_i, \mathbf{e}_s, \mathbf{g}_i)\, d\Omega = \int_{4\pi} |f|^2\, d\Omega, \tag{5.8}$$

where the integration is provided with respect to solid angle Ω.

Let us now turn our attention to the absorbed power. As we already know, the density of the absorbed power is determined by Equation (1.20). Integrating over the scattered volume gives us the power value absorbed by a particle. The normalization of this power per incident wave power flow density determines the cross section of absorption:

$$\sigma_a = k \int_V \varepsilon''(\mathbf{r}') \left| E(\mathbf{r}') \right|^2 d^3\mathbf{r}'. \tag{5.9}$$

The cross section of absorption is small if the scattering inhomogeneity is almost transparent ($\varepsilon'' << 1$) or the particle has very high conductivity ($E \to 0$). The physical difference between σ_s and σ_a is that the cross section of scattering characterizes the spatial redistribution of the incident wave power flow, and the cross section of absorption defines the efficiency of transfer of this energy to heat.

The summed value is the total cross section:

$$\sigma_t = \sigma_s + \sigma_a \tag{5.10}$$

which determines wave attenuation. If the density of the particles is not very high, then we can assume in the first approximation that they scatter independently, which means that the field of scattering of adjacent particles is much smaller than the field of the incident wave; that is, the scattering of fields already scattered does not play a particular role. According to these assumptions, the power of the field scattered by the particle assembly is equal to the sum of the powers scattered by every particle separately, which applies to irregular distribution of particles inside the scattering volume. In fact, it is believed that particles scatter incoherently. Let us consider the volume of a medium having a single section and length ds along the direction of wave propagation. The power flow density at the input of the considered element is S and at the output it is S + dS. The difference between these densities is determined by wave scattering and absorption by particles that are inside the volume; that is, $dS = -SN\sigma_t ds$, where N is the particle density in the volume. The corresponding transfer equation is:

$$\frac{dS}{ds} = -N\sigma_t S\,, \tag{5.11}$$

for which the solution (at σ_t, $N = const$) is:

$$S = S_0 e^{-\Gamma s}, \tag{5.12}$$

where the value:

$$\Gamma = N\sigma_t \tag{5.13}$$

is the *coefficient of extinction*, which defines the degree of wave attenuation in the scattering medium.

The value:

$$\hat{A} = \frac{\sigma_s}{\sigma_t} \tag{5.14}$$

is the scattering albedo and determines the role of scattering in the general balance of propagating wave energy losses.

5.2 SCATTERING BY FREE ELECTRONS

The scattering of electromagnetic waves by electrons is one of the most vivid examples of the scattering process. Due to the incident field, electrons take on an oscillatory motion that is followed by radiation. This reradiation field is the field of electron scattering. Let us, first of all, formulate the equation of electron motion in the field of the incident wave:

$$\frac{d^2\mathbf{r}}{dt^2} + \nu\frac{d\mathbf{r}}{dt} - \gamma\frac{d^3\mathbf{r}}{dt^3} = -\frac{e}{m}\mathbf{E}. \tag{5.15}$$

Here, ν is the frequency of electron collisions (so, in this case, the electrons are not absolutely free, because introducing the frequency of collision takes into account their interaction with other particles), and

$$\gamma = \frac{2e^2}{3mc^3} = 6.25 \cdot 10^{-20} \text{ sec}. \tag{5.16}$$

The term with the third derivative describes the strength of the radiation friction.[36] It is assumed that the processes are not relativistic, so the influence of the wave magnetic field on the electrons is neglected.

In the case of harmonic time dependence, the solution to the movement equation is:

$$\mathbf{r} = \frac{e\mathbf{E}}{m\omega\left[\omega + i\left(\nu + \omega^2\gamma\right)\right]}. \tag{5.17}$$

The inducted dipole moment of an electron is:

$$\mathbf{p} = -e\mathbf{r} = -\frac{e^2\mathbf{E}}{m\omega\left[\omega + i\left(\nu + \omega^2\gamma\right)\right]}. \tag{5.18}$$

The reradiation power is calculated with the help of Equation (1.44), and, being relative to the incident wave power flow density, gives us the following cross section of scattering:

$$\sigma_s = \frac{8\pi}{3} a_e^2 \frac{1}{1 + \left(\nu/\omega + \omega\gamma\right)^2}. \tag{5.19}$$

Here,

$$a_e = \frac{e^2}{mc^2} = 2.82 \cdot 10^{-13} \text{cm} \tag{5.20}$$

is the classical radius of electron. Usually, ν/ω, $\omega\gamma << 1$; the corresponding members can be neglected; and we come to the classical Thomson's formula:

$$\sigma_s = \frac{8\pi}{3} a_e^2. \tag{5.21}$$

The cross section of absorption can be calculated based on the following discussion. The work performed by the field at the electron is the criterion of the wave energy absorption. The work performed in a unit of time is equal to the product

$-1/2\mathrm{Re}(e\mathbf{v}\cdot\mathbf{E})$, where $\mathbf{v}$ is the electron velocity. Dividing this product by the incident wave power flow density gives us the absorption cross section:

$$\sigma_a = \frac{4\pi a_e c\left(\nu+\omega^2\gamma\right)}{\omega^2+\left(\nu+\omega^2\gamma\right)^2} \cong \frac{4\pi a_e c\nu}{\omega^2}. \tag{5.22}$$

We have already said that taking into account electron collisions assumes that the electrons are in an environment of other particles, including charged ones (i.e., plasma). The interaction between charged particles of plasma leads to collective effects. Generally speaking, it is not correct to consider electron motion without taking into account movement in the closest environment. Also, reradiation processes of waves by electrons without partially coherent radiation of the electromagnetic energy by its neighbors cannot be considered. So, the result obtained has an approximate character. The conditions of its validity and the role of collective effects will be further discussed.

5.3 OPTICAL THEOREM

The value of the total cross section is connected with the amplitude of scattering forward in the direction of the incident wave. To show this,[10] we will surround the particle by a sphere of large radius that tends to infinity. Let us consider the energy balance in the volume surrounded by this sphere. Equation (1.17), in which external currents are not considered, will serve as our base. Let us perform the integration over the volume and transform the volume integral from $\nabla\cdot\mathbf{S}$ into a surface one:

$$\int_{4\pi}\left(\mathbf{S}\cdot\mathbf{e}_s\right)r^2 d\Omega = -S_i\sigma_a. \tag{5.23}$$

Because the field is the sum of incident and scattered waves, the Poynting vector of the summary wave is $\mathbf{S} = \mathbf{S}_i + \mathbf{S}_s + \mathbf{S}_{is}$, where the interference term can be written as:

$$\mathbf{S}_{is} = \frac{c}{8\pi}\mathrm{Re}\left\{\left[\mathbf{E}_i\times\mathbf{H}_s\right]+\left[\mathbf{E}_s\times\mathbf{H}_i\right]\right\}.$$

The integral of $\mathbf{S}_i$ is equal to zero, and the integral of $\mathbf{S}_s$ gives us the flow of the scattered power, equal to $S_i\sigma_s$. Thus, we obtain:

$$\sigma_t = -\frac{1}{S_i}\int_{4\pi}\left(\mathbf{S}_{is}\cdot\mathbf{e}_s\right)r^2 d\Omega. \tag{5.24}$$

The expanded view is:

$$\sigma_t = -r\,\mathrm{Re}\int_{4\pi}\Big\{\Big[\big(\mathbf{g}_i\cdot\mathbf{f}^*\big)-\big(\mathbf{e}_s\cdot\mathbf{f}^*\big)\big(\mathbf{g}_i\cdot\mathbf{e}_s\big)\Big]e^{-ikr\left[1-\left(\mathbf{e}_i\cdot\mathbf{e}_s\right)\right]}+$$

$$+\Big[\big(\mathbf{e}_i\cdot\mathbf{e}_s\big)\big(\mathbf{g}_i\cdot\mathbf{f}\big)-\big(\mathbf{e}_s\cdot\mathbf{g}_i\big)\big(\mathbf{e}_i\cdot\mathbf{f}\big)e^{ikr\left[1-\left(\mathbf{e}_i\cdot\mathbf{e}_s\right)\right]}\Big]\Big\}d\Omega.$$

In essence, the integration is performed with respect to spherical angles defining the direction of vector $\mathbf{e}_s$. Because $kr >> 1$, the stationary phase method can be used to calculate the integral. The points of the stationary phase are related to the directions $\mathbf{e}_s = \mathbf{e}_i$ and $\mathbf{e}_s = -\mathbf{e}_i$. The corresponding asymptotic integration leads to the result:

$$\sigma_t = \frac{4\pi}{k}\,\mathrm{Im}\big(\mathbf{g}_i\mathbf{f}(\mathbf{e}_i)\big). \tag{5.25}$$

One should bear in mind, in this case, that the contribution of the stationary point, $\mathbf{e}_s = -\mathbf{e}_i$, is missed because its real part is equal to zero. One should also take into account the product $\mathbf{e}_s \cdot \mathbf{f}(\mathbf{e}_s) = 0$, due to transversality of the scattering field.

The result is referred to as the *optical theorem*, and it demonstrates that extinction is determined by the forward scattering intensity. Note that polarization of the forward scattered radiation cannot be orthogonal to the incident wave polarization, which applies to particles of isotropic material.

Another definition of optical theorem is based on Equation (5.4):

$$\sigma_t = k\,\mathrm{Im}\int_V (\varepsilon-1)\big(\mathbf{g}_i\cdot\mathbf{E}\big)\exp\big[-ik\big(\mathbf{e}_i\cdot\mathbf{r}'\big)\big]d^3\mathbf{r}'. \tag{5.26}$$

The optical theorem allows us to calculate easily the maximum value of the total cross section for nontransparent bodies whose sizes are large compared to the wavelength. The Kirchhoff approximation may be used, in this case, to describe the scattered field. According to the Babinet principle, the scattered field in the fare zone is:

$$\mathbf{E}_s = \frac{ik\mathbf{g}_i}{2\pi}\frac{e^{ikr}}{r}\big(\mathbf{e}_i\cdot\mathbf{e}_s\big)\int_A \exp\big[-ik\big(\mathbf{e}_s\cdot\mathbf{r}'\big)\big]d^2\mathbf{r}'.$$

From here, the amplitude of the forward scattering:

$$\mathbf{f}(\mathbf{e}_i) = \frac{ik\mathbf{g}_i}{2\pi}\int_A \exp\big[-ik\big(\mathbf{e}_s\cdot\mathbf{r}'\big)\big]d^2\mathbf{r}'. \tag{5.27}$$

Now, the mutual orthogonality of vectors $\mathbf{e}_i$ and $\mathbf{r}'$ should be taken into consideration. So,

$$\sigma_t = 2A. \tag{5.28}$$

Here, A is the body section transversal to the direction of the incident wave propagation. Thus, the total cross section of scattering by a large-size body is equal to double the square of its section.

It would be reasonable to ask at what ratio of body size to wavelength we can use Equation (5.28) with reasonable accuracy. It is not possible to answer that question in general terms, as it is necessary to solve the wave diffraction problem and the answers may differ for bodies with different shapes and different physical properties. Analysis of calculation results (e.g., see King[77] for a metallic sphere and a metallic disc with radius a) shows that Equation (5.28) leads to satisfactory results for wavelengths $\lambda = a$ and shorter.

5.4 SCATTERING FROM A THIN SHEET

A thin sheet with thickness d and square A is an example of when the scattering problem can be solved without actually determining an exact solution. If transverse sizes of the sheet are much larger than the wavelength, we can use the solution of wave reflection by a homogeneous layer (see Section 3.3) to calculate the field. Naturally, such an approach is based on the smallness of the edge effect. Use of the term *thin sheet* indicates observance of the condition $kd \ll 1$, and the value $\left|kd\sqrt{\varepsilon}\right|$ can be large. A thin sheet can be a vegetable leaves model. As the scattering takes place at rather low frequencies (corresponding to wavelengths measured in centimeters), the sheet can be semitransparent, and at high frequencies (waves measured in millimeters) it practically does not transmit radiowaves. Let us use the results of Section 3.5 assuming that $\varepsilon_1 = \varepsilon_3 = 1$ and $\varepsilon_2 = \varepsilon$. The calculation can be performed using the Kirchhoff approximation and the theory of diffraction.

For the case of horizontally polarized waves (H-waves), we can use Equation (1.75), which is represented in this case as the sum:

$$\mathbf{E}_i + \mathbf{E}_s = -\frac{1}{2\pi}\frac{\partial}{\partial z}\int\limits_{A_\infty - A} \mathbf{E}_i(\mathbf{r}')\frac{e^{ikR}}{R}d^2\mathbf{r}' - \frac{T(\theta_i)}{2\pi}\frac{\partial}{\partial z}\int\limits_{A} \mathbf{E}_i(\mathbf{r}')\frac{e^{ikR}}{R}d^2\mathbf{r}'.$$

Note that:

$$-\frac{1}{2\pi}\int\limits_{A_\infty - A}(\;)\,d^2\mathbf{r}' = \mathbf{E}_i + \frac{1}{2\pi}\int\limits_{A}(\;)\,d^2\mathbf{r}'.$$

So,

$$\mathbf{E}_s = \frac{1-T(\theta_i)}{2\pi}\frac{\partial}{\partial z}\int_A \mathbf{E}_i(\mathbf{r}')\frac{e^{ikR}}{R}d^2\mathbf{r}'. \tag{5.29}$$

Defining a scattered field in such a way is a generalization of the Babinet principle for semitransparent screens. Equation (5.29) can be rewritten as:

$$\mathbf{E}_s = \frac{ik\left[1-T(\theta_i)\right]}{2\pi}(\mathbf{e}_z\cdot\mathbf{e}_s)\frac{e^{ikr}}{r}\int_A \mathbf{E}_i(\mathbf{r}')e^{-ik(\mathbf{e}_s\cdot\mathbf{r}')}d^2\mathbf{r}'. \tag{5.30}$$

Correspondingly, the amplitude of scattering is:

$$\mathbf{f}(\mathbf{e}_s) = \frac{ik\left[1-T(\theta_i)\right]}{2\pi}(\mathbf{e}_z\cdot\mathbf{e}_s)\int_A \mathbf{E}_i(\mathbf{r}')e^{-ik(\mathbf{e}_s\cdot\mathbf{r}')}d^2\mathbf{r}'. \tag{5.31}$$

At last, using the expression for the incident wave and Equation (5.25), we obtain:

$$\sigma_t = 2A\cos\theta_i \operatorname{Re}\left[1-T(\theta_i)\right]. \tag{5.32}$$

Note that Equation (5.28) applies in the case of an opaque sheet.

In the case of vertical polarization (E-waves), an equation similar to Equation (5.30) must be written for the magnetic field. If the magnetic field is now expressed through an electrical one, then it is easy to show that we again obtain Equation (5.32), where only the coefficient of transmission for vertically polarized waves should be used.

The cross section of absorption is calculated on the basis of simple energetic considerations. The wave power flow incident at the sheet is equal to $S_iA\cos\theta_i$. The flow value of the reflected power is $S_iA\cos\theta_i\left|F(\theta_i)\right|^2$. The flow of the transmitted power is determined by the value $S_iA\cos\theta_i\left|T(\theta_i)\right|^2$. The difference between the power flow of the incident wave and the reflected plus transmitted powers gives us the energy absorbed inside the sheet in a unit of time. This, in turn, determines the cross section of absorption:

$$\sigma_a = \frac{S_iA\cos\theta_i\left(1-|F|^2-|T|^2\right)}{S_i} = A\cos\theta_i\left[1-\left|F(\theta_i)\right|^2-\left|T(\theta_i)\right|^2\right]. \tag{5.33}$$

Now, it is not difficult to calculate the value of the cross section of scattering:

$$\sigma_s = \sigma_t - \sigma_a = A\cos\theta_i\left[\left|F(\theta_i)\right|^2 + \left|1 - T(\theta_i)\right|^2\right]. \tag{5.34}$$

Next, it is important to determine the value of the backscattering differential cross section. By the same reasoning used earlier, we can introduce the amplitude of the backscattered field in the form:

$$\mathbf{f}(\mathbf{e}_s) = \frac{ik\mathbf{g}_i}{2\pi}(\mathbf{e}_z \cdot \mathbf{e}_s)F(\theta_i)\int_A e^{ik(\mathbf{e}_i - \mathbf{e}_s)\cdot\mathbf{r}}d^2\mathbf{r}' \tag{5.35}$$

for the case of H-waves. Because we are considering the case of large plates compared to the wavelength, the main scattering is concentrated in the direction of the specula reflection $\left(\mathbf{e}_s = \mathbf{e}_i - 2(\mathbf{e}_z \cdot \mathbf{e}_i)\mathbf{e}_z\right)$ For this reason, the reflection coefficient in Equation (5.35) is outside the integral sign at an angle equal to the incident one. For the same reason, the integral in this formula is an "acute" function of angles. The cross section of the scattering in the back semisphere is described by the formula:

$$\sigma_d(\mathbf{e}_s) = \frac{k^2A^2}{4\pi^2}\left|F(\theta_i)\right|^2\Psi(\Omega), \tag{5.36}$$

where the function $\Psi(\Omega)$ is determined by Equation (1.117), except that in this case it is called the *indicatrix of diffraction*. Its form is defined by the geometrical shape of the sheet. We can say, generally, that the indicatrix angle spread has a value on the order $\Delta\Omega \cong \lambda^2/A$.

In the case of E-waves, the reflection coefficient for vertically polarized waves and vector $\mathbf{g}_i - 2(\mathbf{e}_z \cdot \mathbf{e}_i)\left[\mathbf{e}_z \times (\mathbf{e}_i \times \mathbf{g}_i)\right)$ instead of vector $\mathbf{g}_i$ must be used in Equation (5.35).

5.5 WAVE SCATTERING BY SMALL BODIES

We often come across cases when natural objects are smaller than the wavelength, including drops of rain, clouds, practically the entire microwave region, vegetation cover, etc. The spatial structure of the electrical field inside these objects is the same as it would be when scatterers (i.e., particles) are placed in the electrostatic field. As we are dealing with particles smaller than the wavelength, then the exponent index in Equation (5.4) approaches zero, and we can write:

$$\mathbf{f}(\mathbf{e}_s) = \frac{k^2(\varepsilon - 1)}{4\pi} \int\limits_V \left[\mathbf{e}_s \times \left(\mathbf{E}^{\text{int}} \times \mathbf{e}_s\right)\right] d^2\mathbf{r}'. \tag{5.37}$$

The superscript (int) is introduced to indicate that we are dealing with the field inside the particle. For the bodies of some shapes, as follows from solutions to the electrostatic problem, the internal field is uniform and

$$\mathbf{f} = \frac{k^2(\varepsilon - 1)\left[\mathbf{e}_s \times \left(\mathbf{E}^{\text{int}} \times \mathbf{e}_s\right)\right]}{4\pi} v, \tag{5.38}$$

where v is the volume of a particle. In particular, such a uniform property applies to the field inside an ellipsoid. Certain forms of particles may be approximated by ellipsoids with some degree of accuracy; therefore, the wave scattering problem of ellipsoidal particles has rather universal importance. We will proceed with consideration of this problem.

Let us introduce Cartesian coordinates ξ, η, and ζ directed along the axes of the ellipsoid. The surface can be described by the equation:

$$\frac{\xi^2}{a^2} + \frac{\eta^2}{b^2} + \frac{\xi^2}{c^2} = 1. \tag{5.39}$$

We will assume the validity of the following relationships between the semi-axes of the ellipsoid: a b c. Further, as we have already agreed, we will assume that $\left|ka\sqrt{\varepsilon}\right| << 1$. Doing so allows us to use the known solution of the ellipsoid polarization problem in a constant electrical field[1] for calculation of the internal field amplitude. We must take into account that, although the electrostatic approximation is used, the dielectric constant of the ellipsoid is considered to be a function of frequency and its value is chosen at the incident field frequency. These calculations show that:[1]

$$\mathrm{E}_j^{\text{int}} = \frac{\mathrm{E}_j}{1 + abc(\varepsilon - 1)^{A_j}/_2}, \tag{5.40}$$

where E_j and $\mathrm{E}_j^{\text{int}}$ are components (along axes ξ, η, ζ) of the external and internal fields. The coefficients are defined as:

$$A_\xi = \int_0^\infty \frac{dt}{\left(t+a^2\right)^{3/2}\sqrt{\left(t+b^2\right)\left(t+c^2\right)}},$$

$$A_\xi = \frac{2}{a^3 \sin^3\varphi}\int_0^\varphi \frac{\sin^2\phi\, d\phi}{\sqrt{1-\kappa^2\sin^2\phi}} = \frac{2}{\kappa^2 a^3 \sin^3\varphi}\left[F(\varphi,\kappa)-E(\varphi,\kappa)\right].$$

$$A_\eta = \int_0^\infty \frac{dt}{\left(t+b^2\right)^{3/2}\sqrt{\left(t+a^2\right)\left(t+c^2\right)}},$$

$$A_\eta = \frac{2}{a^3 \sin^3\varphi}\int_0^\varphi \frac{\sin^2\phi\, d\phi}{\left(1-\kappa^2\sin^2\phi\right)^{3/2}} = \frac{2}{\kappa a^3 \sin^3\varphi}\frac{\partial F}{\partial \kappa}. \tag{5.41}$$

$$A_\zeta = \int_0^\infty \frac{dt}{\left(t+c^2\right)^{3/2}\sqrt{\left(t+a^2\right)\left(t+b^2\right)}}.$$

$$A_\xi + A_\eta + A_\zeta = \frac{2}{abc}.$$

Here, the elliptic integrals are introduced:[40]

$$F(\varphi,\kappa) = \int_0^\varphi \frac{d\phi}{\sqrt{1-\kappa^2\sin^2\phi}},$$

$$E(\varphi,\kappa) = \int_0^\varphi \sqrt{1-\kappa^2\sin^2\phi}\, d\phi, \tag{5.42}$$

$$\kappa^2 = \frac{a^2-b^2}{a^2-c^2},$$

$$\sin\varphi = \sqrt{1-\frac{c^2}{a^2}}.$$

The sphere is a particular case of ellipsoid. In this case, $a = b = c$, and all the coefficients A_j are equal to $A = 2/3a^3$. The internal field is parallel to the incident one and is equal to:

$$\mathbf{E}^{\text{int}} = \frac{3\mathbf{g}_i}{\varepsilon+2}. \tag{5.43}$$

Then, for the amplitude of scattering we have:

$$\mathbf{f} = \frac{k^2 a^3 (\varepsilon - 1)}{(\varepsilon + 2)} \left[\mathbf{g}_i - \mathbf{e}_s (\mathbf{g}_i \cdot \mathbf{e}_s) \right], \tag{5.44}$$

from which the differential cross section of scattering is defined as:

$$\sigma_d = k^4 a^6 \left| \frac{\varepsilon - 1}{\varepsilon + 2} \right|^2 \sin^2 \chi, \tag{5.45}$$

where χ is the angle between the incident wave polarization vector and the direction of scattering. Correspondingly, the cross sections of scattering and absorption are:

$$\sigma_s = \frac{8\pi}{3} k^4 a^6 \left| \frac{\varepsilon - 1}{\varepsilon + 2} \right|^2, \quad \sigma_a = 12\pi\, k a^3 \frac{\varepsilon''}{|\varepsilon + 2|^2}. \tag{5.46}$$

Let us point out some important considerations. The polarization of waves scattered forward and backward by the sphere coincides with the polarization of the incident wave. The cross section of scattering is much smaller than the cross section of absorption due to the smallness of the product *ka*. Accordingly, the total cross section calculated on the basis of the optical theorem is equal to the cross section of absorption. Note that such size relations for scattering and absorption effects are rather typical for wave interaction with small wavelength-scale bodies. Indeed, it follows from Equations (5.4), (5.8), and (5.9) that $\sigma_s \approx v^2/\lambda^4$, but $\sigma_a \approx v/\lambda$. Note also the fact that the intensity of scattering by small particles is inversely proportional to the fourth degree of the wavelength. Such scattering is often referred to as *Rayleigh's* in honor of the person who was first determined this conformity in light wave scattering by gas density fluctuations.

Let us return to the problem of wave scattering by elliptical particles. Let us assume that the incident wave propagates in the z-axis direction and its polarization vector is directed along the x-axis (i.e., $\mathbf{e}_i = \mathbf{e}_z$ and $\mathbf{g}_i = \mathbf{e}_x$). One can connect the coordinate system agreed upon for the scattering ellipse axis with the one introduced by us with the help of Euler's angles Φ, Ψ, and ϑ.[67] These connection formulae can be written down in the form:

$$\begin{aligned} \xi &= \mathrm{x}(\cos\Phi\cos\Psi - \sin\Phi\sin\Psi\cos\vartheta) + \\ &\quad + \mathrm{y}(\cos\Phi\sin\Psi + \sin\Phi\cos\Psi\cos\vartheta) + \mathrm{z}\sin\Phi\sin\vartheta; \\ \eta &= -\mathrm{x}(\sin\Phi\cos\Psi + \cos\Phi\sin\Psi\cos\vartheta) + \\ &\quad + \mathrm{y}(-\sin\Phi\sin\Psi + \cos\Phi\cos\Psi\cos\vartheta) + \mathrm{z}\cos\Phi\sin\vartheta; \\ \zeta &= \mathrm{x}\sin\Psi\sin\vartheta - \mathrm{y}\cos\Psi\sin\vartheta + \mathrm{z}\cos\vartheta. \end{aligned} \tag{5.47}$$

The ranges of change of Euler's angles are given by the intervals $0 \le \Phi \le 2\pi$, $0 \le \Psi \le 2\pi$, and $0 \le \vartheta \le \pi$.

It is easy to establish that the internal field components are equal at the polarization of an incident wave of single amplitude we have chosen:

$$
\begin{aligned}
\mathrm{E}_{\xi}^{\mathrm{int}} &= \frac{\cos\Phi\cos\Psi - \sin\Phi\sin\Psi\cos\vartheta}{1 + abc(\varepsilon - 1)A_{\xi}/2}, \\
\mathrm{E}_{\eta}^{\mathrm{int}} &= -\frac{\sin\Phi\cos\Psi + \cos\Phi\sin\Psi\cos\vartheta}{1 + abc(\varepsilon - 1)A_{\eta}/2}, \\
\mathrm{E}_{\zeta}^{\mathrm{int}} &= \frac{\sin\Psi\sin\vartheta}{1 + abc(\varepsilon - 1)A_{\zeta}/2}.
\end{aligned} \tag{5.48}
$$

It is obvious that depolarization of scattered waves takes place in this case. So, for example, there is the y–component of the field except the x–component at the scattering backward ($\mathbf{e}_s = -\mathbf{e}_z$). We will not carry out the rather complex calculations for the general case; our consideration here will be restricted to only special cases.

The differential cross section of scattering is described by the equation:

$$
\sigma_d^{(x)} = \frac{k^4 \left|\varepsilon - 1\right|^2 V^2}{16\pi^2} G^{(x)}\left(\Phi,\Psi,\vartheta\right)\sin^2\chi_x^{\mathrm{int}} \text{ and } v = \frac{4\pi}{3}abc, \tag{5.49}
$$

where χ_x^{int} is the angle between $\mathbf{e}_s$ and $\mathbf{E}^{\mathrm{in}}$. Here,

$$
\begin{aligned}
G^{(x)}\left(\Phi,\Psi,\vartheta\right) = {} & \frac{\left(\cos\Phi\cos\Psi - \sin\Phi\sin\Psi\cos\vartheta\right)^2}{\left|1 + abc(\varepsilon - 1)A_{\xi}/2\right|^2} + \\
& + \frac{\left(\sin\Phi\cos\Psi + \cos\Phi\sin\Psi\cos\vartheta\right)^2}{\left|1 + abc(\varepsilon - 1)A_{\eta}/2\right|^2} + \frac{\sin^2\Psi\sin^2\vartheta}{\left|1 + abcA_{\eta}/2\right|}.
\end{aligned} \tag{5.50}
$$

The superscript x represents the x-polarization of the incident wave.

The cross section of scattering and absorption are expressed by the formulae:

$$
\sigma_s^{(x)} = \frac{k^4 \left|\varepsilon - 1\right|^2 V^2}{6\pi} G^{(x)}\left(\Phi,\Psi,\vartheta\right) \text{ and } \sigma_a^{(x)} = k\varepsilon'' V G^{(x)}\left(\Phi,\Psi,\vartheta\right). \tag{5.51}
$$

The total cross section calculated on the basis of the optical theorem is, as expected, equal to the cross section of absorption.

In a real situation, the scattering particles may be oriented in a totally random way. In this case, we need to account for the angle-averaged G function about the value:

$$\bar{G}^{(x)} = \frac{4}{3}\left[\frac{1}{\left|2+abc(\varepsilon-1)A_\xi\right|^2}+\frac{1}{\left|2+abc(\varepsilon-1)A_\eta\right|^2}+\frac{1}{\left|2+abc(\varepsilon-1)A_\zeta\right|^2}\right]. \quad (5.52)$$

Similar calculations are performed for y-polarized incident waves. Everywhere the function $G^{(x)}$ should be substituted by the function

$$G^{(y)}(\Phi,\Psi,\vartheta)=\frac{(\cos\Phi\,\sin\Psi+\sin\Phi\,\cos\Psi\,\cos\vartheta)^2}{\left|1+abc(\varepsilon-1)A_\xi/2\right|^2}+\frac{(\sin\Phi\,\sin\Psi-\cos\Phi\,\cos\Psi\,\cos\vartheta)^2}{\left|1+abc(\varepsilon-1)A_\eta/2\right|^2}+\frac{\cos^2\Psi\,\sin^2\vartheta}{\left|1+abc(\varepsilon-1)A_\zeta/2\right|^2}. \quad (5.53)$$

If the incident wave of arbitrary polarization is involved, then the function:

$$G=\frac{|E_x|^2}{|E_x|^2+|E_y|^2}G^{(x)}+\frac{|E_y|^2}{|E_x|^2+|E_y|^2}G^{(y)} \quad (5.54)$$

should be introduced in all formulae for calculating the ellipsoid particle cross sections. Particularly, function G should be substituted by its average value:

$$\langle G\rangle=\frac{1}{2}\left(G^{(x)}+G^{(y)}\right)=\frac{\cos^2\Phi+\sin^2\Phi\,\cos^2\vartheta}{\left|2+abc(\varepsilon-1)A_\xi\right|^2}+\frac{\sin^2\Phi+\cos^2\Phi\,\cos^2\vartheta}{\left|2+abc(\varepsilon-1)A_\eta\right|^2}+\frac{\sin^2\vartheta}{\left|2+abc(\varepsilon-1)A_\zeta\right|^2} \quad (5.55)$$

in the case of randomly polarized wave scattering, when both field components of the incident wave are equal on average. The obvious equality $\langle\bar{G}\rangle=\bar{G}$ is obtained on average over all angles.

Let us now compare the field scattered by a small particle with the radiation field of a dipole (Equation (1.43)). It is easy to establish that the scattering particle is equivalent to the dipole with moment:

$$\mathbf{p} = \frac{(\varepsilon - 1)v}{4\pi}\mathbf{E}^{\text{int}} \tag{5.56}$$

induced by the incident wave of single amplitude. In the case of the spherical particle, the internal field is parallel to the incident one, and the amplitude of the considered induced moment is equal to the polarizability (or perceptivity) of the particle:

$$\alpha = \frac{\varepsilon - 1}{\varepsilon + 2}a^3. \tag{5.57}$$

In general, for arbitrarily shaped particles, the relation between the components of the induced dipole moment and the components of the incident field bears the tensor character:

$$\mathrm{p}_j = \alpha_{jm}\mathrm{g}_{\mathrm{m}}, \tag{5.58}$$

where the value of g_{m} is the polarization vector components of the incident wave. Here, as usual, the summation is taken with respect to repeated subscripts.

5.6 SCATTERING BY BODIES WITH SMALL VALUES OF ε – 1

An example of bodies with small values of $\varepsilon - 1$ is air inhomogeneities inside which the permittivity differs little from that in the environment. If the latter is equal to unity, then $\varepsilon - 1 \ll 1$ inside the inhomogeneity. We already came across such a situation in Chapter 4 when we analyzed the fluctuation phenomenon of radiowave propagation in a turbulent atmosphere. Here, we will consider scattering processes by similar weak inhomogeneities without addressing the phenomenon of stochastic character, and we will assume, in particular, that $\varepsilon - 1 = const$ inside the scattering inhomogeneity.

We can assume in the first approximation that the electrical field strength inside the inhomogeneity being considered is equal to the field of the incident wave. Such an approach for weakly scattering particles is often associated with the name of Born, who used it in the quantum theory of scattering. Thus, the amplitude of scattering can be represented as:

$$\mathbf{f}(\mathbf{e}_s) = \frac{k^2(\varepsilon - 1)\left[\mathbf{e}_s \times (\mathbf{g}_i \times \mathbf{e}_s)\right]}{4\pi}\int_V e^{ik(\mathbf{e}_i - \mathbf{e}_s)\cdot\mathbf{r}'}\,d^3\mathbf{r}' \tag{5.59}$$

in the Born approximation. Hence, using the optical theorem, we obtain:

$$\sigma_t = k\varepsilon'' v. \tag{5.60}$$

Let us point out that, when the scattering volume sizes are large compared to the wavelength, the integral in Equation (5.59) is close to $8\pi^3\delta(\mathbf{e}_i - \mathbf{e}_s)$, which means that for large sizes of inhomogeneities the scattering generally takes place forward. We shall consider for our purposes here the case of a scattering sphere with radius a. Then, the integration in Equation (5.59) leads to the formula:

$$\mathbf{f}(\mathbf{e}_s) = \frac{k^2 a^3 (\varepsilon - 1)\left[\mathbf{e}_s \times (\mathbf{g}_i \times \mathbf{e}_s)\right]}{3} G(u). \tag{5.61}$$

Here,

$$u = ka\left|\mathbf{e}_s - \mathbf{e}_i\right| = 2ka \sin\frac{\theta}{2}, \tag{5.62}$$

where θ is the angle of scattering. The function:

$$G(u) = \frac{3}{u^2}\left(\frac{\sin u}{u} - \cos u\right) \tag{5.63}$$

has been described and tabulated in Van de Hulst[37] and Shifrin.[38]

The cross section of scattering is expressed by the integral:

$$\sigma_s = \frac{\pi}{9} k^4 a^6 \left|\varepsilon - 1\right|^2 \int_0^{\pi} G^2\left(2ka \sin\frac{\theta}{2}\right)\left(1 + \cos^2\theta\right)\sin\theta \, d\theta. \tag{5.64}$$

If $ka >> 1$, then the main integration area is near angle $\theta = 0$, and:

$$\sigma_s = \frac{2\pi}{9} k^2 a^4 \left|\varepsilon - 1\right|^2 \int_0^{\infty} G^2(u)\, u \, du = \frac{\pi}{2} k^2 a^4 \left|\varepsilon - 1\right|^2. \tag{5.65}$$

5.7 MIE PROBLEM

We have analyzed electromagnetic wave scattering by a sphere using approximate ways of computing related to two extreme cases: the radius of a sphere is much smaller than the wavelength or the radius is much larger than the wavelength. In the latter case, we also required maintenance of the condition $\varepsilon - 1 \ll 1$. However, it is often necessary to consider the intermediate case, when the radius of the scattering particle is comparable to the wavelength. In this case, we want an accurate analysis of the radiowave diffraction by a sphere. Such a problem was solved by Mie in 1908; we will not describe this solution in detail here, instead referring the reader to the excellent monographs of Hulst[37] and Shifrin.[38]

First of all, let us introduce the formulae for the total cross section, scattering cross section, and differential cross section of backward scattering. They can be represented in the form:

$$\frac{\sigma_t}{\pi a^2} = \frac{2}{\rho^2} \sum_{l=1}^{\infty} (2l+1) \operatorname{Re}(a_l + b_l), \tag{5.66}$$

$$\frac{\sigma_s}{\pi a^2} = \frac{2}{\rho^2} \sum_{l=1}^{\infty} (2l+1) \left(|a_l|^2 + |b_l|^2 \right), \tag{5.67}$$

$$\frac{\sigma_d(\pi)}{\pi a^2} = \frac{1}{4\pi \rho^2} \left| \sum_{l=1}^{\infty} (-1)^l (2l+1)(a_l - b_l) \right|^2. \tag{5.68}$$

Here, as before, a is the sphere radius and $\rho = ka$. The coefficients a_l and b_l are expressed via cylindrical functions of semi-integer order. Let us now consider the functions:

$$\psi_l(\rho) = \sqrt{\frac{\pi\rho}{2}} J_{l+1/2}(\rho), \quad \zeta_l(\rho) = \sqrt{\frac{\pi\rho}{2}} H^{(1)}_{l+1/2}(\rho), \tag{5.69}$$

where $J_\nu(\rho)$ is the Bessel function, and $H^{(1)}_\nu(\rho)$ is a Hankel function of the first kind.[45] The formulae for a_l and b_l are as follows:

$$\begin{aligned} a_l &= \frac{\psi_l(\rho)\, \psi_l'(\sqrt{\varepsilon}\rho) - \sqrt{\varepsilon}\, \psi_l(\sqrt{\varepsilon}\rho)\, \psi_l'(\rho)}{\zeta_l(\rho)\, \psi_l'(\sqrt{\varepsilon}\rho) - \sqrt{\varepsilon}\, \psi_l(\sqrt{\varepsilon}\rho)\, \zeta_l'(\rho)}, \\ b_l &= \frac{\sqrt{\varepsilon}\, \psi_l(\rho)\, \psi_l'(\sqrt{\varepsilon}\rho) - \psi_l(\sqrt{\varepsilon}\rho)\, \psi_l'(\rho)}{\sqrt{\varepsilon}\, \zeta_l(\rho)\, \psi_l'(\sqrt{\varepsilon}\rho) - \psi_l(\sqrt{\varepsilon}\rho)\, \zeta_l'(\rho)}. \end{aligned} \tag{5.70}$$

Here, ε is the dielectric constant of the sphere. These series begin to converge only when $l > \rho$, so their summation is rather difficult for a sphere of large radius.

Another representation of these coefficients is possible and is more convenient in some cases. It is necessary to take into account that:

$$H_{\nu}^{(1)}(\rho) = J_{\nu}(\rho) + iN_{\nu}(\rho), \tag{5.71}$$

where $N_{\nu}(\rho)$ is Neumann's function.[45] If the function:

$$\chi_l(\rho) = \sqrt{\frac{\pi\rho}{2}} N_{l+1/2}(\rho) \tag{5.72}$$

and angles defined by the relations:

$$\begin{aligned} \tan\delta_l &= \frac{\psi_l(\rho)\psi_l'(\sqrt{\varepsilon}\rho) - \sqrt{\varepsilon}\ \psi_l(\sqrt{\varepsilon}\rho)\psi_l'(\rho)}{\chi_l(\rho)\psi_l'(\sqrt{\varepsilon}\rho) - \sqrt{\varepsilon}\ \psi_l(\sqrt{\varepsilon}\rho)\chi_l'(\rho)}, \\ \tan\gamma_l &= \frac{\sqrt{\varepsilon}\ \psi_l(\rho)\psi_l'(\sqrt{\varepsilon}\rho) - \psi_l(\sqrt{\varepsilon}\rho)\psi_l'(\rho)}{\sqrt{\varepsilon}\,\chi_l(\rho)\ \psi_l'(\sqrt{\varepsilon}\rho) - \psi_l(\sqrt{\varepsilon}\rho)\chi_l'(\rho)}, \end{aligned} \tag{5.73}$$

are taken into account, then

$$\begin{aligned} a_l &= \frac{\tan\delta_l}{\tan\delta_l - i} = \frac{1}{2}\left(1 - e^{-2i\delta_l}\right), \\ b_l &= \frac{\tan\gamma_l}{\tan\gamma_l - i} = \frac{1}{2}\left(1 - e^{-2i\gamma_l}\right). \end{aligned} \tag{5.74}$$

If the particle is small in comparison to the wavelength — more precisely if $\left|\sqrt{\varepsilon}\right|\rho \ll 1$ — then it is sufficient to restrict the summation in these series to the first terms. In this case, we obtain:

$$a_1 = -\frac{2i}{3}\frac{\varepsilon - 1}{\varepsilon + 2}\rho^3.$$

It is easy to be convinced that in this case we have returned to Equations (5.45) and (5.46).

A special case occurs when $\rho \ll 1$ but the product, $\sqrt{\varepsilon}\rho$, is not small; normally, however, $\sqrt{\varepsilon}$ is a sufficiently large value. As an example, let us consider the case of water drops for which the dielectric constant in the microwave region is several tens. In this case, it is necessary to take into account the radiowave absorption in the drop; however, first we will neglect the absorption and assume that ε″ = 0, and we conclude that:

$$\tan\,\delta_l \cong -\frac{(l+1)\rho^{2l+1}}{l(2l-1)!!(2l+1)!!},$$

$$\tan\,\gamma_l \cong -\frac{\rho^{2l+1}}{(2l-1)!!(2l+1)!!}\frac{\psi_{l+1}(\sqrt{\varepsilon}\rho)}{\psi_{l-1}(\sqrt{\varepsilon}\rho)}$$

on the basis of the asymptotic behavior of cylindrical functions. It easy to establish the possibility of peculiar resonance (magnetic) existence at such frequencies when:

$$\psi_{l-1}\left(\sqrt{\varepsilon}\rho\right) = 0. \tag{5.75}$$

We can obtain estimates of this equation from the asymptotic representation:

$$\psi_l(\rho) \cong \sin\left(\rho - \frac{l\pi}{2}\right), \quad \rho >> l\ . \tag{5.76}$$

It follows, then, that (approximately):

$$\sqrt{\varepsilon}\rho_{lm} = (l-1)\frac{\pi}{2} + \mathrm{m}\pi. \tag{5.77}$$

The given estimation is exact for when $l = 1$. In particular, the corresponding value of the wavelength is

$$\lambda_{\mathrm{m}} = \frac{2a\sqrt{\varepsilon}}{\mathrm{m}}, \quad (\mathrm{m} = 1, 2, \ldots). \tag{5.78}$$

Let us suppose that nonresonant members are small in comparison with the resonant one. This supposition follows from the given estimations of the angles δ_l and γ_l. The cross-section resonant values are approximately equal to:

$$\frac{\sigma_{\mathrm{t}}^{\mathrm{res}}}{\pi a^2} = \frac{\sigma_{\mathrm{s}}^{\mathrm{res}}}{\pi a^2} = \frac{8(2l+1)\varepsilon}{\pi^2(2\mathrm{m}+l-1)^2}. \tag{5.79}$$

In particular, when $l = 1$ and $m = 1$,

$$\sigma_{\mathrm{t}}^{\mathrm{res}} = \sigma_{\mathrm{s}}^{\mathrm{res}} = \frac{6\varepsilon a^2}{\pi}. \tag{5.80}$$

We can define this resonance as the main one because the cross-section values decrease with a rise in value of m and l.

Let us now turn to the inclusion of wave absorption inside the scattering sphere. We need to assume in this case that $tg\delta_l$ and $tg\gamma_l$ are complex values. In the process, we also will assume that $\sqrt{\varepsilon} = n' + in''$; note also that $q = \sqrt{\varepsilon}\rho = n'\rho + in''\rho = q' + iq''$. We will also assume that the absorption is small and that $\theta'' << 1$. Then, restricted by the main terms, it is easy to establish that taking into account the absorption does not cause any essential change with regard to the coefficients a_l. As for the coefficients b_l, the expression for them is:

$$b_l = \frac{\nu_l(q')}{\nu_l(q') + 1 - i\psi_{l-1}(q')/q''\psi'_{l-1}(q')}.$$

It is designated here that:

$$\nu_l(q') = -\frac{\rho^{2l+1}}{(2l-1)!!(2l+1)!!}\frac{\psi_{l+1}(q')}{q''\psi'_{l-1}(q')}.$$

The resonant values q' are determined from Equation (5.76), and

$$\nu_l(q'_{res}) = \frac{\rho_{res}^{2l+1}}{\left[(2l-1)!!\right]^2 q'_{res}q''_{res}} = \frac{\left[(l-1)\pi/2 + m\pi\right]^{2l-1}}{\left[(2l-1)!!\right]^2 (n')^{2l} n''}.$$

It was taken into account, in the calculation process, that:

$$\psi_{l+1}(q'_{res}) = -\frac{(2l+1)\psi'_{l-1}(q'_{res})}{q'_{res}}.$$

The cross sections:

$$\frac{\sigma_t^{res}}{\pi a^2} = \frac{8(2l+1)(n')^2 \nu_l(q'_{res})}{\pi^2(2m+l-1)^2\left[1+\nu_l(q'_{res})\right]},$$
$$\frac{\sigma_s^{res}}{\pi a^2} = \frac{8(2l+1)(n')^2 \nu_l^2(q'_{res})}{\pi^2(2m+l-1)^2\left[1+\nu_l(q'_{res})\right]^2} \qquad (5.81)$$

correspond to the coefficient value:

$$b_l^{\text{res}} = \frac{\nu_l\left(q'_{\text{res}}\right)}{1+\nu_l\left(q'_{\text{res}}\right)}. \tag{5.82}$$

At $l = m = 1$,

$$\sigma_{\text{t}}^{\text{res}} = \frac{6\left(n'\right)^2 a^2}{\pi + n''\left(n'\right)^2}, \quad \sigma_{\text{s}}^{\text{res}} = \frac{6\pi\left(n'\right)^2 a^2}{\left[\pi + n''\left(n'\right)^2\right]^2}. \tag{5.83}$$

From here, it is easy to compute the cross section of absorption:

$$\sigma_a^{\text{res}} = \sigma_{\text{t}}^{\text{res}} - \sigma_{\text{s}}^{\text{res}} = \frac{6n''\left(n'\right)^4 a^2}{\left[\pi + n''\left(n'\right)^2\right]^2}. \tag{5.84}$$

The differential cross section of backward scattering is estimated by the value:

$$\sigma_{\text{d}}^{\text{res}}\left(\pi\right) = \frac{\left(2l+1\right)^2 \left|b_l\right|^2 a^2}{4\rho_{\text{res}}^2} = \frac{\left(2l+1\right)\sigma_{\text{s}}^{\text{res}}}{8\pi}. \tag{5.85}$$

Let us point out that this cross section does not reduce but rather increases with increases of l, in contrast to other cross sections. This growth is not infinite and is restricted by the value:

$$\max \sigma_d^{\text{res}} = \frac{4\varepsilon a^2}{\pi^2}. \tag{5.86}$$

Let us emphasize, in conclusion, that our deductions are related to particles that are small compared to the wavelength ($\rho \ll 1$). For this reason, the nonresonant members of the sums defining the unknown cross sections are negligible ($\approx \rho^{4l+2}$). Generally, resonance can be defined by the conditions:

$$\frac{\psi'_l(\sqrt{\varepsilon}\rho)}{\psi_l(\sqrt{\varepsilon}\rho)} = \sqrt{\varepsilon}\,\frac{\chi'_l(\rho)}{\chi_l(\rho)}, \quad \frac{\psi'_l(\sqrt{\varepsilon}\rho)}{\psi_l(\sqrt{\varepsilon}\rho)} = \frac{1}{\sqrt{\varepsilon}}\frac{\chi'_l(\rho)}{\chi_l(\rho)}. \tag{5.87}$$

Equation (5.75) is a particular case of the second of these equations for small-sized particles. Generally, the resonant members of these sums will be the largest, but the sum of the other members will not necessarily be negligible. Because of this fact, the dependence, for example, of the cross sections on the wavelength will not always have a clearly defined resonant character.

5.8 WAVE SCATTERING BY LARGE BODIES

We have already pointed out that the Mie formulae are difficult to use for computing scattered wave parameters for spheres of large radius. Such computations, however, can be done by the numerical method with the help of computers (see, for example, Aivazjan[39]). In order to obtain clear analytical expressions, we must resort to asymptotic methods of calculation;[37,38] however, the obtained results are still rather complex because of the necessity to take into account all the internal wave reflections. The result is essentially simplified in the presence of sufficiently strong wave absorption inside the sphere as, in this case, the internal reflection effects are negligible. The internal reflection being ignored is the inequality $n''\rho \gg 1$, which reflects the fact that waves penetrating inside the sphere do not reach its opposite borders.

In this case it is convenient to use Kirchhoff approximation for computing the scattered field. First, we will consider any form of the scattering body for which the local curvature radii sufficiently exceed the wavelength (one of the conditions of the tangent planes method used in Kirchhoff approximations). The scattering amplitude can be represented in the form:

$$\mathbf{f} = \frac{ik}{4\pi}\oint_S \left\{\mathbf{E}(\mathbf{e}_s\cdot\mathbf{n}) - \mathbf{n}(\mathbf{e}_s\cdot\mathbf{E}) + [\mathbf{e}_s\cdot\mathbf{n}](\mathbf{e}_s\cdot\mathbf{H}) - [\mathbf{e}_s\cdot\mathbf{H}](\mathbf{e}_s\cdot\mathbf{n})\right\} e^{-ik\mathbf{e}_s\cdot\mathbf{r}'} d^2\mathbf{r}' \tag{5.88}$$

on the basis of Equations (1.83) and (1.84).

Thus, the integral is divided into two to cover both the illuminated and shady parts of the surface. According to the Kirchhoff approach, the field on the illuminated part is approximately equal to the one that follows from the laws of wave reflection by the plane interface of two media. The field of the incident wave substitutes for the field of the shady side of the surface according to the Babinet principle. Thus,

$$\mathbf{f}(\mathbf{e}_i,\mathbf{e}_s,\mathbf{g}_i) = \mathbf{f}_1(\mathbf{e}_i,\mathbf{e}_s,\mathbf{g}_i) + \mathbf{f}_2(\mathbf{e}_i,\mathbf{e}_s,\mathbf{g}_i), \tag{5.89}$$

where the first summand corresponds to the integral covering the illuminated side and the second one to the shady side of the surface. It is obvious that the first item describes the wave scattering by the illuminated part of the surface and the second one describes the diffraction phenomenon and the forward scattering connected with it. In the process of integration with respect to the illuminated part of the surface, we should take into consideration the relation $\mathbf{H}_\mathrm{r}^0 = \left[\mathbf{e}_\mathrm{r} \cdot \mathbf{E}_\mathrm{r}^0\right]$ between the magnetic and electric components of the reflected wave. Then,

$$\mathbf{f}_1 = \frac{ik}{4\pi} \int_{S_\mathrm{i}} \Big\{ \left[1 + \left(\mathbf{e}_\mathrm{s} \cdot \mathbf{e}_\mathrm{r}\right)\right]\left(\mathbf{e}_\mathrm{s} \cdot \mathbf{n}\right)\mathbf{E}_\mathrm{r}^0 - \left(\mathbf{e}_\mathrm{s} \cdot \mathbf{E}_\mathrm{r}^0\right)\left[\mathbf{n} + \mathbf{e}_\mathrm{r}\left(\mathbf{e}_\mathrm{s} \cdot \mathbf{n}\right)\right] + \\ + \left[\mathbf{e}_\mathrm{s} \cdot \mathbf{n}\right]\left(\mathbf{E}_\mathrm{r}^0\left[\mathbf{e}_\mathrm{s} \cdot \mathbf{e}_\mathrm{r}\right]\right)\Big\} e^{ik\left(\mathbf{e}_i - \mathbf{e}_\mathrm{s}\right)\cdot \mathbf{r}'} d^2\mathbf{r}'. \tag{5.90}$$

and the same for the shady side:

$$\mathbf{f}_2 = -\frac{ik}{4\pi} \int_{S_\mathrm{sh}} \Big\{ \left[1 + \left(\mathbf{e}_s \cdot \mathbf{e}_i\right)\right]\left(\mathbf{e}_\mathrm{s} \cdot \mathbf{n}\right)\mathbf{g}_\mathrm{i} - \left(\mathbf{e}_\mathrm{s} \cdot \mathbf{g}_i\right)\left[\mathbf{n} + \mathbf{e}_i\left(\mathbf{e}_\mathrm{s} \cdot \mathbf{n}\right)\right] + \\ + \left[\mathbf{e}_\mathrm{s} \cdot \mathbf{n}\right]\left(\mathbf{g}_i \cdot \left[\mathbf{e}_\mathrm{s} \cdot \mathbf{e}_i\right]\right)\Big\} e^{ik\left(\mathbf{e}_i - \mathbf{e}_\mathrm{s}\right)\cdot \mathbf{r}'} d^2\mathbf{r}'. \tag{5.91}$$

Let us now consider scattering by a sphere of large radius a compared with wavelength. Let angles θ and φ be spherical angle coordinates of the observation point. Angle θ is measured from the direction of the sphere illumination. Let α and β be analogous coordinates of the integration point. The values of α in the interval $(0, \pi/2)$ correspond to the shadow zone, and in the interval $(\pi, 2/\pi)$ they correspond to the illuminated area. The phase:

$$\psi = \left[k\left(\mathbf{e}_i - \mathbf{e}_\mathrm{s}\right)\cdot \mathbf{r}'\right] = 2ka \sin\frac{\theta}{2}\left[\cos\alpha \, \sin\frac{\theta}{2} - \sin\alpha \cos\frac{\theta}{2}\cos\left(\beta - \varphi\right)\right]$$

is the fast oscillating function of angles α and β due to the inequality $ka \gg 1$; therefore, the stationary phase method can be used to calculate the function $\mathbf{f}_1$. The point of the stationary phase has the coordinates $\beta = \varphi$ and $\alpha = \pi/2 + \theta/2$ for the illuminated area. In the stationary phase point, the vector of normal is directed in such a way that the equality:

$$\mathbf{e}_\mathrm{s} = \mathbf{e}_\mathrm{r} = \mathbf{e}_i - 2\mathbf{n}\left(\mathbf{n} \cdot \mathbf{e}_i\right) \tag{5.92}$$

is valid. This value of the scattering vector may be put into the pre-exponential factor because it is a slowly changing function. The pre-exponential factor is expressed out of the integral at the arguments' values in the stationary phase point:

$$\mathbf{f}_1\left(\mathbf{e}_\mathrm{s}\right) = \frac{ika^2}{4\pi}\mathbf{E}_\mathrm{r}^0 \sin\theta \int_0^{2\pi}\int_{\pi/2}^{\pi} e^{i\psi} d\beta \, d\alpha. \tag{5.93}$$

Let us emphasize that Equation (5.92) corresponds to the statement that the wave reaches the reception point reflected from the illuminated area of the sphere in agreement with geometrical optics laws (i.e., from the bright point, for which the coordinates are determined by the stationary phase point). The calculation of Equation (5.93) by the stationary phase method gives us:

$$\mathbf{f}_1 = -\frac{a\mathbf{E}_{\mathrm{r}}^0(\pi/2-\theta/2,\,\varphi)}{2}e^{-2ika\sin\theta/2}. \tag{5.94}$$

In the process of calculating the reflected field by using Equation (3.21), for example, it is necessary to remember that the local angle of incidence is equal to $\pi - \alpha = \pi/2 - \theta/2$. The differential cross section of the backward scattering is:

$$\sigma_d^{\leftarrow}(\theta,\varphi) = \frac{a^2\left(\mathrm{E}_{\mathrm{r}}^0\right)^2}{4} = \left[F_{\mathrm{h}}^2\left(\pi/_2 - \theta/_2\right)\sin^2\varphi + F_{\mathrm{v}}^2\left(\pi/_2 - \theta/_2\right)\cos^2\varphi\right]\frac{a^2}{4}. \tag{5.95}$$

Here, Equation (3.20) was used and angle φ is counted from the direction of the polarization. The term *backward scattering* is used by convention, as forward scattering only partially exists here. In particular, at $\theta = 0$, $F_{\mathrm{h}}(\pi/2) = F_{\mathrm{v}}(\pi/2) = -1$, and

$$\sigma_d^{\leftarrow}(0) = \frac{a^2}{4}. \tag{5.96}$$

The cross section of the backward scattering is:

$$\sigma_d^{\leftarrow}(\pi) = \frac{F^2a^2}{4}, \tag{5.97}$$

where

$$F = -F_{\mathrm{h}}(0) = F_{\mathrm{v}}(0) = \frac{\sqrt{\varepsilon}-1}{\sqrt{\varepsilon}+1}. \tag{5.98}$$

Let us now analyze the contribution of the shady side of the sphere to the scattering. Here, the forward scattering (when $\mathbf{e}_{\mathrm{s}} \cong \mathbf{e}_i$) is of primary importance; therefore, the amplitude of scattering is generally defined by the integral:

$$\mathbf{f}_2(\mathbf{e}_{\mathrm{s}}) \cong \frac{ika^2}{2\pi}\mathbf{g}_i\int_0^{2\pi}\int_0^{\pi/2} e^{i\psi}\sin\alpha\cos\alpha\,d\alpha\,d\beta. \tag{5.99}$$

The integration with respect to β gives us the Bessel function:

$$\mathbf{f}_2\left(\mathbf{e}_s\right) = ika^2\mathbf{g}_i \int_0^{\pi/2} J_0\left(ka\sin\theta\sin\alpha\right) e^{2ika\cos\alpha\sin^2\theta/2} \sin\alpha\cos\alpha\, d\alpha.$$

Unity is substituted for the exponential factor because the scattering takes place inside the cone, the angle of spread of which is determined by the equation $\sin\theta/2 \cong 3\pi/2ka$. As a result, we have:

$$\mathbf{f}_2\left(\mathbf{e}_s\right) = ika^2 \frac{J_1\left(ka\sin\theta\right)}{ka\sin\theta}\mathbf{g}_i. \tag{5.100}$$

It is easy to establish that the diffraction field of a large absorptive sphere coincides with the field of diffraction on the aperture of the circle (according to the Babinet principle). In particular, due to forward scattering, $\mathbf{f}_2(\mathbf{e}_i, \mathbf{e}_s, \mathbf{g}_i) = ika^2\mathbf{g}_i/2$, and, according to the optical theorem, $\sigma_t = 2\pi a^2$. The differential forward cross section is equal to:

$$\sigma_d^{\rightarrow}(\theta) = \frac{k^2a^4}{4}\left[\frac{2J_1\left(ka\sin\theta\right)}{ka\sin\theta}\right]^2. \tag{5.101}$$

Let us now compute the differential cross section due to the contributions of both parts of the sphere. This value is defined by the sum:

$$\sigma_d = \sigma_d^{\rightarrow} + \sigma_d^{\leftarrow} + \mathbf{f}_1 \cdot \mathbf{f}_2^* + \mathbf{f}_1^* \cdot \mathbf{f}_2. \tag{5.102}$$

The third interference term is equal to:

$$\sigma_d^{\leftrightarrow} = -\frac{ika^3\mathbf{g}_i}{4}\left[\frac{2J_1\left(ka\sin\theta\right)}{ka\sin\theta}\right]\left(\mathbf{E}_r^{0*}e^{2ika\sin\theta/2} - \mathbf{E}_r^0 e^{-2ika\sin\theta/2}\right). \tag{5.103}$$

In our case, the product is:

$$\left(\mathbf{g}_i \cdot \mathbf{E}_r^0\right) = F_h\left(\pi/2 - \theta/2\right)\sin^2\varphi + F_v\left(\pi/2 - \theta/2\right)\cos\theta\cos^2\varphi$$

on the basis of Equation (3.21). It is worthwhile to speak about the interference only in the case of forward scattering (i.e., at small angles θ). Thus, we will obtain:

$$\sigma_d^{\leftrightarrow} = -\frac{ka^3}{2}\left[\frac{2J_1(ka\sin\theta)}{ka\sin\theta}\right]\sin\left(2ka\sin\frac{\theta}{2}\right). \quad (5.104)$$

Given this, it is rather easy to show that the interference term in Equation (5.102) has an order of smallness estimated by the value $1/ka$ compared to the sum of the others. It is obvious that this term can be neglected so we have:

$$\sigma_d = \sigma_d^{\rightarrow} + \sigma_d^{\leftarrow}. \quad (5.105)$$

This expression indicates that the illuminated and the shady sphere areas scatter waves practically independently.

Let us turn now to the calculation of the cross section of backward and forward scattering. It is convenient to expand the common expression into the following:

$$\sigma_s^{\leftarrow} = \int_{4\pi} \sigma_d^{\leftarrow} d\Omega = \sigma_h + \sigma_v, \quad (5.106)$$

where, as it can be shown,

$$\sigma_h = \pi a^2 \int_0^{\pi/2} \left(\frac{\cos\vartheta - \sqrt{\varepsilon - \sin^2\vartheta}}{\cos\vartheta + \sqrt{\varepsilon - \sin^2\vartheta}}\right)^2 \sin\vartheta\cos\vartheta\, d\vartheta,$$

$$\sigma_v = \pi a^2 \int_0^{\pi/2} \left(\frac{\varepsilon\cos\vartheta - \sqrt{\varepsilon - \sin^2\vartheta}}{\varepsilon\cos\vartheta + \sqrt{\varepsilon - \sin^2\vartheta}}\right)^2 \sin\vartheta\cos\vartheta\, d\vartheta. \quad (5.107)$$

If we make corresponding substitutions in the first and the second integrals:

$$x = -\frac{\cos\vartheta - \sqrt{\varepsilon - \sin^2\vartheta}}{\cos\vartheta + \sqrt{\varepsilon - \sin^2\vartheta}}, \qquad x = \frac{\varepsilon\cos\vartheta - \sqrt{\varepsilon - \sin^2\vartheta}}{\varepsilon\cos\vartheta + \sqrt{\varepsilon - \sin^2\vartheta}}.$$

we will obtain:

$$\frac{\sigma_h}{\pi a^2} = \frac{(\varepsilon - 1)}{8}\int_F^1 (1 - x^2)\, dx\,,$$

$$\frac{\sigma_v}{\pi a^2} = 2\varepsilon^2(\varepsilon - 1)\int_{-1}^F \frac{x^2(1 - x^2)\, dx}{\left[\varepsilon^2(1 - x)^2 - (1 + x)^2\right]^2}.$$

Here, F is defined by Equation (5.98). The first integral is computed elementarily and

$$\frac{\sigma_h}{\pi a^2} = \frac{1}{2}\left(\frac{\sqrt{\varepsilon}-1}{\sqrt{\varepsilon}+1}\right)\left[1-\frac{2}{3\left(\sqrt{\varepsilon}+1\right)}\right]. \tag{5.108}$$

The calculation of σ_v does not involve any fundamental difficulties; however, because it is rather cumbersome to do so, we will not provide its exposition here and instead will show only the ultimate result:

$$\frac{\sigma_v}{\pi a^2} = \frac{1}{2}\left\{\frac{\varepsilon^4+6\varepsilon^2+1}{\left(\varepsilon^2-1\right)^2} - \frac{4\varepsilon^{3/2}\left(\varepsilon^2+1\right)}{\left(\varepsilon^2-1\right)^2} + \frac{2\varepsilon\left(\varepsilon-1\right)^2}{\left(\varepsilon+1\right)^3}\ln\frac{\sqrt{\varepsilon}-1}{\sqrt{\varepsilon}+1} + \right.$$

$$\left. + \frac{8\varepsilon^2\left(\varepsilon^2+1\right)\ln\varepsilon}{\left(\varepsilon^2-1\right)^2\left(\varepsilon+1\right)} - \frac{8\varepsilon^{5/2}}{\left(\sqrt{\varepsilon}+1\right)\left(\varepsilon+1\right)\left(\varepsilon^2-1\right)}\right\}. \tag{5.109}$$

Both expressions can be greatly simplified when $\sqrt{\varepsilon} >> 1$. Here, it is appropriate to note that in all calculations the smallness of the imaginary part of the permittivity is assumed; therefore, no sign of the module is seen in the formulae. Nevertheless, the size of the sphere is set so large that the wave inside the sphere is completely absorbed even in cases of little imaginary part of the permittivity. So, at large values of permittivity:

$$\sigma_s^{\leftarrow} = \pi a^2\left(1-\frac{16}{3\sqrt{\varepsilon}}+\frac{4\ln\varepsilon}{\varepsilon}\right). \tag{5.110}$$

The cross section of the forward scattering is expressed by the integral:

$$\sigma_s^{\rightarrow} = 2\pi a^2\int_0^{\pi/2}\frac{J_1^2\left(ka\sin\theta\right)}{\sin\theta}d\theta = 2\pi a^2\int_0^{\infty}\frac{J_1^2\left(x\right)}{x}dx = \pi a^2 \tag{5.111}$$

as $ka >> 1$.

The total cross section is defined by the sum:

$$\frac{\sigma_s}{\pi a^2} = 1 + \int_0^{\pi/2}\left[F_h^2\left(\vartheta\right)+F_v^2\left(\vartheta\right)\right]\sin\vartheta\cos\vartheta\, d\vartheta. \tag{5.112}$$

Correspondingly, the cross section of absorption is written as:

$$\frac{\sigma_a}{\pi a^2} = 1 - \int_0^{\pi/2} \left[F_h^2(\vartheta) + F_v^2(\vartheta) \right] \sin\vartheta \cos\vartheta \, d\vartheta. \tag{5.113}$$

The last result is highly transparent. The right-hand part of Equation (5.113) describes the energy flow caught by the illuminated part of the sphere and which penetrates inward, where it is absorbed as discussed earlier. In the particular case of great permittivities,

$$\sigma_a \cong \frac{16\pi a^2}{3\sqrt{\varepsilon}} \left(1 - \frac{3\ln\varepsilon}{4\sqrt{\varepsilon}} \right). \tag{5.114}$$

As a second example, we will examine the scattering process by a cylinder with radius a and length l. All of the cylinder sizes are assumed to be large compared to the wavelength. Such cylinders may be considered to be a model of a tree trunk. We will provide the final results without entering into the details. Let us first choose the coordinate system in such a way that the y0z plane is the plane of incidence (Figure 5.1). Let us use θ_i to denote the angle of the wave incident relative to the z-axis and use θ and φ to denote the spherical angles of the scattering direction. Then, by applying the computation procedure used previously, we obtain:

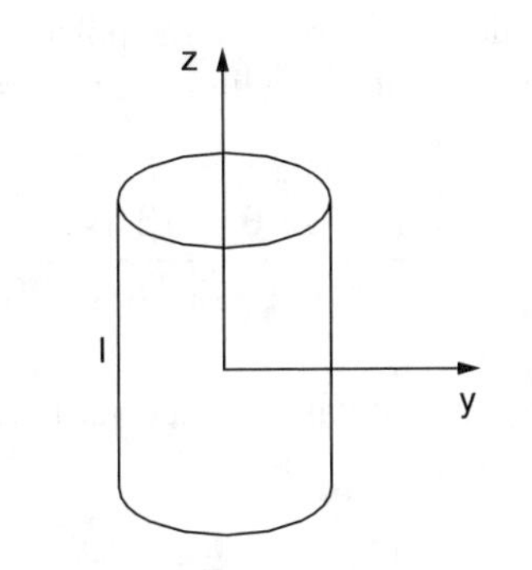

FIGURE 5.1 The coordinate system.

$$\mathbf{f}_1 = \frac{ikal}{2\pi} \sin\theta_i \operatorname{sinc}\left[{}^{kl}\!/_2 \left(\cos\theta_i - \cos\theta \right) \right] \int_{\pi}^{2\pi} \mathbf{E}_r^0(\beta)\, e^{i\psi} \sin\beta \, d\beta,$$

where

$$\psi = -2ka \sin\theta_i \cos\left({}^{3\pi}\!/_4 - {}^{\varphi}\!/_2 \right) \cos\left(\beta - {}^{\varphi}\!/_2 - {}^{3\pi}\!/_4 \right).$$

The last integral is calculated using the stationary phase method. The expression for the differential cross section has the form:

$$\sigma_d^{\leftarrow} = \frac{kal^2}{4\pi}\frac{\sin\theta_i\left|\sin\beta_0\right|\sin c^2(\gamma)}{1-\sin^2\theta_i\sin^2\beta_0}\left[F_{\mathrm{v}}^2(\vartheta)\left(\mathbf{n}\cdot\mathbf{g}_i\right)^2 + F_{\mathrm{h}}^2(\vartheta)\left(\mathbf{g}_i\cdot\left[\mathbf{n}\cdot\mathbf{e}_i\right]\right)^2\right]. \quad (5.115)$$

Here, $\gamma = kl/2(\cos\theta_i - \cos\theta)$ and $\beta_0 = \varphi/2 + 3\pi/4$. We used Equation (3.20) and substituted $\mathbf{n} = -\mathbf{e}_x\cos\beta_0 - \mathbf{e}_y\sin\beta_0$ for $\mathbf{e}_z$. Angle ϑ is defined from:

$$\cos\vartheta = \left(\mathbf{n}\cdot\mathbf{e}_i\right) = -\sin\theta_i\sin\beta_0. \quad (5.116)$$

Let us also take into the account that, in our case,

$$\left[\mathbf{n}\times\mathbf{e}_i\right] = -\mathbf{e}_x\sin\beta_0\cos\theta_i + \mathbf{e}_y\cos\beta_0\cos\theta_i - \mathbf{e}_z\cos\beta_0\sin\theta_i.$$

The following calculations depend on the incident wave polarization. Let us first consider the case when the polarization vector of the incident wave is perpendicular to the axis of the cylinder (i.e., $\mathbf{g}_i = \mathbf{e}_x$). In this case,

$$\sigma_d^{\leftarrow} = \frac{kal^2}{4\pi}\frac{\sin\theta_i\left|\sin\beta_0\right|\sin c^2(\gamma)}{1-\sin^2\theta_i\sin^2\beta_0}\left[F_{\mathrm{v}}^2\cos^2\beta_0 + F_{\mathrm{h}}^2\sin^2\beta_0\cos^2\theta_i\right]. \quad (5.117)$$

In the special case of backward scattering in the plane of incidence ($\varphi = 3\pi/2$),

$$\sigma_d^{\leftarrow} = \frac{kal^2}{4\pi}\sin\theta_i\sin c^2(\gamma)F_{\mathrm{h}}^2\left(\pi/2 - \theta_i\right). \quad (5.118)$$

In some sense, the case when the vector of polarization lies in the plane of the wave incidence is opposite the one considered here, where $\mathbf{g}_i = -\mathbf{e}_x\cos\theta_i + \mathbf{e}_z\sin\theta_i$ and

$$\sigma_d^{\leftarrow} = \frac{kal^2}{4\pi}\frac{\sin\theta_i\left|\sin\beta_0\right|\sin c^2(\gamma)}{1-\sin^2\theta_i\sin^2\beta_0}\left[F_{\mathrm{v}}^2\sin^2\beta_0\cos^2\theta_i + F_{\mathrm{h}}^2\cos^2\beta_0\right] \quad (5.119)$$

As can be seen here, the reflective coefficients for horizontal and vertical polarization have changed places. Particularly, the last formula can be rewritten in the form:

$$\sigma_d^{\leftarrow} = \frac{kal^2}{4\pi}\sin\theta_i\sin c^2(\gamma)F_{\mathrm{v}}^2\left(\pi/2 - \theta_i\right) \quad (5.120)$$

for backward scattering in the plane of incidence.

We will not go into the details but we do want to point out that based on the optical theorem the total cross section is written as:

$$\sigma_t = 4al\sin\theta_i. \tag{5.121}$$

5.9 SCATTERING BY THE ASSEMBLY OF PARTICLES

When not one but many particles are in the wave field, then the scattering process becomes substantially more complicated. First, if the particles move chaotically, then the total scattering field in the general case is of a random character; therefore, we could talk not about the field itself but about its statistical characteristics: the average field, average density of the power flow, etc. Second, in some cases, the effects of multistage scattering at rather high particle densities are significant, such as when the wave scattered by one particle is rescattered by one neighbor, then by another, and so on. For this reason, it is necessary to know the summary incident field when determining the field scattered by any chosen particle which is formed from the field of incidence on the assembled particle waves and scattering fields of other particles. Third, particles of various kinds, different shapes, chaotic orientations, various sizes, etc. might be present. All of these factors should be taken into account when computing scattering by specific systems.

Let us assume a situation where particles occupy the bounded volume. The field of scattering in the general case is the sum:

$$\mathbf{E}_s(\mathbf{r}) = \sum_j \mathrm{E}(\mathbf{r}_j)\, \mathbf{f}_j\left(\mathbf{e}_i^{(j)}, \mathbf{e}_s^{(j)}, \mathbf{g}_i^{(j)}\right) \frac{e^{ik|\mathbf{r}-\mathbf{r}_j|}}{|\mathbf{r}-\mathbf{r}_j|}. \tag{5.122}$$

Here, $\mathbf{r}_j$ is the radius vector of the jth particle, $\mathrm{E}(\mathbf{r}_j)$ is the complex amplitude of the summary wave incident on it, $\mathbf{e}_i^{(j)}$ and $\mathbf{g}_i^{(j)}$ are the vectors of direction and polarization of this wave at the point of the jth particle location, and $\mathbf{e}_s^{(j)}$ is the scattering vector of the chosen particle.

Let us regard an assembly of particles of low density, for which we can assume negligible effects of secondary scattering and that $\mathrm{E}(\mathbf{r}_j)$ is equal to the incident wave field; that is, $\mathrm{E}(\mathbf{r}_j) = \mathrm{E}_i(\mathbf{r}_j)$. In this case, $\mathbf{e}_i^{(j)} = \mathbf{e}_i$ and $\mathbf{g}_i^{(j)} = \mathbf{g}_i$. Then,

$$\mathbf{E}_s(\mathbf{r}) = \sum_j \mathbf{f}_j\left(\mathbf{e}_s^{(j)}\right) \frac{e^{ik|\mathbf{r}-\mathbf{r}_j|}}{|\mathbf{r}-\mathbf{r}_j|} e^{ik(\mathbf{e}_i\cdot\mathbf{r}_j)}.$$

Here, as usual, we have assumed that the incident wave amplitude is equal to unity.

Let us now examine the scattered field in Fraunhofer's zone relative to the scattering volume. In this case, $|\mathbf{r}-\mathbf{r}_j| \cong r - (\mathbf{e}_s\cdot\mathbf{r}_j)$ and $\mathbf{e}_s^{(j)} \cong \mathbf{e}_s$ The particle assembly is defined as one large particle with the scattering amplitude:

$$\mathbf{f}_\Sigma(\mathbf{e}_s) = \sum_j \mathbf{f}_j(\mathbf{e}_s) e^{ik(\mathbf{e}_i - \mathbf{e}_s)\cdot \mathbf{r}_j}. \tag{5.123}$$

The phases of the waves scattered by every particle fluctuate with chaotic movement, the result of which is that the summary scattering amplitude is found to be a random vector. Absent are the fluctuations due to forward scattering movement ($\mathbf{e}_s = \mathbf{e}_i$). All particles scatter coherently in this direction. The summary amplitude of the forward scattering is equal to the sum of the corresponding amplitudes of the individual particles. It is easy to conclude on the basis of the optical theorem that:

$$\sigma_t^{(\Sigma)} = \sum_j \sigma_t^{(j)}, \tag{5.124}$$

that is, the total cross section of the particle assembly, $\sigma_t^{(\Sigma)}$, is equal to the sum of the cross sections of the particles themselves. If the scattering particles have a complex shape and rotate, then $\sigma_t^{(\Sigma)}$ is also found to be a random value because the cross sections $\sigma_t^{(j)}$ are the same. Therefore, the average value of $\sigma_t^{(\Sigma)}$ is taken to be a measure of the scattering by the particle assembly.

Let us suppose that every particle moves randomly around its individual center, $\mathbf{r}_j^0$. We can assume, then, that $\mathbf{r}_j = \mathbf{r}_j^0 + \mathbf{l}_j$, where $\mathbf{l}_j$ is a random vector. In this case, its average value is equal to zero: $\langle \mathbf{l}_j \rangle = 0$. Let $P_1(\mathbf{l}_j)$ be the distribution density of vector $\mathbf{l}_j$. We will now introduce the function $\Phi(\mathbf{e}_i)$ as the random orientation probability of the jth particle. Now, we are able to perform the operation of averaging, which consists of two steps: the averaging according to the orientation of the particles and their location. If we can believe that the function of the location distribution is the same for all of the particles, then we obtain:

$$\langle \mathbf{f}_\Sigma(\mathbf{e}_i, \mathbf{e}_s, \mathbf{g}_i) \rangle = F_1\left[k(\mathbf{e}_i - \mathbf{e}_s)\right] \sum_j \langle \mathbf{f}_j \rangle e^{ik(\mathbf{e}_i - \mathbf{e}_s)\cdot \mathbf{r}_j^0}. \tag{5.125}$$

Here,

$$\langle \mathbf{f}_j \rangle = \int \mathbf{f}_j(\mathbf{e}_i, \mathbf{e}_s, \mathbf{g}_i) \Phi(\mathbf{e}_i) d\mathbf{e}_i.$$

The averaged function obtained depends, in particular, on the direction of the incident wave polarization. The function:

$$F_1(\mathbf{q}) = \int P_1(\mathbf{l}) \, e^{i\mathbf{q}\cdot\mathbf{l}} d^3\mathbf{l} \tag{5.126}$$

is the characteristic function. In the case considered here, it is equal to unity in the direction of the forward scattering, when $\mathbf{e}_s = \mathbf{e}_i$.

Further, for simplicity, let us confine ourselves to studying the assembly of identical particles for which the statistics of their scattering properties is defined by the statistics of their location and the wave interference related with this process. In this case,

$$\left\langle \mathbf{f}_\Sigma\left(\mathbf{e}_i,\mathbf{e}_s,\mathbf{g}_i\right)\right\rangle = F_1\left[k\left(\mathbf{e}_i-\mathbf{e}_s\right)\right]\left\langle \mathbf{f}\right\rangle \sum_j e^{ik\left(\mathbf{e}_i-\mathbf{e}_s\right)\cdot\mathbf{r}_j^0}. \tag{5.127}$$

In particular, the total cross section of the assembly is:

$$\left\langle \sigma_t^{(\Sigma)}\right\rangle = N_\Sigma\left\langle \sigma_t\right\rangle, \tag{5.128}$$

where the averaging is performed with respect to the angle positions of the particles. Here, N_Σ is the total number of particles in the volume. If we introduce the mean density number of the scatterers,

$$\left\langle N\right\rangle = \frac{N_\Sigma}{V}, \tag{5.129}$$

where V is the considered volume, then it is convenient to use the meaning of the cross section of unity volume:

$$\left\langle \sigma_t^0\right\rangle = \left\langle N\right\rangle\left\langle \sigma_t\right\rangle. \tag{5.130}$$

This last equation, as we have already established, defines the wave extinction during their propagation in the scattering media.

Let us now analyze the scattering amplitude fluctuations. The averaged square is:

$$\left\langle \left|\mathbf{f}_\Sigma\left(\mathbf{e}_s\right)\right|^2\right\rangle = \left\langle \left|\mathbf{f}\right|^2\right\rangle \sum_{j,j'}\left\langle e^{ik\left(\mathbf{e}_i-\mathbf{e}_s\right)\cdot\left(\mathbf{r}_j^0-\mathbf{r}_{j'}^0\right)}\right\rangle.$$

If we separate out the terms with $j = j'$ in the given sum, then we can write:

$$\left\langle \left|\mathbf{f}_\Sigma\right|^2\right\rangle = \left\langle \left|\mathbf{f}\right|^2\right\rangle\left\{N_\Sigma + F_2\left[k\left(\mathbf{e}_i-\mathbf{e}_s\right) - k\left(\mathbf{e}_i-\mathbf{e}_s\right)\right]\sum_{j\neq j'} e^{ik\left(\mathbf{e}_i-\mathbf{e}_s\right)\cdot\left(\mathbf{r}_j^0-\mathbf{r}_{j'}^0\right)}\right\}. \tag{5.131}$$

Here,

$$F_2\left(\mathbf{q},\mathbf{q}'\right) = \int P_2\left(\mathbf{l},\mathbf{l}'\right)e^{i(\mathbf{q}\cdot\mathbf{l})+i(\mathbf{q}'\cdot\mathbf{l}')}d^3\mathbf{l}\ d^3\mathbf{l}' \tag{5.132}$$

is the binary characteristic function. The function $P_2(\mathbf{l},\mathbf{l}')$ is the density of the mutual probability distribution of particles located in points $\mathbf{l}$ and $\mathbf{l}'$ relative to their mean positions.

We can simplify the problem by assuming that the particles move freely in the total volume frame independently of each other. Then, it is convenient to choose the coordinate origin in a point inside the scattering volume and assume that $\mathbf{r}_j^0 = 0$. Further, it is necessary to set $P_2(\mathbf{l},\mathbf{l}') = P_1(\mathbf{l})P_1(\mathbf{l}')$ and $F_2(\mathbf{q},-\mathbf{q}') = |F_1(\mathbf{q})|^2$ and we obtain the result:

$$\left\langle |\mathbf{f}_\Sigma|^2 \right\rangle = N_\Sigma \left\langle |\mathbf{f}|^2 \right\rangle \left\{ 1 + (N_\Sigma - 1) \left| F_1\left[k(\mathbf{e}_i - \mathbf{e}_s) \right] \right|^2 \right\}. \tag{5.133}$$

One may note that Equation (5.133) is also the equation for the scattering differential cross section. If we integrate it with respect to a solid angle within 4π, then we will obtain the formula for the cross section of scattering by the particle assembly:

$$\left\langle \sigma_s^{(\Sigma)} \right\rangle = N_\Sigma \left\{ \langle \sigma_s \rangle + (N_\Sigma - 1) \int_{4\pi} \left\langle |\mathbf{f}|^2 \right\rangle \left| F_1\left[k(\mathbf{e}_i - \mathbf{e}_s) \right] \right|^2 d\Omega \right\}. \tag{5.134}$$

If the size of the scattering volume is much less than the wavelength, then the characteristic function put under the integral is approximately equal to unity, and

$$\left\langle \sigma_s^{(\Sigma)} \right\rangle = N_\Sigma^2 \langle \sigma_s \rangle. \tag{5.135}$$

The cross section of scattering is proportional, in this case, to the squared number of particles, which is the sequel of coherency of the wave scattered by single particles. Under these conditions, the sum of the particles may be considered as one particle of small size in the wavelength scale.

In the case, when the size of the scattering volume is large compared to the wavelength, the characteristic function has a sharp maximum in the direction of the incident wave propagation. By giving the scattering amplitude of the single particle a maximum that is not quite as sharp, we can factor out the integral sign at the value $\mathbf{e}_s = \mathbf{e}_i$. So, we now have:

$$\left\langle \sigma_s^{(\Sigma)} \right\rangle = N_\Sigma \left\{ \langle \sigma_s \rangle + (N_\Sigma - 1) \left\langle |\mathbf{f}(\mathbf{e}_i)|^2 \right\rangle \int_{4\pi} \left| F_1\left[k(\mathbf{e}_i - \mathbf{e}_s) \right] \right|^2 d\Omega \right\}. \tag{5.136}$$

A more detailed analysis, which is not presented here, is necessary to explain the physical meaning of this last formula. The first term describes the power of incoherent scattering caused by small-size fluctuations of the density of the particles.

The second term represents the coherent scattering by the volume of particles in view of the body of the corresponding shape. The last term is becoming more clear due to our introducing a definition of the effective permittivity.

The fluctuation intensity of the amplitude of scattering of the particle assembly is:

$$\left\langle \left|\Delta \mathbf{f}_\Sigma\right|^2 \right\rangle = \left\langle \left|\mathbf{f}_\Sigma\right|^2 \right\rangle - \left|\left\langle \mathbf{f}_\Sigma \right\rangle\right|^2 . \tag{5.137}$$

The first summand is determined by Equation (5.133) and the second one is calculated from Equation (5.127), where we can set $\mathbf{r}_j^0 = 0$. As a result,

$$\left\langle \left|\Delta \mathbf{f}_\Sigma\right|^2 \right\rangle = N_\Sigma \left\langle \left|\mathbf{f}\right|^2 \right\rangle + N_\Sigma^2 \left(\left|\mathbf{f}\right|^2 - \left|\left\langle \mathbf{f} \right\rangle\right|^2 \right) \left| F_1\left[k\left(\mathbf{e}_i - \mathbf{e}_s\right)\right]\right|^2 . \tag{5.138}$$

5.10 EFFECTIVE DIELECTRIC PERMITTIVITY OF MEDIUM

If the particles are small in comparison to the wavelength, then the propagation inside the scattering medium can be regarded as wave propagation in a medium with effective dielectric permittivity. In fact, the small scattering particles behave as electrical dipoles for which the moment is induced by the outer field. So, the medium is polarized under the effect of the incident field, and the volumetric density of the dipole moment is equal to:

$$\mathbf{p} = \left\langle N \right\rangle \mathbf{p}_i . \tag{5.139}$$

Here, $\mathbf{p}_i$ is the induced dipole moment. In taking this approach, we assume for the sake of simplicity that all particles are similar. Further, all particles are assumed to be polarized in the direction of the incident field, which means that we are restricted to examination of an isotropic medium. We are dealing here with scattering spheres or chaotically oriented particles. In the last case, the induced dipole moment implied a value averaged with respect to the angles. According to Equations (5.38) and (5.56), it easy to establish that:

$$\mathbf{p}_i = \frac{E_0 \mathbf{f}\left(\mathbf{e}_i\right)}{k^2} = \frac{\mathbf{E}_i \left[\mathbf{g}_i \cdot \mathbf{f}\left(\mathbf{e}_i\right)\right]}{k^2} . \tag{5.140}$$

We have taken into account that during forward scattering the amplitude of scattering is directed along the vector of the electrical field. Also, we are assuming that the particle concentration is small, so the second scattering effect can be neglected, which means that the polarization of scatterers primarily occurs by the incident wave field. Therefore,

$$\mathbf{p} = \mathbf{E}_i \frac{\langle N \rangle \left[\mathbf{g}_i \cdot \mathbf{f}(\mathbf{e}_i) \right]}{k^2}. \tag{5.141}$$

Equation (5.141) allows us to define the medium polarizability as:

$$\beta = \frac{\langle N \rangle \left[\mathbf{g}_i \cdot \mathbf{f}(\mathbf{e}_i) \right]}{k^2} \tag{5.142}$$

and effective permittivity

$$\varepsilon_{\mathrm{e}} = 1 + 4\pi\beta = 1 + \frac{4\pi \langle N \rangle}{k^2} \left[\mathbf{g}_i \cdot \mathbf{f}(\mathbf{e}_i) \right]. \tag{5.143}$$

The second term on the right is small, so, for example, the effective coefficient of refraction is:

$$n_{\mathrm{e}} = \sqrt{\varepsilon_{\mathrm{e}}} = 1 + \frac{2\pi \langle N \rangle}{k^2} \left[\mathbf{g}_i \cdot \mathbf{f}(\mathbf{e}_i) \right], \tag{5.144}$$

and the wave extinction coefficient is:

$$\Gamma = 2kn_{\mathrm{e}}'' = \frac{4\pi}{k} \langle N \rangle \operatorname{Im}\left[\mathbf{g}_i \cdot \mathbf{f}(\mathbf{e}_i) \right] = \langle N \rangle \sigma_{\mathrm{t}} \tag{5.145}$$

in agreement with Equation (5.13).

Let us mention, once more, that the meaning of the effective permittivity is most representative for small particles that behave as dipoles, while we encounter difficulties with the case of large-size particles, where, the formula of mixtures is more sequential. We will return to this subject later.

Let us turn now to Equation (5.136), taking into account that $N_\Sigma = NV >> 1$ and singling out its coherent part:

$$\left\langle \sigma_{\mathrm{s}}^{(\Sigma)} \right\rangle_{\mathrm{cog}} = \left| \langle N \rangle V \left[\mathbf{g}_i \cdot \mathbf{f}(\mathbf{e}_i) \right] \right|^2 \int_{4\pi} \left| F_1 \left[k(\mathbf{e}_i - \mathbf{e}_{\mathrm{s}}) \right] \right| d\Omega.$$

Here, we also took into consideration the fact that $\mathbf{f}(\mathbf{e}_i) = \left[\mathbf{g}_i \cdot \mathbf{f}(\mathbf{e}_i)\right]\mathbf{g}_i$ One should be aware of the expression for the effective permittivity that will give the opportunity to rewrite the last formula as

$$\left\langle \sigma_s^{(\Sigma)} \right\rangle_{cog} = \frac{k^4}{16\pi^2}\left|\varepsilon_e - 1\right|^2 V^2 \int_{4\pi} \left|F_1\left[k\left(\mathbf{e}_i - \mathbf{e}_s\right)\right]\right|^2 d\Omega. \tag{5.146}$$

Let us consider, for example, the case when the scattering particles are concentrated inside a sphere of radius a. It is easy to see in the given case that:

$$F_1\left[k\left(\mathbf{e}_i - \mathbf{e}_s\right)\right] = G\left(2ka \sin \theta/2\right) \tag{5.147}$$

and then, according to the known calculation,

$$\left\langle \sigma_s^{(\Sigma)} \right\rangle_{cog} = \frac{\pi}{2}\left|\varepsilon_e - 1\right|^2 k^2 a^4. \tag{5.148}$$

The comparison of this formula with Equation (5.65) demonstrates their similarity. We can show a similar agreement of results in the general case, also. It should be taken into account for this purpose the research that has been done with regard to wave scattering by bodies with small values of $\varepsilon - 1$ (see Section 5.6). Further, it should be pointed out that the integral in the right-hand part of Equation (5.59) is expressed via a characteristic function that is sharp due to the large size of the scattering volume. Therefore, it also follows from Equation (5.59) that:

$$\mathbf{f}_\Sigma\left(\mathbf{e}_s\right) \cong \frac{k^2\left(\varepsilon - 1\right)\langle N \rangle V}{4\pi} F_1\left[k\left(\mathbf{e}_i - \mathbf{e}_s\right)\right] \tag{5.149}$$

Now it is a simple matter to derive Equation (5.146) by the common computation.

These results mean that the coherent part of the field, scattered by the particle assembly, corresponds to the field of incident wave diffraction on a particle with similar effective permittivity and the shape of the scattering volume. On the basis of this analysis, we can conclude that, while analyzing the interaction of the electromagnetic waves with a scattering medium consisting of small-size scatterers compared to the wavelength, this medium can be examined as a continuous medium with the effective permittivity. The scattering by micro-inhomogeneities of particle concentration is added to the usual processes of wave reflection and diffraction.

5.11 THE ACTING FIELD

In the previous section, the incident wave was assumed to be a field polarizing the particles; thus, we neglected the effect on any chosen particle of the fields scattered by neighboring particles. Such an approach has to be regarded as a first approximation, which is true for low particle density. In reality, any particle is polarized not only by the incident wave effect but also due to the influence of scattering fields reaching the point from other scatterers. The sum of these fields is referred to as the *acting field.* This problem was revealed long ago in the study of electrodynamics and the dielectric polarization phenomenon,[42,43] which led to development of the Lorentz–Lorenz formula.[45,46]

The usual way to derive the acting field formula is based on studying the field inside the spherical cavity around a particle. The cavity radius is assumed to be much smaller than the mean distance between the particles, $r_0 \cong N^{-1/3}$, and must also be much smaller than the wavelength. Then, the cavity can be considered as the polarized particle, inside which the field is equal to:

$$\mathbf{E}_{\mathrm{e}} = \mathbf{E}_i + \frac{4\pi}{3}\mathbf{p}_i \,. \tag{5.150}$$

This is the field acting on the particle. Further, we should refer back to Equation (5.140), where the acting field strength is substituted for the strength of the incident wave. Doing so allows us to tie together the strengths of the acting and the incident fields with the help of the relation:

$$\mathbf{E}_{\mathrm{e}}\left[1 - \frac{4\pi\mathbf{g}_i \cdot \mathbf{f}(\mathbf{e}_i)}{3k^2}\langle N\rangle\right] = \mathbf{E}_i \,. \tag{5.151}$$

It is easy to determine that the polarizability is equal to:

$$\beta = \frac{\langle N\rangle \mathbf{g}_i \cdot \mathbf{f}(\mathbf{e}_i)/k^2}{1 - 4\pi\langle N\rangle \mathbf{g}_i \cdot \mathbf{f}(\mathbf{e}_i)/3k^2}\,, \tag{5.152}$$

and the effective permittivity satisfies the equation:

$$\frac{\varepsilon_{\mathrm{e}} - 1}{\varepsilon_{\mathrm{e}} + 2} = \frac{4\pi}{3}\langle N\rangle \frac{\mathbf{g}_i \cdot \mathbf{f}(\mathbf{e}_i)}{k^2}\,. \tag{5.153}$$

If we introduce the polarizability of particles by the formula:

$$\alpha = \frac{\mathbf{g}_i \cdot \mathbf{f}(\mathbf{e}_i)}{k^2}\,, \tag{5.154}$$

then Equation (5.153) can be rewritten in light of the theory of dielectrics:[42,43]

$$\frac{\varepsilon_e - 1}{\varepsilon_e + 2} = \frac{4\pi}{3}\alpha\langle N\rangle . \tag{5.155}$$

This result is referred to as the Clausius–Mossotti formula (for the refractive index, it is known as the Lorentz–Lorenz formula). As noted previously, this formula is usually derived for medium permittivity affected by polarization of molecules; however, these molecules may be considered to be elementary scatterers,[45] so they do not differ from microscopic particles that are small in comparison to the wavelength.

Equations (5.153) and (5.154) refine the expression for effective permittivity of a rather dense scattering medium. In this case, the notion of a rather dense scattering medium means that the value to the right of Equation (5.153) is not small enough to be neglected. When the value is small enough, Equation (5.153) is naturally reduced to Equation (5.143). The elaboration of Equation (5.153) compared to Equation (5.143) is necessary due to taking into account the role of the scattered fields in particle polarization; that is, the effect of multistage scattering has been included in the consideration.

Let us now perform some estimations by starting with small-radius spheres acting as scatterers. Their susceptibility is determined by Equation (5.57), and the effective permittivity will satisfy the relation:

$$\frac{\varepsilon_e - 1}{\varepsilon_e + 2} = \frac{4\pi}{3}a^3\langle N\rangle\frac{\varepsilon - 1}{\varepsilon + 2}. \tag{5.156}$$

The product $(4\pi/3)a^3\langle N\rangle$ is a part of the volume occupied by the particles in the total volume of the medium. This part does not differ much from unity when the value to the right is not too small, which means that we must consider tightly packed media with small-size scatterers. Natural media are not often considered to be such media, and the multistage scattering effect, in many cases, does not play an important role and can be neglected. However, this consideration can be very important for the resonant scattering case when particle receptivity is high and secondary scattering is appreciable, even at low particle density.

Equation (5.156) expresses the effective permittivity of a mixture of air and particles. With different kinds of particles at hand, Equation (5.156) can be generalized as the Maxwell–Garnet formula:

$$\frac{\varepsilon_e - 1}{\varepsilon_e + 2} = \sum_j \frac{\varepsilon_j - 1}{\varepsilon_j + 2}\rho_j, \quad \sum_j \rho_j = 1 , \tag{5.157}$$

where ρ_j is the volumetric density of the corresponding type of particle. Equation (5.157) agrees with a variant of the mixture theory,[43] and the obtained formula can also be used in some cases for large particles.

5.12 INCOHERENT SCATTERING BY ELECTRONS

In previous sections, we analyzed cases for which it was not necessary, when describing the scattering processes, to take into account correlation in particle polarization, which was implicitly represented while determining the effective field but was not included in the description of the scattering processes. Neglect of the correlation phenomenon was based on the fact that the collective interactions of particles occur on a scale dependent on particle size. These scales are assumed to be much smaller than the wavelength, which means, at least in the first approximation, that we can neglect the collective effects. If electrons appear as the scatterers, then the situation changes significantly. Due to electrostatic interaction, the electron density correlation is much greater than even the value r_0. This scale is determined by the Debye radius:[12]

$$D = \sqrt{\frac{k_b T}{8\pi e^2 N}} = 4.88\sqrt{\frac{T}{N}} \text{ cm.} \tag{5.158}$$

Here, $k_b = 1.38 \cdot 10^{-16}$ erg/K is Boltzmann's constant, and T is the plasma temperature (in Kelvin). The value of the Debye radius is macroscopic (several centimeters in the ionosphere of Earth) and can be in some cases greater than the radiowave length; therefore, the scattering amplitude fluctuations, which define the incoherent scattering intensity, cannot be assumed to be vanishingly small.

The differential cross section of the electron scattering is equal to the fluctuation intensity of the scattering amplitude, which can be determined by Equation (5.138). The difference with the scattering by free electrons indicates that we cannot assume independent particle motion as we did before while deriving Equation (5.138). Further, let us take into account that in this case:

$$\left\langle \left|\mathbf{f}\right|^2 \right\rangle = \left|\left\langle \mathbf{f} \right\rangle\right|^2 = \sigma_d^{(e)} = a_e^2\left(1 - \cos^2\varphi \sin^2\theta\right), \tag{5.159}$$

where φ and θ are the angles of the scattering direction. It is now a simple matter to obtain the relation for the desired differential cross section:

$$\sigma_{d\,\text{ncog}}^{(\Sigma)} = \left\langle \left|\Delta\mathbf{f}_\Sigma\right|^2 \right\rangle = N_\Sigma \sigma_d^{(e)} \left\{1 - N_\Sigma \left|F_1\left[k\left(\mathbf{e}_i - \mathbf{e}_s\right)\right]\right|^2 + \right.$$
$$\left. + N_\Sigma F_2\left[k\left(\mathbf{e}_i - \mathbf{e}_s\right) - k\left(\mathbf{e}_i - \mathbf{e}_s\right)\right]\right\}. \tag{5.160}$$

The mutual function of the distribution is represented by:

$$P_2\left(\mathbf{l}',\mathbf{l}''\right) = P_1\left(\mathbf{l}'\right)P_1\left(\mathbf{l}''\right)\left[1 - h\left(\mathbf{l}' - \mathbf{l}''\right)\right]. \tag{5.161}$$

We will assume that a particle can be located at any point of the scattering volume, and we define the probability density as:

$$P_1(\mathbf{l}) = \frac{1}{V}. \tag{5.162}$$

Under these conditions, the mutual characteristic function:

$$F_2(\mathbf{q},-\mathbf{q}) = \left|F_1(\mathbf{q})\right|^2 - \frac{1}{V}\int h(\mathbf{l})e^{i\mathbf{q}\cdot\mathbf{l}}d^3\mathbf{l}, \tag{5.163}$$

and the differential cross section

$$\sigma_{d\,\text{ncog}}^{(\Sigma)} = N_\Sigma \sigma_d^{(e)}\left[1 - N\int h(\mathbf{l})e^{ik(\mathbf{e}_i-\mathbf{e}_s)\cdot\mathbf{l}}d^3\mathbf{l}\right]. \tag{5.164}$$

For the single charged and equilibrium plasma:[47]

$$h(\mathbf{l}) = \frac{e^2}{k_b T}\frac{e^{-l/D}}{l}. \tag{5.165}$$

Substituting this expression in Equation (5.164) gives us the differential cross section per unit volume:

$$\sigma_d^0 = \frac{1}{V}\sigma_{d\,\text{ncog}}^{(\Sigma)} = \frac{N\sigma_d^{(e)}}{2}\frac{1+8k^2D^2\sin^2\theta/2}{1+4k^2D^2\sin^2\theta/2}. \tag{5.166}$$

Integration over the solid angle gives us the following relation for the scattering cross section:

$$\sigma_s^0 = N\sigma_s^{(e)}\left[1+\frac{3}{4p}+\frac{3}{2p^2}-\frac{3}{4p}\left(1+\frac{2}{p}+\frac{2}{p^2}\right)\ln(1+p)\right], \tag{5.167}$$

where $p = (2kD)^2$ and $\sigma_s^{(e)}$ are determined by Equation (5.21). If the wavelength is smaller than the Debye radius ($p >> 1$), then:

$$\sigma_s^0 = N\sigma_s^{(e)}; \tag{5.168}$$

that is, the electrons scatter independently. This result follows from Thomson's formula, Equation (5.21). In other cases, the effect of collective motion must be considered:

$$\sigma_s^0 = \frac{1}{2} N \sigma_s^{(e)} . \tag{5.169}$$

The differential cross section of backward scattering ($\mathbf{e}_s = -\mathbf{e}_i$, $\theta = \pi$) is equal to:

$$\sigma_d^{(e)} = \frac{N a_e^2}{2} \frac{1 + 8k^2 D^2}{1 + 4k^2 D^2} . \tag{5.170}$$

5.13 RADIOWAVE SCATTERING BY TURBULENT INHOMOGENEITIES

Radiowave scattering by turbulent inhomogeneities exemplifies a random process because of the chaotic origin of medium permittivity pulsation. Such problems may be solved only by approximation methods. For the atmosphere, we assume that the intensity of the permittivity fluctuations is low; therefore, we can use the method of disturbance and following from it the Born approximation (see Section 5.6). Equation (5.59), where the value $\mu(\mathbf{r}) = \varepsilon(\mathbf{r}) - 1$ must be left under the integral sign, may be used as our basis of computation. The properties of the stochastic function $\mu(\mathbf{r})$ are described in Section 4.4. The scattering amplitude can be written as:

$$\mathbf{f}(\mathbf{e}_s) = \frac{k^2 \left[\mathbf{e}_s \times \left(\mathbf{g}_i \times \mathbf{e}_s \right) \right]}{4\pi} \int_V \mu(\mathbf{r}') e^{ik(\mathbf{e}_i - \mathbf{e}_s) \cdot \mathbf{r}'} d^3 \mathbf{r}' . \tag{5.171}$$

This approximation corresponds to the first summand of the Maxwell's equation expansion in terms of the small parameter $\sqrt{\langle \mu^2 \rangle}$. The scattering amplitude is the random vector in this case. We will assume that the volume of scattering is much greater than the correlation radius of the permittivity fluctuations. The integral in Equation (5.171) is a sum of the numerous independent values and, due to the central limit theorem, is subject to the normal law of probability distribution. More exactly, both the real and imaginary parts of the integral satisfy this law. So, it would be legitimate to assume that the scattering amplitude components are Gaussian random values. The mean value of the scattering amplitude is equal to zero due to $\int \mu(\mathbf{r}') = 0$. The square of the scattering amplitude module is also a random value. Correspondingly, the random value is also the differential cross section; therefore, we can use an averaged value of the differential cross section of scattering:

$$\sigma_d(\mathbf{e}_s) = \frac{k^4\left[1-(\mathbf{g}_i\cdot\mathbf{e}_s)^2\right]}{16\pi^2}\iint_V K_\varepsilon(\mathbf{r}'-\mathbf{r}'')e^{ik(\mathbf{e}_i-\mathbf{e}_s)\cdot(\mathbf{r}'-\mathbf{r}'')}d^3\mathbf{r}'d^3\mathbf{r}''. \quad (5.172)$$

Let us set new coordinates of integration: $\mathbf{r}' + \mathbf{r}'' = 2\mathbf{R}$ and $\mathbf{r}' - \mathbf{r}'' = \mathbf{r}$. The integral over **R** is equal to the scattering volume and, with respect to **r**, to the spatial spectrum of the permittivity fluctuations. Therefore, introducing the differential cross section of scattering per volume unity, we have:

$$\sigma_d^0(\mathbf{e}_s) = \frac{\sigma_d}{V} = \frac{\pi k^4}{2}\left[1-(\mathbf{g}_i\cdot\mathbf{e}_s)^2\right]\tilde{K}_\varepsilon\left[k(\mathbf{e}_s-\mathbf{e}_i)\right]. \quad (5.173)$$

Restricting ourselves further to the case of the isotropic turbulence, we can rewrite the last formula as:

$$\sigma_d^0(\mathbf{e}_s) = \frac{\pi k^4}{2}\left(1-\cos^2\varphi\sin^2\theta\right)\tilde{K}_\varepsilon\left(2k\sin\theta/2\right), \quad (5.174)$$

where the angles θ and φ define, as before, the direction of scattering. In the case of small-scale inhomogeneities, when $kl \ll 1$, $\tilde{K}_\varepsilon(2k\sin\theta/2) \cong \tilde{K}_\varepsilon(0)$, and the differential cross section is:

$$\sigma_d^0 = \frac{k^4 l^3\langle\mu^2\rangle}{12\pi}\left(1-\cos^2\varphi\sin^2\theta\right). \quad (5.175)$$

The scattering is almost isotropic in this case. In the critical case of large-scale inhomogeneities ($kl \gg 1$), the scattering generally takes place forward.

After integration over angle φ, the cross section of scattering becomes:

$$\sigma_s^0 = \frac{\pi^2 k^4}{2}\int_0^\pi\left(1+\cos^2\theta\right)\tilde{K}_\varepsilon\left(2k\sin\theta/2\right)\sin\theta\, d\theta. \quad (5.176)$$

For small-scale inhomogeneities,

$$\sigma_s^0 = \frac{4\pi^2 k^4}{3}\tilde{K}_\varepsilon(0) = \frac{2}{9}\langle\mu^2\rangle k^4 l^3. \quad (5.177)$$

For large-scale inhomogeneities,

$$\sigma_s^0 \cong 2\pi^2 k^4 \int_0^\pi \tilde{K}_\varepsilon\left(2k\sin \theta/2\right)\sin \theta/2 \cos \theta/2 \, d\theta$$
$$= \pi^2 k^2 \int_0^{2k} \tilde{K}_\varepsilon(\kappa)\kappa\, d\kappa \cong \pi^2 k^2 \int_0^\infty \tilde{K}_\varepsilon(\kappa)\kappa\, d\kappa. \tag{5.178}$$

In the particular case of the spectrum given by Equation (4.80)

$$\sigma_s^0 = \frac{\pi^2 C_\varepsilon^2 k^2 q_0^2}{2(\nu-1)} = \frac{\langle\mu^2\rangle \Gamma(\nu) k^2}{\pi^{3/2}\Gamma(\nu - 3/2) q_0}. \tag{5.179}$$

It is easy to come to the conclusion that, in this case, $\sigma_s^0 \approx \langle\mu^2\rangle k^2 l$.

5.14 EFFECT OF SCATTERER MOTION

Up to now, we have been discussing scattering processes caused by motionless scatterers. This assumption is not a sound one for many types of natural media. In particular, it is necessary to take into consideration the thermal motion of electrons during wave scattering in the ionosphere. The same concern applies to scattering by hydrometeor particles, elements of land cover (which are susceptible to wind movements), etc. In all of these cases, the scattering takes place with the frequency shift because of the Doppler effect. Scatterers move independently and chaotically, so the spectrum of scattered waves differs from the spectrum of waves radiated by a transmitter, and Doppler widening of the spectrum occurs.

Let us now discuss this process in more detail. We will restrict ourselves to the case of backward scattering by a system of similar particles (e.g., monostatic radar). The field, scattered by the particle clouds, is expressed as:

$$\mathbf{E}(\mathbf{r},t) = k^2 \sum_j \mathbf{p}_j \frac{\exp\left[2ik_0 r_j(t) - i\omega_0 t\right]}{r_j}, \tag{5.180}$$

where $\mathbf{p}_j$ is the dipole moment of the jth scatterer induced by the incident radar radiation; $\mathbf{r}_j$ is the radius vector relative to the radar, the antenna phase center of which is taken for the coordinate origin; and ω_0 is the frequency of radiowaves radiated by the radar, $k_0 = \omega_0/c$. Naturally, the value of the induced dipole moment depends on the radar parameters and the distance to it, so the radar parameters are implicitly kept within the values of the induced dipole moments. It was taken into account when formulating Equation (5.180) that $\mathbf{p}_j \cdot \mathbf{r}_j = 0$.

Let us assume that $\mathbf{r}_j = \mathbf{r} - \mathbf{r}'_j$, where $\mathbf{r}'_j$ is the radius vector of the jth scatterer relative to the coordinate origin chosen inside of the scattering volume. Correspondingly, in the wave zone, we have $r_j \cong r - \mathbf{r} \cdot \mathbf{r}'_j / r$ and we can assume that all induced moments are the same ($\mathbf{p}_j \cong \mathbf{p}$):

$$\mathbf{E}(\mathbf{r},t) \cong \frac{k^2 \exp(2ik_0 r - i\omega_0 t)}{r} \mathbf{p} \sum_j \exp\left[-2ik_0 \frac{\mathbf{r} \cdot \mathbf{r}'_j(t)}{r}\right]. \tag{5.181}$$

Because of the random motion of particles, the field of scattering is a stochastic function obeying the Gaussian statistic (i.e., central limit theorem). It is necessary first to find the correlation function:

$$\hat{K}_E(\mathbf{r},\tau) = \frac{1}{2} \mathrm{Re} \left\langle \mathbf{E}(\mathbf{r},t) \cdot \mathbf{E}^*(\mathbf{r},t+\tau) \right\rangle \tag{5.182}$$

for determination spectrum of the field. We shall take into account that $\mathbf{r}_j(t+\tau) = \mathbf{r}_j(t) + \mathbf{v}_j \tau$, where $\mathbf{v}_j$ is the random velocity of the jth scatterer. Due to the similarity of particles, after averaging we will have:

$$\hat{K}_E(\mathbf{r},\tau) = N_\Sigma \eta(\mathrm{p},\mathrm{r})\, q(\tau), \tag{5.183}$$

where N_Σ is the number of particles in the volume, and $\eta(\mathrm{p},\mathrm{r})$ is the parameter depending on the induced moment value and the distance to the scattering volume,

$$q(\tau) = \left\langle \cos\left[\omega_0 \tau + 2k_0 \left(\frac{\mathbf{r} \cdot \mathbf{v}}{r} \tau\right)\right]\right\rangle. \tag{5.184}$$

Naturally, our further interest will be concentrated on the study of function $q(\tau)$ properties, as other factors of Equation (5.183) do not depend on parameter τ. Correspondingly, the spectrum of the scattered signal can be defined by this function. Let the sounding be directed along the z-axis. Then, Equation (5.184) can be rewritten as:

$$q(\tau) = \left\langle \cos\left[\omega_0 \tau + 2k_0 \mathrm{v}_z \tau\right]\right\rangle. \tag{5.185}$$

Fourier transform of this function gives us the scattered signal spectrum:

$$\hat{q}(\omega) = \frac{c}{4\omega_0}\left[P_z\left(-c\frac{\omega+\omega_0}{2\omega_0}\right) + P_z\left(c\frac{\omega-\omega_0}{2\omega_0}\right)\right]. \tag{5.186}$$

Here, $P_z(\text{v})$ is the particle distribution function relative to the z-component of the velocity. Because positive frequencies are physically significant, it is necessary to assume that the first summand is equal to zero and to double the second one. Thus, we can obtain for the differential cross section of back scattering per unit volume:

$$\tilde{\sigma}_d^0(\omega,\pi) = N\sigma_d(\pi)\frac{c}{2\omega_0}P_z\left(c\frac{\omega - \omega_0}{2\omega_0}\right). \tag{5.187}$$

For example, let us consider the case of scattering by electrons due to thermal movement. It is sufficient to use Maxwell's distribution to describe the velocity distribution:

$$P_z(\text{v}) = \left(\frac{m}{2\pi k_b T}\right)^{1/2}\exp\left(-\frac{m\text{v}^2}{2k_b T}\right) \tag{5.188}$$

Then, for the frequencies, we have:

$$\tilde{\sigma}_d^0(f,\pi) = \frac{N\sigma_d(\pi)c}{2f_0}\sqrt{\frac{m}{2\pi k_b T}}\exp\left[-\frac{mc^2}{8k_b T}\left(\frac{f-f_0}{f_0}\right)^2\right]. \tag{5.189}$$

6 Wave Scattering by Rough Surfaces

6.1 STATISTICAL CHARACTERISTICS OF A SURFACE

Many natural surfaces, such as the soil surface or water surface of the ocean, can be regarded as smooth only in certain circumstances. In general, these surfaces should be considered to be rough, and their interaction with radiowaves should be seen as a scattering process. Whether or not we assume that the surface is rough generally depends on the problem formulation and, particularly, on the ratio of the roughness scales and the wavelength. The nature of the roughness varies depending on the type of surface. Sea surface roughness is a result of the interaction of the wind with the water surface. This interaction has a nonlinear character. A great number of waves with different frequencies and wave numbers are generated as a result, and their mixture leads to oscillations of the sea surface height according to the stochastic function of coordinate and time. However, the velocity of the sea surface movement is small compared to the velocity of light, so time dependence cannot be taken into account in the first approximation. Soil roughness can form as a result of wind erosion, urban activity, and other causes. The soil roughness is also a random function of coordinates. Again, a dependence on time is not considered in the beginning and we are dealing generally with wave scattering by random surfaces. The specific surface will be described by a random function of the elevation $\zeta(\mathbf{s})$, where $\mathbf{s} = \{x,y\}$. The average value $\langle \zeta(\mathbf{s}) \rangle = 0$, so it is assumed that the average surface is given by the plane $z = 0$. The function $\zeta(\mathbf{s})$ is supposed to be statistically homogeneous. It simplifies the problem substantially, as the statistical homogeneity of the real natural rough surfaces take place in the restricted cases. The correlation function:

$$\hat{K}_{\zeta}(\mathbf{s}' - \mathbf{s}'') = \langle \zeta(\mathbf{s}')\zeta(\mathbf{s}'') \rangle \tag{6.1}$$

depends, in this case, on the coordinate difference of the points involved. In many cases, the surface may be assumed to be statistically isotropic. Then, the correlation function depends only on the module $|\mathbf{s}' - \mathbf{s}''|$ (i.e., on the distance between the points of correlation). The correlation function has significant value within the correlation radius which is often defined by the relation:

$$l^2 = \frac{1}{\langle \zeta^2 \rangle} \int K_{\zeta}(\mathbf{s}) d^2\mathbf{s} = \frac{2\pi}{\langle \zeta^2 \rangle} \int_0^{\infty} K_{\zeta}(s) s\, ds. \tag{6.2}$$

It is necessary to represent the correlation scales along the main axis of the anisotropy ellipse in the case of statistical anisotropy.

If we suppose that roughness occupies the bounded surface Σ for whose measure is much larger than the radius of correlation, then the following Fourier expansion is correct:

$$\zeta(\mathbf{s}) = \int \tilde{\zeta}(\mathbf{q})\, e^{i\mathbf{q}\cdot\mathbf{s}} d^2\mathbf{q}\,. \tag{6.3}$$

The spatial spectrum:

$$\tilde{\zeta}(\mathbf{q}) = \frac{1}{4\pi^2} \int_{\Sigma} \zeta(\mathbf{s})\, e^{-i\mathbf{q}\cdot\mathbf{s}} d^2\mathbf{s} \tag{6.4}$$

is a random function with zero mean. Its correlation function is written as:

$$\left\langle \tilde{\zeta}(\mathbf{q}')\tilde{\zeta}^*(\mathbf{q}'')\right\rangle = \frac{1}{16\pi^2} \iint_{\Sigma} \hat{K}_{\zeta}(\mathbf{s}'-\mathbf{s}'')\, e^{-i(\mathbf{q}'\cdot\mathbf{s}'-\mathbf{q}''\cdot\mathbf{s}'')} d^2\mathbf{s}' d^2\mathbf{s}''.$$

By introducing the "gravity center" coordinate $\mathbf{s}'+\mathbf{s}''=2\mathbf{S}$ and the difference $\mathbf{s}'-\mathbf{s}''=\mathbf{s}$, the last integral can be written in the form:

$$\left\langle \tilde{\zeta}(\mathbf{q}')\tilde{\zeta}^*(\mathbf{q}'')\right\rangle = \frac{1}{16\pi^2} \int_{\Sigma} \hat{K}_{\zeta}(\mathbf{s}) \exp\left(-i\frac{\mathbf{q}'+\mathbf{q}''}{2}\cdot\mathbf{s}\right) d^2\mathbf{s} \int_{\Sigma} e^{-i(\mathbf{q}'-\mathbf{q}'')\cdot\mathbf{S}} d^2\mathbf{S}.$$

Let us represent the elevation fluctuation spectrum of the examined surface as:

$$\tilde{K}_{\zeta}(\mathbf{q}) = \frac{1}{4\pi^2} \int \hat{K}_{\zeta}(\mathbf{s})\, e^{-i\mathbf{q}\cdot\mathbf{s}} d^2\mathbf{s}. \tag{6.5}$$

according to the Wiener–Chintchin theorem. The integral in this expression can be spread over infinite limits because the size of the chosen surface was set much larger than the correlation scale. Thus,

$$\left\langle \tilde{\zeta}(\mathbf{q}')\tilde{\zeta}(\mathbf{q}'')\right\rangle = \frac{1}{4\pi^2} \tilde{K}_{\zeta}\left(\frac{\mathbf{q}'+\mathbf{q}''}{2}\right) \int_{\Sigma} e^{-i(\mathbf{q}'-\mathbf{q}'')\cdot\mathbf{S}} d^2\mathbf{S}. \tag{6.6}$$

It follows from here, in particular, that at $\mathbf{q}'=\mathbf{q}''=\mathbf{q}$:

$$\left\langle \left|\tilde{\zeta}(\mathbf{q})\right|^2 \right\rangle = \frac{\tilde{\mathbf{K}}_{\zeta}(\mathbf{q})\Sigma}{4\pi^2}, \tag{6.7}$$

where we have simplified the expression by not distinguishing between the value of the surface and its square. If $\mathbf{q}' \neq \mathbf{q}''$, then the integral in Equation (6.6) can be also taken to infinite limits by maintaining the condition $|\mathbf{q}' - \mathbf{q}''|\sqrt{\Sigma} >> 1$. As a result,

$$\left\langle \tilde{\zeta}(\mathbf{q}')\tilde{\zeta}(\mathbf{q}'') \right\rangle = \tilde{\mathrm{K}}_{\zeta}\left(\frac{\mathbf{q}' + \mathbf{q}''}{2}\right)\delta(\mathbf{q}' - \mathbf{q}''). \tag{6.8}$$

The correlation function of the scalar product is:

$$\left\langle \nabla\zeta(\mathbf{s}')\cdot\nabla\zeta(\mathbf{s})\right\rangle = -\nabla^2\mathbf{K}_{\zeta}(\mathbf{s}) = \int \mathrm{q}^2\tilde{\mathbf{K}}_{\zeta}(\mathbf{q})\, e^{i\mathbf{q}\cdot\mathbf{s}}d^2\mathbf{q}, \quad \mathbf{s} = \mathbf{s}' - \mathbf{s}''. \tag{6.9}$$

In particular,

$$\left\langle (\nabla\zeta)^2 \right\rangle = \int \mathrm{q}^2\tilde{\mathrm{K}}_{\zeta}(\mathbf{q})\, d^2\mathbf{q}. \tag{6.10}$$

Let us now turn to the case of a statistically isotropic surface. We can assume that the main area of integration is situated in the interval $0 \leq \mathrm{q} \leq 2\pi/l$; therefore, the following estimation is valid:

$$\left\langle (\nabla\zeta)^2 \right\rangle \cong 2\pi\tilde{\mathrm{K}}_{\zeta}(0)\int_0^{2\pi/l} \mathrm{q}^3 d\mathrm{q} = \frac{8\pi^3}{l^4}\tilde{\mathrm{K}}_{\zeta}(0).$$

On the other hand, it is easy to determine from Equation (6.2) that $l^2 = 4\pi^2\tilde{\mathrm{K}}_{\zeta}(0)/\langle\zeta^2\rangle$. As a result, we have:

$$\left\langle (\nabla\zeta)^2 \right\rangle \cong 2\pi^3\frac{\left\langle \zeta^2 \right\rangle}{l^2}. \tag{6.11}$$

Further, we will encounter the correlation vector :

$$= i\int \mathbf{q}\tilde{\mathrm{K}}_{\zeta}(\mathbf{q})\, e^{i\mathbf{q}\cdot(\mathbf{s}' - \mathbf{s}'')}d^2\mathbf{q}. \tag{6.12}$$

And, finally, we can calculate the value:

$$\left\langle \zeta(\mathbf{s})\tilde{\zeta}(\mathbf{q})\right\rangle = \frac{1}{4\pi^2}\int_{\Sigma}\left\langle \zeta(\mathbf{s})\zeta(\mathbf{s}')\right\rangle e^{-i\mathbf{q}\cdot\mathbf{s}'}d^2\mathbf{s}' \cong \frac{1}{4\pi^2}\int \hat{\mathbf{K}}_{\zeta}\left(\mathbf{s}-\mathbf{s}'\right) e^{-i\mathbf{q}\cdot\mathbf{s}'}d^2\mathbf{s}'.$$

The last expression is easily transformed to the form:

$$\left\langle \zeta(\mathbf{s})\tilde{\zeta}(\mathbf{q})\right\rangle = \tilde{\mathbf{K}}_{\zeta}(\mathbf{q})e^{-i\mathbf{q}\cdot\mathbf{s}}. \tag{6.13}$$

The previous discussion was concerned with smooth and differentiable surfaces. It is often convenient to eliminate the requirement of differentiability when describing natural surfaces (sea, soil, etc.). To illustrate, we will analyze the structure function of properties of the surface:

$$D_{\zeta}\left(\mathbf{s}'-\mathbf{s}''\right) = \left\langle \left[\zeta\left(\mathbf{s}'\right)-\zeta\left(\mathbf{s}''\right)\right]^2\right\rangle \tag{6.14}$$

assuming statistical isotropy of the surface. It is usual to suppose in the case of a smooth surface that the first derivative of the correlation function at zero is equal to zero ($\hat{\mathrm{K}}'_{\zeta}(0) = 0$). Then, the expansion:

$$D(\mathrm{s}) = \hat{\mathrm{K}}''_{\zeta}(0)\mathrm{s}^2 \tag{6.15}$$

for the structure function is valid close to zero. Such a surface has slopes, which means that the angle is equal to zero and the dispersion (the mean-square slope) is defined as:

$$\left\langle \left(\nabla\zeta\right)^2\right\rangle = -\nabla^2\hat{\mathrm{K}}_{\zeta}(0) = -2\hat{\mathrm{K}}''_{\zeta}(0) = \frac{1}{2}\lim_{s\to 0}\frac{D(\mathrm{s})}{\mathrm{s}^2}. \tag{6.16}$$

The differentiability of the surface is understood in such sense.

Many natural surfaces have a fractal character.[78] Their structure function satisfies the equation:

$$D(\mathrm{s}) = B\mathrm{s}^{\alpha} \tag{6.17}$$

The index α is connected with fractal dimension β (Hausdorff–Besicovitch dimension) by the relation:

$$\alpha = 2\left(2-\beta\right). \tag{6.18}$$

The value obtained lies within the range $1<\beta<2$, which leads to an index interval of $0<\alpha<2$ in Equation (6.17). It would not be correct, in this case, to address the differentiability of the surface. The maximum value of α leads to Equation (6.15), which is typical for smooth surfaces. The value $\beta=1.5$ $(\alpha=1)$, which follows from the theory of Brownian motion, corresponds to the Brownian fractal.

Let us now turn our attention to the properties of rough surfaces. Generally, we cannot expect to develop an exact technique to solve the problem of radiowave diffraction on stochastic surfaces; instead, we must rely on approximation methods that, as a rule, are found effective for asymptotic cases. In our case, the roughness is small compared to the wavelength, which is the opposite of the case of large inhomogeneities. The method of small perturbation is effective in our case, and the Kirchhoff approach is best suited for the second case. Recently, some attempts have been made to find analytical solutions of the latter problem on the basis of an integral equation model (IEM) for electromagnetic fields[79]; however, only some refinement of results have been reported, and the IEM is undergoing improvement.[80]

6.2 RADIOWAVE SCATTERING BY SMALL INHOMOGENEITIES AND CONSEQUENT APPROXIMATION SERIES

Let us assume that the described surface separates into two media. The upper medium has permittivity equal to unity. The permittivity of the lower medium is equal to ε. We assume that a plane wave of single amplitude $\mathbf{E}_i=\mathbf{g}_i e^{ik\mathbf{e}_i\cdot\mathbf{r}}$, $\mathbf{r}=\{x,y,z\}$ is incident from the side of the z-coordinate positive values. To find the scattered field it is necessary to solve Maxwell's equations for both media while maintaining the boundary conditions:

$$\left[\mathbf{n}\times(\mathbf{E}_1-\mathbf{E}_2)\right]_\zeta=0,\quad \left(\mathbf{n}\cdot(\mathbf{E}_1-\varepsilon\mathbf{E}_2)\right)_\zeta=0,\quad \left(\mathbf{H}_1-\mathbf{H}_2\right)_\zeta=0 \tag{6.19}$$

on the surface. The numbers 1 and 2 indicate, respectively, the fields over and under the examined surface. If $\mathbf{e}_z$ is the single normal to the average plane z = 0, then the single normal to the surface $\zeta(\mathbf{s})$ satisfies the equation:

$$\mathbf{n}(\mathbf{s})=\frac{\mathbf{e}_z-\nabla\zeta}{\sqrt{1+(\nabla\zeta)^2}}. \tag{6.20}$$

Let us assume that the roughness is small; that is,

$$k\sqrt{\langle\zeta^2\rangle}<<1. \tag{6.21}$$

This inequality means that the probability is extremely low that deviations of the roughness from the average plane can be more than the wavelength. We also assume that the considered rough surface slopes are small, or $\sqrt{(\nabla\zeta)^2} << 1$. Equation (6.11) allows us to assume that the correlation radius of such a surface is much larger than the height; that is, $l >> \sqrt{\langle\zeta^2\rangle}$.

We can now apply the method of sequential approximations. Let us represent the unknown fields in the form of series:

$$\mathbf{E}_j = \mathbf{E}_j^{(0)} + \mathbf{E}_j^{(1)} + \mathbf{E}_j^{(2)} + \cdots, \quad \mathbf{H}_j = \mathbf{H}_j^{(0)} + \mathbf{H}_j^{(1)} + \mathbf{H}_j^{(2)} + \cdots. \tag{6.22}$$

Here, $j = 1, 2$, and the sum terms represent expansion over the growing degree of ζ and $\nabla\zeta$. Naturally, the sum terms of the same small size satisfy Maxwell's equations.

The fields, however, are expanded into a Taylor series of the form:

$$\mathbf{E}(\zeta) = \mathbf{E}(0) + \frac{\partial \mathbf{E}}{\partial z}\zeta + \frac{1}{2}\frac{\partial^2 \mathbf{E}}{\partial z^2}\zeta^2 + \cdots.$$

Taking into account the approximation $\mathbf{n} = \mathbf{e}_z - \nabla\zeta - \mathbf{e}_z(\nabla\zeta)^2/2 + \cdots$, the border conditions in Equation (6.19) are transferred from surface $\zeta(\mathbf{s})$ on the plane z = 0. Further, the corresponding expansions are continued until the second order of smallness is obtained. Let us set the term of the same order of smallness equal to zero to obtain the boundary conditions for fields of a different order. These boundary conditions for the field of the zero order have the form:

$$\left[\mathbf{e}_z \times (\mathbf{E}_1^{(0)} - \mathbf{E}_2^{(0)})\right]_{z=0} = 0, \quad \left(\mathbf{e}_z \cdot (\mathbf{E}_1^{(0)} - \varepsilon\mathbf{E}_2^{(0)})\right)_{z=0} = 0, \quad \left(\mathbf{H}_1^{(0)} - \mathbf{H}_2^{(0)}\right)_{z=0} = 0. \tag{6.23}$$

From here on, we will omit the subscript z = 0. The boundary conditions for the electromagnetic field of the first order can be written as:

$$\left[\mathbf{e}_z \times \left(\mathbf{E}_1^{(1)} - \mathbf{E}_2^{(1)}\right)\right] = \left[\nabla\zeta \times \left(\mathbf{E}_1^{(0)} - \mathbf{E}_2^{(0)}\right)\right] - \left[\mathbf{e}_z \times \frac{\partial}{\partial z}\left(\mathbf{E}_1^{(0)} - \mathbf{E}_2^{(0)}\right)\right]\zeta \quad (\mathrm{I}),$$

$$\mathbf{H}_1^{(1)} - \mathbf{H}_2^{(1)} = -\zeta\frac{\partial}{\partial z}\left(\mathbf{H}_1^{(0)} - \mathbf{H}_2^{(0)}\right) \quad (\mathrm{II}), \tag{6.24}$$

$$\left[\mathbf{e}_z \times \left(\mathbf{E}_1^{(1)} - \varepsilon\mathbf{E}_2^{(1)}\right)\right] = \left[\nabla\zeta \times \left(\mathbf{E}_1^{(0)} - \varepsilon\mathbf{E}_2^{(0)}\right)\right] - \left[\mathbf{e}_z \times \frac{\partial}{\partial z}\left(\mathbf{E}_1^{(0)} - \mathbf{E}_2^{(0)}\right)\right]\zeta \quad (\mathrm{III}).$$

Analogous expressions can be obtained for the second-order fields. We will not provide them here but refer the reader to Armand,[52] where the second approach is analyzed in detail.

Further actions deal with using Maxwell's equations to solve for every approximation at given boundary conditions. In doing so, the incident plane wave (more exactly, the source of the radiated wave) is the source for the field of the zero approximation. The fields of subsequent approaches are excited by the preceding fields. So, a system of constrained fields is obtained for which it is necessary to solve a succession of Maxwell's equations for fields of different orders. We should bear in mind when doing so that the boundary conditions for the normal to the average plane field components are odd in some sense, but we must keep them in order to minimize calculations when they are indirectly presented. Note that these boundary conditions are analogous to Equations (1.93) and (1.94), which means that our problem is reduced to a problem of fields excited by surface currents. The formal solution of this problem is Equation (1.111) which will be used from here on.

Let us begin with the zero-arch approximation. It does not require any special consideration as it is reduced to a problem of plane wave reflection by the plane interface. This problem has been examined in Chapter 3, but here we will rewrite the formulae in a more convenient form. In particular, the electric field on the surface is written as:

$$\mathbf{E}_1^{(0)} = \mathbf{E}_i + \mathbf{E}_r = 2a_0\left[\frac{\mathbf{g}_i}{a_0+b_0} + \frac{\varepsilon-1}{\varepsilon a_0+b_0}\left(\mathbf{e}_z + \frac{\mathbf{e}_i}{a_0+b_0}\right)\left(\mathbf{e}_z\cdot\mathbf{g}_i\right)\right]e^{i\mathbf{w}_i\cdot\mathbf{s}}. \tag{6.25}$$

Here, $\mathbf{w}_i$ is the wave vector of the incident wave on the plane $z = 0$, and $w_i = k\sin\theta_i$.

Having found the field component of the zero approximation, let us now compute the fields of the first approximation of the perturbation theory. The first step is to rewrite the equivalent surface currents and charges of Equation (6.23) with a more compact right side. We refer the reader to Bass[48] and Armand[52] for more details on the procedure, as we provide only the results here. The first expression of Equation (6.23) can be expressed as:

$$\left[\mathbf{e}_z\times(\mathbf{E}_1^{(1)}-\mathbf{E}_2^{(1)})\right] = -\frac{4\pi}{c}\mathbf{K}_{\mathrm{m}}^{(1)}, \tag{6.26}$$

where the surface magnetic current is:

$$\mathbf{K}_{\mathrm{m}}^{(1)} = \frac{c}{4\pi}\frac{\varepsilon-1}{\varepsilon}\left[\mathbf{e}_z\times\nabla\right]\left(\mathbf{e}_z\cdot\mathbf{E}_1^{0}\right)\zeta. \tag{6.27}$$

The expression obtained for the magnetic current is not only compact but also convenient as it is expressed with regard to only the field over the interface.

The second boundary condition is rewritten as:

$$\left[\mathbf{e}_z \times \left(\mathbf{H}_1^0 - \mathbf{H}_2^0\right)\right] = \frac{4\pi}{c}\mathbf{K}_e^{(1)}, \tag{6.28}$$

where the surface electric current is:

$$\mathbf{K}_e^{(1)} = \frac{ikc(\varepsilon - 1)}{4\pi}\left[\mathbf{e}_z \times \left[\mathbf{e}_z \times \mathbf{E}_1^0\right]\right]\zeta. \tag{6.29}$$

Note that the expression for the surface electric charge has the form:

$$\delta_e^{(1)} = \frac{\varepsilon - 1}{4\pi}\left[\zeta\frac{\partial}{\partial z}\left(\mathbf{e}_z \cdot \mathbf{E}_1^0\right) - \left(\nabla\zeta \cdot \mathbf{E}_1^0\right)\right]. \tag{6.30}$$

Now we can use Equation (6.25) to establish that:

$$\mathbf{K}_m^{(1)} = \Im\left[\mathbf{e}_z \times \nabla\right]\zeta e^{i\mathbf{w}_i \cdot \mathbf{s}}, \tag{6.31}$$

where

$$\Im = \frac{-(\varepsilon - 1)a_0}{2\pi(\varepsilon a_0 + b_0)}\left(\mathbf{e}_z \cdot \mathbf{g}_i\right). \tag{6.32}$$

The electric current is represented as:

$$\mathbf{K}_e^{(1)} = ik\mathbf{i}\zeta(\mathbf{r})e^{i\mathbf{w}_i\mathbf{s}} \tag{6.33}$$

where vector $\mathbf{i}$ is defined as:

$$\mathbf{i} = \frac{c(\varepsilon - 1)a_0}{2\pi}\left\{\frac{\left(\mathbf{e}_z \cdot \mathbf{g}_i\right)}{\varepsilon a_0 + b_0}\left[\mathbf{e}_z - \frac{\varepsilon - 1}{a_0 + b_0}\mathbf{e}_i\right] - \frac{\mathbf{g}_i}{a_0 + b_0}\right\}. \tag{6.34}$$

Now, let us compute the Fourier transform of the introduced surface currents. For the magnetic current:

$$\tilde{\mathbf{K}}_m^{(1)} = \frac{\Im}{4\pi^2}\int_\Sigma e^{-i\mathbf{w} \cdot \mathbf{s}}\left[\mathbf{e}_z \times \nabla\right]\left(\zeta e^{i\mathbf{w}_i \cdot \mathbf{s}}\right) d^2\mathbf{s}. \tag{6.35}$$

Let us take into account that $e^{-i\mathbf{w}\cdot\mathbf{s}}\nabla(\zeta e^{i\mathbf{w}_i\cdot\mathbf{s}}) = \nabla(\zeta e^{i(\mathbf{w}_i-\mathbf{w})\cdot\mathbf{s}}) + i\mathbf{w}\zeta e^{i(\mathbf{w}_i-\mathbf{w})\cdot\mathbf{s}}$. The integral of the first term is transformed into such over the boundary of the surface Σ, and we can set it equal to zero, as the roughness is zero, or the incident field is small on the border of the radiating antenna footprint. The integral of the second summand is transformed to:

$$\tilde{\mathbf{K}}_{\mathrm{m}}^{(1)} = i\Im\left[\mathbf{e}_z \times \mathbf{w}\right]\tilde{\zeta}\left(\mathbf{w} - \mathbf{w}_i\right) \tag{6.36}$$

according to Equation (6.4). Similarly, for the electric current:

$$\tilde{\mathbf{K}}_{e}^{(1)} = ik\mathbf{i}\tilde{\zeta}\left(\mathbf{w} - \mathbf{w}_i\right). \tag{6.37}$$

Now, according to Equation (1.111), the Fourier image of the scattered field is represented in the first approach as:

$$\tilde{\mathbf{E}}_1^{(1)}\left(\mathbf{w}\right) = \frac{4ik\pi}{c\left(\varepsilon\gamma_1 + \gamma_2\right)}\left[\mathbf{q}_1 \times \mathbf{F}\right]\tilde{\zeta}\left(\mathbf{w} - \mathbf{w}_i\right), \tag{6.38}$$

where:

$$\mathbf{F}(\mathbf{w}) = \frac{\varepsilon\Im\left[\mathbf{e}_z \times \mathbf{q}_1\right]}{k} + \frac{\left[\mathbf{q}_1 \times \mathbf{i}\right]}{k} + \frac{k\left(\varepsilon - 1\right)}{\gamma_1 + \gamma_2}\left[\mathbf{e}_z \times \mathbf{i}\right]. \tag{6.39}$$

The field itself is expressed via the integral:

$$\mathbf{E}_1^{(1)}\left(\mathbf{r}\right) = \frac{4ik\pi}{c}\int \frac{\left[\mathbf{q}_1 \times \mathbf{F}\right]\tilde{\zeta}\left(\mathbf{w} - \mathbf{w}_i\right)}{\varepsilon\gamma_1 + \gamma_2} e^{i\mathbf{w}\cdot\mathbf{s} + i\gamma_1 z} d^2\mathbf{w}. \tag{6.40}$$

As the integral extends toward infinite limits, the scattered field satisfies Gaussian probability law. Given a sufficient distance from the surface, the integral can be calculated using the stationary phase method. Let us designate $a = \cos\theta$ and $b = \sqrt{\varepsilon - \sin^2\theta}$ and introduce the vector:

$$\mathbf{p}\left(\theta,\phi\right) = \mathbf{e}_x \sin\theta\cos\phi + \mathbf{e}_y \sin\theta\sin\phi + \mathbf{e}_z \cos\theta, \tag{6.41}$$

and we obtain:

$$\mathbf{E}_1^{(1)}(\mathbf{r}) = \frac{8\pi^2 k^2 \cos\theta_s}{c}\frac{e^{ikr}}{r}\frac{\left[\mathbf{e}_s \times \mathbf{F}\left(\mathbf{e}_s\right)\right]}{\varepsilon a_s + b_s}\tilde{\zeta}\left(\mathbf{w}_s - \mathbf{w}_i\right). \tag{6.42}$$

Here, the subscript s indicates the direction of scattering, and

$$\mathbf{F}(\mathbf{e}_s) = \varepsilon \Im \mathbf{e}_z \times \mathbf{p} + \mathbf{p} \times \mathbf{i} + \frac{\varepsilon - 1}{a + b} \mathbf{e} \times \mathbf{i}. \tag{6.43}$$

Computation of the field scattered into the lower medium is done in the same way.

The formulae for scattered fields allow us to compute easily the scattering amplitudes into upper and into lower semispaces and to determine accordingly the cross section of the scattering. It is necessary to take into account that the scattering amplitude is a stochastic value with a mean value equal to zero. The squared module of the scattering amplitude is also an occasional quantity; therefore, an important definition of the cross section is $\sigma = \langle |\mathrm{f}|^2 \rangle$. It is appropriate to take into account Equation (6.7), which leads us to the conclusion that the scattering section is proportional to the square of the illuminated surface. So, it is reasonable to introduce a definition of the scattering cross section per unit of surface σ^0, the so-called reflectivity. It is a dimensionless value that characterizes the scattering properties of the surface regardless of size:

$$\sigma^0(\mathbf{e}_s) = \frac{16\pi^2 k^4 \cos^2\theta_s}{c^2} \left| \frac{\left[\mathbf{e}_s \times \mathbf{F}(\mathbf{e}_s)\right]}{\varepsilon a_s + b_s} \right|^2 \tilde{\mathrm{K}}_\zeta(\mathbf{w}_s - \mathbf{w}_i). \tag{6.44}$$

Thus, the intensity of the wave scattering is proportional in the first approximation to the power of the surface spatial component with wave vector $\mathbf{k} = \mathbf{w}_s - \mathbf{w}_i$. Its absolute value is:

$$\mathrm{k} = k\sqrt{\sin^2\theta_s + \sin^2\theta_i - 2\sin\theta_s \sin\theta_i \cos(\varphi_s - \varphi_i)}. \tag{6.45}$$

The subscript i represents values related to the incident wave. The result obtained indicates that the incident wave interacts most effectively with only one of the spatial harmonics of the surface. This effect is said about to be a resonance one.

The spatial spectrum of the surface is a rather acute function of angles, so the angle dependence of the scattering intensity (scattering indicatrix) is generally defined by the properties of the function $\tilde{\mathrm{K}}_\zeta(\mathbf{k})$. Particularly, we can assert that maximum scattering occurs in the direction of the specular reflection when $\varphi_s = \varphi_i$ and $\theta_s = \theta_i$.

We must pay special attention to the very important case of backscattering. Particular interest is raised here by the fact that the radar images are formed against a background of wave backscattering. In this case, $\mathbf{e}_s = -\mathbf{e}_i$, and the following expression is obtained for the backscattering reflectivity:

$$\sigma^0(-\mathbf{e}_i) = 4k^4 \left|F_h(\theta_i)\right|^2 \cos^4\theta_i \left[1 + 2(\mathbf{e}_z \cdot \mathbf{g}_i)\left(C + C^* + 2|C|^2 \sin^2\theta_i\right)\right] \tilde{K}_\zeta(2k\sin\theta_i). \tag{6.46}$$

Here,

$$C = \frac{(\varepsilon - 1)\left(\cos\theta_i + \sqrt{\varepsilon - \sin^2\theta_i}\right)\sqrt{\varepsilon - \sin^2\theta_i}}{\left(\varepsilon\cos\theta_i + \sqrt{\varepsilon - \sin^2\theta_i}\right)^2}. \tag{6.47}$$

In the case of horizontal polarization,

$$\sigma_h^0(-\mathbf{e}_i) = 4k^4 \left|F_h(\theta_i)\right|^2 \cos^4\theta_i \tilde{K}_\zeta(2k\sin\theta_i). \tag{6.48}$$

At vertical polarization of the incident wave:

$$\sigma_v^0(-\mathbf{e}_i) = 4k^4 \left|F_h(\theta_i)\right|^2 \cos^4\theta_i \left|1 + 2C\sin^2\theta_i\right|^2 \tilde{K}_\zeta(2k\sin\theta_i). \tag{6.49}$$

At large values of permittivity $C \to 1/\cos^2\theta_i$, and, in general case:

$$\sigma^0(-\mathbf{e}_i) = 4k^4 \left|F_h(\theta_i)\right|^2 \left[\cos^4\theta_i + 4(\mathbf{e}_z\mathbf{g}_i)^2\right] \tilde{K}_\zeta(2k\sin\theta_i). \tag{6.50}$$

Thus, it is a simple matter to write the formulae for horizontal and vertical polarization.

6.3 THE SECOND APPROXIMATION OF THE PERTURBATION METHOD

Before moving on to the analysis of the second approximation, we should explain its importance. Usually, when we attempt to explain the scattering theory using small heterogeneity and apply the related sequential approaches method (perturbation method), the second approximation is not taken into account. To address this problem, let us refer back to Equation (6.22) and analyze the components of different moments. The first moment (coherent part of the field) is equal to:

$$\langle \mathbf{E}_j \rangle = \mathbf{E}_j^{(0)} + \langle \mathbf{E}_j^{(2)} \rangle, \quad \langle \mathbf{H}_j \rangle = \mathbf{H}_j^{(0)} + \langle \mathbf{H}_j^{(2)} \rangle.$$

One can see that the second approximation changes the coherent component of the field. This change proceeds at the second order of smallness and, as a rule, cannot be taken into consideration in mean field calculation.

Another situation appears during calculation of the second moment: average power flow. Here, keeping the summand not greater than the second order of smallness,

$$\mathbf{S}_j = \frac{c}{8\pi}\left\langle\left[\mathbf{E}_j \times \mathbf{H}_j\right]\right\rangle = \mathbf{S}_j^{(0)} + \mathbf{S}_j^{(1)} + \mathbf{S}_j^{(02)}, \tag{6.51}$$

where the interference term:

$$\mathbf{S}_j^{(02)} = \frac{c}{8\pi}\left[\mathbf{E}_j^0 \times \left\langle\mathbf{H}_j^{(2)}\right\rangle\right] + \frac{c}{8\pi}\left[\left\langle\mathbf{E}_j^{(2)}\right\rangle \times \mathbf{H}_j^0\right] \tag{6.52}$$

is analogous to the second summand according to the value order, which corresponds to the power flow density of the first approximation waves. We can now conclude that we cannot neglect waves of the second approach order in the calculation of energetic values.

We will not fully calculate the second approximation fields at permittivity values of a medium with a rough interface. An example of such a calculation can be found in Armand.[52] The analysis given there indicates that we cannot use the perturbation method in the common form to calculate second approximation fields at high permittivity values of the scattering medium. To calculate the field inside the medium, limitation of expansion by the first term of the series is possible by maintaining the condition:

$$k\sqrt{|\varepsilon|\left\langle\zeta^2\right\rangle} << 1 .$$

Thus, the extent of roughness must be smaller than the wavelength in the medium (or skin depth). This requirement may not be valid in some cases — for example, when solving the problem of wave scattering by ripples. For high permittivity values, the problem must be analyzed, from the very beginning, based on the assumption that the field inside the scattering medium is equal to zero, as occurs in the case of ideal conductivity. We can use the Shchukin–Leontovich boundary conditions.[67]

Now we will address the coherent component of the second approach field, $\left\langle\mathbf{E}_1^{(2)}\right\rangle$. As was shown in Leontovich,[67] this component has a plane wave form and propagates in the direction of the specular reflection, thereby interfering with the reflected wave of the zero approximation. Having said this, we will focus primarily on the vertical incident of the original wave and will analyze this problem briefly for ideal conductivity of the scattering surface. Assume we have the following boundary condition:

$$\left[\mathbf{n} \times \mathbf{E}\right]_\zeta = 0 . \tag{6.53}$$

By repeating the procedure for the previous expansions, it is easy to obtain the boundary conditions system on the plane z = 0. These boundary conditions have a form analogous to Equations (6.23) and (6.24), where the fields in the medium are assumed to be equal to zero. We can show that the field on the plane z = 0 is:

$$\mathbf{E}^{(1)}\Big|_{z=0} = 2ik\zeta(\mathbf{s})\mathbf{g}_i \,. \tag{6.54}$$

Then, the field of the first order can be expressed by the integral:

$$\mathbf{E}^{(1)}(\mathbf{r}) = 2ik\mathbf{g}_i \int \tilde{\zeta}(\mathbf{w}) e^{i\mathbf{w}\cdot\mathbf{s}+iz\sqrt{k^2-\mathrm{w}^2}} d^2\mathbf{w} \tag{6.55}$$

in any point of the space. Here, the transversal coordinates of the point of observation are given by vector **s**. We now use Equation (1.75) and a known expression for the field of the spherical wave:

$$\frac{e^{ik|\mathbf{r}-\mathbf{s}'|}}{|\mathbf{r}-\mathbf{s}'|} = \frac{i}{2\pi}\int \frac{e^{i\mathbf{w}\cdot(\mathbf{s}-\mathbf{s}')+i|z|\sqrt{k^2-\mathrm{w}^2}}}{\sqrt{k^2-\mathrm{w}^2}} d^2\mathbf{w}. \tag{6.56}$$

It is easy to obtain from Equation (6.55) that, in the far zone:

$$\mathbf{E}^{(1)} = \frac{k^2\cos\theta_s}{\pi}\frac{e^{ikr}}{\mathrm{r}}\tilde{\zeta}\left(k\frac{\mathbf{s}}{\mathrm{r}}\right)\mathbf{g}_i. \tag{6.57}$$

It follows from here that, in the considered approximation, we have:

$$\sigma^0 = 4k^4\cos^2\theta_s\,\tilde{\mathrm{K}}_\zeta\left(k\frac{\mathbf{s}}{\mathrm{r}}\right). \tag{6.58}$$

These approximations lead to development of a formula describing the power flow of the radiation scattered by a single area:

$$\mathrm{S}_s^{(1)} = \frac{ck^4}{2\pi}\int_0^{\pi/2}\int_0^{2\pi}\tilde{\mathrm{K}}_\zeta(\mathbf{k})\cos^2\theta_s\sin\theta_s d\varphi_s d\theta_s, \quad \mathbf{k} = k\frac{\mathbf{s}}{\mathrm{r}}. \tag{6.59}$$

Let us turn now to calculation of the second-order field (we are interested only in its mean value). We can show that its value on the averaged plane is:

$$\left\langle\mathbf{E}^{(2)}\right\rangle = -\left\langle\zeta\frac{\partial\mathbf{E}^{(1)}}{\partial z}\right\rangle = 2k\mathbf{g}_i\int\tilde{\mathrm{K}}_\zeta(\mathbf{w})\sqrt{k^2-\mathrm{w}^2}d^2\mathbf{w}. \tag{6.60}$$

This expression represents the amplitude of the plane wave describing, on average, the reflected field of the second order.

When calculating the interference term, we also must take into account that the real part of Poynting's vector is important, so:

$$\mathbf{S}_1^{(02)} = -\frac{ck}{2\pi}\mathbf{e}_z \int\limits_{w \le k} \tilde{K}_\zeta(\mathbf{w})\sqrt{k^2 - w^2}\, d^2\mathbf{w}. \tag{6.61}$$

The integration limit inside of a circle of radius $w = k$ corresponds to extraction of the real part of the integral in Equation (6.61).

Let us now turn to integration in the cylindrical coordinate system:

$$\mathbf{S}_1^{(02)} = -\frac{ck}{2\pi}\mathbf{e}_z \int_0^k \int_0^{2\pi} \tilde{K}_\zeta(\mathbf{w})\sqrt{k^2 - w^2}\, w\, dw\, d\varphi$$

$$= -\frac{ck^4}{2\pi}\mathbf{e}_z \int_0^{\pi/2} \int_0^{2\pi} \tilde{K}_\zeta(\mathbf{k}) \cos^2\theta \sin\theta\, d\theta\, d\varphi.$$

It is a simple matter to see that the power flow of the interference summand with the minus sign is equal to the power flow absolute value of scattered waves. So, we have established by a simple example that the energy of scattered waves is the interference component, which reduces the energy of coherently reflected waves by the corresponding value. To be more exact, we should point out that the equality of both power flows is correct with a precision on the order of $1/\sqrt{|\varepsilon|}$. We can obtain this result using the Shchukin–Leontovich boundary conditions for the waves of all orders.

Finally, let us suppose that the topic of interest is not the stochastic but the determined surface with the profile:

$$\zeta(x) = \zeta_0 \cos Kx. \tag{6.62}$$

Let us assume further that the plane wave is incident on this surface in the plane xOy under angle θ_i. The zero approximation field can be written as:

$$\mathbf{E}^0 = -2\mathbf{g}_i e^{ikx\sin\theta_i} \sin\left(kz\cos\theta_i\right).$$

The interface conditions for the field of the second order is:

$$\left[\mathbf{e}_z \times \mathbf{E}^{(2)}\right] = -ikK\zeta_0^2 \left[\mathbf{e}_x \times \mathbf{g}_i\right] e^{ikx\sin\theta_i} \sin 2Kx + \cdots.$$

In the right side of the boundary conditions, we have left only the first nonvanishing term in order to declare a particular circumstance, thus avoiding complete analysis of the scattering process. It is a simple matter to find that the second-order field on the surface will have sinusoidal spatial structure with wave numbers $k_1 = k\sin\theta_i + 2K$ and $k_2 = k\sin\theta_i - 2K$. It is obvious that radiation of the waves will occur under the angles determined from the equations:

$$\sin\theta_1 = \sin\theta_i + \frac{2K}{k}, \quad \sin\theta_2 = \sin\theta_i - \frac{2K}{k}. \tag{6.63}$$

In this case, the inequality $2K < k$ should be observed. Note that the given equalities are nothing other than the famous Bragg equation. Equation (6.63) expresses the conditions of the second-order diffraction maximum. Let us emphasize in this regard that the dependence of the spatial spectrum, Equation (6.45), on the specific vector $\mathbf{k}$ means that these spectral components scatter most intensively in direction $\mathbf{e}_s$, which satisfies the Bragg conditions for first-order diffraction. This spectral component has the wave number $K = 2k\sin\theta_i$ due to backward scattering.

6.4 WAVE SCATTERING BY LARGE INHOMOGENEITIES

Gravitational sea waves are higher, as a rule, than wavelengths of the microwaves region. We will consider sea waves to be surfaces with large inhomogeneities and will examine the small slopes of these surfaces. In doing so, we also assume that the curvature radius of the considered surfaces is much larger than the wavelength, which allows us to use Kirchhoff's approximation to solve the problem of diffraction by large inhomogeneities. It was pointed out in Chapter 1 that the facet model of the surface is not considered for this approximation, which allows us to reduce the scattering process at each point to a process of local reflection from the plane surface.

Let us consider a plane wave incident on a surface with large roughness to calculate the field of scattered waves. Based on Equations (1.82–1.84), the scattering amplitude can be expressed in the form:

$$\mathbf{f} = -\frac{ik}{4\pi}\int_{\Sigma} \mathbf{B}\left[\nabla\zeta(\mathbf{r}')\right] e^{ik(\mathbf{e}_i - \mathbf{e}_s)\cdot\mathbf{r}'} d^2\mathbf{r}', \tag{6.64}$$

and we have:

$$\mathbf{B} = \left(\mathbf{e}_s\cdot\mathbf{E}_r^0\right)\mathbf{n} - \left(\mathbf{n}\cdot\mathbf{e}_s\right)\mathbf{E}_r^0 - \left(\mathbf{e}_s\cdot\mathbf{H}_r^0\right)\left[\mathbf{e}_s\times\mathbf{n}\right] + \left(\mathbf{n}\cdot\mathbf{e}_s\right)\left[\mathbf{e}_s\times\mathbf{H}_r^0\right]. \tag{6.65}$$

The values of the electrical and magnetic fields are determined by Equations (3.21) and (3.22), where the vector of local normal to the surface, $\mathbf{n}$, replaces vector $\mathbf{e}_z$, and the incident field $\mathbf{E}_i^0, \mathbf{H}_i^0$ is replaced by vectors $\mathbf{g}_i$, $[\mathbf{e}_i \times \mathbf{g}_i]$. These last vectors indicate that we are dealing with a plane wave of single amplitude. The reflected field is a function of the local incident angle, whose variations from point to point of the surface are determined by the its slope change. The normal to the surface changes together with slope variations according to Equation (6.20). Thus, vector $\mathbf{B}$ is a function of gradient $\nabla\zeta$.

It is convenient now to turn our attention from integration over the stochastic surface to integration with respect to the mean plane. In this case, the surface elements are connected by the relations $d^2\mathbf{r}' = d^2\mathbf{s}'/(\mathbf{e}_z\mathbf{n}) = \sqrt{1+(\nabla\zeta)^2}\,d^2\mathbf{s}'$. Further, we should take into account that $\mathbf{r}' = \mathbf{s}' + \mathbf{e}_z\zeta(\mathbf{s}')$. Therefore,

$$\mathbf{f} = -\frac{ik}{4\pi}\int_{\Sigma}\mathbf{B}\left[\nabla\zeta(\mathbf{s}')\right]e^{ik(\mathbf{u}_i-\mathbf{u}_s)\cdot\mathbf{s}'-ik(a_i+a_s)\zeta(\mathbf{s}')}\frac{d^2\mathbf{s}'}{(\mathbf{n}\cdot\mathbf{e}_z)}. \qquad (6.66)$$

Here, $\mathbf{u}_i$ and $\mathbf{u}_s$ are projections of vectors $\mathbf{e}_i$ and $\mathbf{e}_z$ on the plane z = 0.

Let us now calculate the average value of the scattering amplitude which describes the coherent component of the scattered field. We should take into account the lack of correlation between the degree of roughness and the slopes measured at the same point. Due to this, vector $\mathbf{B}$ and exponent in the integral of Equation (6.66) are statistically independent. As a result:

$$\langle\mathbf{f}\rangle = -\frac{ik}{4\pi}\left\langle\frac{\mathbf{B}}{(\mathbf{n}\cdot\mathbf{e}_z)}\right\rangle\tilde{P}_\zeta\left[k(a_i+a_s)\right]\int_{\Sigma}e^{ik(\mathbf{u}_i-\mathbf{u}_s)\cdot\mathbf{s}'}d^2\mathbf{s}'.$$

Here,

$$\tilde{P}_\zeta(x) = \int_{-\infty}^{\infty}P_1(\zeta)\,e^{-ix\zeta}d\zeta \qquad (6.67)$$

is the characteristic function of the surface. As sizes of the area Σ are much larger than the wavelength, the integral on the right-hand side is the delta-function, and we have:

$$\langle\mathbf{f}\rangle = -i\pi k\left\langle\frac{\mathbf{B}}{(\mathbf{n}\cdot\mathbf{e}_z)}\right\rangle\tilde{P}_\zeta(2ka_i)\,\delta\left[k(\mathbf{u}_i-\mathbf{u}_s)\right].$$

It is also necessary to consider the smallness of the slope of the surface by averaging vector $\mathbf{B}/(\mathbf{n}\cdot\mathbf{e}_z)$; therefore, we may be restricted by the first term of the sum expanding this vector into a Taylor series. We now have:

$$\left\langle \mathbf{B}(\nabla\zeta)\sqrt{1+(\nabla\zeta)^2}\right\rangle = \mathbf{B}(0) + \frac{1}{2}\nabla_\perp^2\left[\mathbf{B}(0)\sqrt{1+(\nabla\zeta)^2}\right]\left\langle(\nabla\zeta)^2\right\rangle + \cdots,$$

assuming a lack of correlation between the slopes along the orthogonal coordinates. Being able to ignore the second term in this expansion follows from our previous estimation of slope values which means that $\mathbf{n} = -\mathbf{e}_z$, $\mathbf{e}_s = \mathbf{e}_r$ should be taken everywhere, regarding that the normal is the outer one here. The last equality is the result of multiplication of the entire expression by the delta-function, which gives us:

$$\langle \mathbf{f} \rangle = -2i\pi k \cos\theta_i \mathbf{E}_r^0 \tilde{P}_\zeta\left(2k\cos\theta_i\right)\delta\left[k\left(\mathbf{u}_i - \mathbf{u}_s\right)\right]. \tag{6.68}$$

Coherent scattering occurs in the direction of the plane wave reflected by the interface plane ($\mathbf{u}_s = \mathbf{u}_i$). Because the value of the reflected field, $\mathbf{E}_r^0$, is proportional to the coefficient of reflection at the interface of two media, coherent scattering takes place with the following effective coefficient of reflection:

$$F_{\mathrm{h,v}}^{(e)} = F_{\mathrm{h,v}}\tilde{P}_\zeta\left(2k\cos\theta_i\right). \tag{6.69}$$

Also, the process of coherent scattering is reduced to multiplication of the reflection coefficient by the characteristic function of roughness.

Let us consider an example of the Gauss law of height distribution:

$$P_1(\zeta) = \frac{1}{\sqrt{2\pi\langle\zeta^2\rangle}}\exp\left(-\frac{\zeta^2}{2\langle\zeta^2\rangle}\right). \tag{6.70}$$

The characteristic function for it has the view:

$$\tilde{P}_\zeta(x) = \exp\left(-\frac{x^2\langle\zeta^2\rangle}{2}\right). \tag{6.71}$$

Thus, in this case, the reflective coefficient is multiplied by the factor:

$$\tilde{P}_1\left(2k\cos\theta_i\right) = e^{-2k^2\langle\zeta^2\rangle\cos^2\theta_i}. \tag{6.72}$$

It is obvious that the characteristic function is small if $2k^2\langle\zeta^2\rangle\cos^2\theta_i >> 1$, which indicates lack of coherence in the field of scattering. The coherence take place even by large height of roughness only in the case of very small angles of elevation when:

$$\cos\theta_i < 1\Big/ k\sqrt{2\langle\zeta^2\rangle},$$

which allows us to state that the coherent component is practically absent in the scattering of microwaves by natural surfaces.

Thus, the incoherent component of scattering is the primary one, the energy characteristics of which will be the focus of our further calculations. First, we want to find the squared module of the scattering amplitude. This value is equal to:

$$|\mathrm{f}|^2 = \frac{k^2}{16\pi^2}\int\limits_{\Sigma}\int\limits_{\Sigma}\left(\mathbf{B}'\cdot\mathbf{B}''\right)e^{ik\{(\mathbf{u}_i-\mathbf{u}_s)\cdot(\mathbf{s}'-\mathbf{s}'')-(a_i+a_s)[\zeta(\mathbf{s}')-\zeta(\mathbf{s}'')]\}}\frac{d^2\mathbf{s}'d^2\mathbf{s}''}{\left(\mathbf{n}'\cdot\mathbf{e}_z\right)\left(\mathbf{n}''\cdot\mathbf{e}_z\right)}. \tag{6.73}$$

Here, for short, $\mathbf{B}' = \mathbf{B}\left[\nabla\zeta(\mathbf{s}')\right]$ and $\mathbf{B}'' = \mathbf{B}^*\left[\nabla\zeta(\mathbf{s}'')\right]$. This indication also concerns the local normal vector. The exponent under the integrals is a fast oscillating function because the degree of roughness is assumed to be large; therefore, the main area of integration is concentrated near the point $\mathbf{s}' = \mathbf{s}''$ and we can use the expansion $\zeta(\mathbf{s}'') \cong \zeta(\mathbf{s}') + \nabla\zeta(\mathbf{s}')(\mathbf{s}'' - \mathbf{s}')$. Further, we will integrate with respect to $\mathbf{s}' - \mathbf{s}''$ and over $\mathbf{s}'$. The first integration gives us the delta-function, so:

$$|\mathrm{f}|^2 = \frac{k^2}{4}\int\limits_{\Sigma}|\mathrm{B}'|^2\,\delta\left\{k\left[\mathbf{u}_i - \mathbf{u}_s - \left(a_i + a_s\right)\nabla\zeta\left(\mathbf{s}'\right)\right]\right\}\frac{d^2\mathbf{s}'}{\left(\mathbf{n}'\cdot\mathbf{e}_z\right)^2}. \tag{6.74}$$

We will consider the cross section of the scattering to be the average value of the scattering amplitude squared module. To average Equation (6.73) it is necessary to multiply the scattered amplitude squared module by the distribution function $P_1(\nabla\zeta)$ of the slopes and to integrate over the entire region of the $\nabla\zeta$ change. Due to the delta-function properties, we can obtain the value of the distribution function by the argument:

$$\nabla\zeta^\circ = \frac{\mathbf{u}_i - \mathbf{u}_s}{a_i + a_s}. \tag{6.75}$$

The points where the gradient has this value correspond to the surface areas where specular reflection takes place (so-called specular or bright points). The scattering vector $\mathbf{e}_s$ coincides with the local vector of reflection at these points. Therefore, $\mathbf{B} = -2(\mathbf{n}\cdot\mathbf{e}_s)\mathbf{E}_r^0(\theta_i^\circ)$, where θ_i° is understood to be the local angle of incidence corresponding to Equation (6.74). As a result,

$$\sigma = \frac{\left(\mathbf{n}^\circ\cdot\mathbf{e}_s\right)^2}{\left(\mathbf{n}^\circ\cdot\mathbf{e}_z\right)^2\left(\cos\theta_i + \cos\theta_s\right)^2}\left|\mathrm{E}_r^0\right|^2 P_1\left(\nabla\zeta^\circ\right)\Sigma. \tag{6.76}$$

The final formula for the specific cross section is written as:

$$\sigma^0 = \frac{\left(1+\cos\nu\right)^2}{\sin^2\theta_i^\circ\left(\cos\theta_i+\cos\theta_s\right)^4}\times$$
$$\times\left[F_v^2\left(\theta_i^\circ\right)\left(\mathbf{n}^\circ\cdot\mathbf{g}_i\right)^2+F_h^2\left(\theta_i^\circ\right)\left(\mathbf{n}^\circ\cdot\left[\mathbf{e}_i\times\mathbf{g}_i\right]\right)^2\right]P_1\left(\nabla\zeta^\circ\right). \quad (6.77)$$

We used Equation (3.20) to derive this formula. Here, ν is the angle between the directions of the incident and scattered waves, and, correspondingly:

$$\cos\nu=\cos\theta_i\cos\theta_s-\sin\theta_i\sin\theta_s\cos(\varphi_i-\varphi_s).$$

This formula can be rewritten in a rather different way. Let us represent the polarization vector in the local coordinate system as the expansion:

$$\mathbf{g}_i=\cos\eta\,\mathbf{e}_h^{(i)}+\sin\eta\,\mathbf{e}_v^{(i)}, \quad (6.78)$$

as we have done in Section 3.1. Then,

$$\sigma^0=\frac{\left(1+\cos\nu\right)^2}{\left(\cos\theta_i+\cos\theta_s\right)^4}\left[\cos^2\eta F_v^2\left(\theta_i^\circ\right)+\sin^2\eta F_h^2\left(\theta_i^\circ\right)\right]P_1\left(\nabla\zeta^\circ\right). \quad (6.79)$$

In the case of backward scattering, the local angle of incidence is equal to zero, and we have:

$$\sigma^0\left(-\mathbf{e}_i\right)=\frac{F^2\left(0\right)}{4\cos^4\theta_i}P_1\left(\frac{\mathbf{u}_i}{\cos\theta_i}\right). \quad (6.80)$$

We do not consider, in this specific case, the reflection coefficients for horizontal and vertical polarization. At a height distribution that is Gaussian, the slope density distribution also follows the Gaussian law:

$$\sigma^0\left(-\mathbf{e}_i\right)=\frac{F^2\left(0\right)}{4\pi\left\langle\left(\nabla\zeta\right)^2\right\rangle\cos^4\theta_i}\exp\left[-\frac{\tan^2\theta_i}{\left\langle\left(\nabla\zeta\right)^2\right\rangle}\right]. \quad (6.81)$$

Applying Equation (6.11), we obtain:

$$\sigma^0\left(-\mathbf{e}_i\right)=\frac{l^2F^2\left(0\right)}{8\pi^4\left\langle\zeta^2\right\rangle\cos^4\theta_i}\exp\left(-\frac{l^2}{2\pi^3\left\langle\zeta^2\right\rangle}\tan^2\theta_i\right). \tag{6.82}$$

It is apparent that the cross section of backward scattering due to a large degree of roughness depends rather strongly on the angle of incidence with a normal height distribution.

Let us note that we did not take into consideration the shadow effect on some areas of the surface by others. If this effect exists, not all of the areas are of importance in the process of scattering, and the established formulae must be adjusted accordingly. We do not investigate this problem here, instead referring readers to Ishimaru,[49] who has performed such a special study.

Now, we shall regard the case of wave scattering with regard to vertical incidence of the plane wave. For this purpose, let us use Equation (6.78). In this case, $\nu=\theta_s$ and $\theta_i^\circ=\theta_s/2$. In order to simplify the problem, we will restrict ourselves to assuming statistical isotropy of the roughness, and then the slope probability will depend only on $|\nabla\zeta^\circ|=\tan(\theta_s/2)$. Hence,

$$\sigma^0=\frac{P_1\left[\tan\left(\theta_s/2\right)\right]}{\left(1+\cos\theta_s\right)^2}\left[\cos^2\eta F_v^2\left(\frac{\theta_s}{2}\right)+\sin^2\eta F_h^2\left(\frac{\theta_s}{2}\right)\right]. \tag{6.83}$$

Our aim is to calculate the power flow scattered by the surface of the single square. As is already known, this flow is defined by the integral with respect to the solid angle $d\Omega=\sin\theta_s d\varphi_s d\theta_s$. We must take into account, when integrating over the scattering angle θ_s, the small probability of the scattering covering the large angles because the slopes are small. This gives us the opportunity to factor out the integration sign, as the reflecting coefficients at the incident angle value are equal to zero. So, the scattered power flow is equal to:

$$S_s=\frac{c}{4}F^2(0)\int_0^{\pi/2}P_1\left[\tan\left(\theta_s/2\right)\right]\frac{\sin\theta_s d\theta_s}{\left(1+\cos\theta_s\right)^2}=\frac{c}{4}F^2(0)\int_0^1 P_1(x)x dx.$$

The probability that the slope value is greater than unity is extremely low due to the assumptions we have made; therefore, the upper limit in the last integral can be expanded to infinity, which gives us:

$$S_s=\frac{c}{8\pi}F^2(0). \tag{6.84}$$

The value on the right, however, is none other than the power flow density of the reflected wave in the absence of roughness. Thus, the coherently reflected wave practically disappears in the presence of considerable roughness and its energy is pumped over the scattered wave energy.

In conclusion, let us turn to an analysis of wave scattering by fractal surfaces. In the expression under the integral of Equation (6.73), the values $\mathbf{B}/(\mathbf{n}\cdot\mathbf{e}_z)$ depend on slopes but the exponent depends on the roughness spectrum. The roughness spectrum is mainly concentrated in the interval of large scales of inhomogeneities in contrast to the slope spectrum, which gravitates toward the small-scale area (refer to Equation (6.10)). Therefore, the roughness and slopes are found to be weakly correlated for any separation of the correlation points. We have reason, then, to assume a complete absence of the mentioned correlation between roughness and slopes. Moreover, due to the tendency toward the small-scale part of the spectrum, the correlation radius of the slopes is on the order of the inner scale, while the large scales dominates the mechanism of scattering. It is important to remember that Kirchhoff's approximation used here in our research is applicable only in the case of wave scattering by inhomogeneities that are large compared to the wavelength.

Assuming isotropy and Gaussian statistics of roughness, we will now introduce in Equation (6.73) differential coordinates of integration and coordinates of the gravity center, as we have repeatedly done before. Then,

$$|\mathbf{f}|^2 = \frac{k^2}{8\pi}\left[\left\langle\frac{\mathbf{B}(\mathbf{e}_s)}{\mathbf{n}\cdot\mathbf{e}_z}\right\rangle\right]^2 \Sigma\int_0^\infty J_0\left(k|\mathbf{u}_i-\mathbf{u}_s|\,s\right)\exp\left[-k^2(a_i+a_s)^2 D_\zeta(s)\right]s\,ds. \qquad (6.85)$$

We particularly emphasize the dependence of vector $\mathbf{B}$ on the direction of scattering. Because the smallness of the wavelength relative to the surface outer scale is assumed, the wave number k appears as a large parameter (the geometrical optics approximation); therefore, the main area of integration in Equation (6.85) is concentrated close to zero. This gives the opportunity to use the approximation of Equation (6.17) for the structure function to obtain:

$$|\mathbf{f}|^2 = \frac{k^2}{8\pi}\left[\left\langle\frac{\mathbf{B}(\mathbf{e}_s)}{\mathbf{n}\cdot\mathbf{e}_z}\right\rangle\right]^2 \Sigma\int_0^\infty J_0\left(k|\mathbf{u}_i-\mathbf{u}_s|s\right)\exp\left[-k^2(a_i+a_s)^2 E s^\alpha\right]s\,ds. \qquad (6.86)$$

With regard to backscattering,

$$\sigma^0 = \frac{k^2}{8\pi}\left[\left\langle\frac{\mathbf{B}(-\mathbf{e}_i)}{\mathbf{n}\cdot\mathbf{e}_z}\right\rangle\right]^2 \int_0^\infty J_0\left(2ks\sin\theta_i\right)\exp\left[-4k^2Es^\alpha\cos^2\theta_i\right]s\,ds. \qquad (6.87)$$

In the marginal case when $\alpha = 2$ (the differentiable surface), we have Equation (6.81), if we assume:

$$E = \frac{\pi^3 \langle \zeta^2 \rangle}{2l^2}, \quad \left[\left\langle \frac{\mathbf{B}(-\mathbf{e}_i)}{\mathbf{n} \cdot \mathbf{e}_z} \right\rangle \right]^2 = \frac{4F^2(0)}{\cos^2 \theta_i}. \tag{6.88}$$

Equation (6.88) is correct in all cases independent of the fractal dimension value. Thus, we must always consider wave scattering due to the large-scale part of the roughness spatial spectrum (i.e., wave numbers that satisfy the condition $q < k$). The concept of specular points is correct in this case, and we have to assume that $\mathbf{n} = \mathbf{e}_i = -\mathbf{e}_s$ in the expression for $\mathbf{B}(-\mathbf{e}_i)/(\mathbf{n} \cdot \mathbf{e}_z)$. So, we obtain:

$$\sigma_0(-\mathbf{e}_i) = \frac{k^2 F^2(0)}{2\pi \cos^2 \theta_i} \int_0^\infty J_0(2ks \sin \theta_i) \exp\left(-4k^2 E s^\alpha \cos^2 \theta_i\right) s ds. \tag{6.89}$$

In the specific case of Brownian fractals ($\alpha = 1$), the integral is tabulated,[44] and we obtain Hagfor's formula:[81]

$$\sigma_0(-\mathbf{e}_i) = \frac{F^2(0)}{16\pi G\left(\cos^4 \theta_i + G \sin^2 \theta_i\right)^{3/2}}, \quad G = \frac{1}{4k^2 E^2}, \tag{6.90}$$

widely used in the analysis of echo signals obtained by radar for the investigation of the planet.

6.5 TWO-SCALE MODEL

The asymptotic approaches described previously do not fully represent the radiowave scattering processes by natural surfaces. Natural rough surfaces have a wide variety of spectral components, some having scales comparable to the wavelength. This rather narrow scale region is difficult to analyze during radiowave scattering research, and we must resort to numerical computation, which allows us to combine both models (large and small roughness) into one. It is a definite advance in the theoretical analysis of wave scattering by roughness. Combining these models can be done rather easily due to the independence of scattering processes by small and large inhomogeneities. In this case, small roughness is located on the surface along with large roughness. As the latter surfaces have dimensions greater than the wavelength, they may be introduced along with the small roughness. In other words, scattering occurs due to locally plane surfaces with small roughness. If the surface Σ by its size is such that it consists of a lot of large-scale roughness, then the slope changes from point to point, causing local variation in the cross section of small inhomogeneities.

Let us designate the specific cross section of large roughness as σ^0_{mac} and the corresponding cross section of scattering by small roughness as σ^0_{mic}. Then, the combined cross section of scattering is equal to:

$$\sigma^0 = \sigma^0_{\text{mac}} + \frac{1}{\Sigma_0}\int_{\Sigma} \sigma^0_{\text{mic}}\left[\mathbf{n}(\mathbf{r})\right] d^2\mathbf{r}. \tag{6.91}$$

The cross section of scattering by small roughness can be written as a function of the local normal because of the change of slope, revealing its dependence on coordinates on the scattering surface. It is necessary, in this case, to distinguish between the surface Σ_0 of the mean plane and the large-scale rough area under consideration which leans against it. Let us mention that the given formula is true not only for a random surface but also for another one (not random).

The second term in Equation (6.91) is a stochastic value for casual surfaces; therefore, we must perform the second averaging relative to the slope and write:

$$\sigma^0 = \sigma^0_{\text{mac}} + \int \sqrt{1+\mathbf{v}^2}\,\sigma^0_{\text{mic}}\left[\mathbf{n}(\mathbf{v})\right] P_1(\mathbf{v})\, d^2\mathbf{v}. \tag{6.92}$$

Here, we have introduced the designation $\nabla\zeta = \mathbf{v}$. Let us consider in detail the case of backward scattering by the surface with ideal conductivity ($|\varepsilon| \to \infty$). In this case,

$$\begin{aligned}\sigma^0 = {} & \frac{1}{4\cos^4\theta_i} P_1\left(\frac{\mathbf{u}_i}{\cos\theta_i}\right) + \\ & + 4k^4 \int \sqrt{1+\mathrm{v}^2}\left[\left(\mathbf{n}\cdot\mathbf{e}_i\right)^4 + 4\left(\mathbf{n}\cdot\mathbf{g}_i\right)^2\right] \tilde{\mathrm{K}}_\zeta\left(2k\left|\left[\mathbf{n}\times\mathbf{e}_i\right]\right|\right) P_1(\mathbf{v})\, d^2\mathbf{v}.\end{aligned} \tag{6.93}$$

To analyze the second term we can designate it as $\left\langle \sigma^0_{\text{mic}} \right\rangle$; specifically, we will assume a Gaussian distribution of altitudes of large-scale roughness and, correspondingly, a Gaussian distribution of slopes. Thus, we will neglect the quadratic values in Equation (6.93), as doing so will not cause any misunderstanding because of the low probability of large slopes. We will use the model correlation function:

$$\mathrm{K}_\zeta(\mathbf{s}) = \left\langle \zeta^2 \right\rangle \exp\left(-\pi\frac{\mathrm{s}^2}{l^2}\right) \tag{6.94}$$

for small roughness which supposes statistical isotropy of the surface. Equation (6.94) is rather often used for tentative computations. The corresponding spatial spectrum is presented in the form:

$$\tilde{K}_\zeta(\mathbf{w}) = \frac{\langle \zeta^2 \rangle l^2}{4\pi^2} \exp\left(-\frac{w^2 l^2}{4\pi}\right). \tag{6.95}$$

To distinguish the parameters of large and small roughness, we will add the subscripts mac and mic, respectively. The roughness, as already noted above, will be assumed to be statistically isotropic.

Bearing in mind these assumptions, we can write $(\mathbf{n} \cdot \mathbf{e}_i) \cong \cos\theta_i + v\sin\theta_i\cos\gamma$ and $\left|\left[\mathbf{n} \times \mathbf{e}_i\right]\right|^2 \cong \sin^2\theta_i - v\sin 2\theta_i \cos\gamma$. Here, γ is the angle between the vectors $\mathbf{u}_i$ and $\mathbf{v}$. The second term of Equation (6.93) now has the form:

$$\left\langle \sigma^0_{\text{mic}} \right\rangle = \frac{\sigma^0_{\text{mic}}\Big|_{\mathbf{v}=0}}{2\pi\langle v^2 \rangle} \int \left[1 + K(v,\gamma)\right] \exp\left(-\frac{v^2}{2\langle v^2 \rangle} + \frac{k^2 l^2_{\text{mic}}}{\pi} v \sin 2\theta_i \cos\gamma\right) d^2\mathbf{v},$$

where

$$\sigma^0_{\text{mic}}\Big|_{\mathbf{u}=0} = \frac{k^4 \langle \zeta^2_{\text{mic}} \rangle l^2_{\text{mic}}}{\pi^2} \left[\cos^4\theta_i + 4(\mathbf{e}_z \cdot \mathbf{g}_i)^2\right] \exp\left(-\frac{k^2 l^2_{\text{mic}}}{\pi} \sin^2\theta_i\right) \tag{6.96}$$

is the scattering cross section of small roughness in the absence of large roughness, and we have:

$$K(v,\gamma) = \frac{4\left[v\sin\theta_i \cos^3\theta_i - 2(\mathbf{e}_z \cdot \mathbf{g}_i)(\mathbf{g}_i \cdot \mathbf{v})\right]}{\cos^4\theta_i + 4(\mathbf{e}_z \cdot \mathbf{g}_i)^2}. \tag{6.97}$$

Let us point out that for horizontal and vertical polarization, respectively:

$$K(v,\gamma) = 4v\tan\theta_i\cos\gamma, \quad K(v,\gamma) = -\frac{2v\sin 2\theta_i \cos\gamma}{1 + \sin^2\theta_i}. \tag{6.98}$$

The integration gives:

$$\frac{\left\langle \sigma^0_{\text{mic}} \right\rangle}{\sigma^0_{mic}\Big|_{\mathbf{u}=0}} = \left(1 + \frac{2\pi^2 k^2 \langle \zeta^2_{\text{mac}} \rangle l^2_{\text{mic}} \sin 2\theta_i}{l^2_{mac}} \Lambda_{h,v}\right) \exp\left(\frac{\pi k^4 \langle \zeta^2_{\text{mac}} \rangle l^4_{\text{mic}}}{l^2_{\text{mac}}} \sin^2 2\theta_i\right). \tag{6.99}$$

Here, $\Lambda_h = 4\tan\theta_i$ and $\Lambda_v = -2\sin 2\theta_i / \left(1+\sin^2\theta_i\right)$ in accordance with the polarization (horizontal or vertical). In the result, the expression of slope dispersion via dispersion of elevations and the correlation radius was as defined in the formula (6.11).

It is easy to show that the influence of large surface inhomogeneities affects wave scattering by small roughness at the condition $\pi k^4 \left\langle \zeta_{\mathrm{mac}}^2 \right\rangle l_{\mathrm{mic}}^4 \geq l_{\mathrm{mac}}^2$. From this, in accordance with the accepted approximations $2\pi^3 \left\langle \zeta_{\mathrm{mac}}^2 \right\rangle << l_{\mathrm{mac}}^2$, we have the condition $k^4 l_{\mathrm{mic}}^4 >> 2\pi^2$. So, large inhomogeneities have an effect only in the case when the scattering is large scale (meaning that their scale is greater than the wavelength). The inhomogeneities discussed here have a scattering indicatrix with an angle spread on the order of $\lambda / l_{\mathrm{mic}}$.

It is easy to conclude from this that the discussed effect reveals itself at the condition:

$$\lambda / l_{\mathrm{mic}} < \sqrt{\left\langle (\nabla\zeta)^2 \right\rangle} \approx \sqrt{\left\langle \zeta^2 \right\rangle} / l_{\mathrm{mac}}.$$

Let us now compare the backscattering cross sections of large and small roughness. First, however, we can improve the formula for the small-roughness spatial spectrum. The correlation function and the spectrum may be written as:

$$\begin{aligned}
\tilde{\mathrm{K}}_\zeta(\mathbf{w}) &= \left\langle \zeta_{\mathrm{mic}}^2 \right\rangle l_{\mathrm{mic}}^2 \tilde{k}_\zeta\left(l_{\mathrm{mic}}\mathbf{w}\right), \\
\tilde{k}_\zeta(\mathbf{v}) &= \frac{1}{4\pi}\int k_\zeta(\mathbf{q}) e^{i\mathbf{q}\cdot\mathbf{v}} d^2 q, \qquad (6.100) \\
\tilde{\mathrm{K}}_\zeta(\mathbf{s}) &= \left\langle \zeta_{\mathrm{mic}}^2 \right\rangle k_\zeta\left(\mathbf{s} / l_{\mathrm{mic}}\right).
\end{aligned}$$

The advantage of these formulae is their nondimensionality. Let us point out also that $k_\zeta(0) = 1$ and $\tilde{k}_\zeta(0) = 1/2\pi$.

Note that at the nadir radiation $(\theta_i = 0)$, and

$$\Theta = \frac{\sigma_{\mathrm{mic}}^0}{\sigma_{\mathrm{mac}}^0} = 16k^4 \left\langle \zeta_{\mathrm{mic}}^2 \right\rangle l_{\mathrm{mic}}^2 \left\langle (\nabla\zeta)^2 \right\rangle << 1.$$

So, at small zenith angles the intensity of wave scattering from large inhomogeneities exceeds that for small roughness, which seems rather natural; however, due to the smallness of the slope angle dispersion, the angle dependence of the cross section is a rather pointed function, and the intensity of the quasi-specular scattering decreases fast with increase in the zenith angle. The angle dependence from small roughness is not so sharp; therefore, beginning with any zenith angle value, the intensity of this scattering mechanism becomes dominant. Let us estimate the zenith angle value at which the change of the scattering mechanism occurs. The corresponding equation can be written as:

$$\tan^2\theta_i = -2\left\langle (\nabla\zeta)^2 \right\rangle \left\{ \ln\Theta + \ln\left[2\pi\Pi(\theta_i)\,\tilde{k}_\zeta\left(2kl_{\text{mic}}\sin\theta_i\right)\right]\right\}. \quad (6.101)$$

Here, $\Pi(\theta_i) = \cos^8\theta_i$ for horizontal polarization and $\cos^4\theta_i\left(1+\sin^2\theta_i\right)^2$ for vertical polarization. The second summand to the right is a slowly changed function of angle, so, in the first approximation, it can be assumed to have the value $\theta_i = 0$, and

$$\tan^2\theta_i = -2\left\langle (\nabla\zeta)^2 \right\rangle \ln\Theta. \quad (6.102)$$

6.6 IMPULSE DISTORTION FOR WAVE SCATTERING BY ROUGH SURFACES

So far we have investigated the problem of harmonic wave scattering. It is obvious that the scattered field will be an aggregate of sine waves at the sine wave incident on the rough surface. If we turn to the problem of complicated form wave scattering, we will call them impulses for short so the picture will change little. Every spectrum component of the impulse is being scattered at its own individual amplitude and phase. So, the sum of the scattered components of the spectrum will form an impulse of such a shape that, generally speaking, differs from the initial impulse incident on the surface. The situation here is similar to the one which we encountered during consideration of the phenomena of frequency dispersion in Chapter 2.

The spectrum of the scattered impulse is equal to:

$$\tilde{\mathbf{E}}_s(\omega) = \tilde{\mathrm{E}}_i(\omega)\,\mathbf{f}(\omega)\frac{e^{ikr}}{r}, \quad k = \frac{\omega}{c}. \quad (6.103)$$

Here, $\tilde{\mathrm{E}}_i(\omega)$ is the complex amplitude of the corresponding spectral component of the incident impulse. In the scattering amplitude **f**, only its frequency dependence is assigned, which is the main dependence in this case. It is easy to see that the scattering amplitude plays the role of a frequency filter. The filtering approaches of the scattering amplitude are the cause of impulse deformation. The form of the scattered impulse is set by the equality:

$$\mathbf{E}_s(t,\mathbf{r}) = \frac{1}{\mathrm{r}}\int_{-\infty}^{\infty} \tilde{\mathrm{E}}_i(\omega)\,\mathbf{f}(\omega)\,e^{ikr-i\omega t}\,d\omega. \quad (6.104)$$

The scattering amplitude is a random function and follows the random shapes of the scattered impulse. We shall regard the average value of the scattering amplitude to be equal to zero. As we have determined, this assumption is true in the first approximation of the perturbation theory for wave scattering by small-scale roughness and is also correct for large-scale roughness on a scale greater than the wavelength. By virtue of this, the average value of vector $\mathbf{E}_s(t,\mathbf{r})$ is equal to zero, and the primary object of interest is the mean power:

$$\left\langle \left|\mathrm{E}_s(t,r)\right|^2\right\rangle = \frac{1}{r^2}\iint \tilde{\mathrm{E}}_i'\,\tilde{\mathrm{E}}_i''^{*}\left\langle \mathbf{f}'\mathbf{f}''^{*}\right\rangle e^{-i(\omega'-\omega'')(t-r/c)}d\omega'\,d\omega'', \qquad (6.105)$$

where the stroked functions depend on ω' and ω''.

Now it is necessary to calculate the frequency correlation function of the scattering amplitude. We will focus on the case of large roughness, as it is the most interesting one. It follows from Equation (6.66) that:

$$\mathbf{f}(\omega')\cdot\mathbf{f}^{*}(\omega'') = \frac{k^2}{16\pi^2}\int_{\Sigma}\int_{\Sigma}\left(\mathbf{B}'\cdot\mathbf{B}''^{*}\right) e^{i\Psi-i\Phi}\frac{d^2\mathbf{s}'d^2\mathbf{s}''}{\left(\mathbf{n}'\cdot\mathbf{e}_z\right)\left(\mathbf{n}''\cdot\mathbf{e}_z\right)}.$$

Here,

$$\Psi = \left[k\left(\mathbf{s}'-\mathbf{s}''\right)+\frac{K}{2}\left(\mathbf{s}'+\mathbf{s}''\right)\right]\cdot\left(\mathbf{u}_i-\mathbf{u}_s\right),$$

$$\Phi = \left\{k\left[\zeta(\mathbf{s}')-\zeta(\mathbf{s}'')\right]+\frac{K}{2}\left[\zeta(\mathbf{s}')+\zeta(\mathbf{s}'')\right]\right\}\left(a_i+a_s\right),$$

where $2k = k'+k''$ and $K = k'-k''$. Let us change the variables $2\mathbf{S} = \mathbf{s}'+\mathbf{s}''$ and $\mathbf{s} = \mathbf{s}'-\mathbf{s}''$ and perform expansion of the function ζ as we have already done when we derived Equation (6.73). Then,

$$\Psi = k\mathbf{s}+K\mathbf{S}, \quad \Phi = \left[k\mathbf{s}\nabla\zeta(\mathbf{S})+K\zeta(\mathbf{S})\right]\left(a_i+a_s\right).$$

Repeating the method of our previous calculations, we obtain:

$$\left\langle \mathbf{f}(\omega')\mathbf{f}^{*}(\omega'')\right\rangle = \sigma^0\tilde{P}_\zeta\left[K\left(a_i+a_s\right)\right]\int_{\Sigma} e^{iK(\mathbf{u}_i-\mathbf{u}_s)\cdot\mathbf{S}}d^2\mathbf{S}. \qquad (6.106)$$

After we change the variables $2\omega = \omega'+\omega''$ and $\Omega = \omega'-\omega''$, Equation (6.105) becomes:

$$\left\langle \left|\mathrm{E}_s\right|^2 \right\rangle = \frac{\sigma^0}{\mathrm{r}^2} \int\limits_{\Sigma} d^2\mathbf{S} \int d\omega \left(\mathbf{E}_i\left(\omega + \Omega/2\right) \cdot \mathbf{E}_i^*\left(\omega - \Omega/2\right) \right) \times$$

$$\times \int d\Omega \tilde{P}_\zeta \left[\Omega/c \left(a_i + a_s\right) \right] \exp\left[i\Omega/c \left(\left(\mathbf{u}_\mathrm{i} - \mathbf{u}_\mathrm{s}\right) \cdot \mathbf{S} \right) - i\Omega\left(t - r/c\right) \right].$$

It follows from the integral Fourier properties (convolution theorem) that:

$$\int \left(\tilde{\mathbf{E}}_i\left(\omega + \Omega/2\right) \cdot \tilde{\mathbf{E}}_i^*\left(\omega - \Omega/2\right) \right) d\omega = \frac{1}{2\pi} \int \left| \mathbf{E}_i\left(t',0\right) \right|^2 e^{i\Omega t'} dt'. \tag{6.107}$$

This allows us to obtain definitively:

$$\left\langle \left| \mathrm{E}_s\left(\tau\right) \right|^2 \right\rangle = \frac{c\sigma^0}{\left(a_i + a_s\right) r^2} \int\limits_{\Sigma} d^2\mathbf{S} \int dt' \left| \mathrm{E}_i\left(t',0\right) \right|^2 P_1 \left[\frac{c\left(t' - \tau\right) + \left(\mathbf{u}_i - \mathbf{u}_s\right) \cdot \mathbf{S}}{a_i + a_s} \right]. \tag{6.108}$$

Here, as before, $P_1(\zeta)$ is a function of the probability density of the surface elevation distribution. The simplest case corresponds to the direction of the specular reflection. In this case, $\mathbf{u}_s = \mathbf{u}_i$ and we have:

$$\left\langle \left| \mathrm{E}_s\left(\tau\right) \right|^2 \right\rangle = \frac{c\sigma^0 \Sigma}{2r^2 \cos\theta_i} \int \left| \mathrm{E}_\mathrm{i}\left(t',0\right) \right|^2 P_1 \left[\frac{c\left(t' - \tau\right)}{2\cos\theta_i} \right] dt'. \tag{6.109}$$

The law of conservation of the density energy flow is easily established from Equation (6.108):

$$\int \left\langle \left| \mathrm{E}_s\left(\tau\right) \right|^2 \right\rangle d\tau = \frac{\sigma^0 \Sigma}{r^2} \int \left| \mathrm{E}_\mathrm{i}\left(t',0\right) \right|^2 dt'. \tag{6.110}$$

Let us consider now the problem of impulse distortion, assuming the following in view of the Gaussian pulse:

$$\left| \mathrm{E}_i\left(t,0\right) \right|^2 = E_0^2 \exp\left(-\frac{2t^2}{\tau_0^2} \right) \tag{6.111}$$

as well as assuming a Gaussian function of the elevation distribution (Equation (6.70)). Further, we shall restrict ourselves to the case of backward scattering and a rectangular area with sizes a along the x-axis and b along the y-axis. We shall assume that xOy plane is the plane of incidence. Integration over t' gives us:

$$\left\langle \left| \mathrm{E}_s(\tau) \right|^2 \right\rangle = \frac{E_0^2 \sigma^0 b \tau_0}{r^2 \tilde{\tau}} \int_{-a/2}^{a/2} \exp\left[-\frac{2}{\tilde{\tau}^2} \left(\tau - \frac{2x}{c} \sin\theta_i \right) \right] dx, \tag{6.112}$$

where

$$\tilde{\tau}^2 = \tau_0^2 + 16\cos^2\theta_i \frac{\left\langle \zeta^2 \right\rangle}{c^2}. \tag{6.113}$$

Thus, the effect of large inhomogeneities occurs in such a away that the impulse expands and the duration given by Equation (6.113) should be taken into account instead of the initial pulse duration. The physical sense of impulse expansion is clear.

The pulses reflected from the peaks and hollows of rough surfaces come to the receiving point with a mutual delay on the order of 2 cos θ_i $\sqrt{\left\langle \zeta^2 \right\rangle}/c$, which changes the duration of the summary impulse. The effect is not revealed at the condition when $c\tau_0 >> \sqrt{\left\langle \zeta^2 \right\rangle}$. However, for impulses that are long in duration, another phenomenon arises. The problem here concerns pulse edge distortion due to sufficient steepness. Let us regard the case of backward scattering of the pulse with steep envelope by vertical irradiation of the surface $(\theta_i = 0)$. The envelope is represented by the expression:

$$\left| \mathrm{E}_i(t,0) \right|^2 = \begin{cases} 0, & -\infty < t < 0, \\ E_0^2, & 0 \le t < \infty. \end{cases} \tag{6.114}$$

This equation can be represented as:

$$\left\langle \left| \mathrm{E}_s(\tau) \right|^2 \right\rangle = \frac{cE_0^2 \sigma^0 \Sigma}{2r^2} \int_0^\infty P_1\left[\frac{c(t' - \tau)}{2} \right] dt' = \frac{E_0^2 \sigma^0 \Sigma}{r^2} \int_{-c\tau/2}^{\infty} P_1(\zeta) d\zeta. \tag{6.115}$$

If the function of elevation distribution is assumed, as before, to be Gaussian, then:

$$\left\langle \left| \mathrm{E}_s(\tau) \right|^2 \right\rangle = \frac{E_0^2 \sigma^0 \Sigma}{2\mathrm{r}^2} \left[1 + \Phi(q) \right], \quad q = \frac{c\tau}{2\sqrt{2\left\langle \zeta^2 \right\rangle}}. \tag{6.116}$$

The probability integral is now introduced:[45]

$$\Phi(x) = \frac{2}{\sqrt{\pi}} \int_0^x e^{-t^2} dt . \tag{6.117}$$

At $\tau < 0$, before the arrival of the wave reflected from the level of the averaged plane (median wave), we should remember that $\Phi(-|q|) = -\Phi(|q|)$. The probability integral has an asymptotic expansion:

$$\Phi(q) \cong 1 - \frac{1}{\sqrt{\pi} q} e^{-q^2} + \cdots . \tag{6.118}$$

Therefore, long before the median wave arrives,

$$\left\langle \left| \mathrm{E}_s (\tau < 0) \right|^2 \right\rangle \cong \frac{E_0^2 \sigma^0 \Sigma}{2\sqrt{\pi} |q| r^2} e^{-q^2} , \tag{6.119}$$

that is, the power of the scattered wave has a rather small value. At the moment of the median wave arrival ($\tau = 0$),

$$\left\langle \left| \mathrm{E}_s (0) \right|^2 \right\rangle = \frac{E_0^2 \sigma^0 \Sigma}{2r^2} . \tag{6.120}$$

And, eventually, some long time later from this moment, the power of the scattered impulse increases to its maximum value:

$$\left\langle \left| \mathrm{E}_s (\tau > 0) \right|^2 \right\rangle = \frac{E_0^2 \sigma^0 \Sigma}{r^2} . \tag{6.121}$$

In accordance with this analysis, we can argue that the growth of the scattered pulse edge is determined by changes in parameter q within the value interval from -1 to $+1$. It means that the period of the pulse edge growth may be estimated as the value:

$$\hat{\tau} = \frac{4\sqrt{2\langle \zeta^2 \rangle}}{c} . \tag{6.122}$$

It might seem that Equation (6.119) contradicts the causality principle, as the scattered field differs from zero at negative τ. It should be remembered that measuring the time is performed from the moment of wave reflection from the mean plane. The waves, reflected by the roughness, reach the receiving point earlier than the median wave (i.e., negative moments of time). The existence of a remote negative time moment has to be due to the Gaussian elevation distribution being used with the possibility of having large values.

The differential equation:

$$\frac{d\left\langle\left|\mathrm{E}_s(\tau)\right|^2\right\rangle}{d\tau}=\frac{cE_0^2\sigma^0\Sigma}{2r^2}P_1\left(\frac{c\tau}{2}\right) \tag{6.123}$$

follows from Equation (6.115) at any function of elevation distribution. Let us emphasize that this relation is valid only for the stepwise envelope.

We should point out that pulse distortion (more correctly, signals of wide frequency bands) is the basis for various methods of investigating large roughness, particularly sea waves, by a remote sensing technique.

6.7 WHAT IS Σ?

In the previous calculations, we considered wave scattering by a surface of limited size; however, the question of what factors determine the size of this surface was left out of the discussion. We will now discuss this subject in detail. Natural surfaces (e.g., land, water) are not usually regarded, in the context of these problems, as limited in size on the basis of their geometrical characteristics, although limited size is possible with respect to small islands or reservoirs. Setting aside these special cases, we can now pay attention to examples of sea surface or extensive land areas. The boundedness of scattering surfaces, in these cases, is determined, first of all, by the transmitting antenna footprint. However, here we will deal with a spherical wave, not a plane one, so the previous calculations need to be modified.

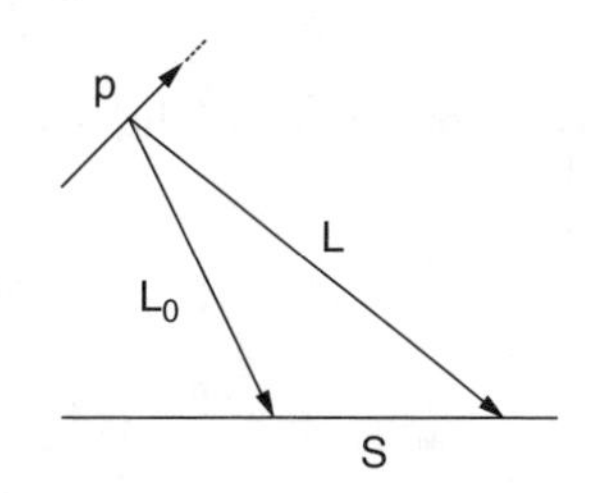

FIGURE 6.1 Field incidence on the surface.

Let us imagine incidence on the surface field as shown in Figure 6.1. The field on the surface is:

$$\mathbf{E}_i(\mathbf{s})=-\frac{ik\mathbf{g}_iE_0e^{ik(\mathrm{L}_0+\mathbf{e}_i\cdot\mathbf{s})}}{2\pi\mathrm{L}_0}\int_A\mathrm{E}_A(\mathbf{p})\exp\left(-ik\frac{\mathbf{p}\cdot\mathbf{s}}{\mathrm{L}_0}\right)d^2\mathbf{p}.$$

In this equation, $\mathbf{E}_A$ is the field in the antenna aperture. The constant E_0 is the subject of definition. Let us direct vector $\mathbf{L}_0$ to the maximum antenna radiation. In order to simplify the following calculations, we will assume that this vector is perpendicular to the antenna plain aperture. We can show that $\mathbf{L}=\mathbf{L}_0+\mathbf{s}$, where $\mathbf{s}$

is the vector of a point on the main plane, and we have an absolute value $\mathrm{L} \cong \mathrm{L}_0 + (\mathbf{e}_i \cdot \mathbf{s})$, where vector $\mathbf{e}_i = \mathbf{L}_0 / \mathrm{L}_0$ now defines the direction of the wave incident. It is assumed that $\mathbf{pL}_0 = 0$. The last representation has reduced the spherical wave to a locally plane one but with changing amplitude along the surface. It was assumed in the previous calculations that a wave of single amplitude was incident on the surface. The amplitude constancy was maintained over the entire surface. Now, this condition does not apply and it becomes necessary to normalize the incident field. For this purpose, let us require the squared amplitude to be equal to unity in the maximum of the pattern (directional diagram) $(\mathbf{s} = 0)$. Thus, we have:

$$E_0^2 = \frac{4\pi^2 \mathrm{L}_0^2}{k^2 \left| \int_A \mathrm{E}_A(\mathbf{p}) d^2\mathbf{p} \right|^2} \tag{6.124}$$

and

$$\mathbf{E}_i(\mathbf{s}) = \mathbf{g}_i T(\mathbf{s}) e^{ik(\mathrm{L}_0 + \mathbf{e}_i \cdot \mathbf{s})}, \tag{6.125}$$

where

$$T(\mathbf{s}) = -\frac{i}{\left| \int_A \mathrm{E}_A(\mathbf{p}) d^2\mathbf{p} \right|} \int_A \mathrm{E}_A(\mathbf{p}) \exp\left(-ik \frac{\mathbf{p} \cdot \mathbf{s}}{\mathrm{L}_0} \right) d^2\mathbf{p}. \tag{6.126}$$

Now let us turn to the main question. We begin with the case of small roughness. We have used the method of few disturbances to describe the scattering characteristics of these inhomogeneities. Let us examine, first of all, the change in representation of equivalent surface currents in terms of the introduced alterations. The magnetic current can be written as (compare with Equation (6.31)):

$$\mathbf{K}_{\mathrm{m}}^{(1)} = \Im\left[\mathbf{e}_z \times \nabla\right] \zeta(\mathbf{s}) T(\mathbf{s})\, e^{i\mathbf{w}_i \cdot \mathbf{s}}, \tag{6.127}$$

and its Fourier image is:

$$\tilde{\mathbf{K}}_{\mathrm{m}}^{(1)} = i\Im\left[\mathbf{e}_z \times \mathbf{w}\right] \tilde{\zeta}_{\mathrm{mod}}\left(\mathbf{w} - \mathbf{w}_i\right), \tag{6.128}$$

where

$$\tilde{\zeta}_{\text{mod}}(\mathbf{w}) = \frac{1}{4\pi^2}\int \zeta(\mathbf{s})T(\mathbf{s})e^{-i\mathbf{w}\cdot\mathbf{s}}d^2\mathbf{s}. \tag{6.129}$$

The integration is performed here over infinite limits contrary to Equation (6.4).

The computations for the electrical current are similar:

$$\mathbf{K}_e^{(1)} = ik\,\mathbf{i}\zeta(\mathbf{s})T(\mathbf{s})e^{i\mathbf{w}_i\cdot\mathbf{s}} \tag{6.130}$$

and

$$\tilde{\mathbf{K}}_e^{(1)} = ik\,\mathbf{i}\tilde{\zeta}_{\text{mod}}(\mathbf{w}-\mathbf{w}_i). \tag{6.131}$$

It is not difficult to see that $\tilde{\zeta}_{\text{mod}}$ must now be used in the expression for the scattered field instead of $\tilde{\zeta}$. The averaged value of this function is equal to zero, and our main interest lies with its mean squared value. While calculating this value, the fourfold integral over $d^2\mathbf{s}'$ and $d^2\mathbf{s}''$ is introduced. Because $\mathbf{s}'+\mathbf{s}''=2\mathbf{S}$ and $\mathbf{s}'-\mathbf{s}''=\mathbf{s}$, we now have:

$$\left\langle\left|\tilde{\zeta}^2_{\text{mod}}(\mathbf{w})\right|^2\right\rangle = \frac{1}{16\pi^2}\iint \mathrm{K}_\zeta(\mathbf{s})T\left(\mathbf{S}+\mathbf{s}/2\right)T^*\left(\mathbf{S}-\mathbf{s}/2\right)e^{-i\mathbf{w}\cdot\mathbf{s}}d^2\mathbf{s}\;d^2\mathbf{S}.$$

The scale of the change in function T is much greater than the scale of the roughness correlation; consequently:

$$\left\langle\left|\tilde{\zeta}^2_{\text{mod}}(\mathbf{w})\right|^2\right\rangle = \frac{\tilde{\mathrm{K}}_\zeta(\mathbf{w})}{4\pi^2}\int\left|T(\mathbf{S})\right|^2 d^2\mathbf{S}. \tag{6.132}$$

The scattering cross section of the rough surface, now infinite in size, will be proportional to this value. From comparison with Equation (6.7), it follows that in this case the value:

$$\Sigma = \int\left|T(\mathbf{S})\right|^2 d^2\mathbf{S} \tag{6.133}$$

should be taken as the square of the scattering surface. From Equations (6.126) and (1.114) it follows that:

$$\left|T(\mathbf{S})\right|^2 = \frac{A_e(\mathbf{S})}{A_e(0)}, \tag{6.134}$$

which represents the section of the antenna pattern by the mean plane of the scattering surface. We neglected the difference from unity of the first factor of Equation (1.114) by assuming sufficient sharpness of the antenna pattern. As a result, we now have a rather obvious outcome:

$$\Sigma = \int \frac{A_e(\mathbf{S})}{A_e(0)} d^2\mathbf{S}, \tag{6.135}$$

reflecting the fact that the power of the field in the receiving point consists of field powers scattered by infinitely small, in the physical sense, elements of the surface. The values of these powers are proportional to the antenna footprint value at the point of scattering.

We can refer to Equation (6.135) for integration over the plane that is perpendicular to vector $\mathbf{L}_0$ and passes through the point where the end of this vector touches the mean plane. This integration is equivalent to integration over a sphere of radius L_0. We obtain:

$$\Sigma = \frac{L_0^2 \lambda^2}{A_e(0)\cos\theta_i} = \frac{4\pi L_0^2}{G(0)\cos\theta_i}, \tag{6.136}$$

by referring back to Equation (1.122). $\lambda^2/A_e(0)$ is the solid angle inside which the main radiating power is concentrated. Being multiplied by $L_0^2/\cos\theta_i$, this angle defines the value of the illuminated area.

It was actually supposed up to now that the receiving antenna is isotropic. If it is not, then the receiving antenna pattern should be taken into consideration, which is easy to do on the basis of our previous analysis. We can use the following expression instead of Equation (6.135):

$$\Sigma = \int \frac{A_t(\mathbf{s}) A_r(\mathbf{s} + \mathbf{s}_0)}{A_t(0) A_r(0)} d^2\mathbf{s}\,. \tag{6.137}$$

Here, the values with the subscript t refer to the transmitting antenna, while the subscript r refers to the receiving antenna. The vector $\mathbf{s}_0$ appears in this expression because the receiving antenna cannot see the point that the transmitting antenna looks at. In other words, this vector connects the crossover points with the mean plane of the maximum directions of the lines of antenna patterns. If the scale of the receiving antenna pattern change is much greater in comparison with that of the transmitting antenna (in other words, the receiving antenna is blunter than the transmitting one), then we must again refer to Equation (6.135). An analogous situation occurs when the transmitting antenna has weak directivity.

For monostatic radar (the transmitting and receiving antennas are combined):

$$\Sigma = \int \left[\frac{A_e(\mathbf{s})}{A_e(0)} \right]^2 d^2\mathbf{s} = \int \left[\Psi(\mathbf{s}) \right]^2 d^2\mathbf{s} = \frac{\mathrm{L}_0^2}{\cos\theta_i} \int_{2\pi} \left[\Psi(\theta,\varphi) \right]^2 d\Omega. \tag{6.138}$$

This representation differs from Equation (6.135). To estimate this difference, let us apply the model provided by Equation (1.123). It is a simple matter to determine that in our case the area defined by Equation (6.138) is less than twice the one computed on the basis of Equation (6.135).

Now, let us turn to consideration of wave scattering by large roughness. First of all, we will address the coherent component of the scattered field using:

$$\langle \mathbf{f} \rangle = 2i\pi k \cos\theta_i \mathbf{E}_r^0 \tilde{P}_\zeta \left[k\left(a_i + a_s\right) \right] \int T(\mathbf{s}) e^{ik\left(\mathbf{u}_i - \mathbf{u}_s\right)\cdot \mathbf{s}} d^2\mathbf{s} \tag{6.139}$$

instead of Equation (6.68) after repeating the calculations carried out in Section 6.4. As the scale of the change in function T is much greater than the wavelength, the integral in Equation (6.139) differs from zero only for scattering in the specular direction. In this case,

$$\langle \mathbf{f} \rangle = 2i\pi k \cos\theta_i \tilde{P}_\zeta \left(2k\cos\theta_i\right) \int T(\mathbf{s})\, d^2\mathbf{s}. \tag{6.140}$$

It is easy to see that:

$$\int T(\mathbf{s})\, d^2\mathbf{s} = -\frac{4i\pi^2 \mathrm{L}_0^2 \mathrm{E}_A(0)}{k^2 \cos\theta_i \left| \int \mathrm{E}_A(\mathbf{p}) d^2\mathbf{p} \right|}. \tag{6.141}$$

As a result,

$$\langle \mathbf{f} \rangle = 2\pi k \cos\theta_i \mathbf{E}_r^0 \tilde{P}_\zeta \left(2k\cos\theta_i\right) \Sigma, \tag{6.142}$$

where

$$\Sigma = \frac{\lambda^2 \mathrm{L}_0^2 \mathrm{E}_A(0)}{\cos\theta_i \left| \int \mathrm{E}_A(\mathbf{p}) d^2\mathbf{p} \right|}. \tag{6.143}$$

Now, instead of the symbolic quantity described by the delta-function in Equation (6.68), we have a real, defined value computed on the basis of knowledge of the field distribution in the antenna aperture or, more accurately, the apparatus antenna function. When the apparatus function is equal to unity, we have:

$$\Sigma = \frac{\lambda^2 L_0^2}{A\cos\theta_i}. \tag{6.144}$$

It is left for us to clarify calculation of the scattering cross section (see Equation (6.76)). Repeating the calculations carried out in Section 6.4 gives us:

$$\sigma = \sigma^0 \Sigma, \tag{6.145}$$

where the scattering area square is defined again by Equation (6.133), which is to be expected. The case of large roughness does not differ, in this respect, from the case of small roughness.

6.8 THE EFFECT OF THE SPHERICAL WAVEFORM ON SCATTERING

Up to now we restricted ourselves to the local plane approximation of a spherical wave with regard to the scattering processes and clearing up the role of limited exposure of the antenna pattern. We will now take into account the spherical waveform in the first approximation. This is particularly important for the problem of short-pulse scattering analysis, especially when the radiation and reception take place at the nadir (a radio altimeter can be an example of such a device). In order to simplify the analysis, we must consider the altimetric mode. The radiation and reception occur in the nadir, in this case, and we have (approximately) $L = \sqrt{L_0^2 + s^2} \cong L_0 + s^2/2L_0$. Taking into account the aforementioned restrictions and approaches, we now have:

$$\mathbf{f}(\omega') \cdot \mathbf{f}^*(\omega'') = \frac{k^2}{16\pi^2} \iint (\mathbf{B}' \cdot \mathbf{B}'') T^2(\mathbf{s}') T^{*2}(\mathbf{s}'') e^{i\Psi - i\Phi} \frac{d^2\mathbf{s}' d^2\mathbf{s}''}{(\mathbf{n}' \cdot \mathbf{e}_z)(\mathbf{n}'' \cdot \mathbf{e}_z)},$$

where

$$\Psi = \frac{1}{L_0}\left[k\left(s'^2 - s''^2\right) + \frac{K}{2}\left(s'^2 + s''^2\right)\right],$$

$$\Phi = 2\left\{k\left[\zeta(\mathbf{s}') - \zeta(\mathbf{s}'')\right] + \frac{K}{2}\left[\zeta(\mathbf{s}') + \zeta(\mathbf{s}'')\right]\right\}.$$

Let us point out that it is necessary to take into consideration the double phase shift from the transmitter to the surface and, backward, from the surface to the receiver. Repeating the calculations of Section 6.6 gives us:

$$(\mathbf{f}'\cdot\mathbf{f}''^{*})=\frac{k^2}{16\pi^2}\iint\left(\mathbf{B}'\cdot\mathbf{B}''\right)T^2\left(\mathbf{S}+\mathbf{s}/2\right)T^{*2}\left(\mathbf{S}-\mathbf{s}/2\right)e^{i\Theta}\frac{d^2\mathbf{S}\,d^2\mathbf{s}}{\left(\mathbf{n}'\cdot\mathbf{e}_z\right)\left(\mathbf{n}''\cdot\mathbf{e}_z\right)}.$$

Here,

$$\Theta=\frac{2k}{\mathrm{L}_0}\left(\mathbf{s}\cdot\mathbf{S}\right)-2k\mathbf{s}\nabla\zeta(\mathbf{S})+\frac{K}{4\mathrm{L}_0}\mathrm{s}^2+\frac{K}{\mathrm{L}_0}\mathrm{S}^2-2K\zeta(\mathbf{S}).$$

The essential area of integration with respect to **s** has a size on the order of $\lambda/\sqrt{\left\langle(\nabla\zeta)^2\right\rangle}$. Therefore, we can assume that $K\mathrm{s}^2/4\mathrm{L}_0 << 1$, and the following:

$$\left\langle\mathbf{f}\left(\omega'\right)\cdot\mathbf{f}^{*}\left(\omega''\right)\right\rangle=\sigma^0\tilde{P}_\zeta\left(2K\right)\int\frac{P_1\left(\nabla\zeta=\mathbf{S}/\mathrm{L}_0\right)}{P_1\left(\nabla\zeta=0\right)}\left|T\left(\mathbf{S}\right)\right|^4\exp\left(\frac{iK\mathrm{S}^2}{\mathrm{L}_0}\right)d^2\mathbf{S}. \tag{6.146}$$

Based on the arguments presented in Section 6.7, we can finally obtain the relation between the shape of the scattered wave and the corresponding shape of the radiated oscillation:

$$\left\langle\left|\mathrm{E}_s(\tau)\right|^2\right\rangle=\frac{c\sigma^0}{2\mathrm{L}_0^2}\int\left|T(\mathbf{S})\right|^4\frac{P_1\left(\nabla\zeta=\mathbf{S}/\mathrm{L}_0\right)}{P_1\left(\nabla\zeta=0\right)}d^2\mathbf{S}\int\left|\mathrm{E}_i\left(t',0\right)\right|^2P_1\left[\frac{c\left(t'-\tau\right)}{2}+\frac{\mathrm{S}^2}{2\mathrm{L}_0}\right]dt'. \tag{6.147}$$

In our case,

$$-c\tau+\frac{\mathrm{S}^2}{\mathrm{L}_0}=-ct+2\mathrm{L}_0+\frac{\mathrm{S}^2}{\mathrm{L}_0}\cong-ct+2\mathrm{L}=-c\tau_e(\mathrm{S}),$$

which corresponds to the sphericity of the waveform inclusion at the arriving time calculation of the waves scattered by the different areas of the surface. Taking into account the slope distribution relates to the fact that the value of the surface square, which effectively scatters, is not always defined by the directionality of the antenna pattern, such as occurs when the pattern angle exposed is smaller than the standard deviation of the surface slopes.

The probability of slopes smaller than the pattern angle exposed is considerable in this case, and the entire surface inside the antenna footprint scatters effectively. Otherwise, because of the low possibility of slopes that exceed their standard deviation, the effective scattering surface is seen from the antenna at a solid angle on the order of $2\pi ar\,\mathrm{ctg}\sqrt{\langle(\nabla\zeta)^2\rangle}$. In other words, in the first case, the integral over the surface in Equation (6.147) is reduced to:

$$\int \left|T(\mathbf{S})\right|^2 d^2\mathbf{S},$$

and in the second case to:

$$\int \frac{P_1\left(\nabla\zeta = \mathbf{S}/\mathrm{L}_0\right)}{P_1\left(\nabla\zeta = 0\right)} d^2\mathbf{S}.$$

The analysis is performed rather easily for the model signal shown in Equation (6.111) by assuming normality of the elevation distribution. The expression for the scattered power is:

$$\left\langle \left|\mathrm{E}_s(\tau)\right|^2 \right\rangle = \frac{E_0^2\tau_0\sigma^0}{\mathrm{L}_0^2\tilde{\tau}_0} \int \left|T(\mathbf{S})\right|^4 \exp\left[-\frac{\mathrm{S}^2}{\mathrm{L}_0^2\left\langle(\nabla\zeta)^2\right\rangle} - \frac{2\tau_e^2}{\tilde{\tau}_0^2}\right] d^2\mathbf{S}. \tag{6.148}$$

Here, $\tilde{\tau}_0$ is defined by Equation (6.113) at $\theta_i = 0$. Furthermore, let us presume, for simplification of the problem, that we are dealing with a circular antenna whose pattern has the model form of Equation (1.123). Then, it is convenient to realize integration on a cylindrical coordinate system. As a result,

$$\left\langle \left|\mathrm{E}_s(\tau)\right|^2 \right\rangle = \frac{\pi\sqrt{\pi}E_0^2 c\tau_0\sigma^0}{2\sqrt{2}\mathrm{L}_0} \exp\left(\beta^2 - \frac{2\tau^2}{\tilde{\tau}_0^2}\right)\left[1 + \Phi(\beta)\right]. \tag{6.149}$$

Here,

$$\beta(\tau) = \sqrt{2}\left(\frac{\tau}{\tilde{\tau}_0} - \frac{c\tilde{\tau}_0}{2\vartheta^2\mathrm{L}_0}\right), \quad \vartheta^2 = \frac{2\theta_0^2\left\langle(\nabla\zeta)^2\right\rangle}{\theta_0^2 + 2\left\langle(\nabla\zeta)^2\right\rangle}. \tag{6.150}$$

Thus, the function $\beta(\tau)$ defines the behavior of the scattered oscillation over time. It is negative at $\tau < 0$, and at its large absolute values the scattered signal form has an asymptotic form:

$$\left\langle \left| E_s(\tau) \right|^2 \right\rangle \cong \frac{\pi E_0^2 c\tau_0 \sigma^0}{2\sqrt{2}\left|\beta\right| L_0} \exp\left(-\frac{2\tau^2}{\tilde{\tau}_0^2} \right). \tag{6.151}$$

Outwardly, the signal shape is practically kept; otherwise, the time of the impulse rising changes because of roughness. A dependence on the pattern angle width and on the slope values is lacking. This takes place because, generally, the head part of the spherical wave front interacts with the surface at this moment, and the scattering has the same character as in the case of a plane wave.

The scattered impulse reaches its maximum not at time moment $\tau = 0$ but at the moment determined from the equation:

$$e^{\beta^2}\left[1 + \Phi(\beta)\right] = \sqrt{\frac{2}{\pi}} \frac{\vartheta^2 L_0}{c\tilde{\tau}_0}. \tag{6.152}$$

Let us assume that the impulse is long if $c\tilde{\tau}_0 \cong c\tau_0 >> \vartheta^2 L_0$. In this case, Equation (6.152) has a solution at large negative values of β. On the basis of the asymptotic representation of Equation (6.118), we now have, instead of Equation (6.152):

$$\beta = -\frac{c\tilde{\tau}_0}{\sqrt{2}\vartheta^2 L_0}$$

or $\tau = 0$.

For short pulses, when $-\tilde{\tau}_0 << \vartheta^2 L_0$, Equation (6.152) has a solution at large positive values of β which gives us the opportunity to reduce the previous equation to a much easier relation:

$$e^{\beta^2} = \frac{\vartheta^2 L_0}{\sqrt{2\pi} c\tilde{\tau}_0}$$

from which we can derive the formula for the time moment of the scattered pulse power maximum:

$$\tau \cong \frac{\tilde{\tau}_0}{2\sqrt{2}} \ln\left(\frac{\vartheta^2 L_0}{\sqrt{2\pi} c\tilde{\tau}_0} \right). \tag{6.153}$$

The scattered power maximum is equal to:

$$\left\langle |E_s|^2_{max} \right\rangle = \frac{\pi \vartheta^2 E_0^2 \sigma^0}{2} \tag{6.154}$$

for long impulses. In this case, the scattering square is defined by the value:

$$\Sigma = \frac{\pi \vartheta^2 L_0^2}{2} . \tag{6.155}$$

The analogous relation for the short impulses has the form:

$$\left\langle |E_s|^2_{max} \right\rangle = \frac{\pi \vartheta^2 E_0^2 \tau_0 \sigma^0}{2\tilde{\tau}_0} \exp\left[-\frac{1}{4} \ln^2 \left(\frac{\vartheta^2 L_0}{\sqrt{2\pi c \tilde{\tau}_0}} \right) \right]. \tag{6.156}$$

This maximum is much smaller than the long impulse of the same peak power. The reason for this difference lies in the considerable widening of the short impulse at the expense of the direct influence of roughness and the indirect influence of the incident wave spherical form. The direct effect is shown by the ratio $\tau_0/\tilde{\tau}_0$, which is less than unity, and the indirect one by the presence of the exponent, which has the reduced angle ϑ.

The pulse-widening (spreading) effect is particularly observed during expansion of the trailing edge. Equation (6.149) is transformed to the form:

$$\left\langle |E_s(\tau')|^2 \right\rangle = \frac{\pi \sqrt{\pi} E_0^2 c \tau_0 \sigma^0}{\sqrt{2} L_0} \exp\left(-\frac{2c\tau'}{\vartheta^2 L_0} - \frac{c^2 \tilde{\tau}_0^2}{2\vartheta^4 L_0^2} \right), \quad \tau' = \tau - \frac{c\tilde{\tau}_0^2}{2\vartheta^2 L_0}, \tag{6.157}$$

when β becomes much greater than unity. This spreading does not play an important role for the long impulse, because it appears only at the tail of this impulse at the very remote time period of $\tau > c\tilde{\tau}_0^2/2\vartheta^2 L_0$, when the intensity is small, as indicated by the second summand in the exponent of Equation (6.157). So, the discussed effect most likely has more of an academic character than a practical one for long impulses.

This phenomenon is crucial for short impulses. The relax time of the pulse trailing edge is estimated by the value $\vartheta^2 L_0/c$, which considerably exceeds the time duration of the pulse itself. This strong effect appears, as already noted, because of the spherical wave form, but it is due to the presence of large roughness. In the absence of roughness (i.e., an ideally plain surface), the effective reflecting area has a size defined by the Fresnel zone $\sqrt{\lambda L_0}$, which is much smaller than the zone of effective scattering, the size of which is on the order of ϑL_0.

Now, we will provide some relations of a general character. First, let us examine the conservation of energy law. The scattered impulse energy is:

$$U_s = \int_{-\infty}^{\infty} \left\langle \left| E_s(\tau) \right|^2 \right\rangle d\tau = \frac{\sigma^0 \Sigma}{L_0^2} U_i, \tag{6.158}$$

where the energy of the incident pulse is defined as:

$$U_i = \int_{-\infty}^{\infty} \left| E_i(t', 0) \right|^2 dt' \tag{6.159}$$

and, in this case, the effective scattering square is:

$$\Sigma = \int \left| T(\mathbf{S}) \right|^4 \frac{P_1\left(\nabla\zeta = \mathbf{S}/L_0 \right)}{P_1\left(\nabla\zeta = 0 \right)} d^2\mathbf{S}. \tag{6.160}$$

Regarding the normal slope distribution and model antenna pattern in Equation (1.123), we arrive at Equation (6.155).

It is convenient to characterize the duration of the scattered impulse by the value whose square is determined by the relation:

$$\bar{\tau}^2 = \frac{\alpha}{U_s} \int_{-\infty}^{\infty} \tau^2 \left\langle \left| E_s(\tau) \right|^2 \right\rangle d\tau. \tag{6.161}$$

It is supposed, now, that the incident pulse shape is an even function of time with duration τ_0. Selection of the constant, α, depends on the pulse shape; for example, $\alpha = 4$ for rectangular pulses and bell-like pulses, which we are discussing here. In the general case, the result of our calculations can be expressed by the formula:

$$\bar{\tau}^2 = \tau_0^2 + \frac{4\alpha}{c^2} \left\langle \zeta^2 \right\rangle + \frac{\alpha}{c^2 L_0^2 \Sigma} \int \left| T(\mathbf{S}) \right|^2 \frac{P_1\left(\nabla\zeta = \mathbf{S}/L_0 \right)}{P_1\left(\nabla\zeta = 0 \right)} S^4 d^2\mathbf{S}. \tag{6.162}$$

The formula then becomes:

$$\bar{\tau}^2 = \tau_0^2 + \frac{4\alpha}{c^2} \left\langle \zeta^2 \right\rangle + \frac{\alpha}{2c^2} \vartheta^4 L_0^2 \tag{6.163}$$

for the models discussed.

6.9 SPATIAL CORRELATION OF THE SCATTERED FIELD

The scattered field, as we have found, has a complex spatial structure of a stochastic character. As the field itself at every point is the result of the summation of the fields scattered by the statistically independent elements of the random surface, the statistic of the field is Gaussian. By studying the problem of large roughness scattering properties, we have determined the mean value of the field and its dispersion. The next step is to find the spatial correlation function.

We will first discuss reception at two points separated in the direction of the radiowave propagation. Setting aside some of the details, we will focus on estimating the correlation radius of the scattered field, which can be considered to be a field induced by random sources situated on the scattering surface. When the correlation radius of these sources is smaller than the radiating area, the van Cittert–Zernike theorem is valid (see Section 1.11). Equation (1.139) is a consequence of this theorem. In our case, we will use $A = \Sigma \cong \lambda^2 z^2 / A_e(0)$. The desired radius of correlation can now be estimated by the relation:

$$s_{\perp} \cong \sqrt{\frac{A_e(0)}{\pi}}\,; \tag{6.164}$$

that is, it is defined by the size of the transmitting antenna.

The case of monostatic radar, when the transmitting and receiving antennae are combined, is more interesting. It is worthwhile to talk, in this case, about correlation of the scattered waves with the radar motion parallel to the mean plane of the rough surface. This is similar to the case when the radar is immovable and only the surface moves. As a result, we are restricted to calculating the average value of the product:

$$(\mathbf{f}' \cdot \mathbf{f}''^{*}) = \frac{k^2}{16\pi^2} \iint \left(\mathbf{B}' \cdot \mathbf{B}''\right) T^2\left(\mathbf{s}'\right) T^{*2}\left(\mathbf{s}''\right) e^{i\Psi - i\Phi} \frac{d^2\mathbf{s}'\, d^2\mathbf{s}''}{\left(\mathbf{n}' \cdot \mathbf{e}_z\right)\left(\mathbf{n}'' \cdot \mathbf{e}_z\right)},$$

where

$$\Psi = 2k\mathbf{u}_i\left(\mathbf{s}' - \mathbf{s}'' - \mathbf{d}\right), \qquad \Phi = 2k\left[\zeta\left(\mathbf{s}'\right) - \zeta\left(\mathbf{s}'' + \mathbf{d}\right)\right].$$

Further, we must change the variables $\mathbf{s}' = \mathbf{S} + \mathbf{s}/2$ and $\mathbf{s}'' = \mathbf{S} - \mathbf{d} - \mathbf{s}/2$ and repeat the calculations of Section 6.8 to obtain:

$$\left\langle \mathbf{f}' \cdot \mathbf{f}''^{*} \right\rangle = \sigma^0\left(-\mathbf{e}_i\right) \int T^2(\mathbf{S}) T^{*2}(\mathbf{S} - \mathbf{d})\, d^2\mathbf{S}. \tag{6.165}$$

Here, **d** is the vector connecting the points sought by the correlation. The expression for the correlation coefficient has the form:

$$\hat{k}_E(\mathbf{d}) = \frac{1}{\Sigma}\int T^2(\mathbf{S})T^{*2}(\mathbf{S}-\mathbf{d})\,d^2\mathbf{S}. \tag{6.166}$$

The square of the effective scattering area is defined by Equation (6.138). The correlation is determined by the degree of overlap of the antenna footprints and depends on the position of the radar. The fact that the value of the correlation coefficient does not depend on the rough surface parameters is an important one.

The situation is rather different for radiation in the nadir (altimeter mode), when it becomes necessary to take into account the spherical waveform. Taking into account our previous calculations, we have:

$$\hat{k}_E = \frac{1}{\Sigma}\int \frac{P_1\left(\nabla\zeta = \mathbf{S}/\mathrm{L}_0\right)}{P_1\left(\nabla\zeta = 0\right)} T^2(\mathbf{S})T^{*2}(\mathbf{S}-\mathbf{d})\,d^2\mathbf{S}. \tag{6.167}$$

where Σ is determined by the same integral at $\mathbf{d} = 0$. Here, the parameters of the roughness are a factor that restricts the size of the effective scattering surface. If we use the model pattern, Equation (1.123), then we obtain the following expression for the correlation coefficient:

$$\hat{k}_E = \exp\left[-\frac{\mathrm{d}^2}{\theta_0^2 \mathrm{L}_0^2}\left(1-\frac{\vartheta^2}{\theta_0^2}\right)\right]. \tag{6.168}$$

6.10 RADIOWAVE SCATTERING BY A LAYER WITH ROUGH BOUNDARIES

In some cases, we encounter the problem of wave scattering by layers where both boundaries are rough. A layer of ice over water or solid soil, a layer of snow on the ground, and a layer of sand on a solid base are examples of this. Undoubtedly, scattering by such layers is quite complex compared to the scattering by a rough surface, as it is requires paying attention to the interference of the wave scattered by both layer interfaces.

The problem is reduced to one of searching the field scattered by the layer with both rough boundaries at the incidence on the field of the plane wave of single amplitude at angle θ_i. Generally, both interfaces of the layer are planes that are parallel to each other (Figure 6.2). The upper boundary is described by the equation $z = \zeta_1(\mathbf{s}_1)$, and the lower one is represented by the equation $z = -d + \zeta_2(\mathbf{s}_2)$. Here, the vectors $\mathbf{s}_1$ and $\mathbf{s}_2$ correspond to the horizontal point coordinates of the upper and lower boundaries. We will assume that roughness ζ_1 and ζ_2 are large in comparison

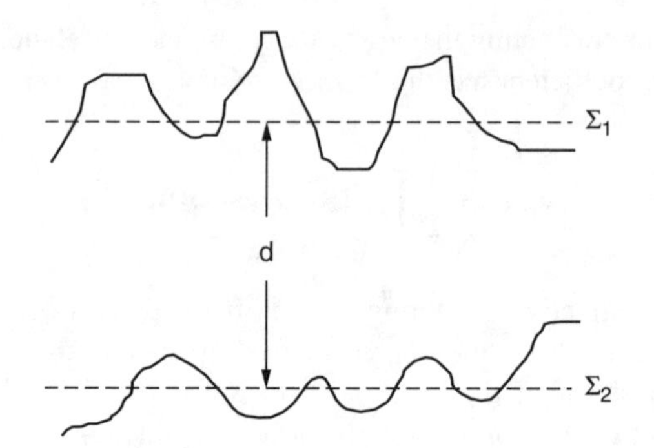

FIGURE 6.2 Upper and lower boundaries of the layer.

to the wavelength and have gradual slopes. In order to simplify the problem, let us regard only backward scattering, which is especially important for radar sounding of layered media. Taking into account the simplifications made in the previous discussions, the formula for the backward scattering can be written as (compare with Equation (6.66)):

$$\mathbf{f}\left(-\mathbf{e}_i\right) = -\frac{ik}{4\pi}\int_{\Sigma_1} \mathbf{B}(0)\, e^{2ik\left(\mathbf{u}_i\cdot \mathbf{s}_1'\right)-2ik\cos\theta_i\zeta_1\left(\mathbf{s}_1'\right)} d^2\mathbf{s}_1'. \tag{6.169}$$

Here, the integral is expanded over the mean plane of the upper interface. The vector $\mathbf{B}(0)$ is given by the formula (compare with Equation (6.65)):

$$\mathbf{B}(0) = \left(\mathbf{e}_z\cdot\mathbf{e}_i\right)\mathbf{E}_r^0 - \left(\mathbf{e}_i\cdot\mathbf{E}_r^0\right) - \left(\mathbf{e}_i\cdot\mathbf{H}_r^0\right)\left[\mathbf{e}_i\times\mathbf{e}_z\right] + \left(\mathbf{e}_z\cdot\mathbf{e}_i\right)\left[\mathbf{e}_i\times\mathbf{H}_r^0\right].$$

As before, the amplitude of the reflected field is expressed via the amplitude of the incident plane wave with the help of Equations (3.21) and (3.22); however, the reflective coefficients are no longer Fresnel ones. They are functions of the coordinates, because of the roughness, and have to be further defined. We should point out here that they are "slow" stochastic coordinate functions in the sense that their correlation radius is of the same order as the roughness correlation radius, and its value is much greater comparable with the wavelength.

Let us first consider the case of horizontal polarization of the incident field. We can show that $\mathbf{B}(0) = -2F_h\left(\mathbf{s}_1'\right)\mathbf{g}_i\cos\theta_i$ in this case, and

$$\mathbf{f} = \mathbf{f}_h\left(-\mathbf{e}_i\right) = \frac{ik}{2\pi}\mathbf{g}_i\cos\theta_i\int_{\Sigma_1} F_h\left(\mathbf{s}_1'\right) e^{2ik\left(\mathbf{u}_i\cdot \mathbf{s}_1'\right)-2ik\cos\theta_i\zeta_1\left(\mathbf{s}_1'\right)} d^2\mathbf{s}_1'. \tag{6.170}$$

Let us now formulate the equation for the reflective coefficient introduced here. For this purpose, we shall use the Stratton–Chu formulae for the field inside the layer. The integration is performed over the boundaries Σ_1 and Σ_2. The general field is parted into two summands: $\mathbf{E} = \mathbf{E}_1 + \mathbf{E}_2$. The first one is defined by an integral over the upper boundary and the second one by an integral over the lower boundary. Let us transform the integration over the rough surfaces into a similar one over the mean planes, taking into account the surface slopes, if necessary, only in the exponent. This means that only the wave phase modulation performed by the rough surface is taken into consideration; the role of depolarization is not regarded in full measure.

The Kirchhoff approximation can be used to describe the fields on the surfaces under consideration because of the large-scale roughness. We can express the field on the inner part of surface Σ_1 via the field of the incident wave with the help of the relations $[\mathbf{n}\times\mathbf{E}] \cong \left[\mathbf{e}_z \times \mathbf{E}\right] = \left(1 + F_h\right)\left[\mathbf{e}_z \times \mathbf{g}_i\right]$, $[\mathbf{n}\times\mathbf{H}] \cong \left(1 - F_h\right)\mathbf{g}_i \cos\theta_i$, and $(\mathbf{n}\cdot\mathbf{E}) \cong \left(\mathbf{e}_z \cdot \mathbf{E}\right) = 0$. It was noted that the incident field is the plane wave of the horizontal polarization and singular amplitude. Let us point out that, in this case, the Green function can be represented in the form:

$$g_1 = \frac{e^{ik\sqrt{\varepsilon_1}R_1}}{R_1}, \quad R_1 = \sqrt{\left|\mathbf{s} - \mathbf{s}_1'\right|^2 + \left[z - \zeta_1\left(\mathbf{s}_1'\right)\right]^2}, \tag{6.171}$$

where ε_1 is the dielectric permittivity inside the layer. The result of this complex calculation gives us:

$$\mathbf{E}_1\left(\mathbf{r}\right) = -\frac{1}{4\pi}\int_{\Sigma_1}\Big\{\Big\{ ik\left[1 - F_h\left(\mathbf{s}_1'\right)\right]g_1\cos\theta_i + \left[1 + F_h\left(\mathbf{s}_1'\right)\right]\left(\mathbf{e}_z \cdot \nabla' g_1\right)\Big\}\mathbf{g}_i -$$
$$-\left[1 + F_h\left(\mathbf{s}_1'\right)\right]\left(\mathbf{g}_i \cdot \nabla' g_1\right)\mathbf{e}_z\Big\} e^{ik\phi_1\left(\mathbf{s}_1'\right)} d^2\mathbf{s}_1',$$
$$\phi_1\left(\mathbf{s}_1'\right) = \left(\mathbf{u}_i\mathbf{s}_1'\right) - \zeta_1\left(\mathbf{s}_1'\right)\cos\theta_i. \tag{6.172}$$

Further, we need an expression for the electrical field on the surface itself. Let us compute the first summand:

$$\mathbf{E}_1'\left(\mathbf{r}_1\right) = -\frac{ik\cos\theta_i}{4\pi}\mathbf{g}_i\int_{\Sigma_1}\left[1 - F_h\left(\mathbf{s}_1'\right)\right]e^{ik\left(R_1' + \phi_1\right)}\frac{d^2\mathbf{s}_1'}{R_1'}.$$

Here, $R_1' = \sqrt{\left|\mathbf{s}_1 - \mathbf{s}_1'\right|^2 + \left[\zeta_1\left(\mathbf{s}_1\right) - \zeta_1\left(\mathbf{s}_1'\right)\right]^2}$. It is obvious that the integration main zone is concentrated close to the point $\mathbf{s}_1' = \mathbf{s}_1$ and has a scale similar to the wavelength. By virtue of this fact, $R_1' \cong \left|\mathbf{s}_1 - \mathbf{s}_1'\right|\sqrt{1 + (\nabla\zeta)^2} \cong \left|\mathbf{s}_1 - \mathbf{s}_1'\right|$. Let us now introduce the new variable $\mathbf{S} = \mathbf{s}_1 - \mathbf{s}_1'$. Then,

$$\mathbf{E}_1'(\mathbf{r}_1) \cong -\frac{ik\left[1-F_h(\mathbf{s}_1)\right]}{4\pi}\mathbf{g}_i e^{ik\phi_1(\mathbf{s}_1)}\int e^{ik[S-(\mathbf{u}_i\cdot\mathbf{S})]}\frac{d^2\mathbf{S}}{S}.$$

The integral included in the last expression is easily calculated in the polar coordinate system:

$$\int e^{ikS-ik(\mathbf{u}_i\cdot\mathbf{S})}\frac{d^2\mathbf{S}}{S} = 2\pi\int_0^\infty J_0\left(k\sin\theta_i\right)e^{ikS}dS = \frac{2\pi}{-ik\sqrt{\varepsilon-\sin^2\theta_i}}.$$

The following tabulated integral was used:[44]

$$\int_0^\infty e^{-\alpha x}J_\nu(\beta x)\,dx = \frac{\left[\sqrt{\alpha^2+\beta^2}-\alpha\right]^\nu}{\beta^\nu\sqrt{\alpha^2+\beta^2}}. \tag{6.173}$$

So,

$$\mathbf{E}_1'(\mathbf{r}_1) = \frac{\left[1-F_h(\mathbf{s}_1)\right]\cos\theta_i}{2\sqrt{\varepsilon_1-\sin^2\theta_i}}\mathbf{g}_i e^{ik\phi_1(\mathbf{s}_1)}.$$

The computation of the second term is rather more complicated as we must overcome the singularity at the point $\mathbf{s}_1' = \mathbf{s}_1$. We will not perform the necessary calculations here, instead referring the reader to the theory of potential,[51] from which it follows that:

$$\mathbf{E}_1'' = -\frac{\mathbf{g}_i}{4\pi}\int_{\Sigma_1}\left[1+F_h\right]\left(\mathbf{e}_z\cdot\nabla' g_1\right)e^{ik\phi}d^2\mathbf{s}_1'$$

$$= \frac{\left[1+F_h\right]\mathbf{g}_i}{4\pi}e^{ik\phi}\int\frac{\partial}{\partial z}\left(\frac{1}{R_1}\right)d^2\mathbf{s}_1' = \frac{\left[1+F_h\right]\mathbf{g}_i}{2}e^{ik\phi}$$

for the case being considered here. All the functions carried out in the integral are done at the argument value $\mathbf{s}_1$.

The third summand also has a singularity; therefore, the corresponding integral should be defined relative to the main value; thus,

$$\mathbf{E}_1'''(\mathbf{r}_1) = \frac{\left[1+F_h\right]\mathbf{e}_z}{4\pi}e^{ik\phi}\int\left(\mathbf{g}_i\cdot\mathbf{S}\right)\frac{1}{R_1'}\frac{dg}{dR_1'}e^{ik\left[R_1'-(\mathbf{u}_i\cdot\mathbf{S})\right]}d^2\mathbf{S}.$$

It is easy to show by integration in the polar coordinate system that the last integral is equal to zero because of the mutual orthogonality of vectors $\mathbf{u}_i$ and $\mathbf{g}_i$ in the case of the horizontal polarization; thus,

$$\mathbf{E}_1(\mathbf{r}_1)=\frac{1}{2}\left\{\frac{\left[1-F_{\mathrm{h}}(\mathbf{s}_1)\right]\cos\theta_i}{\sqrt{\varepsilon_1-\sin^2\theta_i}}+1+F_{\mathrm{h}}(\mathbf{s}_1)\right\}\mathbf{g}_i e^{ik\phi(\mathbf{s}_1)}. \tag{6.174}$$

Now, let us compute the electrical field, $\mathbf{E}_1(\mathbf{r}_2)$, on the lower boundary. It is necessary to assume that $k\mathrm{R}_1 >> 1$. So, we can see that $\nabla' g_1 \cong -ik\sqrt{\varepsilon_1} g_1 \mathbf{R}_1/\mathrm{R}_1$. In addition, we should consider that $d >> |\zeta_1|, |\zeta_2|$, because in the opposite case it is quite probable that both surfaces will cross each other, which would not take place only in the case of their similarity. We are able, in these conditions, to write that $\mathrm{R}_1 \cong \hat{\mathrm{R}}_1+\cos\theta_1\left[\zeta_1(\mathbf{s}_1')-\zeta_2(\mathbf{s}_2)\right]$, where

$$\hat{\mathrm{R}}_1=\sqrt{\left|\mathbf{s}_2-\mathbf{s}_1'\right|^2+d^2},\quad \cos\theta_1=\frac{d}{\hat{\mathrm{R}}_1}. \tag{6.175}$$

It was noted that the z-coordinate has negative values in the space area being analyzed. The expression for the electrical field takes the form:

$$\mathbf{E}_1(\mathbf{r}_2)=-\frac{ik}{4\pi}\int_{\Sigma_1}\left\{\left\{\left[1-F_{\mathrm{h}}(\mathbf{s}_1')\right]\cos\theta_{\mathrm{i}}+\sqrt{\varepsilon_1}\left[1+F_{\mathrm{h}}(\mathbf{s}_1')\left(\mathbf{e}_{\mathrm{z}}\cdot\frac{\hat{\mathbf{R}}_1}{\hat{\mathrm{R}}_1}\right)\right]\right\}\mathbf{g}_i+\right.$$

$$\left.+\sqrt{\varepsilon_1}\left[1+F_{\mathrm{h}}(\mathbf{s}_1')\right]\left(\mathbf{g}_i\cdot\frac{\hat{\mathbf{R}}_1}{\hat{\mathrm{R}}_1}\right)\right\}e^{ik\left[\Psi(\mathbf{s}_1')+\varphi(\mathbf{s}_1')-\sqrt{\varepsilon_1}\cos\theta_1\zeta_2(\mathbf{s}_2)\right]}\frac{d^2\mathbf{s}_1'}{\hat{\mathrm{R}}_1}.$$

Here, $\Psi_1(\mathbf{s}_1')=\sqrt{\varepsilon_1}\hat{\mathrm{R}}_1+(\mathbf{u}_i\cdot\mathbf{s}_1')$ and $\varphi_1(\mathbf{s}_1')=\left(\sqrt{\varepsilon_1}\cos\theta_1-\cos\theta_i\right)\zeta_1(\mathbf{s}_1')$.

To calculate the integral, let us use the stationary phase method. The point of the stationary phase is determined from the equation $\nabla\Psi(\mathbf{s}_1')=0$, or in the expanded view,

$$\frac{\mathbf{s}_1'-\mathbf{s}_2}{\hat{\mathrm{R}}_1}=-\frac{\mathbf{u}_i}{\sqrt{\varepsilon_1}}. \tag{6.176}$$

It follows from this that $\left|\mathbf{s}-\mathbf{s}_1'\right|=d\tan\theta_{\mathrm{t}}$ in the stationary point, where angle θ_{t} is determined by Equation (3.8). So, the stationary phase point can be obtained by the equation:

$$\mathbf{s}_1'=\mathbf{s}_2-\hat{\mathbf{s}},\quad \hat{\mathbf{s}}=\frac{d}{\sqrt{\varepsilon_1-\sin^2\theta_i}}\mathbf{u}_i. \tag{6.177}$$

The main conclusion from the result obtained here is that the field generated by the upper boundary is mainly formed by those surface elements that correspond to the wave refraction lows on the plane interface of the two media. This result is correct only in the case of gradual surface slopes. $\hat{\mathbf{R}}_1/\hat{\mathrm{R}}_1 = \mathbf{e}_t/\sqrt{\varepsilon_1}$ at this condition. We will not carry out the expansion of the phase close to the stationary point but will represent the result as:

$$\mathbf{E}_1(\mathbf{r}_2) = -\frac{ik\mathbf{g}_i}{4\pi}\left\{\left[1 - F_h(\mathbf{s}_2 - \hat{\mathbf{s}})\right]\cos\theta_i + \left[1 + F_h(\mathbf{s}_2 - \hat{\mathbf{s}})\right]\sqrt{\varepsilon_1 - \sin^2\theta_i}\right\}\times$$
$$\times e^{ik\left[\varphi_1(\mathbf{s}_2-\hat{\mathbf{s}}) + (\mathbf{u}_i\cdot\mathbf{s}_2) - \sqrt{\varepsilon_1 - \sin^2\theta_i}\,\zeta_2(\mathbf{s}_2)\right]}\int e^{ik\left[\sqrt{\varepsilon_1}\sqrt{\mathrm{S}^2+d^2} - (\mathbf{u}_i\cdot\mathbf{S})\right]}\frac{d^2\mathbf{S}}{\sqrt{\mathrm{S}^2+d^2}}.$$

Note that $\varphi_1(\mathbf{s}_2 - \hat{\mathbf{s}}) = (\sqrt{\varepsilon_1 - \sin^2\theta_i} - \cos\theta_i)\zeta_1(\mathbf{s}_2 - \hat{\mathbf{s}})$ and $(\mathbf{g}_i\cdot\mathbf{e}_t) = 0$. The last integral is easily computed with the help of Equation (6.56), giving us:

$$\mathbf{E}_1(\mathbf{r}_2) = \mathbf{E}_1\left(\mathbf{s}_2 - \hat{\hat{\mathbf{s}}}\right)\exp\left\{ik\left[\frac{\varepsilon_1 d}{\sqrt{\varepsilon_1 - \sin^2\theta_i}} + \gamma(\mathbf{s}_2)\right]\right\}, \tag{6.178}$$

where $\gamma(\mathbf{s}_2) = \sqrt{\varepsilon_1 - \sin^2\theta_i}\left[\zeta_1(\mathbf{s}_2 - \hat{\mathbf{s}}) - \zeta_2(\mathbf{s}_2)\right]$.

The first factor in Equation (6.178) is defined by Equation (6.174) with the result that the field distribution, given on the surface, transfers as a plane wave in the direction determined by the geometrical optics laws. Note that magnetic field $\mathbf{H}_1(\mathbf{r}_2) = \left[\mathbf{e}_t \times \mathbf{E}_1(\mathbf{r}_2)\right]$.

We shall now turn to the calculation of field $\mathbf{E}_2$, which is described by the integral over Σ_2. One should take into account that, in this case, Green's function on this surface is equal to:

$$g_2 = \frac{e^{ik\sqrt{\varepsilon_1}\mathrm{R}_2}}{\mathrm{R}_2}, \quad \mathrm{R}_2 = \sqrt{\left|\mathbf{s} - \mathbf{s}_2'\right|^2 + \left[z + d - \zeta_2(\mathbf{s}_2')\right]^2}\,. \tag{6.179}$$

Further, $\mathbf{n} \cong -\mathbf{e}_z$. Accordingly to Kirchhoff's approximation, the condition $\mathbf{E} \cong (1 + F_{23}^{(h)})\mathbf{E}_1$ should be assumed. For the magnetic field, $[\mathbf{n}\times\mathbf{H}] \cong \left(1 - F_{23}^{(h)}\right)\left[\mathbf{n}\times\mathbf{H}_1\right]$. Here, $F_{23}^{(h)}$ is the Fresnel coefficient of reflection on the lower interface for the horizontally polarized waves, which is equal to:

$$F_{23}^{(h)} = \frac{\sqrt{\varepsilon_1}\cos\theta_t - \sqrt{\varepsilon_2 - \varepsilon_1\sin^2\theta_t}}{\sqrt{\varepsilon_1}\cos\theta_t + \sqrt{\varepsilon_2 - \varepsilon_1\sin^2\theta_t}}, \tag{6.180}$$

where ε_2 is the permittivity of the hemispace on which the considered layer lies. Thus, we must take into account that the appointed reflective coefficient generally maintains a permanent value, as occurs for the plane interface. We should remember when analyzing the field inside the layer that the following condition must be obeyed:

$$\int_{\Sigma_2} \left\{ ik\left[\mathbf{n}\times\mathbf{H}_1\right] + \left[\left[\mathbf{n}\times\mathbf{E}_1\right]\times\nabla' g_2\right] + \left(\mathbf{n}\cdot\mathbf{E}_1\right)\nabla' g_2 \right\} d^2\mathbf{r}_2' = 0$$

due to the wave equation theory. Actually, no waves are scattered by the lower interface into the upper hemispace at $\varepsilon_2 = \varepsilon_1\left(F_{23}^{(\mathrm{h})} = 0\right)$, which is reflected by the equality above. Because the field past the upper boundary maintains horizontal polarization, it should be assumed, as before, that $(\mathbf{n}\cdot\mathbf{E}_1) \cong (\mathbf{e}_z\cdot\mathbf{E}_1) = 0$.

We need to compute the value of field $\mathbf{E}_2$ on the upper interface of the layer. In this case, we have $\nabla' g_2 \cong -ik\sqrt{\varepsilon_1} g_2 \mathbf{R}_2/\mathrm{R}_2$; as in the previous case, $\mathrm{R}_2 \cong \hat{\mathrm{R}}_2 + \cos\theta_2\left[\zeta_1(\mathbf{s}_1) - \zeta_2(\mathbf{s}_2')\right]$, where $\hat{\mathrm{R}}_2 = \sqrt{|\mathbf{s}_1 - \mathbf{s}_2'|^2 + d^2}$ and $\cos\theta_2 = d/\hat{\mathrm{R}}_2$. As a result,

$$\mathbf{E}_2(\mathbf{r}_1) \cong -\frac{ikF_{23}^{(\mathrm{h})}}{4\pi}\int_{\Sigma_2}\left\{\left[\sqrt{\varepsilon_1 - \sin^2\theta_i} + \sqrt{\varepsilon_1}\left(\mathbf{e}_z\cdot\frac{\hat{\mathbf{R}}_2}{\hat{\mathrm{R}}_2}\right)\right]\mathbf{E}_1(\mathbf{r}_2') - \right.$$

$$\left. -\sqrt{\varepsilon_1}\mathbf{e}_z\left(\frac{\hat{\mathbf{R}}_2}{\hat{\mathrm{R}}_2}\cdot\mathbf{E}_1\right)\right\} g_2\, d^2\mathbf{s}_2' .$$

The stationary phase method can be used again for calculation of the last integral. We will not repeat all of the details of the calculation and show only the result here:

$$\mathbf{E}_2(\mathbf{r}_1) = F_{23}^{(\mathrm{h})} G(\mathbf{s}_1 - 2\hat{\mathbf{s}}) e^{ik\left[(\mathbf{u}_i\cdot\mathbf{s}_1) + \Phi(\mathbf{s}_1) - \cos\theta_i\zeta_1(\mathbf{s}_1 - 2\hat{\mathbf{s}})\right]}, \tag{6.181}$$

and

$$G(\mathbf{s}) = \frac{1}{2}\left\{\frac{\left[1 + F_{\mathrm{h}}(\mathbf{s})\right]\cos\theta_i}{\sqrt{\varepsilon_1 - \sin^2\theta_i}} + 1 + F_{\mathrm{h}}(\mathbf{s})\right\},$$

$$\Phi(\mathbf{s}) = \sqrt{\varepsilon_1 - \sin^2\theta_i}\left[2d - 2\zeta_2(\mathbf{s} - \hat{\mathbf{s}}) + \zeta_1(\mathbf{s}) + \zeta_1(\mathbf{s} - 2\hat{\mathbf{s}})\right].$$

The equality:

$$\mathbf{E}_1(\mathbf{r}_1)+\mathbf{E}_2(\mathbf{r}_1)=\left[1+F_{\mathrm{h}}(\mathbf{s}_1)\right]e^{ik\left[(\mathbf{u}_i\cdot\mathbf{s}_1)-\cos\theta_i\zeta_1(\mathbf{s}_1)\right]}, \tag{6.182}$$

is true on the upper interface and expresses the boundary conditions for the waves with horizontal polarization. Using Equations (6.174) and (6.181), we can obtain the final equation for the reflective coefficient:

$$F_{\mathrm{h}}(\mathbf{s}_1)+F_{12}^{(\mathrm{h})}F_{23}^{(\mathrm{h})}F_{\mathrm{h}}(\mathbf{s}_1-2\hat{\mathbf{s}})e^{ik\Gamma(\mathbf{s}_1)}=F_{12}^{(\mathrm{h})}+F_{23}^{(\mathrm{h})}e^{ik\Gamma(\mathbf{s}_1)}. \tag{6.183}$$

Here

$$F_{12}^{(\mathrm{h})}=\frac{\cos\theta_i-\sqrt{\varepsilon_1-\sin^2\theta_i}}{\cos\theta_i+\sqrt{\varepsilon_1-\sin^2\theta_i}} \tag{6.184}$$

is the coefficient of the horizontally polarized wave reflection by the upper boundary of the layer, and

$$\Gamma(\mathbf{s})=\Phi(\mathbf{s})+\cos\theta_i\left[\zeta_1(\mathbf{s})-\zeta_1(\mathbf{s}-2\hat{\mathbf{s}})\right]. \tag{6.185}$$

Equation (6.183) can be explained from the point of view of geometrical optics and describes successive processes of wave transparency through the upper boundary of the layer; its reflection is the lower one in the corresponding phase shift. It follows, then, that the equation for the reflective coefficient of the vertically polarized waves has to be analogous to Equation (6.183), replacing the reflective coefficients for the horizontal polarization with the same ones for vertical polarization. We now have, in the general case with horizontal and vertical polarization:

$$F(\mathbf{s})+F(\mathbf{s}-2\hat{\mathbf{s}})F_{12}F_{23}e^{ik\Gamma(\mathbf{s})}=F_{12}+F_{23}e^{ik\Gamma(\mathbf{s})}. \tag{6.186}$$

A simple solution to this equation occurs in the case of vertical incidence, when $\hat{\mathbf{s}}=0$ (compare with Equation (3.36a)):

$$F(\mathbf{s})=\frac{F_{12}+F_{23}e^{ik\Gamma_0(\mathbf{s})}}{1+F_{12}F_{23}e^{ik\Gamma_0(\mathbf{s})}},\quad \Gamma_0(\mathbf{s})=2\sqrt{\varepsilon_1}\left[d+\zeta_1(\mathbf{s})-\zeta_2(\mathbf{s})\right]. \tag{6.187}$$

This solution shows that the reflection at every point is the same as from a layer with plane parallel boundaries and of local thickness.

The method of consecutive approaches can be used in the arbitrary case to solve Equation (6.186). It is based on the fact that the modules of the reflective coefficients from the layer boundaries are less than unity. It follows from Equation (6.186) that, for example,

$$F(\mathbf{s}-2\mathbf{s}) = F_{12} + F_{23}e^{ik\Gamma(\mathbf{s}-2\mathbf{s})} - F(\mathbf{s}-4\mathbf{s})F_{12}F_{23}e^{ik\Gamma(\mathbf{s}-2\mathbf{s})}.$$

By substituting this equality into Equation (6.186), we obtain a new equation, where $F\left(\mathbf{s}-6\hat{\mathbf{s}}\right)$ appears. We can rewrite this term with the help of the rule we just formulated to obtain the expansion with respect to the degrees of the reflective coefficients from the boundaries, similar to the expansion shown in Equation (3.49), which describes the processes of consecutive reflections by the layer boundaries. The series obtained converges rather quickly, especially taking into account the absorption, which is important, as a rule, in the natural formations. Therefore, it is acceptable to restrict ourselves to the first approximation in the form:

$$F(\mathbf{s}) = F_{12} + F_{23}\left(1-F_{12}^2\right)e^{ik\Gamma(\mathbf{s})}. \tag{6.188}$$

So, for the amplitude of the backward scattering, we have:

$$\mathbf{f} = \mathbf{f}_{\text{h}}\left(-\mathbf{e}_i\right) = \frac{ik}{2\pi}\mathbf{g}_i\cos\theta_i\int\limits_{\Sigma_1} F_{\text{h}}\left(\mathbf{s}_1'\right)\, e^{2ik\left(\mathbf{u}_i\cdot\mathbf{s}_1'\right)-2ik\cos\theta_i\zeta_1\left(\mathbf{s}_1'\right)}d^2\mathbf{s}_1'.$$

Also,

$$\begin{aligned}\Omega(\mathbf{s}) = {} & 2\sqrt{\varepsilon_1-\sin^2\theta_i}\left[d-\zeta_2(\mathbf{s}-\hat{\mathbf{s}})\right]+ \\ & +\left(\sqrt{\varepsilon_1-\sin^2\theta_i}-\cos\theta_i\right)\left[\zeta_1(\mathbf{s})+\zeta_1(\mathbf{s}-2\mathbf{s})\right].\end{aligned} \tag{6.189}$$

The index $_1$ is now omitted as it is not necessary in the definition of the surface over which the integration is performed or in the definition of the integration variable. In order to simplify further calculations, we will assume that the permittivity imaginary part is small, which allows us to consider Fresnel coefficients as real values and to take into account the imaginary part of permittivity only when calculating the first term in Equation (6.189).

It is a simple matter to establish that the reflectivity scattering consists of three summands. The first one is:

$$\sigma_1^0\left(-\mathbf{e}_i\right) = \frac{k^2\cos^2\theta_i F_{12}^2}{4\pi^2\Sigma}\iint\limits_{\Sigma}\left\langle e^{2ik\cos\theta_i\left[\zeta_1(\mathbf{s}'')-\zeta_1(\mathbf{s}')\right]}\right\rangle e^{2ik\left(\mathbf{u}_i\cdot(\mathbf{s}'-\mathbf{s}'')\right)}d^2\mathbf{s}'d^2\mathbf{s}''. \tag{6.190}$$

This expression describes the wave scattering due to the upper interface of the layer and adds nothing to the previously regarded scattering by a rough surface separating two semispaces. In this case, we should take into account that this calculation can be used only for small incident angles of the plane wave, and, as a result, the calculation over Equation (6.190) will differ in nonessential details from Equation (6.80). Note that significant cross-sectional values of the backward scattering occur only at rather small incident angles (see, for example, Equation (6.81)).

The second summand is written as:

$$\sigma_2^0\left(-\mathbf{e}_i\right)=\frac{k^2\cos^2\theta_i F_{12}F_{23}\left(1-F_{12}^2\right)}{4\pi^2\Sigma}\iint_{\Sigma}\left\langle e^{-ik\left[\cos\theta_i\zeta_1(\mathbf{s}')+\Omega^*(\mathbf{s}'')\right]}+ \right. \\ \left. +e^{ik\left[\cos\theta_i\zeta_1(\mathbf{s}'')+\Omega(\mathbf{s}')\right]}\right\rangle e^{2ik(\mathbf{u}_i\cdot(\mathbf{s}'-\mathbf{s}''))}d^2\mathbf{s}'d^2\mathbf{s}'', \tag{6.191}$$

and the third one is written as:

$$\sigma_3^0\left(-\mathbf{e}_i\right)=\frac{k^2\cos^2\theta_i F_{23}^2\left(1-F_{12}^2\right)^2}{4\pi^2\Sigma}\iint_{\Sigma}\left\langle e^{ik\left[\Omega(\mathbf{s}')-\Omega^*(\mathbf{s}'')\right]}\right\rangle e^{2ik(\mathbf{u}_i\cdot(\mathbf{s}'-\mathbf{s}''))}d^2\mathbf{s}'d^2\mathbf{s}''. \tag{6.192}$$

Let us deal, first, with the third summand. We perform the standard variable substitution $\mathbf{s}'+\mathbf{s}''=2\mathbf{S}$ and $\mathbf{s}'-\mathbf{s}''=\mathbf{s}$ and use the approximation:

$$\Omega\left(\mathbf{s}'\right)-\Omega^*\left(\mathbf{s}''\right)=2id\,\mathrm{Im}\sqrt{\varepsilon_1-\sin^2\theta_i}-2\sqrt{\varepsilon_1-\sin^2\theta_i}\left(\mathbf{s}\cdot\nabla\zeta_2\left(\mathbf{s}-\hat{\mathbf{s}}\right)\right)+ \\ +\left(\sqrt{\varepsilon_1-\sin^2\theta_i}-\cos\theta_i\right)\left[\left(\mathbf{s}\cdot\nabla\zeta_1\left(\mathbf{s}\right)\right)+\left(\mathbf{s}\cdot\nabla\zeta_2\left(\mathbf{s}-2\hat{\mathbf{s}}\right)\right)\right].$$

The integration with respect to $\mathbf{S}$ is equal to the square Σ and the integration with respect to $\mathbf{s}$ is equal to the delta-function. We will assume the statistical independence of the roughness of the upper and lower boundaries. Taking into account these facts and after averaging over the slopes we obtain:

$$\sigma_3^0\left(-\mathbf{e}_i\right)=\frac{\cos^2\theta_i F_{23}^2\left(1-F_{12}^2\right)^2}{4(\varepsilon_1-\sin^2\theta_i)}e^{-2kd\,\mathrm{Im}\sqrt{\varepsilon_1-\sin^2\theta_i}}\left\langle P_1\left[\nabla\zeta_2(\mathbf{s})\right]=\mathbf{a}\right\rangle, \tag{6.193}$$

and

$$\mathbf{a} = \frac{\mathbf{u}_i}{\sqrt{\varepsilon_1 - \sin^2\theta_i}} + \frac{\left(\sqrt{\varepsilon_1 - \sin^2\theta_i} - \cos\theta_i\right)}{2\sqrt{\varepsilon_1 - \sin^2\theta_i}}\left[\zeta_1(\mathbf{s}+\hat{\mathbf{s}}) + \zeta_1(\mathbf{s}-\hat{\mathbf{s}})\right]. \qquad (6.194)$$

The average value of vector **a** is:

$$\langle \mathbf{a} \rangle = \frac{\mathbf{u}_i}{\sqrt{\varepsilon_1 - \sin^2\theta_i}} \qquad (6.195)$$

and the dispersion is:

$$\sigma_{\mathbf{a}}^2 = \left\langle \left[\mathbf{a} - \langle \mathbf{a} \rangle\right]^2 \right\rangle = \frac{\left(\sqrt{\varepsilon_1 - \sin^2\theta_i} - \cos\theta_i\right)^2}{2(\varepsilon_1 - \sin^2\theta_i)} \left\langle (\nabla\zeta_1)^2 \right\rangle \left[1 + \hat{k}_{\nabla\zeta_1}(2\hat{\mathbf{s}})\right], \qquad (6.196)$$

where $\hat{k}_{\nabla\zeta_1}(2\hat{\mathbf{s}})$ is the spatial correlation coefficient of the upper interface slopes.

Let us now consider the case when random slopes of roughness follow the Gaussian law of probability. In this case,

$$P_1\left[\nabla\zeta_2(\mathbf{s}) = \mathbf{a}\right] = \frac{1}{\pi\left\langle (\nabla\zeta_2)^2 \right\rangle} \exp\left[-\frac{a^2}{\left\langle (\nabla\zeta_2)^2 \right\rangle}\right].$$

Because vector **a** is stochastic, the given value of the probability densities is also a stochastic value, and its median value is:

$$\langle P_1 \rangle = \frac{1}{\pi^2 \sigma_{\mathbf{a}}^2 \left\langle (\nabla\zeta_2)^2 \right\rangle} \int \exp\left[-\frac{a^2}{\left\langle (\nabla\zeta_2)^2 \right\rangle} - \frac{\left(\mathbf{a} - \langle \mathbf{a} \rangle\right)^2}{\sigma_{\mathbf{a}}^2}\right] d^2\mathbf{a}.$$

The obtained integral is easily computed:

$$\sigma_3^0(-\mathbf{e}_i) = \frac{\cos^2\theta_i F_{23}^2\left(1-F_{12}^2\right)^2}{4\pi\varepsilon_1\cos^2\theta_t\left[\sigma_{\mathbf{a}}^2+\left\langle\left(\nabla\zeta_2\right)^2\right\rangle\right]}\times$$
$$\times\exp\left\{-2n''kd\cos\theta_t-\frac{\tan^2\theta_t}{\left[\sigma_{\mathbf{a}}^2+\left\langle\left(\nabla\zeta_2\right)^2\right\rangle\right]}\right\}. \tag{6.197}$$

The third summand under consideration describes the power of the waves scattered by the lower boundary.

Finally, we turn our attention back to the second summand, which describes interference of the waves scattered from both layer interfaces. This summand is found to be less than the other two, given our assumptions; this finding is readily understood in light of the Gaussian law of probability. The described component is proportional, in this case, to the value $\exp(-a_1k^2\langle\zeta_1^2\rangle - a_2k^2\langle\zeta_1^2\rangle)$, where the coefficients a_1 and a_2 are of the order of unity. Because we are concerned with large inhomogeneities, we can conclude that the interference term is negligibly small. The scattered waves have strong fluctuating phases, and the phase relations play a primary role in the interference. Because the roughness values of the upper and lower boundaries are statistically independent, phases of the waves scattered by them are also statistically independent, which suggests low coherence of the interfering waves and, consequently, the relatively small interference term.

7 Radiowave Propagation in a Turbulent Medium

7.1 PARABOLIC EQUATION FOR THE FIELD IN A STOCHASTIC MEDIUM

We have already discussed radiowave propagation in a turbulent medium using the geometrical optics approximation. The results obtained in Chapter 3 are limited to small fluctuations of wave parameters (amplitude and phase) that, because of an accumulative effect, do not always occur even at small fluctuations of the medium permittivity. Moreover, the geometrical optics approach is restricted to the value of the distance traveled by the wave in a turbulent medium, as it does not take into account the effects of wave diffraction on the inhomogeneities. It is possible to use the geometrical optics approximation in cases where the Fresnel zone is smaller than the inhomogeneity scale (i.e., when $\overline{l} >> \sqrt{\lambda L}$, where $\overline{l}$ is some scale between the outer and inner scales of the turbulence). For these reasons, we must develop a mathematical instrument to describe in greater detail the processes of wave propagation in a turbulent medium. One such instrument is the parabolic equation method.

The parabolic equation method is based on the following approach. Because we are dealing with a medium with little fluctuation of dielectric permittivity, we can neglect the polarization effects and describe the field by the scalar wave equation written for the radiated component of the electromagnetic field. It is necessary to take into account that turbulent pulsation of the medium generally has dimensions that are significantly greater than the wavelength which means that forward wave scattering by turbulent inhomogeneities is of primary important (see Chapter 5). We can apply the small angle approximation and use a parabolic equation of the type shown in Equation (1.81) for the fluctuating field description. If we insert Equation (1.80) into the wave equation then we can obtain the equation for the fluctuating field:

$$2ik\frac{\partial U}{\partial z} + \nabla_{\perp}^{2}U + k^{2}\mu U = 0 \tag{7.1}$$

which is similar to Equation (1.81). Thus, we can suppose, as before, that permittivity differs little from unity. Substitution of the wave equation with the parabolic one means that we are neglecting waves scattered backward. Note that if we neglect transverse derivatives of the field, then the typical geometrical optics approach follows from Equation (7.1):

$$U_{go}\left(\mathbf{s},z\right)=U_0\exp\left(\frac{ik}{2}\int_0^z\mu\left(\mathbf{s},z\right)dz\right). \tag{7.2}$$

Here **s** is the vector orthogonal to the main wave propagation direction (z-axis). The solution of the obtained parabolic equation is a random function. The first step is to determine the mean field or its coherent component. Let us designate $\langle U\rangle=\bar{U}$, then average Equation (7.1) to obtain:

$$2ik\frac{\partial\bar{U}}{\partial z}+\nabla_{\perp}^2\bar{U}+k^2\langle\mu U\rangle=0. \tag{7.3}$$

To solve this equation is rather difficult because the explicit form of the last term is unknown. The situation is improved when the permittivity fluctuation can be considered Gaussian. The Novikov–Furutzu formula can be applied in this case,[53] and we obtain:

$$\left\langle\mu\left(\mathbf{s},z\right)U\left(\mathbf{s},z\right)\right\rangle=\iint K_{\varepsilon}\left(\mathbf{s}-\mathbf{s}',z-z'\right)\left\langle\frac{\delta U\left(\mathbf{s},z\right)}{\delta\mu\left(\mathbf{s}',z'\right)}\right\rangle d^2\mathbf{s}'dz'. \tag{7.4}$$

Here, we deliberately separated the transverse and longitudinal coordinates. The reason for this separation will become clear later. Using the Novikov–Furutzu formula does not improve the situation without implementing further simplifications related to the necessity to have an explicit view of the functional derivative $\delta U\left(\mathbf{s},z\right)/\delta\mu\left(\mathbf{s}',z'\right)$. We can estimate its value using Equation (7.2), and the functional derivative is:

$$\frac{\delta U_{go}\left(\mathbf{s},z\right)}{\delta\mu\left(\mathbf{s}',z'\right)}=\frac{ik}{2}U_{go}\left(\mathbf{s},z\right)\int_0^z\frac{\delta\mu\left(\mathbf{s},z\right)}{\delta\mu\left(\mathbf{s}',z'\right)}dz.$$

According to the rules of the functional analysis:

$$\frac{\delta\mu\left(\mathbf{s},z\right)}{\delta\mu\left(\mathbf{s}',z\right)}=\delta\left(\mathbf{s}-\mathbf{s}'\right)\Theta\left(z-z'\right), \tag{7.5}$$

where the step function is:

$$\Theta(z-z') = \int_0^z \delta(z-z')dz = \begin{cases} 0 & \text{for } z'>z, \\ 1/2 & \text{for } z'=z, \\ 1 & \text{for } z'<z. \end{cases} \tag{7.6}$$

As a result, we can write:

$$\left\langle \frac{\delta U_{go}(\mathbf{s},z)}{\delta\mu(\mathbf{s}',z')} \right\rangle = \frac{ik}{2}\left\langle U_{go}(\mathbf{s},z)\right\rangle \Theta(z-z')\delta(\mathbf{s}-\mathbf{s}'), \tag{7.7}$$

and it is easy to find that:

$$\left\langle U_{go}(\mathbf{s},z)\right\rangle = U_0 \exp\left[-\frac{k^2}{8}\int_0^z\int_0^z \mathrm{K}_\varepsilon(0,z'-z'')dz'dz''\right]. \tag{7.8}$$

To solve this expression it is necessary to take into account that the integral in Equation (7.2) is subject to Gaussian statistics. If the distance traveled by the wave in the turbulent medium sufficiently exceeds the turbulence scale, then, as we have already shown, the integral in the last formula is estimated by the value of the order $\langle \mu^2 \rangle l z$. It follows, then, that the scale of the mean field change along the wave propagation direction is estimated by the value $1/k^2 \langle \mu^2 \rangle l$. Consequently, the scale of the functional derivative change is the same. This scale is much larger than the turbulence scale as it is assumed that the phase fluctuation shift in the one-inhomogeneity frame is $k\sqrt{\langle \mu^2 \rangle} l \ll 1$. The integration by z′ in Equation (7.4) is concentrated closely to z in the interval of the order l. It means that the functional derivative is not changed within the essential interval of integration. Therefore, we can substitute z for z′ in the functional derivative, Equation (7.4). Then, we can prove that in our case:

$$\left\langle \frac{\delta U(\mathbf{s},z)}{\delta\mu(\mathbf{s}',z)} \right\rangle = \frac{ik}{4}\bar{U}(\mathbf{s},z)\delta(\mathbf{s}-\mathbf{s}'). \tag{7.9}$$

From Equation (7.4) we have:

$$\left\langle \mu(\mathbf{s},z)U(\mathbf{s},z)\right\rangle = \frac{ik}{4}\bar{U}(\mathbf{s},z)A(0) \tag{7.10}$$

and the function (see Equations (4.78) and (4.79)):

$$A(\mathbf{s}) = \int_{-\infty}^{\infty} \mathrm{K}_{\varepsilon}\left(\mathbf{s}, \mathrm{z}\right) dz = 2\pi \int \tilde{\mathrm{K}}_{\varepsilon}\left(\mathbf{q}_{\perp}, 0\right) e^{i\mathbf{q}_{\perp}\cdot\mathbf{s}} d^2\mathbf{q}_{\perp}. \tag{7.11}$$

It should be pointed out that it is not difficult to achieve the same result for the delta-correlation of the permittivity fluctuations in the wave propagation direction, having assumed that:

$$\mathrm{K}_{\varepsilon}(\mathbf{s}, \mathrm{z}) = A(\mathbf{s})\,\delta(\mathrm{z}). \tag{7.12}$$

This rather artificial assumption reflects the fact that the scale of the field change in the wave propagation direction is much larger than the turbulence scale. This also emphasizes the need to use the Markovian approximation for the field computation. Using this approximation, the field is defined in each point by the previous values of permittivity over the z-axis. As a result, the mean field is described by the equation:

$$2ik\frac{\partial \bar{U}}{\partial \mathrm{z}} + \nabla_{\perp}^2 \bar{U} + \frac{ik^3}{4} A(0)\bar{U} = 0 \tag{7.13}$$

assuming Gaussian statistics of permittivity and a Markovian approximation. Let V represent the solution of the parabolic equation at $A(0) = 0$. Then, the general solution, Equation (7.13), can be written down in the form:

$$\bar{U}\left(\mathbf{s}, \mathrm{z}\right) = V\left(\mathbf{s}, \mathrm{z}\right) \exp\left[-\frac{k^2}{8} A(0)\,\mathrm{z}\right]. \tag{7.14}$$

The solution for function V is represented by Equation (1.78). Also, the role of turbulent pulsation of the medium becomes apparent when extinction is considered to calculate the mean (coherent) field. The corresponding extinction coefficient is equal to:

$$\Gamma = \frac{k^2}{4} A(0) = \frac{\pi k^2}{2} \int \tilde{\mathrm{K}}_{\varepsilon}\left(\mathbf{q}_{\perp}, 0\right) d^2\mathbf{q}_{\perp} = \pi^2 k^2 \int_0^{\infty} \tilde{\mathrm{K}}_{\varepsilon}\left(\mathrm{q}_{\perp}, 0\right) \mathrm{q}_{\perp} d^2\mathrm{q}_{\perp}. \tag{7.15}$$

The last formula in this series is obtained by assuming turbulence isotropy. If we compare it with Equation (5.178), then it becomes a simple matter to determine that $\Gamma = \sigma_{\mathrm{s}}^{0}$, which corresponds to the theory of scattering. We recall that σ_{s}^{0} is the wave scattering by unit volume of the turbulent inhomogeneities. It follows, from this comparison, that the extinction coefficient is approximately equal to $\left\langle \mu^2 \right\rangle k^2 l$

(a more specific value is given, for example, by Equation (5.179)). Let us point out in this context that the estimation presented by us, *a priori*, of the rapidity of field attenuation with distance does not differ from the one defined by the extinction coefficient.

7.2 THE FUNCTION OF MUTUAL COHERENCE

In this part, we will derive an equation for the second moment of the field, according to the same assumption as before. First of all, we will consider the function of mutual coherence:

$$\Gamma_U\left(\mathbf{s}_1,\mathbf{s}_2,z\right)=\frac{1}{2}\left\langle U\left(\mathbf{s}_1,z\right)U^*\left(\mathbf{s}_2,z\right)\right\rangle. \tag{7.16}$$

This function corresponds to the definition provided earlier by Equation (1.124), and we derived Equation (7.16) in much the same way as Equation (1.125). As a result, we now have:

$$2ik\frac{\partial\Gamma_U}{\partial z}+\left(\nabla^2_{\mathbf{s}_1}-\nabla^2_{\mathbf{s}_2}\right)\Gamma_U+k^2\left\langle\left[\mu\left(\mathbf{s}_1,z\right)-\mu\left(\mathbf{s}_2,z\right)\right]U\left(\mathbf{s}_1,z\right)U^*\left(\mathbf{s}_2,z\right)\right\rangle=0. \tag{7.17}$$

Here, we have introduced subscripts representing variables with respect to which differentiation in the transverse Laplacian is performed. Further, the reasoning used to derive the equation for the coherent field component can also be applied in this case. Doing so gives us:

$$2ik\frac{\partial\Gamma_U}{\partial z}+\left(\nabla^2_{\mathbf{s}_1}-\nabla^2_{\mathbf{s}_2}\right)\Gamma_U+\frac{ik^3}{2}\left[A\left(0\right)-A\left(\mathbf{s}_1-\mathbf{s}_2\right)\right]\Gamma_U=0. \tag{7.18}$$

The equation for the other second moment is derived in the same way:

$$\tilde{\Gamma}_U\left(\mathbf{s}_1,\mathbf{s}_2,z\right)=\left\langle U\left(\mathbf{s}_1,z\right)U\left(\mathbf{s}_2,z\right)\right\rangle. \tag{7.19}$$

and can be written as:

$$2ik\frac{\partial\tilde{\Gamma}_U}{\partial z}+\left(\nabla^2_{\mathbf{s}_1}+\nabla^2_{\mathbf{s}_2}\right)\tilde{\Gamma}_U-\frac{ik^3}{2}\left[A(0)+A(s_1-s_2)\right]\tilde{\Gamma}_U=0. \tag{7.20}$$

In order to solve Equation (7.17), let us utilize the method used to solve Equation (1.125):

$$k\frac{\partial\tilde{\Gamma}_U}{\partial z}+\left(\mathbf{W}\frac{\partial\tilde{\Gamma}_U}{\partial\mathbf{s}}\right)+\frac{k^3}{4}H(z,\mathbf{s})=0 \tag{7.21}$$

instead of Equation (1.128), represented here, for short, as:

$$H(z,\mathbf{s})=A(0)-A(\mathbf{s})=2\pi\int\left(1-e^{i\mathbf{q}_\perp\cdot\mathbf{s}}\right)\tilde{K}_\varepsilon\left(z,\mathbf{q}_\perp,0\right)d^2\mathbf{q}_\perp. \tag{7.22}$$

We deliberately separated out in this definition the z variable to emphasize the possible dependence of turbulence parameters on the longitudinal coordinate (e.g., altitude in the atmosphere of Earth). Furthermore, we will make the following changes to the variables: $\mathbf{s}-\mathbf{W}z/k=\mathbf{s}'$, $z=z$ (the first of these two equations is the equation of characteristics). As a result of these changes, the ordinary differential equation now can be written as:

$$\frac{d\tilde{\Gamma}_U}{dz}=-\frac{k^2}{4}H\left(z,\mathbf{s}'+\frac{\mathbf{W}z}{k}\right)$$

with the solution:

$$\tilde{\Gamma}_U=M\left(\mathbf{W},\mathbf{s}'\right)\exp\left[-\frac{k^2}{4}\int_0^z H\left(z',\mathbf{s}'+\frac{\mathbf{W}z'}{k}\right)dz'\right]$$

or

$$\tilde{\Gamma}_U=M\left(\mathbf{W},\,\mathbf{s}-\frac{\mathbf{W}z}{k}\right)\exp\left\{-\frac{k^2}{4}\int_0^z H\left[z',\,\mathbf{s}-\frac{\mathbf{W}}{k}\left(z-z'\right)\right]dz'\right\}. \tag{7.23}$$

The function $M\left(\mathbf{W},\mathbf{s}\right)$ is determined from the boundary condition:

$$\Gamma_U(\mathbf{S},\mathbf{s},0)=\Gamma_0(\mathbf{S},\mathbf{s}), \tag{7.24}$$

from which it follows that:

$$M(\mathbf{W},\mathbf{s})=\frac{1}{4\pi^2}\int\Gamma_0(\mathbf{S},\mathbf{s})e^{-i\mathbf{W}\mathbf{S}}d^2\mathbf{S}. \tag{7.25}$$

As a result, we have the following common solution:

$$\Gamma_U = \frac{1}{4\pi^2} \iint \Gamma_0 \left(\mathbf{S}', \mathbf{s} - \frac{\mathbf{W}z}{k} \right) \exp\left\{ i\mathbf{W}(\mathbf{S} - \mathbf{S}') - \right.$$
$$\left. - \frac{k^2}{4} \int_0^z H\left(z, \mathbf{s} - \frac{\mathbf{W}z}{k} \right) dz \right\} d^2\mathbf{S}' \, d^2\mathbf{W} \tag{7.26}$$

or after substituting $\mathbf{W} = k(\mathbf{s} - \mathbf{s}')/\mathrm{z}$ we obtain:

$$\Gamma_U = \frac{k^2}{4\pi^2 \mathrm{z}^2} \iint \Gamma_0(\mathbf{S}', \mathbf{s}') \exp\left\{ i\frac{k}{\mathrm{z}}(\mathbf{S} - \mathbf{S}')(\mathbf{s} - \mathbf{s}') - \right.$$
$$\left. - \frac{k^2}{4} \int_0^{\mathrm{z}} H\left[\mathrm{z}, \mathbf{s} - \frac{z}{\mathrm{z}}(\mathbf{s} - \mathbf{s}') \right] dz \right\} d^2\mathbf{S}' d^2\mathbf{s}'. \tag{7.27}$$

The solution of the equation for the function $\tilde{\Gamma}_U$ unfortunately cannot be found by any analytical form; therefore, it is necessary to resort to approximate methods for its solution.

7.3 PROPERTIES OF THE FUNCTION *H*

Before turning to particular examples, let us discuss in detail some properties of the function $H(\mathrm{z,s})$, which mainly determines the behavior of the coherence function (see Equation (7.22)). Let us assume isotropic turbulence for simplicity. This allows us to obtain:

$$H(\mathrm{z,s}) = 4\pi^2 \int_0^\infty \left[1 - J_0(\mathrm{q}_\perp \mathrm{s}) \right] \tilde{\mathrm{K}}_\varepsilon(\mathrm{z}, \mathrm{q}_\perp, 0)\, \mathrm{q}_\perp d\mathrm{q}_\perp \tag{7.28}$$

by integration in the polar coordinate system. Let us point out once more that, in the case of turbulence isotropy, the function *H* depends only on the distance between the points of correlation. Naturally, this property is retained in the coherence function. To be specific in further calculations, we can use Equation (4.80), and we obtain:

$$H(\mathrm{z,s}) = 4\pi^2 C_\varepsilon^2(\mathrm{z}) \int_0^\infty \left[1 - J_0(\mathrm{q}_\perp \mathrm{s}) \right] \frac{e^{-\mathrm{q}_\perp^2/\mathrm{q}_m^2}\, \mathrm{q}_\perp d\mathrm{q}_\perp}{\left(1 + \mathrm{q}_\perp^2/\mathrm{q}_0^2 \right)^{\nu}}. \tag{7.29}$$

In this case, we can neglect the role of small-scale turbulence. Then, it is convenient to use the relation:

$$\int_0^\infty \frac{J_p(bx)x^{p+1}}{\left(x^2+a^2\right)^{q+1}}dx = \frac{a^{p-q}b^q}{2^q\Gamma(q+1)}K_{q-p}(ab), \tag{7.30}$$

known from the theory of Bessel functions.[44] Here, $K_p(x)$ is the Macdonald function,[44] and

$$H = \frac{2\pi^2 q_0^2 C_\varepsilon^2(z)}{\nu-1}\left[1-\frac{(\nu-1)(q_0 s)^{\nu-1}}{2^{\nu-2}\Gamma(\nu)}K_{\nu-1}(q_0 s)\right]. \tag{7.31}$$

Further, we are particularly interested in the case when $q_0 s \ll 1$. Let us point out, however, that because we have neglected the role of small-scale turbulence it is also necessary to observe the inequality $q_m s \gg 1$. The following approximation is applicable for MacDonald functions at small values of argument:

$$K_p(x) \cong \frac{2^{p-1}\Gamma(p)}{x^p}\left[1-\frac{1}{(p-1)}\left(\frac{x}{2}\right)^2-\frac{\Gamma(1-p)}{\Gamma(1+p)}\left(\frac{x}{2}\right)^{2p}\right]. \tag{7.32}$$

This relationship is true when index p is not an integer number; however, the expansion has a rather different form when the subscripts are integers (which will not be considered here). We can see that the last term of Equation (7.32) should be kept at $p < 1$, and it is acceptable to have the second term in the opposite case. So, for the small separation, we have greater accuracy in those cases for which the correlation distance is much smaller than the turbulence scale:

$$H \cong \frac{2\pi^2 q_0^2 C_\varepsilon^2(z)}{\nu-1}\left[\frac{1}{\nu-2}\left(\frac{q_0 s}{2}\right)^2+\frac{\Gamma(2-\nu)}{\Gamma(\nu)}\left(\frac{q_0 s}{2}\right)^{2\nu-2}\right]. \tag{7.33}$$

At large values of argument, the MacDonald function tends exponentially to zero, and function H is equal to the first factor of Equation (7.31).

Studying the problem in general, we can confirm that the spatial coherence of a wave propagating in a turbulent medium becomes small under the following the condition:

$$\frac{k^2}{4}\int_0^Z H\left[z, s-\frac{z}{Z}(s-s')\right]dz \geq 1. \tag{7.34}$$

7.4 THE COHERENCE FUNCTION OF A PLANE WAVE

Let us now consider a plane wave as an example of calculating the coherence function. In this case, the excitement sources are uniform, in the plane $z = 0$. So, $\Gamma_0(\mathbf{S},\mathbf{s}) = u_0^2/2$ for the plane wave, where u_0^2 is the density of the power source. It is easy to see in this case that integration with respect to $\mathbf{S}'$ gives us $4\pi^2\delta(\mathbf{W})$, and, as a result:

$$\Gamma_U(\mathbf{s},z) = \frac{u_0^2}{2}\exp\left[-\frac{k^2}{4}\int_0^z H(z,\mathbf{s})\,dz\right]. \tag{7.35}$$

We shall assume hereafter that the turbulence spectral index is less than 2. We can only keep the second summand from Equation (7.33). We can now write, assuming constant turbulence parameters along the wave propagation direction,

$$\Gamma_U(\mathbf{s},z) = \frac{u_0^2}{2}\exp\left[-\frac{\sqrt{\pi}\,\Gamma(2-\nu)(q_0 s)^{2\nu-2}\langle\mu^2\rangle k^2 z}{2^{2\nu-1}(\nu-1)\Gamma(\nu-3/2)q_0}\right]. \tag{7.36}$$

Let us now determine the coherence scale by equating the exponent index to unity. The scale represented in such a way is defined from the equation:

$$(q_0 s_{cog})^{2\nu-2} = \frac{2^{2\nu-1}(\nu-1)\Gamma(\nu-3/2)q_0}{\sqrt{\pi}\,\Gamma(2-\nu)\langle\mu^2\rangle k^2 z}. \tag{7.37}$$

At the same time, we must maintain $q_0 s_{cog} \ll 1$ to make Equation (7.36) valid. Obviously, it is possible if the wave has traveled such a large distance that:

$$\langle\mu^2\rangle k^2 z/q_0 \cong \langle\mu^2\rangle k^2 l z \gg 1. \tag{7.38}$$

Let us emphasize that the coherence scale is smaller than the turbulence scale, although it can be greater in those cases when a weak inequality substitutes for a strong one; also, in all of the cases considered here, extinction is appreciable and the field coherent component is small. For the Kolmogorov spectrum, $\nu = 11/6$, and the equation for the coherence scale can be formulated in the simple form:

$$\left(\frac{s_{cog}}{l}\right)^{5/3} = \frac{1.07}{\langle\mu^2\rangle k^2 l L}, \tag{7.39}$$

where the distance traveled (z) was replaced with L.

7.5 THE COHERENCE FUNCTION OF A SPHERICAL WAVE

The sources function for the spherical wave is given as the product:

$$\Gamma_0\left(\mathbf{S}',\mathbf{s}'\right)=\frac{2\pi^2 u_0^2}{k^2}\delta\left(\mathbf{S}'+\frac{\mathbf{s}'}{2}\right)\delta\left(\mathbf{S}'-\frac{\mathbf{s}'}{2}\right). \tag{7.40}$$

The integration in Equation (7.27) is easily performed to give us:

$$\Gamma_U\left(\mathbf{S},\mathrm{s},\mathrm{z}\right)=\frac{u_0^2}{2\mathrm{z}^2}\exp\left\{i\frac{k\left(\mathbf{S}\mathrm{s}\right)}{\mathrm{z}}-\frac{k^2\mathrm{z}}{4}\int_0^1 H\left[\mathrm{z}\left(1-\zeta\right),\mathrm{s}\zeta\right]d\zeta\right\} \tag{7.41}$$

after a simple transform. Here, the dependence on vector **S** appears distinct from the plane wave. In some instances, this dependence is not valid, such as when a spherical wave can be seen in the approach of a plane wave. This dependence disappears when we consider a field on a sphere of constant radius. So, further, we shall restrict ourselves to the case of **S** = 0. Instead of (7.36), we now have the following:

$$\Gamma_U\left(0,\mathrm{s},\mathrm{z}\right)=\frac{u_0^2}{2\mathrm{z}^2}\exp\left[-\frac{\sqrt{\pi}\,\Gamma\left(2-\nu\right)\left(q_0\mathrm{s}\right)^{2\nu-2}\left\langle\mu^2\right\rangle^2 k^2\mathrm{z}}{2^{2\nu-1}\left(\nu-1\right)\left(2\nu-1\right)\Gamma\left(\nu-3/2\right)q_0}\right]. \tag{7.42}$$

The following estimation is true for a spherical wave rather than Equation (7.39):

$$\left(\frac{\mathrm{s}_{\mathrm{cog}}}{l}\right)^{5/3}=\frac{2.85}{\left\langle\mu^2\right\rangle k^2 l L}. \tag{7.43}$$

The coherence radius for the spherical wave is approximately twice as large as that for the plane wave.

8 Radio Thermal Radiation

8.1 EXTENDED KIRCHHOFF'S LAW

The background of the thermal radiation theory of heated bodies will be discussed in this chapter. This radiation appears as the result of random motion of charged particles inside the body. The velocities of this movement are stochastic and, in particular, they change their value and direction occasionally as a result of the interaction (collisions) of particles with each other. The radiated field strength is random, and its intensity depends on the particle energy and, consequently, on the temperature of the body. In this connection, the radiation under discussion is referred to as *thermal*.

We have to imagine that these bodies, and the fluctuation field generated by them, are in a giant thermostat that maintains the thermodynamic balance. This means that the charged particles of the body interact with the given fluctuation field, derive energy from it, reradiate it afresh, and then absorb, reradiate, and so on. In a word, the radiating and absorbing energies are balanced in an equilibrium state for the fluctuation field. The fluctuation field itself can be described as the field radiated by random currents with density **j**(**r**). The mean value of this density is equal to zero, and the spatial correlation function of its frequency spectrum is defined on the basis of the fluctuation–dissipation theorem (FDT):[24,56]

$$\left\langle j_\alpha\left(\omega,\mathbf{r}'\right) j_\beta\left(\omega,\mathbf{r}''\right)\right\rangle = \frac{\hbar\omega^2}{8\pi^2}\coth\left(\frac{\hbar\omega}{2k_bT}\right)\varepsilon''\left(\omega\right)\delta\left(\mathbf{r}'-\mathbf{r}''\right)\delta_{\alpha\beta}. \tag{8.1}$$

Here $\hbar = h/2\pi = 1.05\cdot 10^{-27}\,\text{erg}\cdot\text{sec}$ is Planck's constant. The subscripts α and β represent corresponding coordinate components of the current vector. The FDT, described in such a way, is correct over practically the entire electromagnetic spectrum (at least, for wavelengths that exceed interatomic or intermolecular distances). The energy quantum in the radio region is small (i.e., the inequality $\hbar\omega \ll k_bT$ is true). Therefore, we will use this approach when the averaged energy of quantum oscillator:

$$\Theta(\omega,T) = \frac{\hbar\omega}{2}\coth\left(\frac{\hbar\omega}{2k_bT}\right) \tag{8.2}$$

is substituted for its approximate value $\Theta = k_b T$, and

$$\left\langle j_\alpha(\omega, \mathbf{r}') j_\beta(\omega, \mathbf{r}'') \right\rangle = \frac{\omega}{4\pi^2} \varepsilon''(\omega) k_b T \delta(\mathbf{r}' - \mathbf{r}'') \delta_{\alpha\beta} \tag{8.3}$$

is valid in the radiofrequency band. It is characteristic that orthogonal components of the fluctuation current are not correlated. For the current components themselves, the spatial correlation radius in this case is equal to zero. In fact, it has the scale of particles interaction — for example, interatomic distances in a solid body or free path length in a gas. Because the wavelengths considered here exceed these distances, it is possible to neglect their variation from zero. The most important fact is that the spectral density of the fluctuation current is defined by the imaginary part of the body permittivity (i.e., its ability to absorb electromagnetic waves). Kirchhoff's law applies here implicitly, as it connects radiating and absorbing properties of the body. Let us point out in this connection, that we are dealing with nonmagnetic materials, so the magnetic losses default, and we do not need to represent the magnetic fluctuation currents.

In order to calculate the fields generated by fluctuation currents, we need to know Green's function of the considered body — that is, the diffraction field excited inside the body by a single current source:

$$\mathbf{j}_G(\mathbf{r}') = \mathbf{e}\delta(\mathbf{r} - \mathbf{r}'), \tag{8.4}$$

where $\mathbf{e}$ is a single vector, generally speaking, of arbitrary direction. The field $\mathbf{E}_d, \mathbf{H}_d$, excited by this current, is the diffraction field. To determine the fluctuation field, we can use the mutuality theorem in the form of Equation (1.64). The fluctuation current and field is represented by $\mathbf{j}_1, \mathbf{E}_1$, while $\mathbf{j}_2, \mathbf{E}_2$ represents the current (Equation (8.4)) and the diffraction field generated by it. Omitting unnecessary subscripts, we now have the following equation for the fluctuation field:

$$\mathbf{e} \cdot \mathbf{E}(\mathbf{r}) = \int_V \mathbf{j}(\mathbf{r}') \cdot \mathbf{E}_d(\mathbf{r}, \mathbf{r}', \mathbf{e}) d^3\mathbf{r}'. \tag{8.5}$$

Also, we have the expression for the calculation of the fluctuation field component, oriented in the direction of vector $\mathbf{e}$. Its average value is equal to zero, as the average value of the fluctuation current is also equal to zero. In this connection, let us point out that the diffraction field is the determining value. Let us also emphasize another very important fact. The imaginary part of the dielectric permittivity in Equations (8.1) and (8.3) can be a function of the coordinates. Particularly, it can be equal to zero if, for example, part of the considered volume V occupies a vacuum. So, the volume can include as the actual heated body, which serves as a fluctuation field source, any part of space up to infinity. It is important that point $\mathbf{r}$ of the dipole, existing in the diffraction field, is situated inside the volume, but it can be outside of the material body. In particular, it can be removed from the radiating body so far that the incident on the body wave is practically a plane wave.

Due to the isotropy of the fluctuation currents and statistical properties and based on the FDT, we now have the following for the mean intensity of the field component:

$$\left\langle \left| \mathbf{e} \cdot \mathbf{E}(\mathbf{r}) \right|^2 \right\rangle = \frac{\omega k_b T}{4\pi^2} \int_V \varepsilon''(\omega, \mathbf{r}') \left| \mathbf{E}_d(\mathbf{r}, \mathbf{r}', \mathbf{e}) \right|^2 d^3\mathbf{r}'. \tag{8.6}$$

If we now recall Equation (1.20), describing the density of losses of electromagnetic energy in a substance, then the last result can be rewritten as:

$$\left\langle \left| \mathbf{e} \cdot \mathbf{E}(\mathbf{r}) \right|^2 \right\rangle = \frac{2 k_b T}{\pi} \int_V Q(\mathbf{r}, \mathbf{r}', \mathbf{e}) \, d^3\mathbf{r}'. \tag{8.7}$$

So, the intensity of the fluctuation field is determined by the value of the thermal losses of the diffraction field excited in the body by a unit current applied at the point where the fluctuation field is being studied and directed according to the vector of the fluctuation field polarization.

The result obtained is sometimes referred to as the extended Kirchhoff's law. It is called *extended* because the law initially formulated by Kirchhoff was restricted to the case of a body large in size compared to the wavelength (i.e., the geometrical optics problem). No such limit is stated in the relations being considered here, so the extended Kirchhoff's law applies.

We should point out the dependence of the integrands in the previous formulae on vector **e** and the diffraction field dependence on the auxiliary dipole polarization; thus, the polarization is dependent on the radiation of the heated bodies. Let us suppose now that the radiation is detected by a receiver responding to only one linear polarization. We can assume that the receiver is rather distant and detects the waves with polarization orthogonal to the line connecting the receiver and the center of gravity of the radiating body. We can direct the z-axis along this line. The discussion in this case is about the reception of waves, the polarization of which is oriented in the plane {x, y}. Let us consider the case of a receiver detecting the x-polarization. In this case, it will react not only to the x-polarization waves but also to waves polarized in the plane. The difference between these waves and the x-polarized waves is that the power of their fluctuations will be detected by the receiver with the weight $\cos^2 \eta$, where η is the angle between vectors $\mathbf{e}$ and $\mathbf{e}_x$. In the case of statistical independence of waves of different polarization, the fluctuation intensity of the detected x-polarized field will be equal to the weighted sum of intensities of all the fields polarized in the {x, y} plane. In other words,

$$\left\langle \left| E_x \right|^2 \right\rangle = \frac{2 k_b T}{\pi} \int_0^{2\pi} \int_V Q(\mathbf{r}, \mathbf{r}', \eta) \cos^2 \eta \, d\eta \, d^3\mathbf{r}'. \tag{8.8}$$

In the case of the receiver responding only to the y-component of waves, we have:

$$\left\langle \left| \mathrm{E}_{\mathrm{y}} \right|^2 \right\rangle = \frac{2k_{\mathrm{b}}T}{\pi} \int_0^{2\pi} \int_V Q(\mathbf{r},\mathbf{r}',\eta) \sin^2 \eta \, d\eta \, d^3\mathbf{r}'. \tag{8.9}$$

If the receiver detects both orthogonal polarizations, then the power of the resulting field will be equal to the sum of Equations (8.8) and (8.9).

Let us now compute the spectral density of Poynting's vector z-component for different polarizations. To do so, we must take into account that the spectral densities given by Equation (8.7) and others are two-way (i.e., they are applicable over the entire real axis of frequencies from – to +. We, however, are interested in a one-way spectral density that is associated with the positive half-axis of frequencies. This means that the previous expressions should be multiplied by 2:

$$\begin{Bmatrix} \tilde{\mathrm{S}}_{\mathrm{x}}(\omega) \\ \tilde{\mathrm{S}}_{\mathrm{y}}(\omega) \end{Bmatrix} = \frac{c}{2\pi} \begin{Bmatrix} \left\langle \left| \mathrm{E}_{\mathrm{x}} \right|^2 \right\rangle \\ \left\langle \left| \mathrm{E}_{\mathrm{y}} \right|^2 \right\rangle \end{Bmatrix} = \frac{ck_{\mathrm{b}}T}{\pi^2} \int_0^{2\pi} \int_V Q(\mathbf{r},\mathbf{r}',\eta) \begin{Bmatrix} \cos^2\eta \\ \sin^2\eta \end{Bmatrix} d\eta \, d^3\mathbf{r}'. \tag{8.10}$$

8.2 RADIO RADIATION OF SEMISPACE

Let us now consider the radiation of semispace. We will first consider the case when the receiving antenna is situated in a vacuum and directed perpendicularly to the plane boundary of the radiating medium. In this case, the diffraction field and, consequently, the volume density of absorption do not depend on the polarization. The integration in Equation (8.10) can be performed over the angle to give the factor π. Further, we are interested in the spectral density of the power flow that is detected by the receiver for any linear polarization:

$$\tilde{W}_{\mathrm{x}}(\omega) = \frac{\omega c k_{\mathrm{b}} T \varepsilon''(\omega)}{8\pi^2} \int A_e(\mathbf{s}) d^2\mathbf{s} \int_0^{\infty} \left| \mathrm{E}_{\mathrm{d}}(\mathbf{s},\mathrm{z}) \right|^2 d^2\mathrm{z}\,. \tag{8.11}$$

Here, A_e is the antenna area, and integration with respect to **s** represents integration over the plane that is perpendicular to the z-axis. Writing the expression in such a form shows that we have implicitly used the geometrical optics approach. Equation (8.11) assumes that the radiation field in the view of plane waves coming from different directions reaches the antenna and summarizes their intensities with a weight given by the antenna area value. One can point out, in this connection, that this approximation requires the position of the point in the field being searched to be at a distance from the interface much greater than the wavelength. Equation (8.11) implies at the same time that the antenna reacts only to polarization orthogonal to the z-axis, which occurs only in the case of a highly directional antenna. Therefore, the main area of integration with respect to **s** in Equation (8.11) is concentrated close to the coordinate origin. From here, using the geometrical optics approximation, the diffraction field can be expressed as:

$$\mathrm{E_d}(\mathbf{s},\mathrm{z}) \cong \mathrm{E_d}(0,\mathrm{z}) \cong \mathrm{E}_i^0\left[1+F(0)\right]e^{ik\sqrt{\varepsilon}\mathrm{z}},$$

where E_i^0 is the field of auxiliary dipole on the interface and $F(0)$ is the Fresnel reflective coefficient at zero incident angle. After integration over z and simple transforms, we obtain:

$$\tilde{W}_\mathrm{x} = \frac{c^2 k_\mathrm{b} T\left(1-|F(0)|^2\right)}{8\pi^2}\int\left|\mathrm{E}_i^0\right|^2 A_\mathrm{e}(\mathbf{s})d^2\mathbf{s}.$$

The field on the surface is easily calculated if we take into account that it is generated by an electrical dipole with moment equal to the value $\mathrm{p} = -1/i\omega$ (compare with the first term of Equation (1.40)). The dipole field on the surface being considered close to point $\mathbf{s} = 0$ is:

$$\mathbf{E}_i^0 = \frac{k^2}{i\omega}\frac{e^{ik\mathrm{r}}}{\mathrm{r}}\mathbf{e}_\mathrm{p} \tag{8.12}$$

according to Equation (1.38). Further, we should keep in mind that $d^2\mathbf{s}/\mathrm{r} = d\Omega$ (the solid angle element) and use Equation (1.122). As a result,

$$\tilde{W}_\mathrm{x}(\omega) = \frac{k_\mathrm{b} T}{2}\left[1-|F(0)|^2\right]. \tag{8.13}$$

Just the same result will apply to a y-field component, so the spectral density of the total power flow can be written as:

$$\tilde{W}(\omega) = \left[1-|F(0)|^2\right]k_\mathrm{b} T. \tag{8.14}$$

This result is obvious and reflects the detailed balance between emitted and reflected energy flows.

It is convenient to express the power detected by an ideal receiver (i.e., by a receiver with perfectly matched circuits) in the temperature scale. Such temperature is called brightness and is equal to:

$$T_{\circ}(\omega) = \left[1 - |F(0)|^2\right] T. \tag{8.15}$$

Note that our method of calculation leads to the-so called antenna temperature; however, there is no difference between brightness and antenna temperature in the considered case of semispace.

The black-body reflective coefficient is equal to zero and, in this case, the brightness temperature is simply equal to the temperature of the black body. So, the brightness temperature is the temperature of a black body at which it radiates with the same intensity as the heated body at a given polarization and frequency. In the example discussed here, the following value is the emissivity (coefficient of emission):

$$\kappa(\omega) = 1 - |F(0)|^2 \tag{8.16}$$

and it is more convenient to write:

$$T_{\circ}(\omega) = \kappa(\omega) T. \tag{8.17}$$

This last expression is considerably more widely used than the particular case from which it was obtained. At the inclined observation of a plane-stratified medium, the emissivity is equal to:

$$\kappa(\omega, \theta) = 1 - \frac{1}{2}\left[|F_{\mathrm{h}}(\theta)|^2 + |F_{\mathrm{v}}(\theta)|^2\right]. \tag{8.18}$$

Here, the second summand is equal to half the sum of the reflective coefficients for the horizontal and vertical polarized waves. These coefficients for a plane-layered medium differ, in general, from the Fresnel ones. The angle θ is the zenith angle of observation in this case. Let us emphasize the fact that Equation (8.18) concerns media inside which radiation that has penetrated is fully absorbed. So, for example, Equation (8.18) has to be modified by adding the transmission coefficient in the case of a limited-thickness layer. In the general case,

$$\kappa(\omega, \theta_i) = 1 - \frac{1}{2}\left[F_{\mathrm{h}}(\theta_i) + F_{\mathrm{v}}(\theta_i) + T_{\mathrm{h}}(\theta_i) + T_{\mathrm{v}}(\theta_i)\right]. \tag{8.19}$$

Let us point out two circumstances regarding the radiation of a semispace filled by a transparent, or more exactly, a weakly absorbing medium. First, the brightness temperature is equal to the temperature of the body observed on vertical polarization at the Brewster angle. This means that the medium is like a black body (the emissivity is equal to unity) under the specified conditions. If we were to perform the vertical polarization emission measurement and change, in the process, the angle of observation, then the body temperature is determined at the angle of maximum radiation, and the dielectric constant is calculated by this angle value.

The second circumstance is connected with Equation (3.25), from which it follows that the temperature of an emitting semispace can be easily determined by observing both polarizations at $\theta = 45°$. It is computed by the formula:

$$T = \frac{\left(T_{\circ}^{(h)}\right)^2}{2T_{\circ}^{(h)} - T_{\circ}^{(v)}}. \tag{8.20}$$

The emissivity is a function of frequency. This frequency dependence is twofold due to the permittivity frequency dependence and to the interference and resonance phenomena that are described by the diffraction field.

The temperature of natural media cannot be constant over the space. This situation is typical, for example, for soil, which is not uniformly heated by solar radiation. In these cases, the formulae defining the brightness temperature demand elaboration. One should take into account when performing the corresponding calculations that, strictly speaking, a medium that is not uniformly heated is not in equilibrium, even with the heat transfer process; however, if spatial temperature gradients are rather small, then the medium can be assumed to be locally in equilibrium. The definition of small gradients is not formulated in the general case and always requires elaboration, taking into account the peculiarities of the problem being studied. One can assume, in our cases, that the demand of small spatial gradients of temperature is always fulfilled.

The temperature spatial variations are followed by a spatial change of the heated medium permittivity, as the permittivity is a temperature function. The spatial changes, in this case, should also be taken into account. Further, we will summarize the problem and take into consideration permittivity spatial changes caused by various factors, not only temperature. Among these are spatial variation of the medium density or concentration changes in impurities.

Let us examine, for instance, the case of a semispace that is not uniformly heated (e.g., soil in the morning). The permittivity of the medium will be assumed to be a function only of depth, and we will concentrate on observation at the nadir. Instead of Equation (8.11), we now have:

$$T_{\circ} = \frac{\omega c}{4\pi^2} \int A_e(\mathbf{s}) d^2\mathbf{s} \int_0^{\infty} \varepsilon''(z) T(z) \left|\mathrm{E}_{\mathrm{d}}(\mathbf{s}, z)\right|^2 dz. \tag{8.21}$$

Let us assume that the temperature and permittivity change slightly on a scale of the order of the wavelength. We can use, in this case, the Wentzel–Kramers–Brillouin (WKB) approximation for the diffraction field. According to this, the reflection inside the medium cannot be taken into consideration, and we should analyze only the wave reflection on the medium–vacuum interface, which is characterized by the reflective coefficient:

$$F = \frac{1-\sqrt{\varepsilon_0}}{1+\sqrt{\varepsilon_0}}, \quad \varepsilon_0 = \varepsilon(0).$$

As a result,

$$\left|E_d\right|^2 = \frac{k^4\left|1+F\right|^2}{\omega^2 r^2}\sqrt{\frac{\left|\varepsilon_0\right|}{\left|\varepsilon(z)\right|}}\exp\left[-\int_0^z \Gamma(\zeta)\,d\zeta\right], \tag{8.22}$$

where the coefficient of absorption is:

$$\Gamma(z) = 2\gamma = \frac{k}{i}\left[\sqrt{\varepsilon(z)} - \sqrt{\varepsilon^*(z)}\right] = 2kn''(z). \tag{8.23}$$

After rather simple transforms, we obtain:

$$T_\circ = \frac{1}{2}\left|1-F^2\right|\int_0^\infty \frac{\sqrt{\varepsilon(z)}+\sqrt{\varepsilon^*(z)}}{\sqrt{\left|\varepsilon(z)\right|}}T(z)\Gamma(z)\exp\left[-\int_0^z \Gamma(\zeta)\,d\zeta\right]. \tag{8.24}$$

Let us now regard two extreme cases. The first one refers to weak absorption in the sense that the imaginary permittivity part is much smaller than its real one; that is $\varepsilon'' \ll \varepsilon'$. Then, we will neglect the imaginary permittivity part everywhere it is reasonable to do so to obtain:

$$T_\circ = \left(1-F^2\right)\int_0^\infty T(z)\Gamma(z)\exp\left[-\int_0^z \Gamma(\zeta)\,d\zeta\right]dz. \tag{8.25}$$

In the case of strong absorption, when the real and imaginary parts of the permittivity are comparable, the integral from the absorbing coefficient changes quickly on the wavelength scale. Other cofactors in Equation (8.24) can be assumed to be "slow" functions, which gives us the opportunity, using integration by parts, to obtain an expansion with respect to the reverse degree of $\Lambda\Gamma(0)$, where Λ is the scale of the temperature or permittivity change. Because $\Lambda\Gamma(0) \approx 1/\lambda$ in this case, the series terms decrease quickly. The sum of the first two is represented in the form:

$$T_{\circ} = \left(1-|F|^2\right) T(0)\left\{1+\frac{1}{\Gamma(0)}\frac{d}{dz}\left[\frac{T(z)\left(\sqrt{\varepsilon}+\sqrt{\varepsilon^*}\right)}{\sqrt{|\varepsilon|}}\right]_{z=0}\right\}. \tag{8.26}$$

It follows from Equation (8.26), that, at strong absorption, the temperature spatial variability is not appreciably displayed in the intensity of thermal radiation. It is understandable, because in this case the essential contribution to the radiation is brought by the fluctuation currents that are situated close to the medium interface.

Let us now consider the case of an eroded boundary of radiating medium to see what corrections the unsharpness of border introduces into the emissivity. We will assume that thickness d of the transient layer is small in comparison with the wavelength. In order to calculate the emissivity at the zero incident angle, we will use Equation (3.119) to obtain:

$$\kappa(\omega) = \left(1-F_f^2\right)\left\{1-k^2\left[2F_f I+\left(1-F_f\right)^2 J_1^2(0)\right]\right\}. \tag{8.27}$$

We also must use the model for the permittivity depth distribution:

$$\varepsilon(z) = 1+3\left(\varepsilon_{\mathrm{d}}-1\right)\left(\frac{z}{d}\right)^2 - 2\left(\varepsilon_{\mathrm{d}}-1\right)\left(\frac{z}{d}\right)^3. \tag{8.28}$$

The advantage of this model is that the function describing the permittivity depth profile is integrated with the points z = 0, d and has zero derivatives there. For this reason, it "is smoothly connected" with the permittivity values on the boundaries of the transient layer. Integration over the equations provided in Section 3.8 gives us:

$$\kappa(\omega) = \left(1-F_{\mathrm{f}}^2\right)\left[1+\frac{\left(\sqrt{\varepsilon_d}-1\right)^2 k^2 d^2}{20}\right]. \tag{8.29}$$

It is easy to see that, as was expected, the correction is small at the accepted approaches; however, we can also see that it increases the emissivity. It takes place because of better matching between the vacuum and the emitting medium.

The transient layers improve the matching rather appreciably in some cases, as can be seen by analyzing Equation (3.41). Without going into the details, let us consider the case of dielectric media under the conditions of $1<\varepsilon_2<\varepsilon_3$, where $\psi_{12}=\psi_{23}=\pi$. Let us now set the conditions at which the reflective coefficient of the layer system converts to zero. First, the condition $2\psi=\pi$ should be observed in order to change the cosine in the numerator of Equation (3.41) to –1; in fact, this will take place at any odd numbers of π. Here, however, we will restrict ourselves to a very simple case. Under these conditions, we are dealing with a quarter-wave layer of thickness $d = \lambda/4\sqrt{\varepsilon}$ (a quarter of the wavelength in the layer). Because we are discussing layers for which the thickness is on the order of the wavelength as well as weakly absorbing media, then we should assume that $\tau = 0$ in Equation (3.41). Then, the second requirement of converting the reflective coefficient to zero and, correspondingly, the emissivity to unity will be satisfied by the equality $|F^{23}| = |F^{12}|$, which leads to the necessary validity of the relation $\varepsilon_3 = \varepsilon_2^2$.

We should emphasize the resonant character of the effect described, as the monochromatic emission is under discussion. Because reception of thermal radiation takes place in the frequency bandwidth, which is often a rather wide one, then the mentioned resonance can be eroded and full conversion of the emissivity to unity does not occur. This erosion is found to be weak in the case of a narrow bandwidth. The corresponding analysis should be carried out using, for example, Equation (3.57). In the case considered here, $\beta_2 - \beta_1 \ll 1$. Then, by expansion in a series, the following is obtained:

$$\bar{\kappa} = 1 - \bar{F}^2 \cong 1 - \frac{\pi^2 (\varepsilon_2 - 1)^2}{192\varepsilon_2} \left(\frac{\Delta f}{f_0} \right)^2 . \tag{8.30}$$

Here, f_0 is the central frequency and Δf is the bandwidth of the receiving frequencies. It would appear that the difference between the quarter-wave layer emissivity and unity seems to be small.

Let us consider, finally, the radiation of a weakly reflected layer. The diffraction field can be described by the WKB approximation in this case; therefore, the expression for the brightness temperature is obtained from Equation (8.25), where the reflective coefficient is set equal to zero (i.e., no sharp jump of permittivity on the border). So,

$$T_{\circ} = \int_0^d T(z)\Gamma(z) \exp\left[-\int_0^z \Gamma(\zeta)\, d\zeta \right] dz . \tag{8.31}$$

Here, d is the layer thickness. In the case of constant temperature inside the layer,

$$T_{\circ} = T\left(1 - e^{-\tau}\right), \quad \tau = \int_0^d \Gamma(\zeta) d\zeta . \tag{8.32}$$

If the integral absorbing coefficient (optical thickness) τ is large, then the brightness temperature is equal to the temperature of the layer, and the latter temperature becomes similar to that of the black body. In the case of small absorption,

$$T_{\circ} = \tau T;$$

that is, the brightness temperature is proportional to the integral absorption in the layer.

8.3 THERMAL RADIATION OF BODIES LIMITED IN SIZE

Bodies limited in size are those having angle sizes much smaller than the antenna pattern width. In this case, the bodies have to be in the wave (Fraunhofer) zone of the antenna, and the antenna itself is in the Fraunhofer zone of the emitting body; therefore, the field of auxiliary dipole in the area of this body coincides to a high degree of accuracy with the plane wave field. Then, if one compares Equations (5.9), (8.6), and (8.7), it is a simple matter to obtain:

$$\int_V Q\, d^3\mathbf{r}' = \frac{c}{8\pi}\left|\mathrm{E}_i^0\right|^2 \sigma_a^{(x,y)} = \frac{\pi}{2c\lambda^2 r^2}\sigma_a^{(x,y)}. \tag{8.33}$$

taking into account that, if the field in Equation (5.9) is generated inside the body by a plane wave of single amplitude, then the amplitude of the incident on the body plane wave in Equation (8.6) is equal to $2\pi/(-\lambda r)$, in agreement with Equation (8.12). Based on Equation (8.10), we now have:

$$\tilde{\mathrm{S}}_{x,y}(\omega) = \frac{k_b T \sigma_a^{(x,y)}}{4\pi\lambda^2 r^2}. \tag{8.34}$$

Multiplying the power flows by the antenna effective square gives us the spectral power density for each polarization, which is easily expressed through the brightness temperatures:

$$T_{\circ}^{(x,y)} = \frac{T\, A_e(0)\,\sigma_a^{(x,y)}}{2\pi\lambda^2 r^2}. \tag{8.35}$$

The value of zero in the antenna effective area means that the receiving antenna is looking at the radiating body; that is, the directivity pattern is oriented toward this body.

We will now introduce the black body concept for this case. Let us project the radiating body onto the plane (Figure 8.1) that is perpendicular to the line linking the radiating body and the antenna (in our case, the z-axis). We will use $\Sigma(\mathbf{e}_i)$ to represent the square of the projected body. Obviously, the value of this projected square depends on the radiation direction. In this case, the black body is the element of the plane surface, projected as discussed above, that is perpendicular to the propagation direction and is able to fully absorb the radiation incident on it of a given frequency independent of the polarization. Thus, the body emissivity at the given frequency and the designated direction can be identified as:

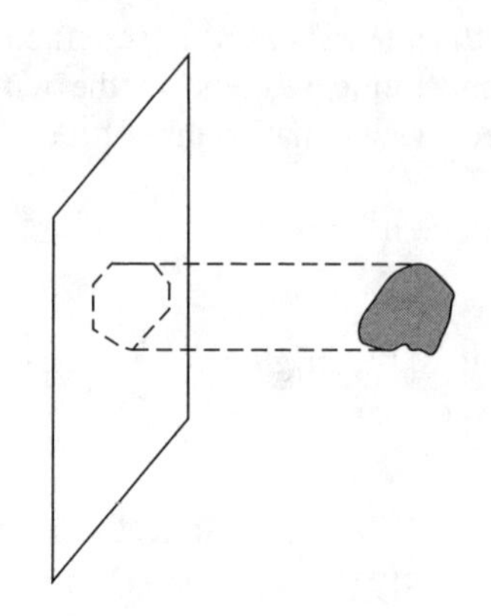

FIGURE 8.1 Projecting a body on the plane.

$$\kappa^{(x,y)}\left(\omega, \mathbf{e}_i, \eta\right) = \frac{\sigma_a^{(x,y)}\left(\omega, \eta\right)}{\Sigma\left(\mathbf{e}_i\right)}. \tag{8.36}$$

Further, note that $\Sigma\left(\mathbf{e}_i\right)/\mathrm{r}^2 = \Omega\left(\mathbf{e}_i\right)$ is the solid angle at which the considered body can be seen from the center of the antenna at this direction, and we have:

$$A_e(0) = \frac{\lambda^2}{\Omega_A}, \tag{8.37}$$

where Ω_A is the angle spread of the antenna pattern. As a result, we now have:

$$T_\circ^{(x,y)}\left(\omega, \mathbf{e}_i\right) = T\frac{\Omega\left(\mathbf{e}_i\right)}{\Omega_A}\kappa^{(x,y)}. \tag{8.38}$$

Let us now study the problem of radiation by particles occupying a layer in the space. Natural examples of such layers include clouds or crowns of trees. We will assume that theses particles emit independently in the sense that the second scattering does not play an appreciable role. In doing so, it is also assumed that densely packed media are not under discussion. Further, let us assume, for simplicity, similarity of the absorbing cross sections for all particles and the fact that they all have the same temperature. The concentration N is also assumed to be constant over the space. The brightness temperature of the layer, then, can be written as:

$$T_\circ^{(x,y)} = \frac{TN\sigma_a^{(x,y)}}{\lambda^2}\int\limits_{4\pi} A_e\left(\Omega\right) d\Omega \int\limits_{z_0}^{z_0+d} e^{-\Gamma\left(z-z_0\right)} dz\,.$$

The integrals on the right-hand side perform the summation of emitted particles over the volume with regard to the fact that their radiation falls within different areas of the antenna pattern. We have taken into account the particle radiation attenuation due to extinction upon its propagation in the scattering medium. This is the reason why our statement about the particle emission independence was not absolute, as it is partly taken into account by introducing the extinction. Further, it is necessary to mention that part of the emitted energy is reflected backward, as we have established that the medium has effective permittivity (Equation (5.143)). As the result, we now have:

$$T_{\circ}^{(x,y)} = \frac{TN\sigma_{a}^{(x,y)}}{2\Gamma}\left[1-\left|F(0)\right|^{2}\right]\left(1-e^{-\tau}\right) \tag{8.39}$$

after the integration. The extinction coefficient is here determined by Equation (5.13). The reflective coefficient, introduced here, is described by the formula $\left(1-\sqrt{\varepsilon_{e}}\right)/2$ in the first approximation, which follows from the expression for the Fresnel reflective coefficient on the assumption that the permittivity differs little from unity. Keeping in mind the definitions of the extinction coefficient (Equation (5.13)) and albedo (Equation (5.14)), then we can define the emissivity of the layer filled by the particles as:

$$\kappa^{(x,y)} = \left[1-\left|F(0)\right|^{2}\right]\left(1-\hat{A}^{(x,y)}\right)\left(1-e^{-\tau}\right). \tag{8.40}$$

The emergence of the albedo should be noted, as it was absent in our consideration of a continuous medium. If the scattering layer is sufficiently thick in such a way that we can set $\tau =$ with high accuracy, then Equation (8.40) does not have to differ from Equation (8.16), as a scattering medium is equivalent to a continuous medium with effective permittivity. The difference is explained by the fact that wave scattering due to fluctuations in particle density is taken into account (sometimes implicitly) in the scattering medium description but does not occur in a continuous medium.

8.4 THERMAL RADIATION OF BODIES WITH ROUGH BOUNDARIES

We have already pointed out that, in the region of microwave frequencies, earth surfaces should often be regarded as rough ones. Because the diffraction field inside a body with a rough surface differs from that for a body with a smooth interface, the problem of radiation of bodies with roughness requires special consideration. It is natural, as the scattering processes add up to the reflection and they make their own contribution to the energy exchange between a body and a field.

We do not intend to study the general problem but instead will limit our discussion here to the case of a semispace with a rough interface. In this case, we will assume that plane z = 0 is the mean surface of the interface. The stochastic surface itself is assumed to have properties similar to those that were defined in Chapter 6 when we explored the wave scattering problem. It is easy to see that both problems are closely connected. Following the Chapter 6 approach, we will consider two asymptotic cases: small roughness and large roughness. To estimate the emissivity value, we will use a technique based on the problems of wave scattering that were solved in Chapter 6 which allows us to determine the amount of energy absorbed by the semispace. In order to do this, it is sufficient to calculate the amount of energy reflected and scattered by the interface. The remaining energy enters the body and is absorbed there. The body absorbability is estimated in such a way that leads to estimation of the emissivity on the basis of Equation (8.7).

The corresponding calculation for the small roughness case is very simple. We will restrict ourselves to studying the vertical incidence of the single amplitude plane wave. The reflected energy is determined by the power flow $c\left|F(0)\right|^2/8\pi$. From this, we should subtract the flow of the second approximation coherent component and add the flow of the first approximation scattered field. We have not provided the relevant computations for the general case, so the reader is referred to Armand,[52] which indicate that the flow of the coherent component is equal to:

$$\mathbf{S}_1^{(02)} = -\frac{ck^2\left\langle\zeta^2\right\rangle(\varepsilon+1)\left(\sqrt{\varepsilon}-1\right)^2}{2\pi\left(\sqrt{\varepsilon}+1\right)^3}\mathbf{e}_z \tag{8.41}$$

for small slopes and when the correlation radius exceeds the wavelength. The corresponding calculation for the noncoherent component of the first approximation is described by:

$$S_s^{(1)} = \frac{ck^2\left\langle\zeta^2\right\rangle\left(\sqrt{\varepsilon}-1\right)^2}{2\pi\left(\sqrt{\varepsilon}+1\right)^2}. \tag{8.42}$$

The emissivity can be written as:

$$\kappa = 1-\left|F(0)\right|^2 - 4k^2\left\langle\zeta^2\right\rangle\sqrt{\varepsilon}\left|F(0)\right|^3, \quad k\sqrt{\left\langle\zeta^2\right\rangle\varepsilon} \ll 1, \quad \varepsilon'' \ll \varepsilon'. \tag{8.43}$$

A comparison with Equation (8.16) reveals that small roughness reduces the emissivity by the value:

$$\Delta\kappa = 4k^2\left\langle\zeta^2\right\rangle\sqrt{\varepsilon}\left|F(0)\right|^3, \tag{8.44}$$

which is rather small, and corresponding changes in the brightness temperature are a fraction of the degree Kelvin.

Analysis of thermal radiation for large-scale roughness is much more complicated. We will base our discussion here on the facet model approximation (or geometrical optics approximation). It is reasonable, in this approximation, to separate, at every point, the incident field into the horizontal components $E_h^{(i)}$ and vertical components $E_v^{(i)}$. Let us note that this separation has a local character and changes from point to point. The incident field is set by Equation (8.12).

We can now introduce a local coordinate system consisting of the vector of the wave incident ($\mathbf{e}_i$); the orthogonal to $\mathbf{e}_i$ and tangent to the surface vector $(\mathbf{n}\times\mathbf{e}_i)\Big/\sqrt{1-(\mathbf{n}\times\mathbf{e}_i)^2}$; and the orthogonal to $\mathbf{e}_i$ and lying in the local plane of the incident vector $\left[\mathbf{e}_i\times(\mathbf{n}\times\mathbf{e}_i)\right]\Big/\sqrt{1-(\mathbf{n}\times\mathbf{e}_i)^2}$.

Vector $\mathbf{n}$, as before, is the vector of the local normal. We will assume that the auxiliary dipole is rather far from the surface in such a way that the incident field in every point of the surface might be regarded as the field of the plane wave. Then, it is easy to see that:

$$E_h^{(i)}=\frac{k^2}{i\omega r}\frac{\left[\mathbf{e}_p\times(\mathbf{n}\times\mathbf{e}_i)\right]}{\sqrt{1-(\mathbf{n}\times\mathbf{e}_i)^2}},\quad E_v^{(i)}=\frac{k^2}{i\omega r}\frac{(\mathbf{e}_p\times\mathbf{n})}{\sqrt{1-(\mathbf{n}\times\mathbf{e}_i)^2}}.$$

So, the contribution of a single surface to formation of the spectral intensity of a fluctuation field oriented along vector $\mathbf{p}$ is equal to:

$$d\left\langle E_p^2\right\rangle=\frac{-k_bT}{\lambda^2 c r^2\left[1-(\mathbf{n}\times\mathbf{e}_i)^2\right]}\Big[\left(1-|F_h|^2\right)\left[\mathbf{e}_p\times(\mathbf{n}\times\mathbf{e}_i)\right]^2+$$

$$+\left(1-|F_v|^2\right)(\mathbf{e}_p\times\mathbf{n})^2\Big](\mathbf{n}\times\mathbf{e}_i).$$

Let us now define a new coordinate system $\{x',y',z'\}$ at the point of reception; these unit vectors are related to the unit vectors of the former coordinate system by the equations:

$$\begin{aligned}\mathbf{e}_{x'}&=\mathbf{e}_x\cos\theta_i+\mathbf{e}_z\sin\theta_i,\\ \mathbf{e}_{y'}&=\mathbf{e}_y,\\ \mathbf{e}_{z'}&=-\mathbf{e}_x\sin\theta_i+\mathbf{e}_z\cos\theta_i=-\mathbf{e}_i.\end{aligned}\tag{8.45}$$

Let us represent the incident wave polarization vector in the form:

$$\mathbf{e}_{\mathrm{p}} = \mathbf{e}_{x'} \cos\eta + \mathbf{e}_{y},$$
$$\sin\eta = \mathbf{e}_{x} \cos\eta \cos\theta_i + \mathbf{e}_{y} \sin\eta + \mathbf{e}_{z} \cos\eta \sin\theta_i. \tag{8.46}$$

The fluctuation field received is divided into the field of horizontal polarization, $E_h = E_y$, and the vertical polarized field, $E_v = E_z$. The corresponding contributions of the single surface element are equal; for example, in the vertical polarization case:

$$d\left\langle \left|E_v\right|^2 \right\rangle_{\mathrm{p}} = -\frac{k_b T}{\lambda^2 c r^2 \left[1 - \left(\mathbf{n}\times\mathbf{e}_i\right)^2\right]^2} \left\{ \left[\mathbf{e}_\xi \times \left(\mathbf{n}\times\mathbf{e}_i\right)\right]^2 \left[\mathbf{e}_{\mathrm{p}} \times \left(\mathbf{n}\times\mathbf{e}_i\right)\right]^2 \left(1 - \left|F_h\right|^2\right) + \right.$$
$$\left. + \left(\mathbf{e}_\xi \times \mathbf{n}\right)^2 \left(\mathbf{e}_{\mathrm{p}} \times \mathbf{n}\right)^2 \left(1 - \left|F_v\right|^2\right) \right\} \left(\mathbf{n}\times\mathbf{e}_i\right).$$

This expression describes the spectral density of the field component at a fixed direction of the auxiliary dipole. Further, it is necessary to summarize the contribution of the entire emission polarization; that is, we must calculate the value:

$$d\left\langle \left|E_v\right|^2 \right\rangle = \int_0^{2\pi} d\left\langle \left|E_v\right|^2 \right\rangle_{\mathrm{p}} d\eta.$$

We will take into account Equation (6.20) for this purpose, and we obtain:

$$\left(\mathbf{e}_{\mathrm{p}} \times \mathbf{n}\right) = \frac{\sin\theta_i - \gamma_x \cos\theta_i}{\sqrt{1+\gamma^2}} \left(\cos\eta - \mu \sin\eta\right). \tag{8.47}$$

We also have:

$$\gamma_x = \left(\mathbf{e}_x \times \nabla\zeta\right) = \frac{\partial\zeta}{\partial x},$$
$$\gamma_y = \left(\mathbf{e}_y \times \nabla\zeta\right) = \frac{\partial\zeta}{\partial y}, \tag{8.48}$$
$$\gamma^2 = \gamma_x^2 + \gamma_y^2 = \left(\nabla\zeta\right)^2,$$

and

$$\mu = \frac{\gamma_y}{\sin\theta_i - \gamma_x \cos\theta_i}. \tag{8.49}$$

Analogously,

$$\left[\mathbf{e}_p \times \left(\mathbf{n} \times \mathbf{e}_i\right)\right] = \frac{\sin\theta_i - \gamma_x \cos\theta_i}{\sqrt{1+\gamma^2}}\left(\sin\eta + \mu\cos\eta\right). \tag{8.50}$$

After integration over angle η, we obtain:

$$d\left\langle \left|\mathrm{E}_v\right|^2 \right\rangle - = \frac{\pi k_b T}{\lambda^2 c r^2} \frac{\left[\mathbf{e}_x \times \left(\mathbf{n}\times\mathbf{e}_i\right)\right]^2\left(1-\left|F_h\right|^2\right) + \left(\mathbf{e}_x \times \mathbf{n}\right)^2\left(1-\left|F_v\right|^2\right)}{1-\left(\mathbf{n}\times\mathbf{e}_i\right)^2}\left(\mathbf{n}\times\mathbf{e}_i\right).$$

Now, we have to multiply this value by $cA_e(\mathbf{s})/2\pi$ and integrate over the entire interface to obtain the expression for the spectral density of the power of the vertically polarized waves detected by the receiving antenna. As the result, we will obtain the following for the brightness temperature of the vertical polarization:

$$T_\circ^{(v)} = -\frac{T}{2\lambda^2}\int \frac{1-\left|F_v\right|^2 + \mu^2\left(1-\left|F_h\right|^2\right)}{1+\mu^2} A_e(\mathbf{s})\left(\mathbf{n}\times\mathbf{e}_i\right)\sqrt{1+\gamma^2}\,\frac{d^2\mathbf{s}}{r^2}. \tag{8.51}$$

The integration here is performed over the main plane. Analogously, for the horizontal polarization:

$$T_\circ^{(h)} = -\frac{T}{2\lambda^2}\int \frac{1-\left|F_h\right|^2 + \mu^2\left(1-\left|F_v\right|^2\right)}{1+\mu^2} A_e(\mathbf{s})\left(\mathbf{n}\times\mathbf{e}_i\right)\sqrt{1+\gamma^2}\,\frac{d^2\mathbf{s}}{r^2}. \tag{8.52}$$

We will assume that the antenna directivity pattern is rather pointed, and factor the distance r that can carry out the integral sign. Then, we will use Equation (8.37) for the antenna effective square and will restrict the integration to the bounded surface element Σ_A, which is illuminated by the antenna. Then, we will take into account that $\cos\theta_i \Sigma_A = \Omega_A r^2$. Thus, we have:

$$\kappa_{\mathrm{v}} = \frac{1}{\cos\theta_i \Sigma_A} \int_{\Sigma_A} \frac{1-\left|F_{\mathrm{v}}\right|^2 + \mu^2\left(1-\left|F_{\mathrm{h}}\right|^2\right)}{1+\mu^2} \cos\alpha\sqrt{1+\gamma^2}\, d^2\mathbf{s},$$

$$\kappa_{\mathrm{h}} = \frac{1}{\cos\theta_i \Sigma_A} \int_{\Sigma_A} \frac{1-\left|F_{\mathrm{h}}\right|^2 + \mu^2\left(1-\left|F_{\mathrm{v}}\right|^2\right)}{1+\mu^2} \cos\alpha\sqrt{1+\gamma^2}\, d^2\mathbf{s}. \tag{8.53}$$

Here, $\cos\alpha = -(\mathbf{n}\cdot\mathbf{e}_i)$, where $\alpha(\mathbf{s})$ is the local incident angle.

Let us now consider some particular cases. If the observation is performed at the nadir ($\theta_i = 0$), then the vertical polarization will be directed along the x-axis and the horizontal polarization along the y-axis. Thus, we will have:

$$\kappa_{\mathrm{v}} = \frac{1}{\Sigma_A} \int_{\Sigma_A} \left[\gamma_{\mathrm{x}}^2\left(1-\left|F_{\mathrm{v}}\right|^2\right) + \gamma_{\mathrm{y}}^2\left(1-\left|F_{\mathrm{h}}\right|^2\right)\right] \frac{d^2\mathbf{s}}{\gamma^2},$$

$$\kappa_{\mathrm{h}} = \frac{1}{\Sigma_A} \int_{\Sigma_A} \left[\gamma_{\mathrm{x}}^2\left(1-\left|F_{\mathrm{h}}\right|^2\right) + \gamma_{\mathrm{y}}^2\left(1-\left|F_{\mathrm{v}}\right|^2\right)\right] \frac{d^2\mathbf{s}}{\gamma^2}. \tag{8.54}$$

Further simplifications are based on the assumption of small slopes. It is easy to show, in this case, by expansion in the Taylor series that:

$$\left|F_{\mathrm{v,h}}\right|^2 \cong \left|F(0)\right|^2\left[1 \mp \beta\right], \quad \beta = \gamma^2\left(\frac{1}{\sqrt{\varepsilon}} + \frac{1}{\sqrt{\varepsilon^*}}\right). \tag{8.55}$$

The relations:

$$\kappa_{\mathrm{v,h}} = 1 - \left|F(0)\right|^2 \pm \Delta\kappa,$$

$$\Delta\kappa = \left(\frac{1}{\sqrt{\varepsilon}} + \frac{1}{\sqrt{\varepsilon^*}}\right)\frac{\left|F(0)\right|^2}{\Sigma_A} \int_{\Sigma_A} \left(\gamma_{\mathrm{x}}^2 - \gamma_{\mathrm{y}}^2\right) d^2\mathbf{s} \tag{8.56}$$

are obtained for the emissivities in this approximation.

With regard to the case of a regular interface of a sinusoidal wave (e.g., a simplified model of sea waves):

$$\zeta(\mathrm{x,y}) = \zeta_0 \sin\left(K\mathrm{x}\cos\vartheta + K\mathrm{y}\sin\vartheta\right). \tag{8.57}$$

It will be assumed further that the size of the illuminated surface element (antenna footprint) is considerably larger than the roughness scale. Then, by omitting the small terms of the order $(K^2\Sigma_A)^{-1})$, we obtain:

$$\Delta\kappa = \frac{|F(0)|^2}{2}\left(\frac{1}{\sqrt{\varepsilon}} + \frac{1}{\sqrt{\varepsilon^*}}\right)(K\zeta_0)^2 \cos 2\vartheta. \tag{8.58}$$

This result displays the polarization of radiation of a semispace bounded by a surface with regular roughness and applies to the case of a statistically anisotropic surface. In this case, integration over the square is equivalent to the statistical averaging; therefore,

$$\Delta\kappa = |F(0)|^2\left(\frac{1}{\sqrt{\varepsilon}} + \frac{1}{\sqrt{\varepsilon^*}}\right)\left(\langle\gamma_x^2\rangle - \langle\gamma_y^2\rangle\right). \tag{8.59}$$

Naturally, additions to the emissivity disappear in the case of statistical isotropy of the surface with observation occurring in the nadir. The same result can be obtained based on the analysis of surface scattering characteristics using, in particular, Equation (6.79).

Observation that occurs at any angle (which should not be too large in order to avoid shadowing) can be easily analyzed at large values of permittivity (i.e., when the reflective coefficients for both polarizations are close to unity). For the latter, the following approximations apply:

$$F_{\mathrm{h}}(\alpha) \cong -1 + \frac{2\cos\alpha}{\sqrt{\varepsilon}}, \quad F_{\mathrm{v}}(\alpha) \cong 1 - \frac{2}{\sqrt{\varepsilon}\cos\alpha}. \tag{8.60}$$

The expressions for emissivity have the form:

$$\begin{aligned} \kappa_{\mathrm{v}} &= \frac{2}{\cos\theta_i\,\Sigma_A}\left(\frac{1}{\sqrt{\varepsilon}} + \frac{1}{\sqrt{\varepsilon^*}}\right)\int_{\Sigma_A} \frac{1+\mu^2\cos^2\alpha}{1+\mu^2}\sqrt{1+\gamma^2}\,d^2\mathbf{s}, \\ \kappa_{\mathrm{h}} &= \frac{2}{\cos\theta_i\,\Sigma_A}\left(\frac{1}{\sqrt{\varepsilon}} + \frac{1}{\sqrt{\varepsilon^*}}\right)\int_{\Sigma_A} \frac{\cos^2\alpha+\mu^2}{1+\mu^2}\sqrt{1+\gamma^2}\,d^2\mathbf{s}. \end{aligned} \tag{8.61}$$

Further, we can assume small slopes and perform the necessary expansions and approximations to obtain:

$$\kappa_{\mathrm{v,h}} = 1 - \left|F_{\mathrm{v,h}}(\theta_i)\right|^2 + \Delta\kappa_{\mathrm{v,h}}, \tag{8.62}$$

The corresponding corrections for roughness differ for the two types of polarization. For vertical polarization,

$$\Delta\kappa_{\mathrm{v}} = \frac{\Delta\kappa}{\cos\theta_i}, \tag{8.63}$$

where $\Delta\kappa$ has been defined by Equation (8.56). For horizontal polarization,

$$\Delta\kappa_{\mathrm{h}} = -\frac{\Delta\kappa}{\cos\theta_i} + \frac{\left(2\langle\gamma_x^2\rangle + \langle\gamma^2\rangle\right)\sin^2\theta_i}{\cos\theta_i}\left(\frac{1}{\sqrt{\varepsilon}} + \frac{1}{\sqrt{\varepsilon^*}}\right). \tag{8.64}$$

Typically, the emission polarization takes place at the inclined observation even in the case of a statistically isotropic interface, in which case,

$$\Delta\kappa_{\mathrm{v}} = 0, \quad \Delta\kappa_{\mathrm{h}} = \frac{2\langle\gamma^2\rangle\sin^2\theta_i}{\cos\theta_i}\left(\frac{1}{\sqrt{\varepsilon}} + \frac{1}{\sqrt{\varepsilon^*}}\right). \tag{8.65}$$

Finally, let us consider the case of radiation by a semispace bounded by a plane but with a change in the horizontal direction reflective coefficient, by which we mean a medium with changes in the horizontal direction permittivity. Thus, the scales of change are so great that the geometrical optics approximation can be used for sufficiently short waves. This means that local radiowave reflections occur in the same way as for a homogeneous medium with particular local characteristics. In this sense, we can talk about the local reflective coefficient and, correspondingly, about local emissivity. If areas with different emissivity values fall within the antenna footprint, then we can show that the summary emissivity is equal, in the first approach, to the mean weighted integral emissivity:

$$\kappa_{\Sigma} = \frac{1}{\Sigma_A}\int_{\Sigma_A} \kappa(\mathbf{s})\, d^2\mathbf{s}. \tag{8.66}$$

9 Transfer Equation of Radiation

9.1 RAY INTENSITY

This chapter presents another method to describe wave propagation that is not based on Maxwell's equations and the wave concepts that followed from them, but rather on energy considerations. As is becoming evident, such a description is best suited for wave propagation that is chaotic in both value and direction. We have established that such fields occur, in particular, due to scattering and thermal radiation. Naturally, the emission description must be based on the determination of the averaged field properties. Coherent components of the field at intensive scattering or upon thermal radiation are extremely scarce or do not exist at all. Squared characteristics are the primary values used to describe regularities of the radiation propagation.

Poynting's vector, $\mathbf{S}(\mathbf{r},\mathbf{s})$ is found to be a random function of coordinates $\mathbf{r}$ and direction $\mathbf{s}$. Its statistical properties are characterized by the probability density $P_1(\mathbf{S}) = P_1(\mathrm{S},\Omega)$. This, in particular, defines the probability of having a Poynting's vector value in the interval (S, S+dS) and to be inside the solid angle $d\Omega$ in the direction given by vector $\mathbf{s}$. In the case being considered here, Poynting's vector is not a very suitable energy characteristic because its main value might be equal to zero due to its vector character. This does not mean that the field power is also equal to zero. Simply due to the chaotic character of wave propagation, Poynting's vector turns out to be, on average, equal to zero; therefore, it is more convenient to regard Poynting's vector characteristics in a particular direction. So, we can define the concept of ray intensity as:

$$I(\mathbf{r},\mathbf{s}) = \int_0^\infty \mathrm{S}P_1(\mathrm{S},\Omega)\,\mathrm{S}^2 d\mathrm{S}, \tag{9.1}$$

which is the average value of Poynting's vector in a single interval of the solid angles in direction $\mathbf{s}$.

Let us develop this idea further. Because our discussion is about not only harmonic oscillations but also those for which the power is distributed in a spectral interval, we will take the word *power* to mean spectral density so frequency ω has to be included in the arguments used to define the values of interest. We will not do so in all cases, however, to avoid overloading the formulae with various sorts of designations. It is enough to remember that, apart from noted exceptions, the discussion will concentrate on spectral densities.

Let us now consider the power spectral density received by the antenna, which is equal to $\tilde{W}(\omega) = S(\mathbf{r},\mathbf{s})A_e(\Omega)$ and includes the antenna effective area in direction **s**. Thus, it is assumed that the antenna receiving properties do not depend on the polarization of the incident radiation. The received spectral density is a random value due to the chaotic character of Poynting's vector:

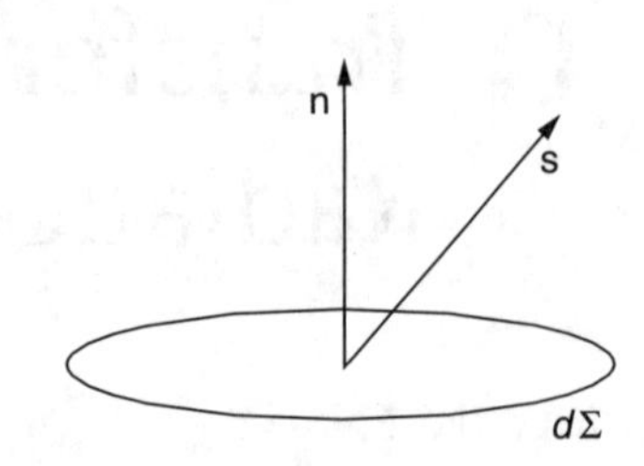

FIGURE 9.1 Oriented elementary area.

$$\left\langle \tilde{W}(\omega) \right\rangle = \int\limits_{4\pi} A_e(\Omega)\, d\Omega \int\limits_0^\infty P_1(S,\Omega)\, S^3 dS = \int\limits_{4\pi} A_e(\Omega)\, I(\mathbf{r},\mathbf{s})\, d\Omega. \tag{9.2}$$

It might seem odd that the integration over the solid angle is produced within 4π, but integration is performed within the angle given by the antenna directivity pattern.

Because A_e has the dimensionality of a square, another definition of ray intensity, as the spectral density of the power flow at the unit solid angle, follows from the last expression. This definition is more traditional. Let us consider the arbitrarily oriented elementary area $d\Sigma$ (Figure 9.1). The orientation direction is determined by the single vector **n**. The power spectral density passing through the introduced surface element in the direction of the single vector **s** is defined by the expression:

$$d\tilde{W}(\omega) = I(\mathbf{r},\mathbf{s})\left(\mathbf{n}\cdot\mathbf{s}\right) d\Sigma\, d\Omega. \tag{9.3}$$

The ray intensity differs from the spectral density of the power flow (Poynting's vector values) in that the latter value is normalized per unit square while the ray intensity is additionally normalized per unit solid angle. This leads to a difference in the coordinate dependence. For example, the power flow decreases with distance r^{-2} as it recedes from the point source, while the ray intensity remains constant.

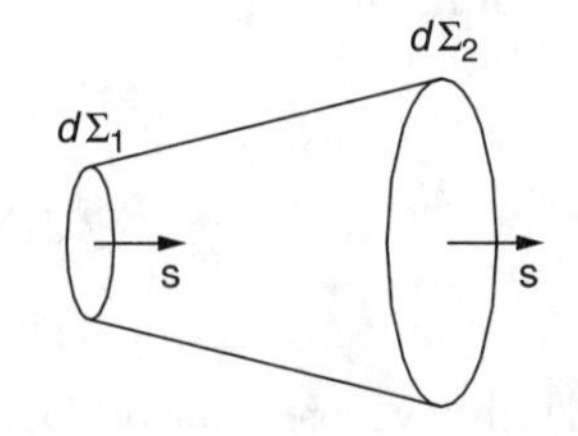

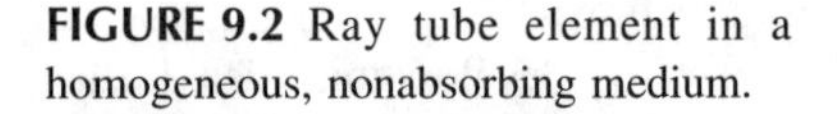

FIGURE 9.2 Ray tube element in a homogeneous, nonabsorbing medium.

Let us now assign some properties of the ray intensity by examining a ray tube element in a homogeneous nonabsorbing medium (Figure 9.2), restricted at the ends by the surface elements $d\Sigma_1$ and $d\Sigma_2$, which are perpendicular to vector **s**. The power flow through the second surface is defined relative to the power flow through the first surface by the relation $I\left(\mathbf{r}_1,\mathbf{s}\right)d\Sigma_1 d\Omega_1$ By virtue of the energy conservation law, however, it must be equal to $I\left(\mathbf{r}_2,\mathbf{s}\right)d\Sigma_2 d\Omega_2$. Because $d\Sigma_2 d\Omega_2 = d\Sigma_2 d\Sigma_1 / r^2 = d\Sigma_1 d\Omega_1$ where $r = |\mathbf{r}_2 - \mathbf{r}_1|$, we obtain the equality:

$$I(\mathbf{r}_2) = I(\mathbf{r}_1), \tag{9.4}$$

which expresses the ray intensity constancy along the ray. In differential form, this property of the ray intensity can be written as the transfer equation:

$$\frac{dI}{ds} = (\mathbf{s} \cdot \nabla I) = 0 \tag{9.5}$$

in a homogeneous, nonabsorbing medium.

Let us now study the changes in ray intensity with wave reflection and refraction on the interface of transparent media. The energy conservation law requires the equality $I_i \cos\theta_i \, d\Sigma \, d\Omega_i = I_r \cos\theta_r \, d\Sigma \, d\Omega_r + I_t \cos\theta_t \, d\Sigma \, d\Omega_t$ on the interface. The sense of the subscripts introduced here is the same as that used for the wave reflection problems in Chapter 3. Further, we will take into account that $\theta_r = \theta_i$ and $\sqrt{\varepsilon_1} \sin\theta_i = \sqrt{\varepsilon_2} \sin\theta_t$ (Snell's law), as well as the relations $\delta\Omega_r = \delta\Omega_i = \sin\theta_i d\theta_i d\varphi_i$, $d\Omega_t = \sin\theta_t d\theta_t d\varphi_t$, and $d\varphi_t = d\varphi_i$. The differentiation result follows from Snell's law: $\sqrt{\varepsilon_1} \cos\theta_i d\theta_i = \sqrt{\varepsilon_2} \cos\theta_t d\theta_t$. Therefore,

$$\frac{I_i}{\varepsilon_1} = \frac{I_r}{\varepsilon_1} + \frac{I_t}{\varepsilon_2}. \tag{9.6}$$

The ray intensity of the reflected waves is connected with the ray intensity of the incident waves by the obvious relation:

$$I_r = \left|F(\theta_i)\right|^2 I_i. \tag{9.7}$$

Here, the reflective coefficient corresponds to any polarization and is defined by the half sum of reflective coefficients in the case of nonpolarized waves (compare with Equation (8.18)). Thus, we obtain the following expression for the ray intensity of reflected waves:

$$I_t = \frac{\varepsilon_2}{\varepsilon_1}\left(1 - \left|F(\theta_i)\right|^2\right) I_i \tag{9.8}$$

At a negligibly small coefficient of reflection, we find that $I_t/\varepsilon_2 = I_i/\varepsilon_1$. The last expression is generalized by the equation for the value:

$$J(\mathbf{r}, \mathbf{s}) = \frac{I(\mathbf{r}, \mathbf{s})}{\varepsilon(\mathbf{r})} = const \tag{9.9}$$

along the ray for a medium with permittivity that changes slowly (at the wavelength scale). Note that the stated constancy follows from the geometrical optics approximation. Let us state once more that we are discussing transparent media — more accurately, the imaginary part, for which the permittivity is much less than for the real part. The transparency of the type of medium considered here is particularly emphasized by the fact that the reflection angle is implicitly assumed to be a real value for calculation of the solid angle elements.

Note that invariant Equation (9.9) takes place when the geometrical optics conditions are not observed but the medium or bordering media are in a state of the thermal equilibrium (see, for example, Bogorodsky and Kozlov[57] or Aspresyan and Kravztov[58]).

9.2 RADIATION TRANSFER EQUATION

Ray intensity satisfies an equation known as the radiation transfer equation, which is of the integer–differential type. It derives from consideration of the energy balance within an elementary volume, represented as a cylinder of length ds with unit square of transverse cross section. Equation (9.5) is an example of such an equation, and it is correct for a homogeneous nonabsorbing and nonscattering medium. These processes, together with changes in the ray tube cross section, must be taken into consideration. The change in ray tube cross section is described by Equation (9.9), which, in this case, is convenient to rewrite in the form of the differential relation $dI = I\,d(\ln\varepsilon)$. It describes amplification or attenuation of the ray intensity on the elementary segment of ray ds dependent on the permittivity gradient sign. The radiation will be weakened over the same distance due to absorption in the medium itself (for example, in air) and due to absorption and scattering by particles entering the medium (for example, drops of clouds). The change in radiation intensity caused by volume sources has to be addressed. Thus, we obtain the following differential equation:

$$\frac{dI}{ds} = \left(\frac{d\ln\varepsilon}{ds} - \Gamma - N\sigma_t\right)I + E. \tag{9.10}$$

Here, Γ is the absorption coefficient of the medium, N is the particle concentration, and E is the volume density of radiation sources.

Let us regard two processes that are important radiation sources. One of them is the process of rescattering due to particles of radiation coming upon the considered volume from other (side) directions. The density of the flow incident within the solid angle element in the direction $\mathbf{s}'$ is equal to $S_i = I\left(\mathbf{r},\mathbf{s}'\right)d\Omega'$. Multiplying by the value $\sigma_d\left(\mathbf{s}',\mathbf{s}\right)/R^2$ gives us the power flow density scattered in direction $\mathbf{s}$ at a distance R from the particle. We can obtain the value of the ray intensity reradiated in direction $\mathbf{s}$ by integrating over all directions and summing up the contribution of all the particles inside the volume being considered:

$$dI(\mathbf{r},\mathbf{s}) = N(\mathbf{r})\,ds\int_{4\pi}\sigma_d\left(\mathbf{s}',\mathbf{s}\right)I\left(\mathbf{r},\mathbf{s}'\right)d\Omega'.$$

The second source of volume radiation is the proper thermal emission of the medium. We will establish the spectral density value by a rather unusual method. First, we will consider the thermal radiation of the particles that are in the medium. The corresponding spectral density is equal to:

$$dI(\mathbf{r},\mathbf{s}) = \tilde{S}(\omega)R^2 = \frac{N(\mathbf{r})\sigma_a(\mathbf{r})k_b T(\mathbf{r})\varepsilon(\mathbf{r})}{2\pi\lambda^2}ds$$

on the basis of Equation (8.34). The combined wave power of both polarizations was taken into account when writing this expression, so the polarization effects are not discussed here. The appearance of permittivity of the medium as a factor is rather unexpected; however, it is an expression of Clausius' law, which connects the equilibrium intensity in a transparent isotropic medium with the same in a vacuum. It is now more convenient to examine the spectral density in the oscillation frequency scale f. In this case, 2π has to be reduced (recall that $\omega = 2\pi f$).

Now, it is easy to think about how to calculate the radiation of the medium itself. For this purpose, it is enough to replace the product $N\sigma_a$ with absorption coefficient Γ in the previous formula. This substitution is a reflection of Kirchhoff's law for equilibrium radiation in the ray definition. Thus, we have the ratio:

$$\frac{E_\omega}{\Gamma_\omega n_\omega^2} = \Theta_\omega(T). \tag{9.11}$$

Here, $\Theta_\omega(T)$ is the universal function of temperature and frequency. We substituted the permittivity for the refractive index, emphasizing the ray character of Equation (9.11). The subscript with the frequency is introduced to note the spectral character of the given relation. The universal function indicated here, which is independent of the physical nature of the substance, is expressed through the Planck or Rayleigh–Jeans function for the case of radio frequencies (see Chapter 8). As a result, we obtain the equation for the value defined by Equation (9.9):

$$\left(\mathbf{s}\cdot\nabla J_f\right)+\left(\Gamma+N\sigma_t\right)J_f = N\int_{4\pi}\sigma_d\left(\mathbf{s}',\mathbf{s}\right)J_f\left(\mathbf{r},\mathbf{s}'\right)d\Omega'+\left(\Gamma+N\sigma_a\right)\frac{k_bT}{\lambda^2}. \tag{9.12}$$

The subscript f is introduced here in order to emphasize again that we are discussing spectral density on the scale of the oscillation frequency but not the circular one; however, the subscript will be omitted from now on. We note also that λ is the radiation wavelength in a vacuum.

Let us express the differential cross section in terms of the scattering indicatrix with the help of the relation:

$$\sigma_d\left(\mathbf{s}',\mathbf{s}\right)=\sigma_s\psi\left(\mathbf{s}',\mathbf{s}\right)=\hat{A}\,\sigma_t\psi\left(\mathbf{s}',\mathbf{s}\right),\quad \int_{4\pi}\psi\left(\mathbf{s}',\mathbf{s}\right)d\Omega'=1. \tag{9.13}$$

The transfer equation can be rewritten as:

$$(\mathbf{s}\cdot\nabla J)+(\Gamma+N\sigma_t)J=N\hat{A}\,\sigma_t\int_{4\pi}\psi(\mathbf{s}',\mathbf{s})J(\mathbf{r},\mathbf{s}')\,d\Omega'+\left[\Gamma+(1-\hat{A})N\sigma_t\right]\frac{k_bT}{\lambda^2}. \tag{9.14}$$

It is necessary at this point to make some remarks. The transfer equation we derived obviously has a geometric–optical character that reveals itself in the inclusion of the dependence of all parameters on the only coordinate counted over the ray, in the inclusion of the propagation medium permittivity, and in the label itself — the ray intensity of the main energetic value. By the way, other labels have been used for value *I* in the literature: intensity, spectral brightness, simple brightness, energetic brightness, etc. The deduction itself is not based on the wave and statistical concept of radiation propagation. In particular, wave interference due to scattering by an assembly of particles is not considered, but we have introduced, by the simple summation of the scattered power, the idea of a scattered wave incoherence. Moreover, the waves coming to any point from different directions are also considered to be incoherent. The integral term and the extinction component coefficient in Equation (9.14) are written, obviously, using the single scattering approach; consequently, the equation concerns the case of rarefied media. The situation is rather improved by the substitution of the product $N\sigma_t$ for the total cross section per unit volume; however, it is difficult to determine up to what level of media density the given substitution works. In the case of small particles (compared to the wavelength), for which dipole scattering is the primary type, the matter is reduced to the introduction of dielectric permittivity based on the Lorentz–Lorenz formula. This formula takes into account the mutual polarization of particles as the first approximation with respect to the density, and the scattering itself is described as scattering by the density fluctuations.

The most logical deduction of the transfer equation is based on the analysis of spatial spectral properties of the coherency function. In particular, the ray intensity is defined as a Fourier transform of this function over the differential coordinates. We will not examine this problem in detail but refer interested readers to the appropriate literature.[58] The validity of the transfer equation has been proven by the fact that the solutions of many problems based on it agree with the experimental data.

Let us point out, as well, that the transfer equation is analogous to the Boltzmann kinetic equation for the stationary case. Particularly, its integral term is similar to the collision integral and describes the scattering of light particles (photons) by heavy particles. It explains why in our case the collision integral is linear with respect to the ray intensity. Obviously, many methods developed to solve the kinetic equation can be used to solve radiation transfer problems.[61]

Let us now introduce the optical thickness using the formula:

$$\tau(\mathrm{s})=\int_0^{\mathrm{s}}\left[\Gamma(\mathrm{s})+N(\mathrm{s})\sigma_t(\mathrm{s})\right]ds \tag{9.15}$$

with the following definition:

$$c(\mathrm{s}) = \frac{\hat{\mathrm{A}}\, N(s)\,\sigma_t(\mathrm{s})}{\Gamma(\mathrm{s}) + N(\mathrm{s})\,\sigma_t(\mathrm{s})}. \tag{9.16}$$

Then, the transfer equation can be written down in a dimensionless view:

$$\frac{dJ}{d\tau} + J - c\int_{4\pi} \psi(\mathbf{s}',\mathbf{s})\,J(\tau,\mathbf{s}')\;d\Omega' = (1-c)\frac{k_b T}{\lambda^2}. \tag{9.17}$$

For transfer problems, we do not raise the question of emission polarization. This problem demands a special approach that considers scattering effects. The polarization of scattered radiation can differ from the polarization for incident radiation, and, generally speaking, it should be taken into account by adding the intensities of differently polarized waves with the same weight, as was done with Equation (9.12). The situation gets even more complicated when radiation transfer in anisotropic plasma is taken into account. In this case, it is possible to have partial transformation of waves of one type to the other (for example, ordinary into extraordinary) upon scattering. In order to describe this, we would have to put together a system of linked transfer equations, including pointed transformations. We will not go into the details of this problem but instead refer readers to the appropriate literature.[58] Here, we will note only that polarization phenomena do not play a crucial role in our further discussions; therefore, we will consider the formulated equations above to be quite sufficient for our aims.

9.3 TRANSFER EQUATION FOR A PLANE-LAYERED MEDIUM

The approach for a plane-layered medium is appropriate for many of the cases we will examine further. We will suppose that all the medium properties (absorption, scattering) depend on the z-coordinate. We will also assume that all rays are straight at an angle θ to the z-axis. For the atmosphere of Earth, for example, this means that we can ignore its spherical stratification and neglect the ray bending due to refraction. We are entitled to assume that $s = \mathrm{z}/\cos\theta$ and to write:

$$\mu\frac{dJ}{dz} + J - c(\mathrm{z})\int_{4\pi} \psi(\mathrm{z},\mathbf{s}',\mathbf{s})\,J(z,\mathbf{s}')\,d\Omega' = \left[1 - c(\mathrm{z})\right]\frac{k_b T}{\lambda^2}. \tag{9.18}$$

Here,

$$z(\mathrm{z}) = \int_0^{\mathrm{z}} \left[\Gamma(\mathrm{z}') + N(\mathrm{z}')\,\sigma_t(\mathrm{z}')\right]dz', \quad \mu = \cos\theta. \tag{9.19}$$

We also represented the coordinate dependence of the scattering indicatrix, emphasizing the possible types of particle changes and their shape from the layer to layer.

Let us now consider a simple example of solving the transfer equation — an absorbing atmosphere with no scattering particles. The albedo is equal to zero, so $c = 0$. We assume that the radiation comes from the semispace toward a receiving point on Earth ($z = 0$). In this case, the transfer equation can be written in the form:

$$-\mu \frac{dI}{dz} + I = \frac{k_b T}{\lambda^2}, \quad I(z,\mu)_{z\to\infty} \to 0. \tag{9.20}$$

The minus before the first summand on the left-hand side is here because we are concerned with radiation propagating in the direction of negative values of z. The solution of this equation at the formulated boundary condition is:

$$I(z,\mu) = \frac{k_b}{\lambda^2} e^{z/\mu} \int_{z(z)}^{z_\infty} T e^{-z'/\mu} dz', \quad z_\infty = \int_0^\infty \Gamma(z) dz . \tag{9.21}$$

At $z = 0$, the spectral density of the ray intensity that we obtained should be multiplied by the effective area of the antenna directed at angle θ to the zenith and integrated over all possible directions. The antenna itself is assumed to have a narrow pattern, which permits us to neglect the μ change during the integration. As a result, we obtain the spectral density of power and then the brightness temperature. In fact, it will be the antenna temperature value, but both temperature values, as was already mentioned, practically coincide at a rather acute antenna pattern. Thus,

$$T_\circ(\mu) = \int_0^{z_\infty} T e^{-z/\mu} dz = \frac{1}{\mu} \int_0^\infty \Gamma(z) T(z) \exp\left[-\frac{1}{\mu} \int_0^z \Gamma(z') dz' \right] dz. \tag{9.22}$$

Let us now study the radiation of a layer of thickness d situated in the altitude interval $\left[z_0, z_0 + d \right]$. The temperature inside the layer is assumed to be constant, and we will also assume an absence of absorption in the layer environment of the frequency waves being studied. It follows from Equation (9.22) that, in this case:

$$T_\circ(\mu) = \frac{T}{\mu} \int_{z_0}^{z_0+d} \Gamma(z) \exp\left[-\frac{1}{\mu} \int_{z_0}^{z} \Gamma(z') dz' \right] dz = T\left[1 - e^{-\tau(d,\mu)} \right], \tag{9.23}$$

where the optical thickness is defined by the integral within the radiating layer:

$$\tau(d,\mu) = \frac{1}{\mu} \int_{z_0}^{z_0+d} \Gamma(z')\, dz'.$$

This equation agrees with Equation (8.32) and conveys the fact that the reflection effects on the interfaces are neglected.

Let us investigate further the radiation of a semispace by considering a soil with constant temperature T_g radiating into an absorbing atmosphere. To simplify the problem, we will concentrate on the analysis of radiation propagation in the vertical direction relative to the boundary; that is, we will assume that $\mu = 1$. Equation (9.21) can be used to determine the radiation ray intensity of the soil at the surface; however, we should not ignore the fact that the soil permittivity differs substantially from unity, which is why it is necessary to derive an equation for value J that describes the radiation inside the soil. Thus, we will have:

$$J_g(0) = \frac{I_g(0)}{\varepsilon_g} = \frac{k_b T_g}{\lambda^2}. \tag{9.24}$$

Equations (9.24) and (9.8) define the boundary condition for the ray intensity of atmospheric radiation. The transfer equation gives us the solution:

$$I_a\left[z(z)\right] = \frac{k_b}{\lambda^2} e^{-z} \left(T_0 + \int_0^z T_a e^{z'} dz' \right). \tag{9.25}$$

The constant T_0 is fixed from the boundary condition:

$$I_a(0) = \frac{I_g(0)}{\varepsilon_g}\left(1 - |F|^2\right) = \frac{k_b T_g}{\lambda^2}\left(1 - |F|^2\right). \tag{9.26}$$

The expression for the ray intensity of the soil and atmosphere at altitude z above the interface is obtained. Let us suppose that the radiation is received onboard a satellite at an altitude much greater than the thickness of the atmosphere. This gives us the opportunity to regard $z \to \infty$ in our formulae, and, accordingly, $z \to z_\infty$. The corresponding expression for the brightness temperature gives us:

$$T_\circ = \kappa_g T_g \exp\left[-\int_0^\infty \Gamma(z)\, dz \right] + \int_0^\infty T_a(z) \Gamma(z) \exp\left[-\int_z^\infty \Gamma(z')\, dz' \right] dz.$$

The first summand describes the soil radiation attenuated by absorption in the atmosphere, and the second one describes the sum of the proper radiation of the atmospheric layer, also attenuated by absorption; however, the represented formula is not complete, as it excludes the atmospheric summand directed downward toward the soil border, reflected upward, and then attenuated by absorption in the atmosphere. The intensity of descending radiation on the interface is easily defined from Equation (9.21) as follows:

$$I_{a\downarrow}(0) = \frac{k_b}{\lambda^2}\int_0^{\infty} T(z)\Gamma(z)\exp\left[-\int_0^{z}\Gamma(z')\,dz'\right]dz'.$$

Now, it is easy to take into account the resulting contribution in the brightness temperature and to obtain as a result:

$$T_{\circ} = \left\{\kappa_g T_g + \int_0^{\infty} T_a(z)\Gamma(z)\left[e^{z(z)} + |F|^2 e^{-z(z)}\right]dz\right\}e^{-z_{\infty}}. \tag{9.27}$$

The known equation is obtained for the soil brightness temperature in the absence of atmospheric absorption. The solution to the total problem could be obtained at once if we were to use the more complete boundary condition:

$$I_{a\uparrow}(0) = \frac{k_b}{\lambda^2}\left[\kappa_g T_g + |F|^2\int_0^{\infty} T(z)\Gamma(z)e^{-z(z)}dz\right] \tag{9.28}$$

instead of Equation (9.26).

9.4 EIGENFUNCTIONS OF THE TRANSFER EQUATION

Let us now consider a medium in which particles put into it influence predominantly radiation propagation. This means that the absorption in the medium itself is rather small (i.e., assume $\Gamma = 0$). To simplify the problem, we also assume a small variance in permittivity from unity; hence, it is sufficient to assume that $\varepsilon = 1$. In this case, the function $c(z) = \hat{A}$ is a constant value which means that the scattering particles are invariable in all space and the albedo does not depend on spatial coordinates. So, the transfer equation can now be written as:

$$\mu\frac{dI}{dz} + I - \hat{A}\int_{4\pi}\psi(\mathbf{s}',\mathbf{s})I(z,\mathbf{s}')\,d\Omega' = \left(1-\hat{A}\right)\frac{k_b T}{\lambda^2}. \tag{9.29}$$

Equation (9.29) is an integer–differential equation in which the right-hand side plays the role of an impressed force. Let us now consider the question: What are the eigenfunctions of the homogeneous equation (i.e., the equation with an impressed force equal to zero)? Naturally, these functions will depend on the scattering indicatrix type. To simplify the problem, we will assume the simplest case of isotropic scattering indicatrix; that is,

$$\psi(\mathbf{s}',\mathbf{s}) = \frac{1}{4\pi}. \tag{9.30}$$

In this case, the ray intensity will depend on only the z-coordinate and angle θ or value μ. Then, it is possible to perform the integration in Equation (9.29) with respect to azimuth angle $\varphi'(d\Omega' = \sin\theta' d\theta' \, d\varphi' = d\mu' \, d\varphi')$, and the homogeneous transfer equation can be written in the form:

$$\mu \frac{dI(z,\mu)}{dz} + I(z,\mu) - \frac{\hat{A}}{2}\int_{-1}^{1} I(z,\mu') \, d\mu' = 0 \,. \tag{9.31}$$

We will now search for the solution in terms of the separation of variables method:[60]

$$I_\nu(z,\mu) = \phi_\nu(\mu) e^{-z/\nu}. \tag{9.32}$$

The functions $\phi_\nu(\mu)$ are eigenfunctions and ν represents eigenvalues. As eigenfunctions are determined accurate to constant multiplier, it is convenient to introduce the normalization:

$$\int_{-1}^{1} \phi_\nu(\mu) \, d\mu = 1. \tag{9.33}$$

Then, it follows from the transfer equation that:

$$(\nu - \mu)\phi_\nu(\mu) = \frac{\hat{A}\nu}{2}. \tag{9.34}$$

The direction cosine μ varies within the interval [–1,+1]. If the eigenvalue ν lies outside this interval, then the unknown eigenfunctions are represented in the form:

$$\phi_\nu(\mu) = \frac{\hat{A}\nu}{2}\frac{1}{\nu - \mu}. \tag{9.35}$$

The eigenvalues themselves are defined from the condition of the normalization, Equation (9.33), and we obtain the following equation:

$$\frac{\hat{A}\nu}{2}\ln\frac{\nu+1}{\nu-1}=1. \tag{9.36}$$

This other view of the equation form is more transparent for our analysis:

$$\frac{\tanh\xi}{\xi}=\hat{A}, \qquad \xi=\frac{1}{\hat{A}\nu}. \tag{9.37}$$

Because the albedo value is less than unity and based on the hyperbolic tangent behaviors, we can state that this equation has two real roots differing only in sign. Let us designate them as $\pm\nu_0$. We exclude from our consideration the roots on the imaginary axis because doing so would require a more detailed analysis.[60]

It is a simple matter to estimate the values of the roots in extreme cases. If $\hat{A}<<1$, then $\xi >> 1$, and

$$\frac{1}{\nu_0}\cong\tanh\frac{1}{\hat{A}}\cong 1-2e^{-2/\hat{A}}. \tag{9.38}$$

The value ξ is small in the other extreme case of the albedo approaching unity $(1-\hat{A} << 1)$. The following equation is obtained from the hyperbolic tangent expansion:

$$\frac{1}{\nu_0}\cong\sqrt{3(1-\hat{A})}. \tag{9.39}$$

Also, we have established the existence of two modes:

$$\phi_{\pm\nu_0}(\mu)=\pm\frac{\hat{A}\nu_0}{2}\frac{1}{\pm\nu_0-\mu} \tag{9.40}$$

from the discrete spectrum of eigenfunctions. The eigenvalues of this spectrum are situated on the real semi-axis [–1,–] and [+1,+].

The situation becomes rather more complicated when we include [–1,+1] in the discussion. First, Equation (9.34) cannot be divided by $\nu-\mu$ in all cases. Second, we must take into account the functional equality $x\delta(x)=0$, which allows us to add the summand $\Lambda(\nu)\delta(\nu-\mu)$ to the obvious solution of Equation (9.34). Thus, the continuous spectrum modes from the class of generalized functions[61,62] are added to the already found functions of the discrete spectrum. In general, they can be written in the form:

$$\phi_\nu(\mu) = \frac{\hat{A}\nu}{2} P \frac{1}{\nu - \mu} + \Lambda(\nu)\delta(\nu - \mu), \tag{9.41}$$

where P represents the principal value. Let us recall that generalized functions have meaning not per se but under the integral sign; it is important to understand the principal value sign and delta-function in this regard. The normalization condition leads to the definition:

$$\Lambda(\nu) = 1 - \frac{\hat{A}\nu}{2} P \int_{-1}^{1} \frac{d\mu}{\nu - \mu} = 1 - \frac{\hat{A}\nu}{2} \ln \frac{1+\nu}{1-\nu}. \tag{9.42}$$

Let us point out that the principal representation of function $\Lambda(\nu)$ corresponds to the representation of the second summand via the integral. Its representation through the logarithm of Equation (9.42) is related to the case when ν lies inside the integration interval; in this case, we should use the rules of principal value calculations. In other cases, it is necessary to introduce the substitution:

$$\ln \frac{1+\nu}{1-\nu} \Rightarrow \ln \frac{\nu+1}{\nu-1},$$

which means that $\Lambda(\nu_0) = 0$.

These eigenfunctions possess the properties of orthogonality and completeness over the entire segment $(-\infty < z < +\infty,\ -1 \le \mu \le +1)$. The orthogonality is defined with weight μ; that is,

$$\int_{-1}^{1} \mu \phi_\nu(\mu) \phi_{\nu'}(\mu)\, d\mu = 0, \quad \nu' \ne \nu. \tag{9.43}$$

The norms of discrete modes are connected by the equality $N(-\nu_0) = -N(\nu_0)$ and the value:

$$N(\nu_0) = \int_{-1}^{1} \mu \phi_{\nu_0}^2(\mu)\, d\mu = \frac{\hat{A}\nu_0 \left[1 - (1-\hat{A})\nu_0^2\right]}{2\left(\nu_0^2 - 1\right)}. \tag{9.44}$$

Calculation of the norm in the case of a continuous spectrum becomes more complex due to the peculiarities of eigenfunctions, particularly because they are not functions with the integrated square. For completeness, any "good" function can be represented as:

$$f(\mu)=\tilde{f}_{+0}\,\phi_{\nu_0}(\mu)+\tilde{f}_{-0}\,\phi_{-\nu_0}(\mu)+\int_{-1}^{1}\tilde{f}(\nu')\,\phi_{\nu'}(\mu)\,d\nu' . \tag{9.45}$$

The components of the discrete spectrum are easily found. For example, the first of them is defined from the obvious equality:

$$\int_{-1}^{1}\mu\, f(\mu)\,\phi_{\nu_0}(\mu)\,d\mu=\tilde{f}_{+0}\int_{-1}^{1}\mu\,\phi_{\nu_0}^2(\mu)\,d\mu=\tilde{f}_{+0}\,N(\nu_0) . \tag{9.46}$$

By analogy, the spectral density has to satisfy the relation:

$$\tilde{f}(\nu)N(\nu)=\int_{-1}^{1}\mu\, f(\mu)\,\phi_{\nu}(\mu)\,d\mu=\int_{-1}^{1}\tilde{f}(\nu')\,\phi_{\nu'}(\mu)\,d\nu'\int_{-1}^{1}\mu\,\phi_{\nu}(\mu)\,d\mu . \tag{9.47}$$

Further calculations are more complex due to the fact that the result of integration depends on the order of integration due to the singularity of the integrands. Substitution in Equation (9.47) of the obvious expressions of the eigenfunctions provided in Equation (9.41) gives us:

$$N(\nu)=\nu\,\Lambda^2(\nu)-\frac{\hat{A}^2\nu}{4\tilde{f}(\nu)}P\int_{-1}^{1}\frac{d\mu}{\mu-\nu}P\int_{-1}^{1}\frac{\mu\,\nu'\tilde{f}(\nu')}{\nu'-\mu}\,d\nu'+$$

$$+\frac{\hat{A}^2\nu}{4\tilde{f}(\nu)}\int_{-1}^{1}d\nu'\int_{-1}^{1}P\frac{1}{\mu-\nu}P\frac{\mu\,\nu'\tilde{f}(\nu')}{\nu'-\mu}\,d\mu .$$

The Poincare–Bertrand formula now becomes very useful:[63]

$$P\int_L\frac{dt}{t-t_0}P\int_L\frac{\varphi(t,t_1)}{t_1-t}=-\pi^2\varphi(t_0,t_0)+\int_L dt_1\int_L P\frac{1}{t-t_0}P\frac{\varphi(t,t_1)\,dt}{t_1-t} . \tag{9.48}$$

Here, L is a smooth contour on the plane of complex variable t. The points t_0 and t_1 are on this contour. This formula now gives us:

$$N(\nu) = \nu\left[\Lambda^2(\nu) + \frac{\pi^2\hat{A}^2}{4}\nu^2\right] = \nu\left[\left(1 - \frac{\hat{A}\nu}{2}\ln\frac{1+\nu}{1-\nu}\right)^2 + \frac{\pi^2\hat{A}^2}{4}\nu^2\right]. \quad (9.49)$$

9.5 EIGENFUNCTIONS FOR A HALF-SEGMENT

The eigenfunctions of the transfer equation discussed previously apply to the case of infinite space; that is, we considered the case when the radiation propagation direction could be any (–1 μ 1). However, the case of semispace occurs frequently when the direction of radiation propagation is restricted by the limits (0 μ 1). The opposite case of negative directions is similar. The transfer to the half-segment (0 μ 1) demands, generally speaking, reworking the problem because the eigenfunctions examined in the previous part were defined on a full segment (–1 μ 1). Luckily, we can prove that the functions defined by us have the property of completeness for the half-segment, too.[61] Hence, it is not necessary to introduce new functions, but the orthogonality problem does require some special attention. It is necessary to introduce another weight function $W(\mu)$, which satisfies the singular equation:

$$\frac{\hat{A}\nu}{2}P\int_0^1 \frac{W(\mu)\,d\mu}{\nu-\mu} + \Lambda(\nu)\,W(\nu) = \frac{\hat{A}\nu}{2} \quad (9.50)$$

and the condition:

$$\int_0^1 \frac{W(\mu)d\mu}{\nu_0-\mu} = 1. \quad (9.51)$$

By using:

$$W(\mu) = (\nu_0 - \mu)V(\mu) \quad (9.52)$$

we can reduce the previous equation to the uniform equation of Carleman:

$$a(v)V(v) - \hat{A}P\int_0^1 \frac{V(\mu)d\mu}{\mu - v} = 0,\ a(v) = \frac{2\Lambda(v)}{v}. \quad (9.53)$$

The solution to Equation (9.53) must satisfy the condition:

$$\int_0^1 V(\mu)\,d\mu = 1. \quad (9.54)$$

The theory of singular integral equations is covered in detail in the well-known monograph of Musheleshvily.[63]

We can now see that, according to our method, the solution has the form:

$$V(\mu)=\frac{Ce^{\tau(\mu)}}{(1-\mu)\sqrt{a^2(\mu)+\pi^2\hat{A}^2}},\quad \tau(\mu)=\frac{1}{\pi}P\int_0^1\frac{\Theta(\nu)\,d\nu}{\nu-\mu},\quad \tan\Theta(\nu)=\frac{\pi\hat{A}}{a(\nu)}. \tag{9.55}$$

We can also write the solution in the form:

$$W(\mu)=\frac{\hat{A}\mu^{3/2}(\nu_0-\mu)X(\mu)}{2\sqrt{N(\mu)}}, \tag{9.56}$$

where

$$X(z)=\frac{1}{1-z}\exp\left[\frac{1}{\pi}P\int_0^1\frac{\Theta(z')\,dz'}{z'-z}\right]. \tag{9.57}$$

Integration by part allows us to state that:

$$X(z)=\exp\left[-\frac{\hat{A}}{2}\int_0^1\left(1+\frac{\hat{A}z'^2}{1-z'^2}\right)\frac{z'\ln\left(|z-z'|\right)\,dz'}{N(z')}\right]. \tag{9.58}$$

The tables for this function can be found in Case and Zweifel.[60]

The relations

$$X(z)X(-z)=\frac{\hat{\Lambda}(z)}{(1-\hat{A})(\nu_0^2-z^2)},$$
$$X(\nu_0)X(-\nu_0)=-\frac{1-(1-\hat{A})\nu_0^2}{2(1-\hat{A})\nu_0^2(\nu_0^2-1)} \tag{9.59}$$

are found to be useful where the function:

$$\hat{\Lambda}(z)=1-\frac{\hat{A}z}{2}\int_{-1}^1\frac{d\mu}{z-\mu} \tag{9.60}$$

is determined as the function of complex variable z (compare with Equation (9.42)), and

$$\int_0^1 \frac{W(\mu)d\mu}{\nu-\mu} = 1-\left(\nu_0-\nu\right)X(\nu). \tag{9.61}$$

Let us point out that the first expression in Equation (9.59) is valid on the entire complex plane z with the section along the real axis on the segment from zero to the unity. It can be written as:

$$X(\mu)\,X(-\mu) = \sqrt{\frac{N(\mu)}{\mu}}\,\frac{1}{(1-\hat{A})\left(\nu_0^2-\mu^2\right)} \tag{9.62a}$$

on the section itself. The last result is derived with the help of Sohotsky–Plemelj formula.[63]

The table of normalizing integrals is cited in Reference 60. Equation (9.61) is very helpful for its deduction. The unknown integrals can be represented as:

$$\int_0^1 \phi_\nu(\mu)\phi_{\nu'}(\mu)W(\mu)d\mu = W(\nu)\frac{N(\nu)}{\nu}\delta\left(\nu-\nu'\right),$$
$$\int_0^1 \phi_{+\nu_0}(\mu)\phi_\nu(\mu)W(\mu)d\mu = 0, \tag{9.62b}$$

$$\int_0^1 \phi_{-\nu_0}(\mu)\phi_\nu(\mu)W(\mu)d\mu = \hat{A}\nu\,\nu_0 X\left(-\nu_0\right)\phi_{-\nu_0}(\nu), \tag{9.62c}$$

$$\int_0^1 \phi_{\pm\nu_0}(\mu)\phi_{+\nu_0}(\mu)W(\mu)d\mu = \mp\left(\frac{\hat{A}\nu_0}{2}\right)^2 X\left(\pm\nu_0\right),$$
$$\int_0^1 \phi_{-\nu}(\mu)\phi_{+\nu_0}(\mu)W(\mu)d\mu = \frac{\hat{A}\nu\,\nu_0}{4}X(-\nu), \tag{9.62d}$$

$$\int_0^1 \phi_{-\nu}(\mu)\phi_{\nu'}(\mu)W(\mu)d\mu = \frac{\hat{A}\nu'}{2}\phi_{-\nu}\left(\nu'\right)\left(\nu_0+\nu\right)X(-\nu)\,. \tag{9.62e}$$

9.6 PROPAGATION OF RADIATION GENERATED ON THE BOARD

Let us now consider the problem of radiation by sources given on the layer board as a function $I_0(\mu)$. This means that the coefficients of the ray intensity expansion:

$$I(z,\mu) = b_0\phi_{\nu_0}(\mu)e^{-z/\nu_0} + \int_0^1 b(\nu)\phi_\nu(\mu)e^{-z/\nu}d\nu \tag{9.63}$$

are found from the boundary condition:

$$I(0,\mu) = b_0\phi_{\nu_0}(\mu) + \int_0^1 b(\nu)\phi_\nu(\mu)d\nu = I_0(\mu). \tag{9.64}$$

The standard procedure leads to the relations:

$$\begin{aligned} b_0 &= -\frac{4}{A^2\nu_0^2X(\nu_0)}\int_0^1 I_0(\mu')W(\mu')\phi_{\nu_0}(\mu')d\mu', \\ b(\nu) &= \frac{\nu}{W(\nu)N(\nu)}\int_0^1 I_0(\mu')W(\mu')\phi_\nu(\mu')d\mu'. \end{aligned} \tag{9.65}$$

In particular, note that at zero albedo (i.e., no scattering) $\phi_{\nu_0}(\mu) = 0$, $\phi_\nu(\mu) = \delta(\nu - \mu)$, $N(\nu) = \nu$, and $b(\nu) = I_0(\nu)$, and the ray intensity is:

$$I(\mu) = I_0(\mu)e^{-z/\mu}. \tag{9.66}$$

The obtained formulae are rather complex. They can be simplified a little in the directed radiation source given by the formula $I_0(\mu) = \delta(\mu - \mu_0)$; however, even in this case the general analysis is not simple, so we will limit ourselves to the study of an extreme case.

The first summand in Equation (9.63) becomes dominant at a sufficient distance from the source, in this case from the interface, as $\nu_0 \geq 1$. Then, at z >> 1:

$$I(z,\mu) \cong -\frac{V(\mu_0)}{X(\nu_0)(\nu_0 - \mu)}e^{-z/\nu_0}. \tag{9.67}$$

It is easy to determine that the distance over which the radiation propagates can be obtained by the equation:

$$\int_0^{z_0} N(z')\sigma_t(z')\,dz' = \nu_0. \tag{9.68}$$

We note that here N represents the particle concentration. For a given concentration and cross section, this distance is equal to:

$$z_0 = \frac{\nu_0}{N\sigma_t}. \tag{9.69}$$

When the albedo tends to unity (in the case of no absorption), the discrete eigenvalue tends to infinity ($\nu_0 \to \infty$), which means that the medium is filled up with radiation for which the intensity does not depend on coordinates z and μ at a sufficient distance from the source and tends to the value $I(z,\mu) \to V(\mu_0)$. True, infinite time is needed to reach this value in the whole volume.

9.7 RADIATION PROPAGATION IN A FINITE LAYER

The problem of radiation propagation in the semispace occupied by scattering elements was discussed in the previous section. Closer to reality is the problem of radiation propagation in a layer of finite thickness d. The difference from the previous consideration is that the condition:

$$I(z_d,\mu) = 0, \quad \mu < 0, \quad z_d = \int_0^d N(z')\sigma_t(z')dz' \tag{9.70}$$

should be added to the boundary condition of Equation (9.64). The appearance of this second boundary condition allows us to include backward radiation in our analysis. Thus, the solution of the transfer equation should be represented as the sum:

$$\begin{aligned} I(z,\mu) &= b_{0+}\phi_{\nu_0}(\mu)e^{-z/\nu_0} + b_{0-}\,\phi_{-\nu_0}(\mu)e^{z/\nu_0} + \\ &+ \int_0^1 b(\nu')\phi_{\nu'}(\mu)e^{-z/\nu'}d\nu' + \int_0^1 b(-\nu')\,\phi_{-\nu'}(\mu)e^{z/\nu'}\,d\nu'. \end{aligned} \tag{9.71}$$

The source of radiation will be assumed, for the sake of simplicity, to be in the form of a delta-function. We then obtain two equations, adding up and subtracting the boundary conditions:

$$\delta\left(\mu-\mu_0\right)=B_{0\pm}\Phi_{\nu_0\pm}(\mu)+\int_0^1 B_\pm\left(\nu'\right)\Phi_{\nu'\pm}(\mu)d\nu' , \tag{9.72}$$

which are equivalent to the boundary conditions in Equations (9.64) and (9.70). We can introduce the following functions for brevity:

$$\begin{aligned} &B_{0\pm}=b_{0+}\pm b_{0-}\,e^{z_d/\nu_0},\quad B_\pm(\nu)=b(\nu)\pm b(-\nu)e^{z_d/\nu},\\ &\Phi_{\nu_0,\nu\pm}(\mu)=\phi_{\nu_0,\nu}(\mu)\pm\phi_{-\nu_0,\nu}(\mu)e^{-z_d/\nu_0,\nu} \end{aligned} \tag{9.73}$$

and the following relations are useful:

$$\begin{aligned} &\int_0^1 V(\mu)\phi_{\pm\nu_0}(\mu)\,d\mu=\mp\frac{\hat{A}\nu_0}{2}X\left(\pm\nu_0\right),\quad \int_0^1 V(\mu)\phi_\nu(\mu)d\mu=0,\\ &\int_0^1 V(\mu)\phi_\nu(-\mu)\,d\mu=\frac{\hat{A}\nu}{2}X\left(-\nu\right),\quad \nu\geq 0. \end{aligned} \tag{9.74}$$

They follow from Equation (9.62a) for normalizing integrals. If we take into account the properties of eigenfunction due to changes in the argument and subscript sign, then the following relations can be obtained:

$$B_{0\pm}=-\frac{\frac{2}{A}V\left(\mu_0\right)\mp\int_0^1 B_\pm\left(\nu'\right)X\left(-\nu'\right)e^{-z_d/\nu'}\nu'd\nu'}{\nu_0\left[X\left(\nu_0\right)\mp X\left(-\nu_0\right)e^{-z_d/\nu_0}\right]} . \tag{9.75}$$

The equations of orthogonality lead to the relations:

$$\begin{aligned} B_\pm(\nu)\frac{W(\nu)N(\nu)}{\nu}&=W\left(\mu_0\right)\phi_\nu\left(\mu_0\right)\mp B_{0\pm}\frac{\hat{A}^2\nu\nu_0^2X\left(-\nu_0\right)}{2\left(\nu_0+\nu\right)}e^{-z_d/\nu_0}\mp\\ &\mp\frac{\hat{A}^2\nu}{4}\int_0^1\frac{\nu_0+\nu'}{\nu+\nu'}X\left(-\nu'\right)B_\pm\left(\nu'\right)e^{-z_d/\nu'}\nu'\,d\nu'. \end{aligned} \tag{9.76}$$

If Equation (9.75) is substituted here, then we obtain inhomogeneous Fredholm integral equations of the second order with respect to the functions $B_{\pm}(\nu)$, and the method of consecutive approximations is one to solve them. It is especially effective in the case of a thick layer, when $z_d >> 1$, when the integral terms are small and can be dropped in the approximation. The zero approximation gives us the following equations:

$$b_{0+} = -\frac{2V(\mu_0)X(\nu_0)}{\nu_0 U(\nu_0)} e^{z_d/\nu_0}, \quad b_{0-} = -\frac{2V(\mu_0)X(-\nu_0)}{\nu_0 U(\nu_0)} e^{-z_d/\nu_0}, \tag{9.77}$$

Here,

$$U(\nu_0) = X^2(\nu_0)e^{z_d/\nu_0} - X^2(-\nu_0)e^{-z_d/\nu_0}. \tag{9.78}$$

Analogously,

$$b(\nu) = \frac{\nu V(\mu_0)}{W(\nu)N(\nu)}\left[(\nu_0 - \mu_0)\phi_\nu(\mu_0) + \frac{\hat{A}^2\nu\nu_0 X^2(-\nu_0)}{(\nu_0+\nu)U(\nu_0)} e^{-z_d/\nu_0}\right],$$
$$b(-\nu) = \frac{\hat{A}^2\nu^2\nu_0 X(\nu_0)X(-\nu_0)V(\mu_0)}{(\nu_0+\nu)U(\nu_0)W(\nu)N(\nu)} e^{-z_d/\nu}. \tag{9.79}$$

We will now restrict ourselves to the case of an albedo close to unity. Then, we can assume that $\nu_0 >> 1$ and use:

$$X(\nu_0) \cong -\frac{1}{\nu_0}, \quad U(\nu_0) \cong \frac{2}{\nu_0^2}\sinh\left(\frac{z_d}{\nu_0}\right). \tag{9.80}$$

In this approximation, we have:

$$b_{0+} = \frac{V(\mu_0)}{\sinh\left(z_d/\nu_0\right)} e^{z_d/\nu_0}, \quad b_{0-} = -\frac{V(\mu_0)}{\sinh\left(z_d/\nu_0\right)} e^{-z_d/\nu_0}. \tag{9.81}$$

It is easy to see that we arrive at the previous result in the case of a layer of infinite thickness ($z_d >> \nu_0$).

For the amplitudes of the continuous spectrum, the summand with $b(-\nu)$ should be taken into account only close to the right boundary of the regarded layer. We can show that in this area the corresponding summand has the ν_0^{-1} order of smallness.

The second summand in the first expression of Equation (9.79) is also small; therefore in this case, we can assume that $b(-\nu) = 0$ in zero-order approximation, and

$$b(\nu) \cong \frac{\nu_0 \nu V(\mu_0)}{W(\nu)\, N(\nu)} \phi_\nu(\mu_0) \cong \frac{\nu V(\mu_0)}{V(\nu)\, N(\nu)} \phi_\nu(\mu_0). \tag{9.82}$$

As before, we can neglect the components of the continuous spectrum far away from the sources and write the ray intensity as:

$$I(z,\mu) \cong V(\mu_0) \frac{\sinh\left[\left(z_d - z\right)/\nu_0\right]}{\sinh\left(z_d/\nu_0\right)}. \tag{9.83}$$

The ray intensity turns to zero, in the given approximation, at z = z_d. The approximation is exactly correct for $\mu < 0$, reflecting the absence of radiation entering from outside. For $\mu > 0$, describing radiation coming out, it is true if we neglect the values of the order of at least ν_0^{-1} or z_d^{-1}.

Now, let us define the value of radiation coming out at the boundary z = 0. For this purpose, it is necessary to calculate the ray intensity $I(0,\mu < 0)$, which in this case is equal to:

$$I(0,\mu < 0) = V(\mu_0)\left[1 + \nu_0 \int_0^1 \frac{\phi_\nu(\mu < 0)\phi_\nu(\mu_0)\, \nu\, d\nu}{W(\nu) N(\nu)}\right].$$

As pointed out in Reference 60, this last integral is computed analytically and is equal to:

$$\int_0^1 \frac{\phi_\nu(\mu < 0)\phi_\nu(\mu_0)\nu\, d\nu}{W(\nu) N(\nu)} = \frac{1}{\nu_0 - \mu}\left[\frac{1}{(\nu_0 - \mu_0) X(\nu_0)} + \frac{1}{(\mu_0 - \mu) X(\mu)}\right]. \tag{9.84}$$

Taking into account the adopted approximations, we have:

$$I(0,\mu < 0) = \frac{V(\mu_0)}{(\mu_0 - \mu) X(\mu)}. \tag{9.85}$$

9.8 THERMAL RADIATION OF SCATTERERS

Now, let us assume thermal radiation of the scatterers occupying the semispace. The peculiarity here is that the processes of volume radiation are accompanied by the processes of volume scattering. We need to solve the equation:

$$\mu \frac{dI(z,\mu)}{dz} + I(z,\mu) - \frac{\hat{A}}{2}\int_{-1}^{1} I(z,\mu')d\mu' = \left(1-\hat{A}\right)\frac{k_b T}{\lambda^2}. \tag{9.86}$$

To simplify the problem, one can assume that the temperature of the particle is constant. This allows us to search for the problem solution in view of the sum:

$$I(z,\mu) = \frac{k_b T}{\lambda^2} + b_0\,\phi_{\nu_0}(\mu)e^{-z/\nu_0} + \int_0^1 b(\nu)\phi_\nu(\mu)e^{-z/\nu}d\nu. \tag{9.87}$$

It is easy to see that the first summand in the written solution satisfies Equation (9.86) on the right-hand side. The two other summands define the solution of the homogeneous transfer equation with the condition that it tends to zero at $z \to \infty$. The whole solution has to turn to zero on the boundary of the semispace, which reflects the fact of the radiation coming back from the left. The condition:

$$b_0\,\phi_{\nu_0}(\mu) + \int_0^1 b(\nu)\phi_\nu(\mu)\,d\nu = -\frac{k_b T}{\lambda^2}$$

occurs on this basis. The standard procedure for use of the orthogonality conditions allows us to obtain:

$$b_0 = \frac{2k_b T}{\hat{A}\nu_0 X(\nu_0)\lambda^2}, \quad b(\nu) = -\frac{k_b T}{\lambda^2}\frac{\nu}{W(\nu)N(\nu)}\int_0^1 W(\mu')\phi_\nu(\mu')\,d\mu'. \tag{9.88}$$

The integral in the second formula can be computed with the use of Equations (9.40) and (9.61); however, it is not necessary to do so, as further on we will be interested in the value $I(0,\mu<0)$, which describes the intensity of the radiation coming out, thereby determining the emissivity of the system of scatterers (medium) under consideration. It follows from our formulae that:

$$\kappa(\mu<0) = 1 + \frac{1}{\left(\nu_0 + |\mu|\right)X(\nu_0)} - \int_0^1 W(\mu')d\mu' \int_0^1 \frac{\phi_\nu(\mu<0)\phi_\nu(\mu')\nu\,d\nu}{W(\nu)N(\nu)}.$$

The inner integral is calculated according to Equation (9.84). It is necessary to take into account Equations (9.51) and (9.61) in the further calculations, and we obtain:

$$\kappa(\mu<0)=\frac{1}{(\nu_0+|\mu|)X(-|\mu|)}. \tag{9.89}$$

Note that the obtained formula satisfies the extreme conditions. At albedo equal to zero, $\nu_0=1$ and

$$X(z)=\frac{1}{1-z}, \tag{9.90}$$

and $\kappa(\mu)=1$. This medium behaves as a black body. When the albedo is equal to unity, $\nu_0\to\infty$, and the emissivity becomes equal to zero, which is natural. It is necessary to point out that the absence of reflections from the interface is assumed implicitly in our consideration, which means that the emissivity becomes equal to unity.

9.9 ANISOTROPIC SCATTERING

The model we used for isotropic particle scattering is an idealization of natural processes and should not be regarded too critically; however, it does give us the opportunity to describe correctly many versions of the phenomenon on the basis of a rather simple model. It is natural that further refinement of the model requires the inclusion of scattering anisotropy. We will briefly analyze the process of taking into account anisotropy for the rather simple case of particles for which the scattering indicatrix depends only on the angle between the directions of incident and scattered waves. This angle is defined by the scalar product $(\mathbf{s}'\cdot\mathbf{s})=\cos\gamma$. The ray intensity will not depend on azimuth angle φ for this assumption. The homogeneous transfer equation can be written as:

$$\mu\frac{dI(z,\mu)}{dz}+I(z,\mu)-\frac{\hat{A}}{2}\int_{4\pi}\psi(\mathbf{s}'\cdot\mathbf{s})I(z,\mu')d\Omega'=0. \tag{9.91}$$

In this approach, the scattering indicatrix is represented as the expansion over the Legendre polynomial:

$$\psi(\mathbf{s}'\cdot\mathbf{s})=\psi(\cos\gamma)=\sum_{n=0}^{\infty}a_nP_n(\cos\gamma). \tag{9.92}$$

The following equality is valid in the chosen coordinate system:

$$\cos\gamma = \cos\theta\cos\theta' + \sin\theta\sin\theta'\cos\left(\varphi - \varphi'\right), \tag{9.93}$$

Here, the primed angles correspond to the direction of the incident wave and the nonprimed angles to the scattering direction. In an addition theorem,[44]

$$P_n(\cos\gamma) = P_n(\mu)P_n\left(\mu'\right) + 2\sum_{m=1}^{n}\frac{(n-m)!}{(n+m)!}P_n^{(m)}(\mu)P_n^{(m)}\left(\mu'\right)\cos m\left(\varphi - \varphi'\right), \tag{9.94}$$

where $P_n^{(m)}$ are Legendre associated polynomials, and $\mu = \cos\theta, \mu' = \cos\theta'$. Let us substitute expansion Equation (9.92) in Equation (9.91) and take into account Equation (9.94). Then, it is easy to perform the integration over φ', using the property of axis symmetry for ray intensity. As a result, the sum connected with Legendre-associated polynomials will disappear, and the transfer equation is as follows:

$$\mu\frac{dI\left(z,\mu\right)}{dz} + I\left(z,\mu\right) - \frac{\hat{A}}{2}\sum_{n=0}^{\infty}a_n P_n\left(\mu\right)\int_{-1}^{1}I\left(z,\mu'\right)P_n\left(\mu'\right)d\mu' = 0. \tag{9.95}$$

We will now search for a solution again in the form of Equation (9.32). Let us introduce for short the notation:

$$q_{\nu,j} = \int_{-1}^{1}\phi_\nu\left(\mu'\right)P_j\left(\mu'\right)d\mu'. \tag{9.96}$$

Then, we will have:

$$\left(\nu - \mu\right)\phi_\nu(\mu) - \frac{\hat{A}\,\nu}{2}\sum_{n=0}^{\infty}a_n\,q_{\nu,n}\,P_n(\mu) = 0. \tag{9.97}$$

Let us first establish the equations for the coefficients of Equation (9.96). For this purpose, the last equation is multiplied by $P_l(\mu)$ and is integrated according to angle μ. Thus, we must take into account the following relations:[44]

$$(2l+1)\mu\,P_l(\mu) = (l+1)P_{l+1}(\mu) + l\,P_{l-1}(\mu) \tag{9.98}$$

and the orthoganality equations of Legendre polynomials:[44]

$$\int_{-1}^{1} P_n(\mu)P_l(\mu)d\mu = \frac{2}{2l+1}\delta_{nl}. \tag{9.99}$$

As a result,

$$(l+1)\, q_{\nu,l+1} + l\, q_{\nu,l-1} = \nu\, q_{\nu,l}\left(2l+1-a_l\hat{A}\right). \tag{9.100}$$

While solving this system of recurrent equations, we will take into account that a_0 has to be equal to 1 due to normalization of the scattering indicatrix, Equation (9.13). Besides, it follows from Equation (9.100) that all $q_{\nu,l}$ are proportional to $q_{\nu,0}$. It is reasonable to assume that $q_{\nu,0} = 1$ without violation of generality. Then, it follows that:

$$q_{\nu,0} = 1,\quad q_{\nu,1} = \nu\left(1-\hat{A}\right),\qquad 2q_{\nu,2} = \nu^2\left(1-\hat{A}\right)\left(3-a_1\hat{A}\right)-1. \tag{9.101}$$

Let us represent the sum in Equation (9.97) by:

$$M(\mu,\nu) = \sum_{n=0}^{\infty} a_n\, q_{\nu,n}\, P_n(\mu). \tag{9.102}$$

The solution from the area of discrete eigenvalues $\left(|\mathrm{Re}\,\nu| > 1\right)$ is represented in the form:

$$\phi_\nu(\mu) = \frac{\hat{A}\,\nu M(\mu,\nu)}{2(\nu-\mu)}. \tag{9.103}$$

The normalization conditions in Equation (9.33) open the way to determining an equation for the definition of discrete eigenvalues. Neumann's formula is useful for deduction of this equation:[44]

$$Q_n(\nu) = \frac{1}{2}\int_{-1}^{1} \frac{P_n(\mu)}{\nu-\mu}d\mu. \tag{9.104}$$

Here, Q_n is a Legendre function of the second order. In the result, the dispersion equation of the discrete eigenvalue spectrum has the form:

$$\hat{A}\,\nu\sum_{n=0}^{\infty} a_n\, q_{\nu,n}\, Q_n(\nu)=1\,. \tag{9.105}$$

This convenient equation is effective for a limited number of expansion terms (Equation (9.92)) and usually applies except for the pencil-beam scattering indicatrix. In this special case, it is more reasonable to use another approximation. Let us consider, for example, the case when all $a_n = 0$ for $n > 1$. Note that $|\mathrm{Re}\,\nu| > 1$ and

$$Q_0(\nu)=\frac{1}{2}\ln\frac{\nu+1}{\nu-1},\quad Q_1(\nu)=\frac{\nu}{2}\ln\frac{\nu+1}{\nu-1}-1. \tag{9.106}$$

After simple transforms, we obtain:

$$\frac{\tanh p(\xi)}{\xi}=\hat{A},\qquad p(\xi)=\hat{A}\xi\,\frac{\hat{A}\,\xi^2+(1-\hat{A})\,a_1}{\hat{A}^2\,\xi^2+(1-\hat{A})\,a_1},\qquad \xi=\frac{1}{\hat{A}\,\nu}. \tag{9.107}$$

Because $p(\xi)$ is an even function, then it follows from the stated equation that discrete eigenvalue $-\nu_j$ accompanies the discrete eigenvalue $+\nu_j$. We can show that the roots of Equation (9.107) are real and that only two of them exist, as in the case of isotropic scattering. The squared roots of Equation (9.107) lie in the interval $0<\xi^2<(1-\hat{A})|a_1|/\hat{A}$ in the case of a negative value of a_1 (a rather exotic case).

At a small albedo, we can show that the asymptotic eigenvalue is approximately:

$$\nu_0 \cong 1+2\exp\left[-\frac{2}{(1+a_1)\hat{A}}\right] \tag{9.108}$$

and for an albedo that is about unity:

$$\nu_0 \cong \sqrt{\frac{\hat{A}}{(1-\hat{A})(3-a_1\hat{A})}}. \tag{9.109}$$

The expression for functions of a continuous spectrum can be written as:

$$\phi_\nu(\mu)=\frac{\hat{A}\,\nu}{2}P\frac{M(\nu,\mu)}{\nu-\mu}+\Lambda(\nu)\delta(\nu-\mu)\,, \tag{9.110}$$

where now

$$\Lambda(\nu) = 1 - \frac{\hat{A}\,\nu}{2} P\int_{-1}^{1} \frac{M(\nu,\mu)}{\nu - \mu} d\mu = 1 - \hat{A}\,\nu \sum_{n=0}^{\infty} a_n\, q_{\nu,n}\, Q_n(\nu). \qquad (9.111)$$

When we turn to the half-segment, the result is a singular integral equation for the weight function, as in the case of isotropic scattering. Its solution depends on the form of scattering indicatrix. As a whole, the problem gets more complicated, so further details are not provided here and the interested reader is referred to the relevant literature.[60]

9.10 DIFFUSION APPROXIMATION

It was stated previously that only the discrete component of a spectrum is kept at some distances from the radiation sources and the interfaces of the scattering layer, especially in the case of a particle albedo that is about unity. So the ray intensity inside the layer is:

$$I(z,\mu) \approx b_{0+} \frac{\hat{A}\nu_0}{2(\nu_0 - \mu)} e^{-z/\nu_0} + b_{0-} \frac{\hat{A}\nu_0}{2(\nu_0 + \mu)} e^{z/\nu_0},$$

where the coefficients $b_{0\pm}$ are determined from the terms of the problem. Thus, we should assume that $\nu_0 >> 1$, and the solution is approximately described by the formula:

$$I(z,\mu) \approx \frac{\hat{A}}{2}\left[b_{0+} e^{-z/\nu_0} + b_{0-} e^{z/\nu_0} - \frac{\mu}{\nu_0}\left(b_{0+} e^{-z/\nu_0} - b_{0-} e^{z/\nu_0} \right) \right]. \qquad (9.112)$$

It is necessary to note that the thickness of the scattering layer has to be greater than the value defined by Equations (9.68) and (9.69).

We can now find an approximate solution to the described transfer equation, especially relative to its diffusion component. We will focus here on an axially symmetric distribution of radiation intensity, which is described by Equation (9.95). Inside the medium, the general solution can be represented as the sum of the attenuated emission incident on the boundary (or boundaries) and the diffusion component of the ray intensity whose occurrence is due to the scattering of incident radiation by particles. Based on these arguments, we may search for a solution in the form:

$$I(z,\mu) = I_0(\mu) e^{-z/\mu} + I_d(z,\mu)\,. \qquad (9.113)$$

Here, the first summand corresponds to the attenuated incident emission (compare with Equation (9.66)). After substitution of this expression into Equation (9.95), we obtain the equation for the diffusion component:

$$\mu \frac{dI_d(z,\mu)}{dz} + I_d(z,\mu) - \frac{\hat{A}}{2}\sum_{n=0}^{\infty} a_n P_n(\mu) \int_{-1}^{1} I_d\left(z,\mu'\right) P_n\left(\mu'\right) d\mu' = \frac{\hat{A}}{2}\sum_{n=0}^{\infty} a_n L_n(z)\, P_n(\mu). \tag{9.114}$$

The right part of this equation plays the role of the radiation volume source. Here,

$$L_n(z) = \int_{0}^{1} I_0\left(\mu'\right) e^{-z/\mu'} P_n\left(\mu'\right) d\mu' . \tag{9.115}$$

Making a generalization of what was declared before, we can argue that the transfer equation solution inside the scattering layer of a plane-layered medium has the form:

$$I(z,\mu) = U(z) + \frac{3}{4\pi}\mu\, F(z) \tag{9.116}$$

when the value of the albedo is large enough. Equation (9.113) represents *diffusion*. The value:

$$U(z) = \frac{1}{2}\int_{-1}^{1} I(z,\mu) d\mu \tag{9.117}$$

is the *mean diffusion intensity*, and the function:

$$F(z) = 2\pi \int_{-1}^{1} I(z,\mu)\mu d\mu \tag{9.118}$$

determines the mean diffusion flow.

Let us substitute Equation (9.116) into Equation (9.114). Reference to Equation (9.114) indicates, in particular, that we are assuming that the indicatrix is described by a function such as Equation (9.92). After substitution, we will have:

$$\mu\left(\frac{dU}{dz} + \frac{3 - a_1 \mathrm{A}}{4\pi} F\right) + \mu^2 \frac{3}{4\pi}\frac{dF}{dz} + \left(1 - \mathrm{A}\right)U = \frac{\hat{\mathrm{A}}}{2}\sum_{n=0}^{\infty} a_n L_n(z) P_n(\mu) . \tag{9.119}$$

Let us integrate this equation over μ in the interval [–1,+1], and then multiply by μ and again integrate in the same interval. We will get, as result, two equations:

$$\frac{dF}{dz}+4\pi\left(1-\hat{A}\right)U=2\pi\,\hat{A}L_0(z)\,,\qquad \frac{dU}{dz}+\frac{3-a_1A}{4\pi}F=\frac{a_1\hat{A}}{2}L_1(z)\,,\qquad (9.120)$$

and it can be seen that both satisfy the diffusion equation. We will write it out only for mean diffusion intensity, which can be written as:

$$\frac{d^2U}{dz^2}-q^2U=\frac{\hat{A}}{2}\left[3-a_1(1-\hat{A})\right]L_0(z),\qquad q^2=\left(1-A\right)\left(3-a_1A\right)\cong\frac{1}{\nu_0^2}\,.\qquad (9.121)$$

The equation for diffusion flow is similar, differing only on the right-hand side.

The deduced equations require formulation of boundary conditions. This problem is not a simple one, as the equations are correct in the depth of the medium but the conditions should be examined for the zones where they are unrealizable; therefore, approximate boundary conditions should be discussed. We can formulate the absence of the intensity flow, inwards the medium on the medium boundaries. For example, let us consider the case of a semi-infinite medium. The lack of full intensity flow at the boundary z = 0 means the following equality is true:

$$\int_0^1 I(0,\mu)\mu d\mu=0\qquad (9.122)$$

or

$$\frac{dU(0)}{dz}-\frac{3-a_1\hat{A}}{2}U(0)=\frac{a_1\hat{A}}{2}L_1(0)\,.\qquad (9.123)$$

The problem in general seems to be rather more complicated, and we refer readers to the relevant literature.[49]

9.11 SMALL-ANGLE APPROXIMATION

Up to now we have considered a scattering indicatrix, which differs slightly from an isotropic one. We will now turn our attention to the pencil-like indicatrix — that is, cases when the scattering occurs within small angles. Simplifications are possible under these circumstances. In the case of a pencil-like indicatrix, in the zero approximation we have:

$$\int_{4\pi}\psi\left(\mathbf{s}',\mathbf{s}\right)I\left(\mathbf{r},\mathbf{s}'\right)d\Omega'\approx I(\mathbf{r},\mathbf{s})\int_{4\pi}\psi\left(\mathbf{s}',\mathbf{s}\right)d\Omega'=I(\mathbf{r},\mathbf{s})$$

and the transfer equation transforms into a simple differential equation, which simplifies the problem considerably.

Let us suppose that the source and the receiver are situated on the z-axis. In this case, the emission is concentrated in the cone with a small expansion angle, the axis of which will be the z-axis. Note that the radiation direction vector is equal to:

$$\mathbf{s} = \sin\theta\cos\varphi\mathbf{e}_x + \sin\theta\sin\varphi\mathbf{e}_y + \cos\theta\mathbf{e}_z \cong \mathbf{s}_\perp + \mathbf{e}_z.$$

We will designate $\mathbf{r} = \mathbf{r}_\perp + \mathbf{e}_z$ and $\psi(\mathbf{s}',\mathbf{s}) = \psi(|\mathbf{s}' - \mathbf{s}|)$ based on the assumption of scattering indicatrix axial symmetry. It is reasonable to use the following expansion for the ray intensity:

$$I(z,\mathbf{r}_\perp,\mathbf{s}'_\perp) \cong I(z,\mathbf{r}_\perp,\mathbf{s}_\perp) + (\mathbf{s}_\perp - \mathbf{s}'_\perp)\cdot\nabla_\perp I + \frac{1}{2}\left[(\mathbf{s}_\perp - \mathbf{s}'_\perp)\cdot\nabla_\perp\right]^2 I + \cdots$$

Further, we will take into account that:

$$\int_{4\pi}\psi(|\mathbf{s}''|)\,(\mathbf{s}''_\perp)\,d\Omega' = 0,\quad \int_{4\pi}\psi(|\mathbf{s}''|)(\mathbf{s}''_\perp\cdot\nabla_\perp)^2 I\,d\Omega' = \frac{\langle\gamma^2\rangle}{2}\nabla_\perp^2 I$$

at adopted assumptions of indicatrix properties. Here, $\mathbf{s}'' = \mathbf{s}' - \mathbf{s}$, and the average value $\langle\gamma^2\rangle$ characterizes the angle width of the scattering indicatrix, which is defined by the integral:

$$\langle\gamma^2\rangle = \frac{1}{2}\int_0^\pi \psi(\cos\gamma)\sin^3\gamma\,d\gamma. \tag{9.124}$$

The use of the ray intensity expansion is the main point of the small-angle approximation, which leads to the parabolic equation:

$$\frac{\partial I}{\partial z} + (\mathbf{s}_\perp\cdot\nabla_\perp I) + N\sigma_a I - \frac{N\sigma_s\langle\gamma^2\rangle}{4}\nabla_\perp^2 I = 0. \tag{9.125}$$

Note that extinction is now determined not by the total cross section of scattering but only by the cross section of absorption.

We will search for a solution to this equation by using the Fourier integral:

$$I(z,\mathbf{r}_\perp,\mathbf{s}) = \int J(z,\mathbf{s},\mathbf{q})e^{i\mathbf{q}\cdot\mathbf{r}_\perp}d^2\mathbf{q}. \tag{9.126}$$

To simplify our approach, we will omit the subscript $\perp$ on the vector of direction, and we now have the ordinary differential equation:

$$\frac{d\tilde{I}}{dz}+\left[i(\mathbf{s}\cdot\mathbf{q})+N\sigma_a+\frac{1}{4}N\sigma_s\langle\gamma^2\rangle q^2\right]\tilde{I}=0.$$

The solution of this equation is obvious:

$$\tilde{I}(\mathrm{z},\mathbf{r}_\perp,\mathbf{s})=\tilde{I}_0(\mathbf{s},\mathbf{q})\exp\left\{-\left[N\sigma_a+i(\mathbf{s}\cdot\mathbf{q})+\frac{1}{4}N\sigma_s\langle\gamma^2\rangle q^2\right]\mathrm{z}\right\},$$

where the first factor is defined from the boundary condition:

$$\tilde{I}_0(\mathbf{s},\mathbf{q})=\frac{1}{4\pi^2}\int I(0,\mathbf{r}'_\perp,\mathbf{s})e^{-i\mathbf{q}\cdot\mathbf{r}'_\perp}d^2\mathbf{r}'_\perp.$$

After rather simple transforms, the solution has the form:

$$I(\mathrm{z},\mathbf{r}_\perp,\mathbf{s})=\frac{e^{-N\sigma_a\mathrm{z}}}{\pi N\sigma_s\langle\gamma^2\rangle\mathrm{z}}\int I(0,\mathbf{r}_\perp,\mathbf{s})\exp\left\{-\frac{(\mathbf{r}_\perp-\mathbf{r}'_\perp-\mathbf{s}\mathrm{z})^2}{N\sigma_s\langle\gamma^2\rangle\mathrm{z}}\right\}d^2\mathbf{r}'_\perp. \quad (9.127)$$

Another representation of the solution can be written down in the form:

$$I(\mathrm{z},\mathbf{r}_\perp,\mathbf{s})=\frac{e^{-N\sigma_a\mathrm{z}}}{\pi}\int I\left(0,\mathbf{r}_\perp-\mathbf{s}\mathrm{z}+\mathbf{p}\sqrt{N\sigma_s\langle\gamma^2\rangle\mathrm{z}},\mathbf{s}\right)e^{-\mathrm{p}^2}d^2\mathbf{p}. \quad (9.128)$$

The parameter:

$$\rho=N\sigma_s\langle\gamma^2\rangle\mathrm{z} \quad (9.129)$$

characterizes the radiation divergence in addition to the initial one for wave propagation in a scattering medium. Where it is small, this additional divergence is not perceived and the ray intensity is:

$$I(\mathrm{z},\mathbf{r}_\perp,\mathbf{s})=I(0,\mathbf{r}-\mathbf{s}\mathrm{z})e^{-N\sigma_a\mathrm{z}}, \quad (9.130)$$

which corresponds to the radiation at the input with the inclusion of the attenuation due to natural absorption by particles.

The distance:

$$z_{div} = \frac{1}{N\sigma_s \left\langle \gamma^2 \right\rangle} \tag{9.131}$$

defines the range in the medium after which the divergence due to scattering will begin to be perceived; however, the condition:

$$\sigma_a < \sigma_s \left\langle \gamma^2 \right\rangle \quad \text{or} \quad \frac{1\text{-}\hat{A}}{\hat{A}} < \left\langle \gamma^2 \right\rangle \tag{9.132}$$

has to be observed, because, in the opposite case, the radiation will be absorbed without approaching the distance (Equation (9.132)). Naturally, discussion about the divergence ceases to have its conventional meaning. It is easy to determine that the last condition takes place only at an albedo close to unity.

The ray intensity is taken as the sum:

$$I = \frac{k_b T}{\lambda^2} + I', \tag{9.133}$$

for studying the proper thermal radiation. Here, the primed component satisfies the parabolic equation with the boundary condition:

$$I'\left(0, \mathbf{r}_\perp, \mathbf{s}\right) = -\frac{k_b T}{\lambda^2}. \tag{9.134}$$

The general problem solution is represented as:

$$I\left(z, \mathbf{r}_\perp, \mathbf{s}\right) = \frac{k_b T}{\lambda^2}\left(1 - e^{-N\sigma_a z}\right). \tag{9.135}$$

The emissivity of the layer with thickness z_d has the usual form:

$$\kappa = 1 - e^{-N\sigma_a z_d}. \tag{9.136}$$

Note the fact that proper radiation is seen, in this solution of the problem, only of the right as we study emission propagating in the direction of positive values of the z-axis; moreover, it does not depend on the observation direction (i.e., no dependence on vector **s**). For this reason, in particular, the radiation of the left is missing. To obtain it, we must rewrite the problem where $z \Rightarrow -z$ and set the boundary conditions at the right boundary. As it turns out, the direct and backward waves are not related with each other, and we obtain a Kirchhoff's law abnormality because the radiation is absorbed ($\sigma_a \neq 0$) in this scattering model, but radiation in the reverse direction of the incident wave is lacking. All of this is the result of our rather artificial assumption about the absence of backward scattering.

We can now study the scattering characteristics of a large-radius sphere (see Section 5.8). This scatter has an approximately isotropic indicatrix for backward scattering and a pencil-like one for forward scattering. In the first approximation, the scattering indicatrix can be chosen in the form:

$$\psi(\mathbf{s}',\mathbf{s}) = \frac{\psi_0}{4\pi} + (1-\psi_0)\delta(\mathbf{s}-\mathbf{s}'). \tag{9.137}$$

This choice of indicatrix leads us to the known equation:

$$\mu\frac{dI}{dz} + I - \frac{\hat{A}_e}{2}\int_{-1}^{1} I(z,\mu')\,d\mu' = \left(1-\hat{A}_e\right)\frac{k_b T}{\lambda^2}, \tag{9.138}$$

where the equivalent albedo is:

$$\hat{A}_e = \frac{\psi_0 \hat{A}}{1-\hat{A}+\psi_0\hat{A}}, \tag{9.139}$$

and the dimensionless distance is:

$$z = N\sigma_t\left(1-\hat{A}+\psi_0\hat{A}\right)\mathrm{z}. \tag{9.140}$$

So, we arrive again at the well-known problem.

10 General Problems of Remote Sensing

The second part of this book is dedicated to the background of remote sensing by radio methods. The notion of remote sensing of the environment is usually understood as the determination of characteristics of a medium by devices that are far from the object being studied. The concept of *environment* includes all objects (both natural and of anthropogenic origin) that form man's habitat. These are the natural objects (soil, vegetation, atmosphere, etc.) around us on Earth, as well as near Earth and in outer space. These objects also include people themselves and animals. Some elements of the environment are represented schematically in Figure 10.1.

The goal of measurements carried out for remote sensing is to define various environmental parameters that can be used to obtain a deeper understanding of natural processes, to improve economic activities through the realization of the preventive actions necessary to protect the environment, and to discover and monitor extraordinary natural and anthropogenic situations.

The time and spatial scales of observed characteristics have a very wide range (from part of a second to centuries for time and from units of meters to units of a global scale for space). The measurers can be mounted on ground and air platforms, on rockets, and on space craft. Some of these platforms are also shown in Figure 10.1.

Environmental remote sensing assumes the practical absence of disturbance in the studied medium during measurements. This is achieved by electromagnetic application or remote sensing acoustic waves. The wide application includes electromagnetic, microwave, and ultrahigh-frequency waves, all of which interact effectively with natural media. It is supposed that the interaction of electromagnetic waves with the environment, defined by the electrophysical and geometrical parameters of the researched objects, is closely connected with the structure, thermal regime, geophysical characteristics, and other parameters of these objects. Radiowave interaction with natural media was described in the first part of this book (Chapters 1 to 9), which was devoted to radio propagation theory in various media. This theory is the physical background of radio methods for remote sensing of natural media. The devices for research, as well as the development of processing technology for experimental data, are created on this basis. In the following chapters, we consider devices that are used for remote sensing and some methods for processing experimental data. In this chapter, which may be considered as an introduction to remote sensing, some problems of environmental remote sensing are covered from the position of radio methods:

- Formulation of the remote sensing problem
- Radiowave bands applied to remote sensing
- Main principles of processing remote sensing experimental data

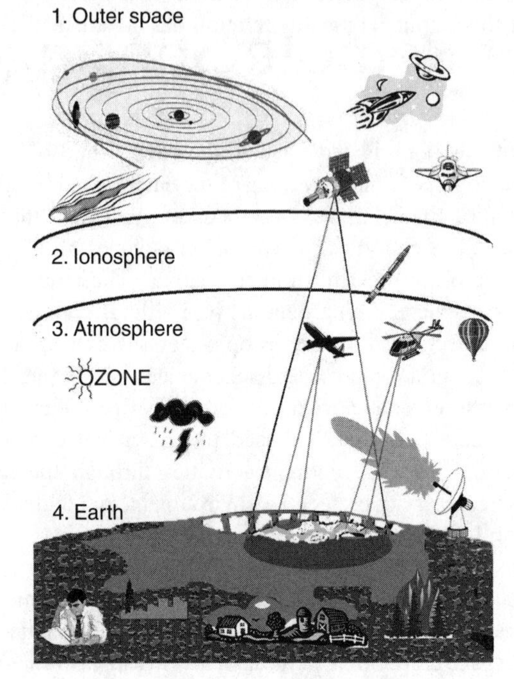

FIGURE 10.1 Schematic representation of the environment.

10.1 FORMULATION OF MAIN PROBLEM

The main goal of remote sensing is, as was already mentioned, to obtain various kinds of data about the environment. In this book, we will consider only radiowaves as the source of such information. Radiowaves are generated from both artificial and natural sources. The methods applied to artificially generate waves are often called *active* as opposed to *passive* approaches based on using naturally generated waves. It is necessary to point out that active methods are generally connected with coherent waves, while incoherent waves are typical for passive methods.

The high frequency power gathered by an antenna at the receiver input is amplified (often with a frequency decrease due to heterodyning). As a result, one or several voltages are formatted at the receiver output. Each of them is linearly related to the field strength entered the measuring system input. Sometimes this relation has a functional character. Also, the receiving–amplifying part of a device contributes the complementary noise, the power of which is defined by the receiver noise temperature (T_n). The sources of interference may have another origin, particularly with regard to extraneous waves at the antenna input. As a rule, interference is supposed to be additive, although this does not hold in all cases.

The signal from the receiving/amplifying component enters the processing device, where the required measurement parameters (e.g., amplitude, phase, frequency, delay time) are separated. The processing operation is optionally linear. As a result, the instrument can be mathematically represented as a set of operators (A_1, A_2, ..., A_i) converting the characteristics of input strengths E_{in} at the antenna into the voltages V_i at the output. Thus, this relation has a statistical character:

$$V_i(t) = A_i\left\{E_{in}(t)\right\} + \Delta V_i\,, \tag{10.1}$$

where ΔV_i is the errors generated by the noises of i-th channel of the measuring instrument or the measuring system.

We will now provide simple examples of the relations of output voltages with measured quantities of parameters for some instruments. In order to do this, we will briefly describe the main points of operators (discussed further in Chapter 11) for the most frequently used instruments of remote sensing.

10.1.1 Radar

At least two operators, A_1^{rl}, and A_2^{rl}, correspond with this instrument. They are associated with output voltages V_1^{rl} and V_2^{rl}, corresponding to two characteristics of received radiation. One of them is proportional to the time delay between radiated and received signals and the second one to the received signal power:

$$\begin{aligned} \left[V_1^{rl}(t)\right]_{ij} A_1^{rl}\left\{E_{in}(t)\right\} &\sim \tau(t) = 2\int_0^R \frac{dR}{c(R)}, \\ \left[V_2^{rl}(t)\right]_{ij} A_2^{rl}\left\{E_{in}(t)\right\} &\sim P_j^{\text{rec}} = \frac{P_i^{\text{rad}}\, G_i^{\text{rad}}\, \sigma_{ij}\, A_j^{\text{rec}}}{16\pi^2 R^4}, \end{aligned} \tag{10.2}$$

where P_j^{rec} is the power of the j-th polarization of the wave in the receiving antenna input, A_j^{rec} is the effective area of the receiving antenna at the j-th polarization, P_i^{rad} is the transmitter power at the i-th polarization, G_i^{rad} is the gain of the transmitting antenna at the i-th polarization, σ_{ij} is the radar cross-polarization section of the target backward scattering, R is the distance from the target to the radar, and $c(R)$ is the radiowave velocity. Signal processing may be more varied. In particular, the operators of polarization and spectral analyses would be added to the two mentioned above, which are most commonly used.

10.1.2 Scatterometer

The scatterometer is a variant of a radar where the power of the received signal is the only object of measurement. The operator A^{sct} associates the output voltage V_{ij}^{sct} with a quantity equal to the ratio of the power P_j^{rec} at the receiving antenna input to the power P_i^{rad} at the transmitting antenna output (i and j are the corresponding polarization):

$$V_{ij}^{\text{sct}}(t) = A^{\text{sct}}\left\{E_{in}(t)\right\} \Rightarrow \frac{P_J^{\text{rec}}}{P_i^{\text{rad}}} \sim \frac{1}{16\pi^2 R^2}\int dl \int_{2\pi} D_i^{\text{rad}}(\Omega)\,\sigma_{ij}(l,\Omega)\,A_j^{\text{rec}}(\Omega)\,d\Omega\,, \tag{10.3}$$

where D_i^{rad} is the transmitting antenna directional coefficient at the i-th polarization. On the right side is the integral with respect to depth l and over the solid angle as a distributed object of our research (e.g., cloud drops, ionospheric electrons, sea-surface irregularities). Therefore, in the considered case, $\sigma_{ij}(l,\Omega)$ is the cross section per volume unit. It is supposed that the target is distributed in some volume; thus, we have an integral with respect to l. It is assumed further that the layer thickness is much less than the distance to the radar, and the integration over Ω is mainly concentrated within the major lobe of a pencil-beam antenna. This gives us the opportunity to put distance R outside the integral sign. If we deal with a surface target (sea ripples, for example), it is necessary to assume that $\sigma_{ij}(l,\Omega) = \sigma_{ij}^0(\Omega)\,\delta(l-R)$, where $\sigma_{ij}^0(\Omega)$ is a dimensionless value (cross section per area unit or backscattering reflectivity). When the backscattering reflectivity is a constant, Equation (10.3) is quite simplified and, at the matched polarization:

$$A^{\text{sct}}\left\{E_{\text{in}}(t)\right\} \Rightarrow \frac{P_j^{\text{rec}}}{P_j^{\text{rad}}} \sim \frac{\sigma^0 A_j^{\text{rec}}(0)}{4\pi R^2}. \tag{10.4}$$

10.1.3 Radio Altimeter

The radio altimeter is also a functionally simplified radar. The main interest here is the arriving time of the signal; the operator A^{alt} relates output voltage V^{alt} to the time interval (τ) between the radiated and received radio pulses:

$$V^{\text{alt}}(t) = A^{\text{alt}}\left\{E_{\text{in}}(t)\right\} \Rightarrow \tau(t) = 2\int_0^h \frac{dh}{c(h)}, \tag{10.5}$$

where h is the altimeter altitude above a reflecting surface, and $c(h)$ is the radiowave velocity depending on altitude.

10.1.4 Microwave Radiometer

The operator A^{rm} associates the output voltage with a quantity that is proportional to the brightness temperature of an object:

$$V_i^{\text{rm}}(t) = A_i^{\text{rm}}\left\{E_{\text{in}}(t)\right\} \Rightarrow P_j^{\text{rec}} \sim T_{\circ}\,. \tag{10.6}$$

The operators mentioned above will be refined later when we describe specific instruments used for remote sensing (Chapter 11). Information about the primary techniques for calibration will also be given in Chapter 11. This calibration allows us to estimate the coefficients of proportionality and permanent biases that are negligible in Equations (10.2) to (10.6).

10.2 ELECTROMAGNETIC WAVES USED FOR REMOTE SENSING OF ENVIRONMENT

Remote sensing of the natural environment is realized within a wide range of electromagnetic waves — from ultraviolet to radio (see Figure 10.2). Each section of the range has its own merits and demerits; therefore, the most effective approach is the application of different areas of the electromagnetic spectrum as appropriate. We consider in this book only part of the radio region: millimetric, centimetric, decimetric, and, particularly, ultrahigh frequency (UHF). The advantage of using this spectral part of the region as opposed to the optical or infrared is connected with the depth of penetration that can be achieved in a medium which allows us to detect variation in medium parameters related to the depth of the structure. Using vehicle-borne instruments, radiowaves are absorbed weakly in the atmosphere and clouds. This creates the conditions for all weather observations of Earth's surface. In addition, the application of radio instruments, as opposed to optical ones, does not require illumination of the area being studied by solar light, which allows us to carry out investigations regardless of the time of day. Also, some spectral intervals in this region interact effectively with the ionosphere, atmosphere, and atmospheric formations, as well as with elements of ground and sea surfaces. This gives us the opportunity to use them to investigate these media.

The main drawback of using the radio region is the rather low (in comparison to the optical and infrared regions) spatial resolution, especially by passive sounding (see Equation (1.120)). Only synthetic aperture radars overcome this difficulty and achieve spatial resolution comparable with optical and infrared devices (see Chapter 11).

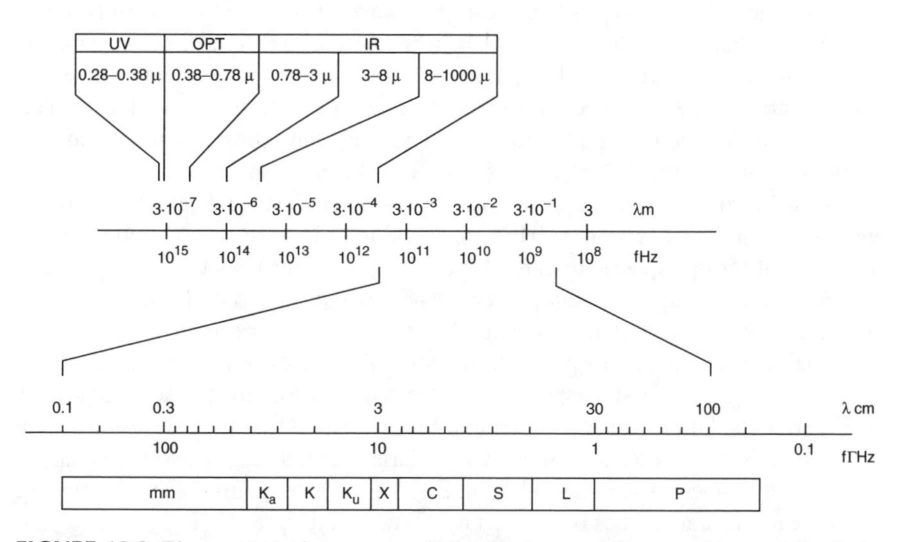

FIGURE 10.2 Electromagnetic waves, which can be used for remote sensing of the environment.

Effective application of radiowaves to investigate natural objects depends on the required spatial resolution and specific peculiarities of radio propagation in the experimental conditions. The problem of various objects interacting with electromagnetic waves is discussed in Chapters 12 to 15.

In the case of sounding from space through the ionosphere, the lower limit of the frequency region (f_{min}) is determined by the maximum of the ionospheric plasma frequency (f_p) connected with the maximum of electron concentration N_{max} (see Equation (2.31)). It was pointed out in Chapter 2 that the value of f_p in the electron concentration maximum is on the order of 10 MHz. The limitations connected with wave propagation in the ionosphere are naturally no longer relevant to the use of airborne instruments; however, they appear again if, for example, we are dealing with upper ionosphere observations (see Chapter 3).

The upper frequency border of the sounding region from space is defined by the atmospheric absorption of electromagnetic waves. The main absorbing components are water vapor and oxygen. In the radio band, oxygen has a series of absorption lines at a wavelength of 0.5 cm and a separate line at a wavelength of 0.25 cm. Water vapor has absorbtion lines corresponding to wavelengths 1.35 and 0.163 cm, and also a series of absorption lines at waves shorter then 1 mm. As absorption at frequency $3 \cdot 10^{11}$ Hz is of the order at 10 db this frequency is assumed to be the upper border frequency region for the radio sensing of Earth from space. Hence, the electromagnetic region of sounding waves from space is determined by the inequality

$$0{,}1 < \lambda < 10^3 \text{ cm} .$$

The transparency windows of the millimetric wave region lie at the wave bands of 0.2, 0.3, 0.8, and 1.25 cm (Figure 10.3) in the absence of clouds, snow, rain, etc. One has to take into account when planning experiments the help of both aerospace-borne instruments and devices mounted on the ground. Meteorology radar, in particular, is a common example. It is fitted to take into consideration radiowave scattering and absorption by hydrometeors (clouds, rains, snow).

In underground sounding, an important consideration is the depth of penetration into the researched layers, and UHF is the band used in this case. A similar band is also applied for ionospheric research for other reasons (see Chapters 3, 13, and 15).

Frequencies lying at the transparency windows and at regions of selective atmospheric absorption, depending on the problem being studied, are applied for the study of the atmosphere and atmospheric formations. The waves of millimetric, centimetric, and decimetric bands, depending on the requirements for the sounding depth and spatial resolution, are also preferable for the study of biological objects.

Remote sensing with radiowave help is based, as indicated earlier, on changes in the wave characteristic as a result of interaction with the environment. The change in radiowave characteristics is detected by the receiving systems. The output signals then allow us to obtain the position, form, and geophysical parameters of natural formations.

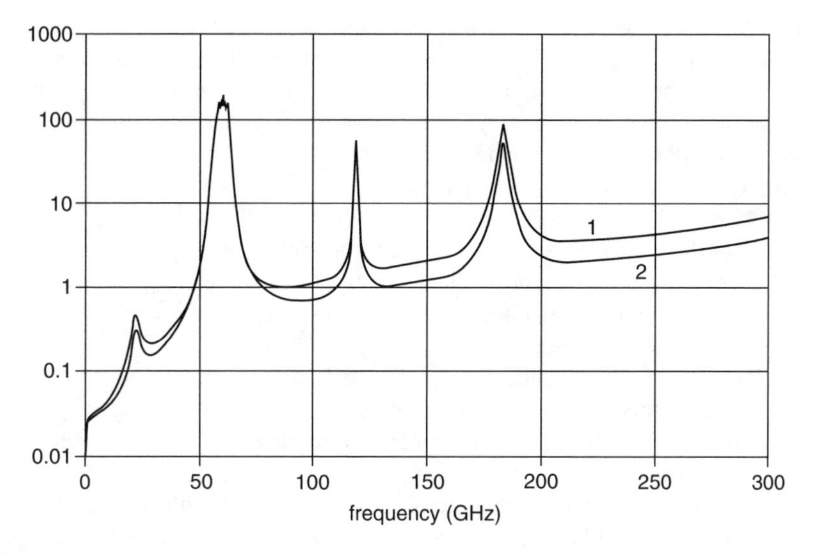

FIGURE 10.3 Microwave absorption due to atmospheric gases: 1, normal humidity (7.0 g/m^3); 2, humidity (4 g/m^3).

Listed below are the main radiowave characteristics determined by remote sensing:

- Amplitude, intensity, and power flow of the electromagnetic field
- Time of propagation
- Direction of the radiowave propagation
- Phase properties of radiowaves
- Frequency and frequency spectrum of receiving signal
- Polarization characteristics of received signal
- Change of the pulse shape

In order to obtain information about the geometry, physico-chemical properties, structure, state, and dynamics of a natural formation, we must formulate an inverse problem to study the change of these values in space and time and use *a priori* information about the investigated object itself and about the characteristics of its interaction with the electromagnetic field.

10.3 BASIC PRINCIPLES OF EXPERIMENTAL DATA PROCESSING

The main goal for thematic processing of experimental data obtained through environmental remote sensing is to define the characteristics of a medium in space and time. As a rule, such characteristics are the values related to its physico-chemical properties, structure, etc. In order to reach this goal, we must solve a wide range of problems that are referred to as inverse ones from the point of view of causal and investigatory connections. However, it is an inverse problem in only some cases —

namely, those having a great number of unknown parameters (where the state of an object is described by some coordinate function); we will discuss those problems further toward the end of this chapter. The other inverse problems have been given such labels as problems of classification, factorization, parameter estimation, model discrimination.[83,84] We have divided these problems into three groups according to the requirements for remote sensing data processing:

- Classification problems are related to defining the type of object being observed and its qualitative characteristics (e.g., space observation of land areas where it is difficult to distinguish forest tracts from open soil or ice plots from open water).
- Parameterization problems are connected with the numerical estimation of parameters of studied objects (e.g., not a question of what we see during a flight above the ocean, but rather determining the surface temperature of the water or the seawave intensity).
- Inverse problems of remote sensing are associated with the creation of continuous profile distributions for various parameters of the researched objects (e.g., height profiles of tropospheric temperature, height profiles of ionospheric electron concentration).

The problems of classification deal with the selection of object groups having approximately similar parameters with regard to interaction with electromagnetic waves and, consequently, as one may expect, comparable physico-chemical and structural characteristics. One can subdivide a body of mathematics for classification based on different directions of cluster (grouping close results of multidimensional measurements) and structure (grouping of spatio-temporary areas with structures of close multidimensional measurements) analyses, as well as multidimensional scaling (limitation by magnitude).[84]

The classification problem is generally solved by multichannel methods; however, before turning to them, let us say a few words about some of the possible single-channel methods. The simplest one is associated with the establishment of boundaries for the functional quantities of instrument output voltages (parameters of interaction) within limits, where the investigated objects may be related to a particular class. The simplest kind of such functionals can be maximum and minimum values, medians, dispersion, correlation coefficients of experimental, *a priori* data, etc. Obviously, the boundaries themselves are established on the basis of *a priori* information (from theory or previous experimental data often obtained by *in situ* methods) (see Figure 10.4a). The characteristics of the distribution function and especially its having multiple modes can be used for classification (Figure 10.4b). The elements of the textured analyses can be applied in the case of sufficient *a priori* information. These elements may relate to the specific form of signal from defined elements of the sounding environment and with the contours of two-dimensional images.

The technique of multidimensional scaling is seldom applied for multichannel measurements (thresholds are established from *a priori* data similarly to the one-channel case). More often, in this case, we resort to different methods of cluster

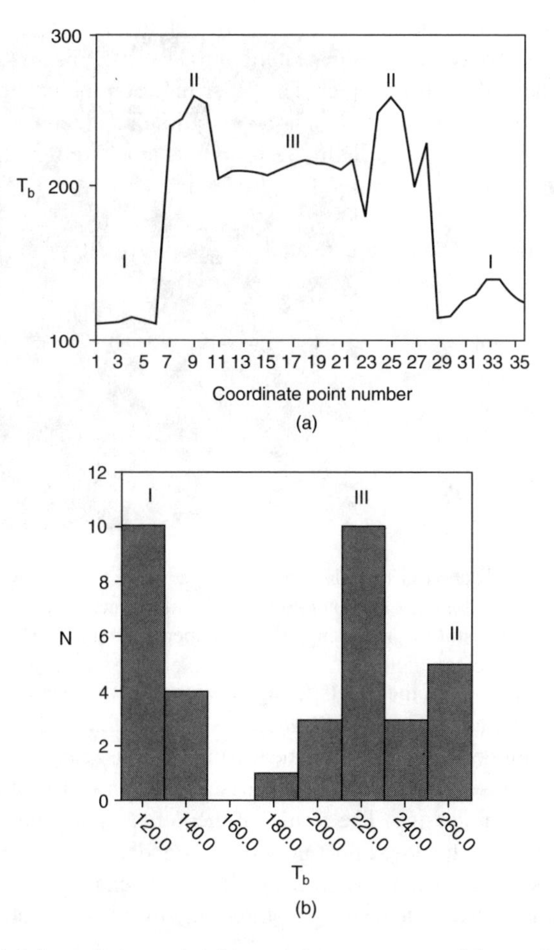

FIGURE 10.4 (a) Schematic image brightness temperature around Antarctica; (b) histogram of this temperature. I, sea; II, sea ice; III, continental ice.

analyses. As a rule, three types of information are taken into consideration: multidimensional data of measurements, data about closeness after processing the experimental materials, and data about classes obtained as a result of experimental and *a priori* data processing multidimensional data chosen from the train of data obtained from different measurement channels. The closeness criterion here is defined by the parameters of discrepancy or similarity for the separated sets (clusters) of the experimental data, such as intercorrelation data in different measurement channels, the intersection of data, or other similar parameters (e.g., the Euclidean distance between two similar objects or some other functional closeness).

For classification purposes, the ensemble of experimental points (comparable according to some feature) is intercepted in the measurement space. This process is known as *clusterization*. The set boundaries are defined by the expected credibility value of the obtained results. From this point of view, the intuition of the researcher plays no small role here. These boundaries may be ascertained in the process of

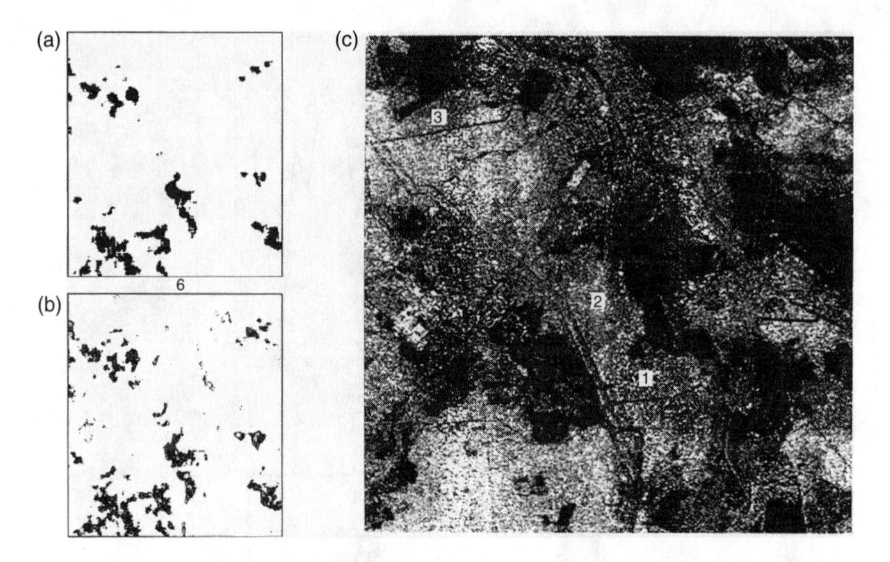

FIGURE 10.5 (a) Application of two classification stages of forest types with a usage texture parameter; (b) application for classification of a trizonal artificial neural network; (c) image of a fir forest obtained as a result of processing synthetic aperture radar (SAR) data.

establishing the relations of these sets with the elements of the studied environment. This process, known as *cluster identification*, is usually realized by teaching and is carried out for unknown objects by measuring various elements of the known environment and subsequently comparing these measurement results with the outcome of the cluster processing. The results of theoretical and experimental research can be also used for the identification. Many standard computer programs are available for cluster analysis of experimental data. The example of ice field clusterization on the basis of remote sensing at three microwave channels is discussed in Livingstone et al.[136]

The texture methods, as compared to cluster methods, are associated with another type of classification. If the cluster techniques classify objects by single elements of the spatial resolution of an instrument, then the texture methods do so according to the structure of the fields of the observed objects. Continuous fields are usually considered, but it is also possible to examine noncontinuous fields. The body of mathematics regarding this area is extensive, it is well algorithmized, and numerous computer programs are available for texture analyses. Figure 10.5 shows the results of the texture procedure for the selection of forest tracts.[137,138]

Certainly, other more complicated methods of pattern recognition are available, but the techniques described briefly above have gained the widest application for remote sensing. It is necessary to point out once more that the need to address these methods is conditioned by the complicated structure of many natural objects and the practical impossibility of computing exactly the results of their interaction with electromagnetic waves. Therefore, these methods do not assume knowledge of the relations between some parameters of the environment and the characteristics of their interaction with electromagnetic fields; however, knowledge of interaction

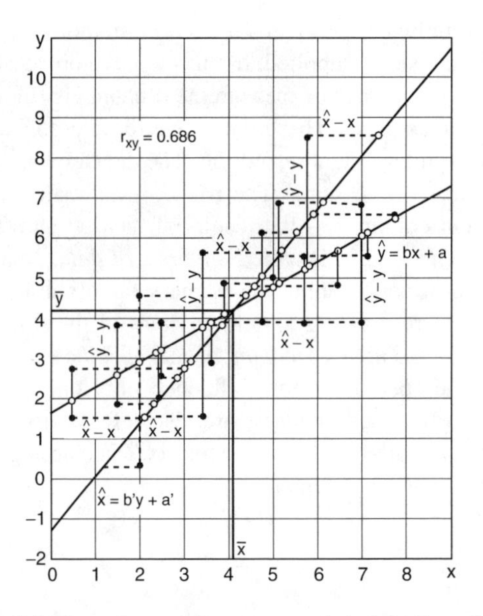

FIGURE 10.6 (a) Straight line of regression *y* on *x* and straight line of regression *x* on *y*; (b) regression for the same field of a correlation, where $\overline{x}$ and $\overline{y}$ are average values of the variables.

models facilitates both the clusterization and identification of separated clusters and cluster spatial structures.

Before turning to the second group of problems (problems of parameterization), let us consider briefly the factorial approach to remote sensing problems. This approach is associated with both the classification and the parameterization of natural formations. Parameters such as atmospheric humidity, water content and temperature of clouds, temperature of the sea surface, soil moisture, vegetation biomass, and many other characteristics (described in Chapters 12 to 16) can be considered as the factors. The simplest factorial problems (e.g., assessing the influence of a small number of known causes) are solved, as a rule, by regressive analysis technique.[139] In regressive analyses, we graph the regressive curves reflecting the statistical relation between numerical values of factors (e.g., soil moisture, biomass of vegetation) and parameters of the radiowave interaction with the medium being researched, such as brightness temperature or the scattering cross section. An example of a one-dimensional linear regression of two variables, *x* and *y*, is provided in Figure 10.6. The regressive line $\left(\hat{y} = bx + a\right)$ is plotted by the experimental points y_j based on the condition $\sum_j \left(y_j - \hat{y}_j\right)^2$. Figure 15.12 (brightness temperature with regard to dependence on the subsoil water level) give an example of the regressive line use in remote sensing. The regressive lines inclination angles may be used in some cases for the identification (classification) of factors.

To solve more complicated problems related to unknown causes, different variants of the factorial analyses are applied. In this case, the processing of experimental data obtained by a large number of measurement channels (more than the number of expected factors) takes place. These data have to be associated with the terrain coordinates and have similar spatial resolution. The data are joined in the rectangular matrix **Y** for the factorial processing. The rows of this matrix determine the measurement channels and columns — the results of measurements along the definite curve on the terrain. This matrix is called a *matrix of data.* Analysis of this matrix allows us to obtain information about the primary factors influencing the variation of experimental data corresponding to defined areas of the studied terrain.

These factors are classified as *common* and *specific* ones by their effect on experimental data. The specific factors influence only one channel; the common factors that affect all processed channels are also referred to as *general.* The data are normalized for the factorial analysis, and matrix **Y** is rearranged into the so-called standardized matrix **Z** with the elements:

$$z_{ij} = (y_{ij} - \bar{y}_i) / \sigma_i \, , \tag{10.7}$$

where $\bar{y}_i$ is the main signal value in the *i*-th channel, and σ_i is the standard deviation in the same channel.

Factorial analysis is practically reduced to standardized data presentation as a linear combination of hypothetical variables or factors:

$$z_{ij} = a_{i1}p_{1j} + a_{i2}p_{2j} + \ldots + a_{ir}p_{rj} + \zeta_i \, . \tag{10.8}$$

Here, a_{ij} are coefficients (determined during factor analysis) that define the influence grade of the *j*-th common factor; $p_{1j} \ldots p_{rj}$ are factor scores (the numerical value of influencing characteristics) at the *j*-th sample; and ζ_i is the common effect of the unique factors of *i*-th channel. This equality expresses the basic model of factor analysis. Thus, it is supposed that the matrix of standardized data is defined only by common factors, and by applying the matrix form of notation we obtain:

$$\mathbf{Z} = \mathbf{AP} \, . \tag{10.9}$$

Matrix **A** is the *factor pattern* and its elements, a_{ij}, are factorial loadings. Matrix **P** represents by itself the matrix of numerical quantities (parameters) of the factors p_{ij}.

The fundamental theorem of factorial analysis maintains that matrix **A** is related to the correlation matrix **R**, the elements of which are the correlation coefficients between rows (channels) of standardized matrix **Z**. In the case of uncorrelated factors,

$$\mathbf{R} = \mathbf{AA}' \, , \tag{10.10a}$$

where $\mathbf{A}'$ is the transposed matrix of the factorial loads, and

$$\mathbf{R} = \mathbf{ACA}' \tag{10.10b}$$

in the case of correlation of factors. **C** is the correlative matrix reflecting the relations between factors.

Matrix **C** is computed on the basis of *a priori* information about the physical connections between factors. Matrix **A** is defined by solving Equation (10.10a). The method of main components or the centroidal method is often applied for these purposes.

Different models of factorial analysis are used depending on the accepted *a priori* assumptions. We can separate these models into two groups. For the first group, we assume that the number of common factors is known. Then, the factor loads a_{ij} and the numerical quantities of the factors p_{ij} are determined from Equation (10.10a). In the process, the summarized dispersion added by the negligible factors in the general data dispersion of each channel, is minimized.

For models of the second group, we must first determine the number of common parameters required to provide affinity of experimental and calculated correlation matrixes. To do so, we use the sequential approach technique, from one to n common factors. The computation is stopped when the differences between elements of the experimental and calculated matrix reach the same order as the measurement and computation errors. It is useful to point out that, in this case, computation of the common factors is performed by applying Equation (10.10a) where the reduced matrix $\mathbf{R}_{\mathrm{h}}$ is substituted for the correlation matrix **R**. Matrix $\mathbf{R}_{\mathrm{h}}$ differs from matrix **R** by its diagonal terms, which are called the *commonalities* in this case. The commonalities give us an estimation of the contribution of the common factors to the common data dispersion in the processed segment. The commonalities estimation is a separate problem of factorial analysis. A rough estimation is sufficient in the case of a great number of channels; for example, the maximal value of nondiagonal terms of the chosen row can be used for the diagonal term. The qualitative side of such classification is demonstrated by Figure 10.7.

The first group of factorial analysis models is more appropriate for problems of parameter estimation; the second group, for classification problems. The factorial models perform linearization of experimental data on the given segment of processing and estimate the quantity and the intensity of the factors impacting the output signal change. Factorial analysis is especially useful for the preliminary simultaneous processing of a great number of channels.

Parameterization problems belong to the main class of remote sensing problems. They are connected with quantitative estimation of the parameters of the natural object being studied. It is supposed in the process of problem solving that the model function relates the instrument displays with the structure and physico-chemical properties of the objects. This relation depends upon the accuracy of the parameters; *a priori* model functions may be refined and modified during specific studies. Some these functions were addressed in the first part of this book and will be examined in following chapters with regard to significant objects of the environment. Here,

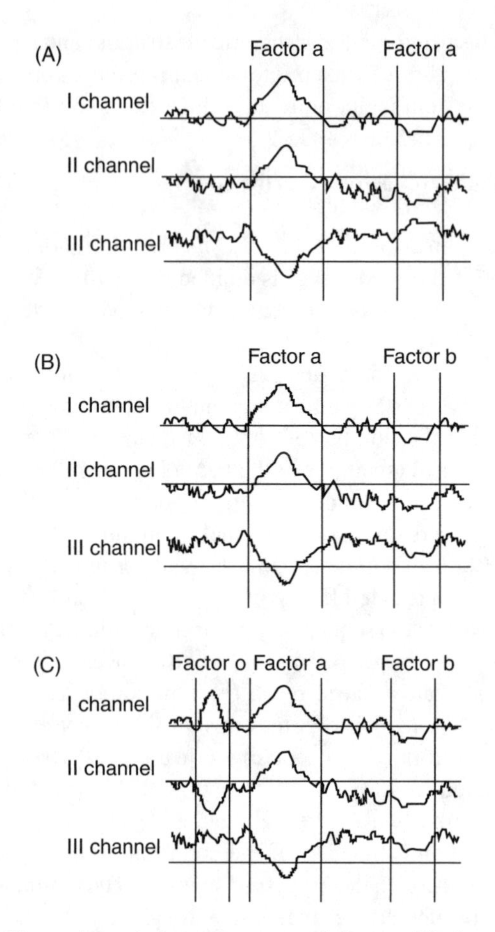

FIGURE 10.7 (A) Three channels have one general factor; (B) three channels have two general factors; (C) three channels have one common factor (o) and two general factors (a and b).

we will consider only some general problems of estimating the parameters of a function.

Suppose that model functions F_i connect measured electromagnetic wave parameters I_i of the i-th channel with the studied objects characteristics, x_j. Then, we can write the following system for calculating the parameters of the medium:

$$\begin{aligned} I_i &= F_i(x_1, x_2, ..., x_n) + \Delta_i \\ i &= 1, 2, ..., m, \quad m \geq n \end{aligned}, \tag{10.11}$$

including in the consideration the measurement errors, $\Delta_i^{(1)}$, and the model conception uncertainties, $\Delta_i^{(2)}$. Here, i is the measurement channel number, n is the number of parameters to be determined, and $\Delta_i = \Delta_i^{(1)} + \Delta_i^{(2)}$ are summarized errors of the

i-th channel. It is often supposed that the error distribution function is known, even if only approximately, and it is taken into account in the solution process.

The considered system is not quite an ordinary one solvable by higher algebra technique. On the one hand, if x_j and Δ_i are regarded as unknown values, then the system is not definite because the number of unknown values is always more than the number of equations. On the other hand, we can neglect the measurement errors by assuming them to be equal to zero. In this case, the number of equations is usually more than the number of unknown parameters and the system becomes contradictory. The solution of Equation (10.11) is divided into two steps:[140] (1) definition of unknown parameters using the minimum of data, and (2) definition of unknown parameters by using redundant data. These two steps are closely connected, and the processing may be confined by one of them.

By processing using a minimum of data selection from Equation (10.11), then the number of equations being equal to the number of unknown parameters is $n = m$, assuming the errors to be equal to zero. We usually solve the obtained nonlinear (in the general case) system of equations by applying the iterative procedure. In most cases, this procedure results in a sufficiently good initial approximation of the unknown parameters $x_1^0, x_2^0, ..., x_n^0$ by using the techniques of reassembly and rearrangement of equations and by varying the initial conditions and other combinations. This can be done even in the absence of *a priori* data regarding the environmental parameters being studied. After this, it is convenient to linearize Equation (10.11) by the following expansion:

$$I_i = F_i(x_1, x_2, ..., x_n) + \Delta_i = F_i(x_1^0, x_2^0, ..., x_n^0) + \Delta_i + \\ + \sum_{j=1}^{n} \partial F_i / \partial x_j \big|_{x_j = x_j^0} \delta x_j, \quad i = 1, 2, ..., m \tag{10.12}$$

where δx_j distinguishes parameter x_j from the starting approach, x_j^0.

The system of linear equations in Equation (10.12) may be solved by changing δx_j by the methods for the solution of redundant equations (see later) on the assumption that the errors (Δ_i) are equal to zero. In this way, we are *processing the linear model by the minimum of data*. The problem is then reduced to solution of the linear equations ($\Delta_i = 0$):

$$\begin{Vmatrix} I_1 \\ . \\ . \\ I_n \end{Vmatrix} = \begin{Vmatrix} a_{11} & .. & a_{1n} \\ . & & \\ . & & \\ a_{n1} & .. & a_{nn} \end{Vmatrix} \begin{Vmatrix} x_1 \\ . \\ . \\ x_n \end{Vmatrix}$$

or

$$\mathbf{I} = \mathbf{AX}.$$

The solution will have the form:

$$\mathbf{X} = \mathbf{A}^{-1}\mathbf{I} \tag{10.13}$$

where the matrix $\mathbf{A}^{-1}$ is the inverse one. The inverse matrix appears when the matrix determinant differs from zero; however, if the matrix elements are given approximately, then the question about the determinant of matrix $\mathbf{A}$ differing from zero is irrelevant. This observation relates particularly to cases when a change of coefficients in the frame of accuracy can change the determinant sign. A system having such a matrix cannot be solved with sufficient accuracy. The matrix is a stable one when small changes of basic matrix elements lead to small changes of inverse matrix elements. If the inverse matrix is unstable, then the basic matrix is an ill-conditioned one. It is clear that we must choose experimental conditions that will lead to a stable matrix of the linear equations.

The determinant of the basic matrix must not be too small to provide stability of the inverse matrix. It is difficult to define precisely the notion of "too small" because multiplication of the matrix by any quantity changes the determinant but does not change the inverse matrix stability.[141] Hadamard's inequality can be used as a diagnostic criterion for the determinant value estimation:

$$|\mathbf{A}| \le \sqrt{\prod_{i=1}^{n}\sum_{j=1}^{n}\left|a_{ij}\right|^2}. \tag{10.14}$$

Hadamard's inequality has to be close to equality for inverse matrix stability. Obviously, the experimental conditions must be chosen in such a way as to give us the opportunity to obtain the maximum of the matrix $\mathbf{M}$ determinant; in particular, it can be done by choosing data for these channels that lead to value maximization of the determinant.

The next step of analysis is *redundant data processing*. The random errors are assumed to be basic and other ones can be neglected. If the last condition is not fulfilled, the considered methods give biases. We can select from the two major groups of methods to obtain estimations by redundant measurements.[83,140] We must know the distribution function of observed values to apply the first group of methods; the maximal likelihood and Bayes' methods are commonly used, and they will be covered below. The regressive method, the method of minimal squares, and the method of minimal modules are related to the second group. They allow estimations close or equal to the maximal likelihood ones on the basis of the formal computing techniques without knowledge of the distribution function for the observed values.

Let us consider the *method of maximal likelihood* to define the environment parameters on the basis of measuring data in m channels. In the process, $m > n$. The density of joint distribution (f) can be represented in the form:

$$f(I_1,I_2,\ldots,I_m,x_1,x_2,\ldots,x_n)=f(I_1,\,x_1,x_2,\ldots,x_n)f(I_2,x_1,x_2,\ldots,x_n).f(I_m,x_1,x_2,\ldots,x_n)$$
$$=g_m(I_1,I_2,\ldots,I_m\,|x_1,x_2,\ldots,x_n)=g_m(\overline{I}\,|\,\overline{x}) \tag{10.15}$$

for the case of independence of individual measurements.[83,140] The function $g_m(\overline{I}\,|\,\overline{x})$ is the *function of likelihood*. It is a joint distribution function for the observed quantities and is regarded as a function of unknown parameters x_i. The observed environment parameter quantities obtained by the likelihood function maximum are referred to as *reliable estimations* or *estimations of maximal likelihood*. Maximal likelihood estimations are distributed by the normal law under very common conditions. They are mutually effective and therefore consistent. The mathematical expectation of the estimation tends to approach the true value with an increasing number of measurements.

Taking into consideration the exponential view of the distribution function, the maximization of the logarithm of the likeliness function usually is used; that is,

$$L_m(\overline{I}\,/\,\overline{x})=\ln g_m(\overline{I}\,/\,\overline{x})=\sum_{j=1}^{m}\ln f(I_j,\,x_1,\,x_2,\ldots,\,x_n). \tag{10.16}$$

The value L can be maximized relative to x by setting the partial derivatives equal to zero; thus,

$$\frac{\partial L_m}{\partial x_1}=\sum_{j=1}^{m}\frac{\partial \ln f(I_j,\,x_1,\,x_2,\ldots,\,x_n)}{\partial x_1}=0;$$
$$\frac{\partial L_m}{\partial x_n}=\sum_{j=1}^{m}\frac{\partial \ln f(I_j,\,x_1,\,x_2,\ldots,\,x_n)}{\partial x_n}=0. \tag{10.17}$$

The solution of these equations gives us the unknown estimations. We have achieved processing using the minimum of data, as the number of equations is equal to the number of the unknowns. Equation (10.17) is reduced to a linear one when the model functions are linear and the distribution function is normal. The method of the reverse matrix may be used in this case to solve the equation system.

Estimation by Bayes' method is based on maximization of a posterior probability distribution for the investigated parameters which are considered as random values. The *a posteriori* probability distribution is:

$$g_0(\bar{x}\,|\,\bar{I}) = g(\bar{I},\bar{x})\,/\,g_1(\bar{I}) = g_0(\bar{I}\,|\,\bar{x})\,g_2(\bar{x})\,/\,g_1(\bar{I})\,, \tag{10.18}$$

where $g_0(\bar{x}\,|\,\bar{I})$ is a posterior distribution of probability (i.e., the density of the probability distribution for $\bar{x}$ when observations $\bar{I}$ are obtained by this time); $g(\bar{I},\bar{x})$ is the mutual density of distribution for $\bar{I}$ and $\bar{x}$; $g_0(\bar{I}\,|\,\bar{x})$ is the function of likelihood; $g_2(\bar{x})$ is the *a priori* density of distribution $\bar{x}$; $g_1(\bar{I})$ is the *a priori* density of distribution for the observation results $\bar{I}$.

All values of x are supposed to be equally probable if *a priori* information about the distribution function is lacking. In this case, Bayes' estimations coincide with the maximal likelihood ones. We may expect refinement of the maximal likelihood method when *a priori* distribution $g_2(\bar{x})$ is known to be due to the additional information. This refinement takes place for a limited number of channels. The contribution of the additional information becomes negligibly small as a result of an unlimited increase in the number of channels, and the maximal likelihood estimations coincide asymptotically with the Bayes' method ones.

Now, let us to turn to the second group of processing techniques. The first one we will consider here is *the regressive method*.[142] We can write down the system of m equations for n unknown values in the matrix form ($m > n$)"

$$\mathbf{I} = \mathbf{AX} = \underset{m}{\|\mathbf{I}\|} = \underset{m\times n}{\|\mathbf{A}\|}\underset{n}{\|\mathbf{X}\|}\,.$$

We know that $\mathbf{X} = \mathbf{M}^{-1}\mathbf{I}$ at $m = n$. We can obtain a good linear estimation for $\mathbf{X}$ when $m > n$ by using a similar formula:

$$\bar{\mathbf{X}} = \mathbf{M}^{-1}\mathbf{Y} \tag{10.19}$$

where $\mathbf{M}$ is Fisher's information matrix:

$$\mathbf{M} = \sum_{j=1}^{m} \frac{1}{\sigma_j^2}\mathbf{F}_j\mathbf{F}_j'; \qquad \mathbf{Y} = \sum_{j=1}^{m} \frac{1}{\sigma_j^2} I_j\mathbf{F}_j\,, \tag{10.20}$$

σ_j^2 is the errors dispersion in j-th channel, and $\mathbf{F}_i = \|a_{i1}, a_{i2}, \ldots, a_{in}\|$. Thus, the dispersion matrix equals $\mathbf{M}^{-1}$.

This estimation minimizes the sum of the squares of the weighted deviations on the right and left sides of the equations. If the measurements in channels are uniformly precise and their dispersion is unknown, we can substitute unity for the dispersion. The estimation, in this case, minimizes not the weighted deviations but simply the sum of deviations squared. As a result of estimating this parameter, we obtain:

$$\sigma^2 = (m-n)^{-1}\sum_{j=1}^{m}\left[I_j - \mathbf{X}'\mathbf{F}_j\right]^2 , \quad (10.21)$$

as well as the estimation of dispersion. A similar regressive method is applied for the solution of the nonlinear equation. The iterative procedure is needed for realization of this technique.

The *criterion function method*[140] is applied to solve the system of equations:

$$I_i = F_i(x_1, x_2, ..., x_n), \quad i = 1, 2, ..., m \quad (m > n) , \quad (10.22)$$

where I_i are the measurement results, F_i are the model functions, and x_j are parameters of the model functions being defined. We can substitute the quantities of any parameter $\left(x_j^0\right)$ into these equations. The differences $\left(\Delta_1, \Delta_2, ..., \Delta_m\right)$ between the measured quantities and values generated by the model function are referred to as the *discrepancy.* Those parameter quantities are assumed to be the system solution that correspond to the minimum of some objective function for the discrepancies $\Phi\left(\Delta_1, \Delta_2, ..., \Delta_m\right)$.

It is necessary to choose the objective function in such a way as to obtain estimations as close as possible to those of maximum likelihood. In practice, the most applicable are the objective functions for the discrepancy of the sum of weighted squares:

$$\Phi_1 = \sum_{i=1}^{m} w_i\left[I_i - F_i\left(x_1, x_2, ..., x_n\right)\right]^2 , \quad (10.23)$$

where w_i are weighting coefficients. If the model errors are stochastic and the accuracy of the measurements is not sufficiently high, then we must assume that the errors are distributed not by Gaussian law but by Laplace's law. In this case, we need to minimize the objective function that is the sum of the discrepancy modules:

$$\Phi_2 = \sum_{i=1}^{m} w_i\left|I_i - F_i(x_1, x_2, ..., x_n)\right| . \quad (10.24)$$

If, in practice, the distribution function differs from the one assumed for analysis, then the estimation obtained will have dispersions larger than the maximal likelihood. The ratio of this dispersions is referred to as the *effectiveness of estimation.*

Table 10.1 provides data regarding the effectiveness of four distribution functions (f_1, normal; f_2, Laplace; f_3, gamma; f_4, Cauchy) and four minimizing functionals:

$$\Phi_1 = \sum_{i=1}^{m} w_i \left[I_i - F_i(x_1, x_2, \ldots, x_n) \right]^2$$

$$\Phi_2 = \sum_{i=1}^{m} w_i \left| I_i - F_i(x_1, x_2, \ldots, x_n) \right|$$

$$\Phi_3 = \sum_{i=1}^{m} w_i \frac{\left[I_i - F_i(x_1, x_2, \ldots, x_n) \right]^4}{r^4} \tag{10.25}$$

$$\Phi_4 = \sum_{i=1}^{m} w_i \ln\left\{ k^2 + \left[I_i - F_i(x_1, x_2, \ldots, x_n) \right]^2 \right\},$$

where r and k are parameters of the corresponding distributions.

We assumed before that the model function established the relations between the results of the remote measurements and that the characteristics of the studied natural object are known *a priori* with the accuracy to the parameters which are the objects of determination; however, occasionally in remote sensing we have to choose among several models that can be used for parameter estimation. Choosing the best model option is aided by applying the following criteria:[83]

- Simplest form (e.g., linear) combined with reasonably acceptable errors
- Minimum of coefficients by acceptable errors
- Reasonable physical background
- Minimal sum of squares for the discrepancies between predicted and experimental quantities
- Minimal value of standard deviation σ_y getting by the model adjustment

TABLE 10.1
Effectiveness of Parameter Estimation by the Objective Functions of Discrepancy

Functional Minimization	Density of Distribution of Measurement Error			
	f_1	f_2	f_3	f_4
Φ_1	1.00	0.50	0.74	0.00
Φ_2	0.64	1.00	0.31	0.81
Φ_3	0.60	0.05	1.00	0.00
Φ_4	0.07	0.79	0.05	1.00

The following procedure is applied for selecting the best model. All models, after assessment of the physical foundation (the third criterion), can be arranged by complexity (first and second criteria), then the parameters of these models can be estimated. After that, the correctness of the models is analyzed. The correctness means that the variance of the experimental data relative to the quantities predicted by the model does not exceed a value determined by the accuracy of measurements. Usually, we divide the experimental data dispersion into two summands. The first summand is connected with measurements errors s_e^2 and the second with the diversity of the experimental data from the model (s_r^2). The ratio s_r^2 / s_e^2 is analyzed. The distribution function of this ratio is called the *F*-distribution. Tables of this distribution are provided in the literature (e.g., Himmelblau[83]). If this ratio exceeds a tabulated table value for the given verification $(1 - \alpha)$, then the chosen model is incorrect and must be excluded from consideration.

Though satisfaction of the *F*-criterion signifies the correctness of the model, it is still possible to observe an essential distinction between the real and computed models. This distinction can be determined by analysis of the so-called remainders (i.e., the deviation between the interval-averaged experimental values $\overline{y}_i$ and those predicted by the models at the similar interval $\hat{y}_i$). In any case, these remainders must not contradict the main assumptions of regression analysis: independence of observed errors, constancy of the dependent variable dispersion, and the normal law for errors. One of the requirements for the remainders is a stochastic distribution relatively to $\hat{y}$. Its absence means that the model cannot be used.

We analyze the remainders by five main features that give us the opportunity to choose and sometimes improve a model: detection of peaks, detection of trend, detection of the violent level shift, detection of change in errors dispersion (usually supposed to be constant), and research of the remainders on the normality. Along with analysis of the remainders, stepped regression is applied based on a sequence of including and excluding some variables and determining their influence and significance. Other, more complicated methods are also available.[83]

10.3.1 Inverse Problems of Remote Sensing

The procedure of thematic processing of remote sensing data is required to reproduce the altitude profiles of the parameters (humidity, temperature, etc.) of the natural medium. The technique of reproduction is based on processing the interaction characteristics for electromagnetic waves with the medium by different frequency, angle of observation, etc. Dependent on *a priori* information, this problem can refer either to preliminary ones (the profile is known with some accuracy of the parameters) or to problems regarded in the present part when the information about a profile has a sufficiently general character: continuity of the profile itself and its derivatives, limitation of the profile variation, belonging to a known assembly of the stochastic function, etc.

The problem is generally formulated in the following manner: Determine function z defined in metric space F (the space of the natural object parameters) by measured function u, which is defined in another metric space, U (the parameter space of electromagnetic wave interaction with the environment). Thus, operator A, which translates functions from one space to another, is supposed to be known:

$$u = Az, \quad u \in U, \quad z \in F . \tag{10.26}$$

We can distinguish correctly formulated problems from those that are ill posed. According to Hadamard,[23] the correctly formulated problem requires the following conditions:

- Solution z from space F occurs for any element $u \in U$ (condition of existence).
- The solution is unique (condition of uniqueness).
- Small variations of u lead to small variations of z (condition of stability in the corresponding spaces).

We must point out that the incorrectness definition is only related to the given pair of metric spaces. The problem can appear correctly formulated in other metrics.

We will consider only ill-posed problems, as they take place in remote sensing mainly in the form of the Fredholm equation of the first order:

$$\int_a^b K(x,s)\, z(s)\, ds = u(x), \quad c \leq x \leq d , \tag{10.27}$$

where $K(x,s)$ is the equation kernel that is assumed to be a continuous function with continuous partial derivatives $\partial K/\partial x$ and $\partial K/\partial s$; $z(s)$ is an unknown function from space F where $a \leq s \leq b$; and $u(x)$ is the given function from space U where $c \leq x \leq d$.

The stability problem is the most complicated problem in the Fredholm equation for the first-order solution. Indeed, let us assume that δu represents the accuracy of function u, knowledge of which is supposed to be small. Correspondingly, $\delta z(s)$ is an addition to the exact solution of the problem for Equation (10.27). Obviously, the new function satisfies the integral equation:

$$\int_a^b K(x,s)\,\delta z(s)\, ds = \delta u(x),$$

which is similar to the input equation. The addition, $\delta z(s)$, cannot be a small module even when function $\delta u(x)$ is small; only the integral at the left must be small. However, the integral can be small when the module of $\delta z(s)$ is not small, but the function itself frequently changes the sign inside the integration interval with a period that is much smaller than the kernel scale change. For these reasons it can be seen that small errors in the experimental data will not lead to small variations of solution in the inverse problem. This is the reason why an ill-posed problem cannot be solved without having additional *a priori* information about the unknown solution. In our case, this *a priori* information is based, as a rule, on the physical properties of the studied medium parameters (e.g., monotony, continuity, limitation of quantities). The *a priori* information defines the algorithm of the approximate solution obtained for the ill-posed problem. Here, we will regard only some typical cases.

In many cases, the additional information has a quantity character that allows us to narrow the range of possible solutions. The range of possible solutions, U, can be reduced to be compact. In this case, the methods of fitting, quasi-solution, quasi-inversion, etc. can be applied. Most of these methods are based on generalizations of the discrepancy method considered above. For example, in the fitting method, we choose a z_0 on the compact U that minimizes the function:

$$\rho_{L_2}\left(Az_0, u_\delta\right) = \left\{ \int_c^d \left[Az_0 - u_\delta(x)\right]^2 dx \right\}^{1/2} . \tag{10.28}$$

Here, u_δ is the input experimental data with errors (i.e., $u_\delta = u + \delta u$).

When the additional information has a general qualitative character, the range of possible solutions cannot be reduced to a compact one, in which case *a priori* information can be used to define the regularizing algorithm that allows us to find the solution closest to the exact one. This information can require the deterministic or statistical properties of an unknown solution. We will briefly consider only two cases.

In reality, as was said, experimental data are known but with some errors, which means that we deal not with the exact Equation (10.27) but with the following equation:

$$\int_a^b K(x,s)\, z(s)\, ds = u_\delta(x) . \tag{10.29}$$

Now, we can find the approximate solution (z_δ) of the studied equation. The variance of the experimental data relative to the accurate quantities is given by the squared metric (metric U):

$$\rho_{L_2}\left(u, u_\delta\right) = \left\{ \int_c^d \left[u(x) - u_\delta(x)\right]^2 dx \right\}^{1/2} \tag{10.30}$$

and the solution variance in the uniform metric (metric F):

$$\rho_C\left(z, z_\delta\right) = \max_{s \in [a,b]} \left| z(s) - z_\delta(s) \right| . \tag{10.31}$$

In order to overcome the problem of instability, the stabilizing functional is introduced according to Tikchonov:

$$\Omega[z] = \int_a^b \left[p(x)(z')^2 + q(x)z^2 \right] dx . \tag{10.32}$$

The functions $p(x)$ and $q(x)$ are positive inside the interval $[a, b]$, and they determine the requirement for the function $z(x)$ module and its derivative inside the mentioned interval. Simply speaking, the discussion is about the limitations for the function and its derivative oscillations, the so-called requirement for solution smoothness. The selection principle mentioned earlier allows us to state that function $z_\delta(s)$ from space F leading to minimization of the stabilizing functional $\Omega[z]$ is a desired approximate solution.[23] The following question is then raised: What is the accuracy of a solution obtained in this way? Certainly, the solution accuracy must be no worse than the accuracy of the input data. So, among the chosen functions from set F only those are appropriate that satisfy the condition:

$$\rho_{L_2}\left(Az, u_\delta\right) = \delta , \tag{10.33}$$

where δ is the mean squared error of measurements inside the interval $[a, b]$. So, we have now come to the problem of the conditional extremity. It is necessary, in this case, to look for the function that minimizes the smoothing functional:

$$M^\alpha\left[z, u_\delta\right] = \rho_{L_2}^2\left(Az, u_\delta\right) + \alpha\Omega[z] . \tag{10.34}$$

The parameter α is defined by the discrepancy quantity:

$$\rho_{L_2}\left(Az^\alpha, u_\delta\right) = \delta, \tag{10.35a}$$

or by solving the following equation numerically:[111]

$$\rho(\alpha) = \left[\rho_{L_2}\left(Az_\delta^\alpha, u_\delta\right)\right]^2 = \delta^2, \tag{10.35b}$$

which has a unique solution $\alpha = d(\delta) > 0$.

The procedure of Equation (10.34) minimization leads to Euler's integro-differential equation:[23]

$$\alpha\left\{q(s)z-\frac{d}{ds}\left[p(s)\frac{dz}{ds}\right]\right\}+\int_a^b \bar{K}(s,t)z(t)dt=g(s)\,, \tag{10.36}$$

where

$$\bar{K}(s,t)=\int_c^d K(x,s)\,K(x,t)\,dx,\quad g(s)=\int_c^d K(x,s)u_\delta(x)dx\ . \tag{10.37}$$

This equation has to satisfy the boundary conditions making the function z or its derivative vanish at the ends of the interval $[a, b]$. The unknown solution cannot satisfy these conditions. Then, if the real conditions — for example, $z(a)$ and $z(b)$ — are known, then the introductory function:

$$\tilde{z}(s)=z(s)-\frac{z(a)}{b-a}(b-s)-\frac{z(b)}{b-a}(s-a) \tag{10.38}$$

gives us an equation relative to this new function similar to Equation (10.29) but with a different right side of the equation. An analogous rearrangement can be done in the case of the boundary conditions relative to derivatives of the solution.[23]

Now, let us consider a technique of statistical regularization. These methods are often useful for media parameter profiles measured repeatedly by other methods (e.g., *in situ*), and they have reliable statistical characteristics. An example of this approach is being able to reconstruct atmospheric parameters by using experimental data obtained with the help of weather balloons. The reconstruction of atmospheric humidity with this method is represented in Figure 10.8.

In order to understand the main point of the statistical regularization method,[143] let us reduce Equation (10.27) to the algebraic equation system:

$$\sum_{i=1}^{n} K_{ji}z_i=u_j,\quad j=1,2,\ldots,m \tag{10.39}$$

and z_i and u_j are linear functionals from the functions $z(s)$ and $u(x)$ — namely, their values at points of control or the coefficients of these function expansions in an orthogonal function series. The number of equations (m) is greater than the number of unknowns. The preliminary information, in the form of an *a priori* distribution function of the unknown solution, and the distribution function of the measurements errors are introduced into the solution process. The introduction of the *a priori* distribution function, $P(\mathbf{Z})$, permits us to bound the assembly function that can be

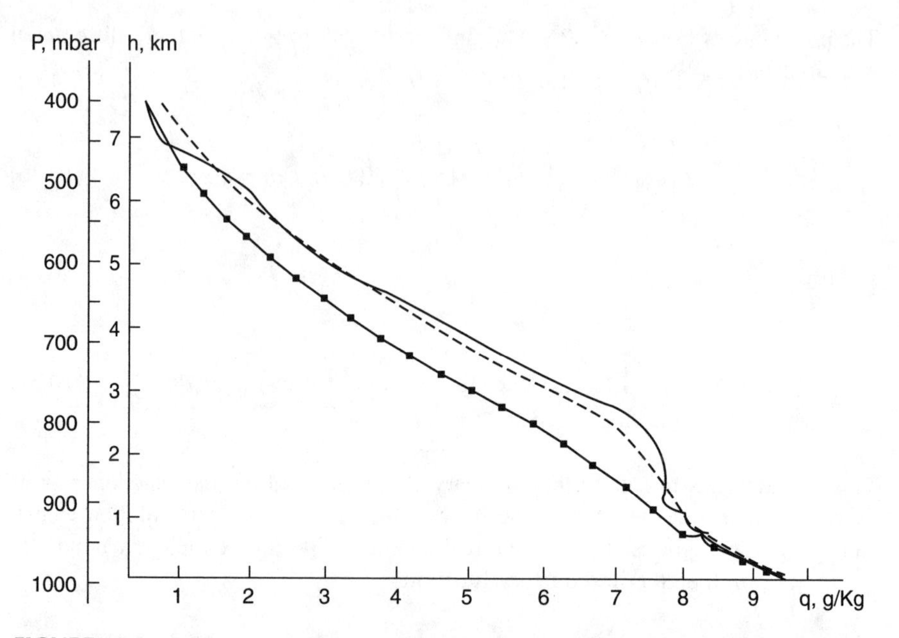

FIGURE 10.8 Matching of the recovered profiles of dampness at simulation of radiometric measurements in a 1.35-cm wave band (dotted line) with profiles obtained by measurement by probes (full curve) and by interpolation of ground values (triangles).

chosen as our equation solution. We will briefly consider two ways to represent *a priori* statistical information.

Taking into account the measurements errors, we can rewrite Equation (10.39) in the matrix view:

$$\mathbf{K}\,\mathbf{Z} + \delta = \mathbf{U}\ , \tag{10.40}$$

where δ is the matrix column or the m-dimensional vector of measurement errors which has the distribution function $P_{\delta}(\delta)$. The density of the conditional probability for vector $\mathbf{U}$ by the given vector $\mathbf{Z}$ is determined by the relation:

$$P\left(\mathbf{U}\,|\,\mathbf{Z}\right) = P_{\delta}\left(\mathbf{U} - \mathbf{KZ}\right). \tag{10.41}$$

Now, we can obtain an *a posteriori* probability:

$$P(\mathbf{Z}\,|\,\mathbf{U}) = \frac{P(\mathbf{Z})\,P(\mathbf{U}\,|\,\mathbf{Z})}{\int P(\mathbf{Z})\,P(\mathbf{U}\,|\,\mathbf{Z})\;d\mathbf{Z}}\ . \tag{10.42}$$

The unknown function mathematical expectation determined by the *a posteriori* probability is a very regularized solution, which is sought for the formulated problem. The mathematical expectation of any function $F(\mathbf{Z})$ is:

$$\langle F(\mathbf{Z})\rangle = \int F(\mathbf{Z})P(\mathbf{Z}\,|\,\mathbf{U})d\mathbf{Z} = \frac{\langle F(\mathbf{Z})P(\mathbf{U}\,|\,\mathbf{Z})\rangle}{\langle P(\mathbf{U}\,|\,\mathbf{Z})\rangle}, \tag{10.43a}$$

where the average on the right-hand side is accomplished by an *a priori* distribution $P(\mathbf{Z})$. Correspondingly, the regularized solution is:

$$\mathbf{Z}_r = \langle \mathbf{Z}\rangle = \int \mathbf{Z}P(\mathbf{Z}\,|\,\mathbf{U})d\mathbf{Z} = \frac{\langle \mathbf{Z}P(\mathbf{U}\,|\,\mathbf{Z})\rangle}{\langle P(\mathbf{U}\,|\,\mathbf{Z})\rangle}. \tag{10.43b}$$

We will consider this Bayes strategy for two kinds of *a priori* statistical information: a solution in an assembly of bounded samples and a solution in an assembly given by the correlation matrix.

Let us assume that an *a priori* assembly of available solutions is given by its vectors $\mathbf{Z}_\nu$, where $\nu = 1, 2, \ldots, N$. These vectors are obtained by a random sample satisfying the *a priori* distribution $P(\mathbf{Z})$. In this case, averaging over the sample can be substituted in Equation (10.43) for averaging over the assembly. Hence,

$$\langle F(\mathbf{Z})\rangle = \frac{\sum_{\nu=1}^{N} F(\mathbf{Z}_\nu)P(\mathbf{U}\,|\,\mathbf{Z}_\nu)}{\sum_{\nu=1}^{N} P(\mathbf{U}\,|\,\mathbf{Z}_\nu)} = \frac{\sum_{\nu=1}^{N} w_\nu F(\mathbf{Z}_\nu)}{\sum_{\nu=1}^{N} w_\nu}, \tag{10.44}$$

where w_ν are weighting coefficients depending on the measured values $\mathbf{U}$ and the conditional probabilities $P(\mathbf{U}|\mathbf{Z}_\nu)$. Assuming independence of the measurement errors for the various u_j leads to the expressions:

$$P(\mathbf{U}\,|\,\mathbf{Z}^\nu) = \prod_{j=1}^{m} (2\pi s_j^2)^{-1/2} \exp\left\{-\frac{1}{2s_j^2}\left[u_j - \sum_{i=1}^{n} K_{ji} z_i^\nu\right]^2\right\}, \tag{10.45}$$

$$w_\nu = \exp\left\{-\sum_{j=1}^{m} \frac{1}{2s_j^2}\left[u_j - \sum_{i=1}^{n} K_{ji} z_i^\nu\right]^2\right\} \tag{10.46}$$

in the case of a normal distribution with zero mathematical expectation and the dispersions s_j^2. We can obtain the regularized solution and its error assuming that $F(\mathbf{z}) = z_i$ and $F(\mathbf{Z}) = (z_i - \langle z_i\rangle)^2$.

Now, let us consider the situation when the correlation matrix:

$$C_{ij} = \langle z_i z_j \rangle = \int z_i z_j P(\mathbf{Z})\, d\mathbf{z} \tag{10.47}$$

is known on the basis of previous information. The function $P(\mathbf{Z})$, which satisfies the formulated condition and minimizes the entropy functional:

$$H[\mathbf{Z}] = \int P(\mathbf{Z}) \ln P(\mathbf{Z})\, d\mathbf{Z} \tag{10.48}$$

is chosen as an *a priori* probability distribution function. It has been shown that this function is:

$$P(\mathbf{Z}) = \alpha \exp\left\{ -\frac{1}{2} \left(\mathbf{Z} \cdot (\mathbf{C}^{-1}\mathbf{Z}) \right) \right\}, \tag{10.49}$$

where α is a constant and $\mathbf{C}^{-1}$ is the inverse correlation matrix. The result is obtained by using. Bayes' formula, Equation (10.42), which allows us to obtain the expression for an *a posteriori* probability function:

$$P(\mathbf{Z} \,|\, \mathbf{U}) = b \exp\left\{ -\frac{1}{2s^2} \left[\left(\mathbf{Z} \cdot \left(\left(\mathbf{L}'\mathbf{L} + s^2\mathbf{C}^{-1} \right) \mathbf{Z} \right) \right) - 2\left((\mathbf{L}'\mathbf{g}) \cdot \mathbf{Z} \right) \right] \right\} \tag{10.50}$$

where b is constant and $\mathbf{L}'$ is the transpose of the matrix. Here,

$$s = \sqrt[m]{\prod_{j=1}^{m} s_j^2}$$

is the mean geometric error of the measurements, $L_{ji} = s/s_j K_{ji}$, and $g_j = s/s_j u_j$. As in the previous case, the regularized solution and the mean square of its recovery errors are defined from the equations:

$$\left(\mathbf{L}'\mathbf{L} + s^2\mathbf{C}^{-1} \right) \mathbf{Z} = \mathbf{L}'\mathbf{g}, \quad \sigma_i^2 = s^2 \left(\mathbf{L}'\mathbf{L} + s^2\mathbf{C}^{-1} \right)_{ii}^{-1}. \tag{10.51}$$

Some other means of statistical regularization are given in Turchin et al.[143]

11 Radio Devices for Remote Sensing

Two classes of devices and corresponding methods are used for remote sensing of the environment. Devices that radiate radiowaves and receive them after their interaction with media belong to the first class and are referred to as the active devices. Devices of the second class, as a rule, receive intrinsic thermal radiation of natural objects and are referred to as passive devices. The different types of radars, altimeters, scatterometers, and radio occultation instruments are examples of active devices. The different types of microwave radiometers are classified as the passive devices.

11.1 SOME CHARACTERISTICS OF ANTENNA SYSTEMS

All radio devices for remote sensing are equipped with antenna systems for the reception of radiowaves (and for their radiation, in the case of active instruments). This is the reason why we will consider again the properties of an antenna. We have established already in the Chapter 1 that transmitting antenna are characterized by directivity $D(\theta,\varphi)$ and pattern $\Psi(\theta,\varphi)$. The definition of directivity was based on the angular distribution of the radiated power relative to the power at the antenna input. The energy losses (e.g., taking place in the feeding systems) were not taken into consideration; therefore, we need to include in our analysis the antenna efficiency:

$$\eta = \frac{P_i}{P_t}, \qquad (11.1)$$

which reflects the ratio of radiated power P_i to input power P_t. So, the gain, $G(\theta,\varphi) = \eta D(\theta,\varphi)$, often figures into engineering computations.

The effective area of the antenna was used to characterize the receiving properties of the antenna. Normally, this notion relates to the geometrical antenna area via the aperture efficiency :

$$\eta_a = \frac{A_e^{\max}}{A}. \qquad (11.2)$$

Here, the maximum effective area occurs in the numerator. Equation (1.114) helps us to define the aperture efficiency as:

$$\eta_a = \frac{\left| \int_A F(\mathbf{r}') d^2\mathbf{r}' \right|^2}{A \int_A \left| F(\mathbf{r}') \right|^2 d^2\mathbf{r}'} . \tag{11.3}$$

The normalized amplitude of the aperture field (the apparatus function $F(\mathrm{r}')$) is substituted here for the field. The utilization factor is equal to unity in the case of a uniform aperture field distribution (the apparatus function equals unity). In this case, the antenna pattern of the circular antenna can be written as shown in Equation (1.118). We must point out that a uniform distribution is an idealization of practical situations. In reality, the utilization factor is less than unity.

The side lobes of both kinds of antennae (transmitting and receiving) are of interest, as very often they are sources of interference. One of the main reasons for the high side lobe level is the sharp change of apparatus function at the edge of the antenna; therefore, uniform apparatus function leads to a high level of the side lobes. The first side lobe of the circular antenna, in this case, is 17.6 dB less compared to the major one; therefore, in order to lower this level, we want the apparatus function at the edge of the antenna to tend to zero. As a modeling example, we can choose the apparatus function for the circular antenna in the view $F(\mathrm{r}) = 1 - (\mathrm{r}/a)^2$. The antenna pattern will be:

$$\Psi(\theta) = \frac{4 J_2^2 (ka \sin\theta)}{(ka \sin\theta)^4} , \tag{11.4}$$

and the first side lobe level will be 24.6 dB;[4] however, in this case, the beamwidth is $0.64\lambda/a$ whereas it is $0.51\lambda/a$ for the uniform apparatus function. The utilization factor is 0.75.[4] The stray factor:

$$\beta = 1 - \frac{\iint_{\Omega_0} F_n(\theta,\varphi)\, d\Omega}{\iint_{4\pi} F_n(\theta,\varphi)\, d\Omega} \tag{11.5}$$

is applied for the side lobe role general assessment. Here, Ω_0 is the solid angle of the major lobe estimated by the first zero line of the antenna pattern. In the case of the patterns represented by Equation (1.118), it is reduced to $\beta = J_0^2(j_{1,2})$, where $j_{1,2}$ is the second root of the first-order Bessel function, and we have the corresponding quantity $\beta = 0.16$. This value is defined by a similar formula as in Equation (11.4), except that in this case the root of the second-order Bessel function will be the zero-order function argument $j_{2,2} = 5.136$ and $\beta = 0.017$.

The level of the side lobes, as a rule, is not very important for radar. More important is the signal-to-noise ratio; therefore, the standard approach is to have maximal aperture efficiency and, correspondingly, maximum directivity. For radiometry, however, a low level of side lobes is one of the most important parameters. So, we may have to sacrifice sensitivity due to some deterioration of the spatial resolution.

11.2 APPLICATION OF RADAR DEVICES FOR ENVIRONMENTAL RESEARCH

Radar devices for remote sensing of the environment are mounted on the surface of the Earth, on ships and riverboats, and on air- and spacecraft. Earth-mounted radar is used for the study of the atmosphere and ionosphere and to observe coastal sea zones and inland reservoirs, as well as various types of land and the structure of the surface layers. Ship radar is used mostly for water surface monitoring and atmospheric research, as well as coastal area supervision. All elements of the environment are the objects of observation for air- and spaceborne radars.

The main purposes of such research are analysis of radiowave intensity and polarization, the study of the shape and time delay of signals, etc. These data are used for radar mapping, altimetry, scatterometric research, and subsurface sounding. The main areas of application are geology (e.g., structure and lithology), hydrology (e.g., determining soil moisture; mapping river basins, inner reservoirs, floods, snowcover), farming (e.g., mapping crops, monitoring vegetation growth and maturation, determining field borders, observing soil moisture dynamics), forestry (e.g., estimating merchantable wood, monitoring deforestation, detecting forest fires, assessing fire hazard situations), oceanology (e.g., sea-wave zone monitoring, measuring near-water wind velocity, determining the direction in which a pollution area is moving, studying sea currents and geoid forms), atmospheric research (e.g., sounding clouds and rain areas, determining temperature height profiles, defining atmospheric turbulence, detecting minor gaseous constituents), study of ionospheric dynamics (e.g., altitude profiles of electron concentrations, turbulence motion, electron and ion temperature measurement), cartography (e.g., topographical survey of regions that are difficult to access, land use mapping), polar area observations (e.g., monitoring and investigating sea ice, mapping continental ice covers, observing iceberg formation and their motion, researching changes in glaciers).

Generally, four types of radar devices are used for remote sensing: panoramic radars, scatterometers, radio altimeters, and subsurface radars. Panoramic radar involves sectoring or a circular scanning view and side-looking radar; the latter type includes radar with a real aperture and synthetic-aperture radars (SAR). Scatterometers are classified according to the methods for selecting a given area. The main applications are angle–angle (the antenna pattern is narrow on both planes), angle–time (the pattern is narrow only on the azimuthal plane, thus the signal is modulated by narrow pulses), angle–frequency (the pattern is narrow on the azimuthal plane and is directed along the platform flight line; signal discrimination is achieved by Doppler selection), frequency–time (SAR with pulse modulation). Altimeters are classified by the modulation method: pulse, frequency, or phase modulation. Subsurface radars differ in where the antenna is mounted: remote or applied (the antenna is put on the surface of the studied object).

11.3 RADIO ALTIMETERS

The radio altimeter is in theory one of the simplest radar systems. Applications of remote sensing using radio altimeters include measuring water levels, defining geoid forms, estimating seawave intensity, monitoring glacier growth, mapping land surface topographic, determining upper cloud levels, etc. Pulse-modulated or chirp signals are normally used for altimetric research. For the traveling time of the signal (t_{r-t}) over the path, transmitter–surface–receiver serves as a device for measuring the altitude of the altimeter platform above the studied surface. The common relation is:

$$H = \frac{t_{r\text{-}t} v}{2}, \tag{11.6}$$

where v is the mean velocity of radiowave propagation in the atmosphere of Earth. The frequency spectrum of altimetric systems is chosen in such a way that the influence of the ionosphere is small. Usually, the C, X, and Ku-bands are preferable for these devices.

The error in the altitude determination is:

$$\delta H = \frac{v}{2}\delta t_{r-t} + \frac{t_{r-t}}{2}\delta v, \tag{11.7}$$

where δt_{r-t} and δv are errors in the delay time and mean velocity, respectively. The first summand contribution is defined by the leading edge steepness of the optimally processed pulse and the signal-to-noise ratio at the altimeter output. Generally,

$$\delta t_{r-t} = \frac{\tau_i}{\sqrt{2(S/N)}}, \tag{11.8}$$

where τ_i is the leading edge duration after reflection (scattering), and S/N is the signal-to-noise ratio.[89] Changes in the leading edge as a result of scattering by a rough surface are described in Chapter 6.

The second term in Equation (11.7) depends on the propagation conditions. In the case of spaceborne altimeters, the contributions of both the troposphere and ionosphere must be taken into account. The tropospheric contribution, corresponding to Equation (4.57), can be written as:

$$\frac{t_{\text{r-t}}}{2}\delta v = \frac{(\varepsilon_0 - 1)H_T}{2} \cong 2.5 \text{ m}. \tag{11.9}$$

Equation (4.69) permits us to estimate the ionospheric contribution as:

$$\frac{t_{\text{r-t}}}{2}\delta v = \frac{1.21 \cdot 10^{21}}{f^2}. \tag{11.10}$$

For the X-band, it will be about 10 cm. For greater accuracy when defining altitude, it is necessary to develop procedures for correcting the radio propagation. In the case of the troposphere, atmospheric models are used. Microwave radiometers are added to measure water vapor content, which gives us a more accurate calculation of the mean value of the refractive index. Correcting for ionospheric influences requires the design of a two-frequency altimeter or the use of very high frequencies, which generates new problems (scattering by clouds, for example). In order to overcome these, continuous analysis of signals is needed using, for example, a strobing technique.

The accuracy of the altimeter also depends on the antenna beamwidth and its positional stability. These effects become most apparent in the case of strong reflection from the surface edge coinciding with the footprint board. As was noted in Chapter 6, surface roughness leads to a change in the duration of the scattered signal.[89] Examples of backscattered signals are given in Figure 11.1. This is one reason why estimating the time delay using the arrival of the leading edge of the pulse gives us a more accurate altitude measurement. Simultaneously, pulse distortion analysis allows the study of the roughness parameters. In particular, this technique is used for sea waves intensity monitoring.

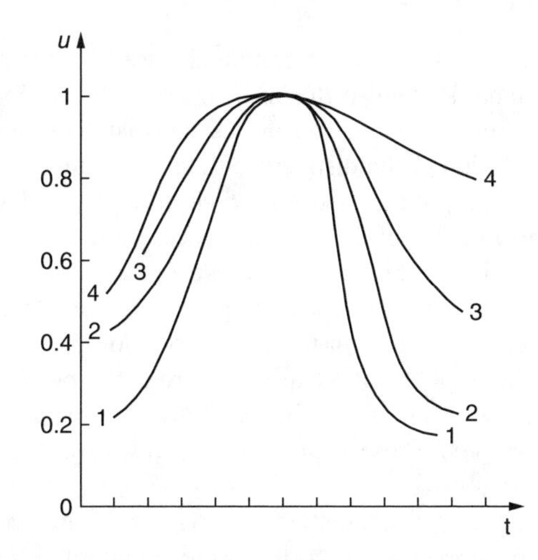

FIGURE 11.1 The approached pulse shape of normalized altimeter impulses: (1) salt desert; (2) and (3), agricultural holdings; (4) surface of lake.

The first spaceborne altimeter was tested during the SKYLAB mission. It operated at a frequency of 13.9 GHz, and the pulse duration was of the order of 25 nsec. The measurements taken showed that we can determine the topographic relief under the satellite trace with an accuracy of ±15 meters for the case of a relatively smooth surface without mountains and steep slopes. The altitude above a quiet water surface could be determined with an accuracy of about ±1 meter.[89]

The investigation of ocean problems requires a more accurate altimeter. Research has shown that errors of altitude determination must be less than 10 cm. Very short pulses are needed in this case, which explains why chirp modulation is useful in modern altimeters. In particular, pulses with a duration of the order of 100 μsec and a frequency deviation inside ΔF = 320 MHz are used in modern altimeters. The compression procedure provides the altitude definition with an accuracy that is similar to pulses of 3.1-nsec duration.

A spaceborne altimeter of very high accuracy is applied in the U.S.–French mission TOPEX/POSEIDON, which is designed to research water surface characteristics.[145] The C- and Ku-bands (frequencies of 5.3 and 13.6 GHz, respectively) are used. The two-frequency system allows us to correct the ionospheric influences. Microwave radiometers, operating at frequencies of 18, 21, and 37 GHz, provide the tropospheric correction. A good system of 1.5-m antenna orientation minimizes errors connected with main lobe deviation from the nadir. There is a system based on the ground for the satellite coordinates determination with heightened accuracy.

A complicated procedure of corrections has been developed for altimeter data processing and interpretation. Depending on the seawave intensity and the data averaging time, the accuracy of the mean level of the oceanic surface definition lies within a range of several centimeters.

11.4 RADAR SYSTEMS FOR REMOTE SENSING OF THE ENVIRONMENT

A different type of panoramic radar is commonly used for remote sensing of the environment. These radar systems differ in their principles of operation and in the technique of observation. The received pulses are distinguished by arrival time, which allows determination of the distance to the illuminated object (in the case of monostatic radar), and amplitude, both of which allow us to define the scattering properties (reflectivity) of this object. In this process, the direction of the antenna major lobe is fixed, and the direction of the investigated target is determined. The scattering properties of targets can depend on polarization of illuminated waves and the targets themselves, as we saw in Chapter 5 (Section 5.6). The targets usually scatter the waves with a polarization different from the polarization of incident waves. So, the polarization matrix or Stokes matrix are the subject of interest in more complicated systems. These data open the way for mapping the scattering properties of the environment.

The radar systems applied to remote sensing are separated roughly into two classes: radar for atmospheric research (troposphere and ionosphere) and radar for monitoring the surface of Earth (land and water). Ionospheric radar will be discussed in Chapter 13; here, let us describe briefly tropospheric radar, or weather radar (WR), which is used primarily for the investigation of hydrometeors. Scanning space is made possible with the help of a pencil-beam antenna. The main parameter of measurement is the reflectivity, which is determined via the intensity of the received pulses. The power of a signal received from a point scatterer is defined by the radar equation:

$$W(L) = \frac{PDA_e\sigma}{4\pi L^4} = \frac{PA_e^2\sigma}{L^4\lambda^2} \; . \tag{11.11}$$

Here, W is the received power; P is the pulse transmitted power; D and A_e are the radar antenna directivity and its effective area, respectively; σ is the target differential cross section of backscattering; L is the distance between the radar and the target; and λ is the radiowavelength. In contrast to the traditionally applied formulae, we use the physical definition of the cross section, which is different from the radar one by the factor 4π (see Chapter 5). Among other factors, we do not take into account different engineering coefficients describing the feeder system efficiency. We also assume the absence of wave attenuation in the environment. In the future, we will generally restrict ourselves to the case of clouds as the subject of the study. They are distributed targets; therefore, the power scattered by a cloud layer of thickness l can be defined as:

$$W(L) = \frac{P\sigma_d^0(\pi,L)l}{L^2\lambda^2}\int A_e^2(\mathbf{r})\frac{dr}{L^2} = \frac{P\sigma_d^0(\pi,L)l}{L^2\lambda^2}\int A_e^2(\Omega)d\Omega \; , \tag{11.12}$$

where $\sigma_d^0(\pi,L)$ is the differential cross section of the backscattering per unit volume of the cloud. The coordinate integration in Equation (11.12) is realized over the plane transverse to the wave propagation direction. This integration is easily transformed to integration over the solid angle. To calculate this integral, we can use the model antenna formula (Equation (1.123)) to obtain:

$$I = \int A_e^2(\Omega)d\Omega = 2\pi\int_0^{\pi/2} A_e^2(\theta)\sin\theta d\theta \cong 2\pi\int_0^{\infty} A_e^2(\theta)\theta d\theta = \frac{\lambda^2 A_e(0)}{2} \; . \tag{11.13}$$

With the result

$$W(L) = \frac{P\sigma_d^0(\pi,L)l A_e(0)}{2L^2} \; . \tag{11.14}$$

If we had used the antenna pattern (Equation (1.118)), then instead of Equation (11.13), we would have obtained:

$$I = 8\lambda^2 A_e(0) \int_0^\infty \frac{J_1^4(x)\, dx}{x^3} \cong 0.456\lambda^2 A_e(0) .$$

One can easily see that the difference is insignificant, and that Equation (11.14) is quite good for future analysis.

The cloud drops are very small compared to radiowavelengths, and the Rayleigh approximation for the cross section of scattering can be used. Equation (5.45) is suitable for this case and gives us:

$$\sigma_d^0(\pi, L) = N k^4 a^6 \left| \frac{\varepsilon - 1}{\varepsilon + 2} \right|^2 , \tag{11.15}$$

where N is the drop concentration. Equation (11.15) is valid on the assumption that all the drops are similar. In reality, their radius a is the stochastic value, and we need to apply to the main value:

$$\langle a^6 \rangle = \int_0^\infty a^6 f(a)\, da , \tag{11.16}$$

where $f(a)$ is the distribution function of drop radius. The parameter:

$$\Theta = N \langle a^6 \rangle = N \int_0^\infty a^6 f(a) da \tag{11.17}$$

is referred to as the reflectivity, which we discussed before;[89] hence,

$$W(L) = \frac{8\pi^4 P A_e(0) l}{\lambda^4 L^2} \left| \frac{\varepsilon - 1}{\varepsilon + 2} \right|^2 \Theta(L) . \tag{11.18}$$

The layer thickness is determined by the pulse duration τ_p:

$$l = \frac{c\tau_p}{2} . \tag{11.19}$$

The relations obtained represent the method of defining the reflectivity of the clouds.

Different distribution functions are used for calculating the $\langle a^6 \rangle$.[39] One of the most universal is the distribution proposed by Deirmendjian:[86]

$$f(a) = \frac{\gamma}{\Gamma(\beta)} \left(\frac{\alpha}{\gamma a_{\mathrm{m}}^{\gamma}} \right)^{\beta} a^{\alpha} \exp\left[-\frac{\alpha}{\gamma} \left(\frac{a}{a_m} \right)^{\gamma} \right], \quad \beta = \frac{1+\alpha}{\gamma}. \tag{11.20}$$

Here, a_{m} is the modal radius (i.e., most probable one). The parameters α and γ depend on the type of hydrometeors. The averaged value is:

$$\langle a^6 \rangle = \frac{\Gamma(\beta + 6/\gamma)}{\Gamma(\beta)} \left(\frac{\gamma}{\alpha} \right)^{6/\gamma} a_{\mathrm{m}}^6 . \tag{11.21}$$

The data represented in Table 12.2 allow us to determine combinations of parameters for various types of hydrometeors. Spectral analysis of the scattered signals allows us to study the internal motion in clouds. Similar formulae can be used for snow and hail clouds. It is sometimes necessary to use more exact formulae for the scattering cross section, especially in the case of hail, whose size can be comparable with the wavelength. The reflectivities of precipitation are connected with its intensity J (mm/hour) by the empirical formula:[89]

$$\Theta = AJ^b . \tag{11.22}$$

Parameters $A = 200$ and $b = 1.6$ are used for rain, and $A = 2000$ and $b = 2.0$ are used for snow in moderate latitudes.[89] Weather radar applications are used in meteorological services and aviation. Recently, they have also found application in space research along with the help of so-called rain radars.[91,92]

Radar is sometimes applied to tropospheric turbulence research. The specific cross section of the backward scattering is, in this case (accordingly to Equation (5.173)):

$$\sigma_{\mathrm{d}}^0(\pi) = \frac{\pi k^4}{2} \tilde{\mathrm{K}}(2k) \cong \frac{\pi q_0^{2\nu} C_\varepsilon^2}{2^{1+2\nu}} k^{4-2\nu} . \tag{11.23}$$

One can see that $\sigma_{\mathrm{d}}^0(\pi) \propto \lambda^{-1/3}$ in the case of Kolmogorov–Obukhov turbulence. The calculations show that this cross section is small, and powerful radar is needed for the detection of tropospheric turbulence.[89]

Pencil-beam antennae are seldom used for remote sensing of land, but they do find application in airborne navigation systems; however, they have poor space resolution due to difficulties encountered by having a large-size antennae on board. This is the reason why side-looking radar (SLR) and synthetic aperture radar are widely used. We will first discuss SLR. An antenna with a narrow pattern in one direction can be rather easily mounted on the flying platform — for example, a leaky waveguide antenna mounted along an airplane fuselage. So, the spatial resolution of the radar along the flight direction can be written as:[90]

$$\delta X = \frac{\lambda}{d} R = \beta_{\text{h}} R \,, \tag{11.24}$$

where d is the horizontal antenna size, and R is the slant range to the illuminated element of the land surface. The radar radiates short pulses which permit discrimination of the reflections from the land elements separated by distance:

$$\delta Y = \frac{c\tau_{\text{p}}}{2\sin\theta} \,, \tag{11.25}$$

where θ is the angle between vertical and the direction to the illuminated element (see Figure 11.2a). Equation (11.25) determines the spatial resolution across the direction of flight. More exactly, lines of similar ranges are circles and lines of similar angle positions are lines (see Figure 11.2b). Thus, it can be said that Equation (11.24) defines the resolution by azimuth and Equation (11.25) by one-in range.

The SLR antenna looks sideways, thus its name. The operation of such radar is based on backward scattering of rough surfaces; for example, SLR does not "see" smooth water surfaces. The backward scattering depends on inclinations of the land or water elements (see Chapter 6) and is the basis of mapping a landscape by SLR. The latter one is realized during the platform flight when the range selection gives the scan line, and the frame scan is realized in the flight process. It is similar to formation of a television image. In the first airborne SLR, the radar image was displayed by an electron-ray tube and then photographed on film. The film served as the memory for information storage. Now, such storage is a function of the computer and associated devices.

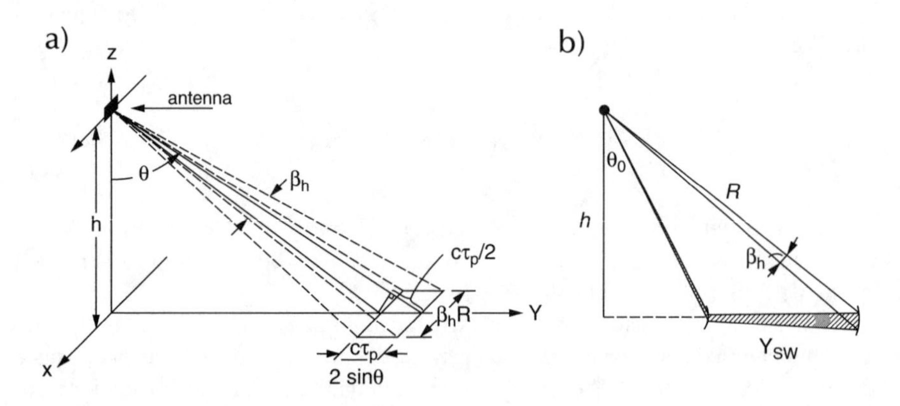

FIGURE 11.2 Side-looking radar operation.

Equation (11.11) can be used for the calculation of the received signal power. It is necessary only to substitute slant range R for distance L. The target cross section is now defined by the sizes of the space resolution element (pixel):

$$\sigma = \sigma^0 \delta X \cdot \delta Y \cong \sigma^0 \frac{\lambda}{d} R \cdot \delta Y. \qquad (11.26).$$

Here, d is the antenna size along the flight direction. More exactly, as was done in the case of weather radar (see Equation (11.12)), we had to calculate with respect to the antenna pattern along the x-direction instead of using the pixel size in the form of Equation (11.24). The following equation can be used for our estimation:

$$\Psi(\varphi) = \frac{\sin^2(kd/2\sin\varphi)}{(kd/2\sin\varphi)^2}. \qquad (11.27)$$

A pattern of this kind occurs with a linear antenna with a uniform distribution of sources.[4] In this case,

$$\delta X \cong \frac{\lambda}{\pi D} R \int_{-\infty}^{\infty} \frac{\sin^4 x}{x^4} dx = \frac{2\lambda}{3d} R.$$

The difference is not considerable; therefore, we will use Equation (11.26) for the sake of simplicity, taking into account that it is accurate to within a coefficient of the order of unity. So,

$$W(R) = \frac{PA_e^2(0,\theta)\sigma^0 \cdot \delta Y}{\lambda d R^3}. \qquad (11.28)$$

Note the antenna area dependence on the nadir angle, θ, which means that the magnitude of the reflected signal depends on the pixel position. Certainly, the specific cross section σ^0 also depends on the pixel position. This dependence is the basis of the radar image.

The swath of the radar image is determined by the pulse time repetition (T):

$$Y_{sw} \cong \frac{cT}{2\sin\theta_0}, \qquad (11.29)$$

where θ_0 is the nadir angle (see Figure 11.2b) of an SLR observation near line. It is assumed, then, that $Y_{SW} << R_{min}$. The antenna beam angle in the vertical plane has to be of the order of:

$$\gamma \cong \frac{Y}{R_{\min}} \cos\theta_0 \cong \frac{cT \cot\theta_0}{2R_{\min}} . \tag{11.30}$$

In this case, all of the observed points will be sufficiently illuminated. On the other hand, the spurious signals corresponding to reflections from points outside the swath will be essentially weakened. These signals appear due to the periodic repetition of the pulses radiation.

The time repetition cannot be too long. It is necessary to illuminate any pixel at least once during the flight; therefore, the following inequality has to be performed:

$$T < \frac{\delta X}{v} = \frac{\lambda R}{vd} , \tag{11.31}$$

where v is the platform velocity.

The analysis of SLR images has some peculiarities. One of them is connected with the fact that the pixel position is fixed in the slant range scale, which leads to the distortion of images. To correct this, we must remember the relation $l = \sqrt{R^2 - H^2}$, which is valid when we assume the land to be smooth and plain. This formula needs to be changed slightly to take into consideration the spherical character of the surface of Earth. The relief heights also distort scales and form radar shadows (Figure 11.3).

The spaceborne SLR for the Russian *Kosmos*-1500 mission[114] and subsequent *Ocean* satellites is an example of the effective application of such instruments for remote sensing. This 3.1-cm SLR operates at a 650-km orbit and provides images

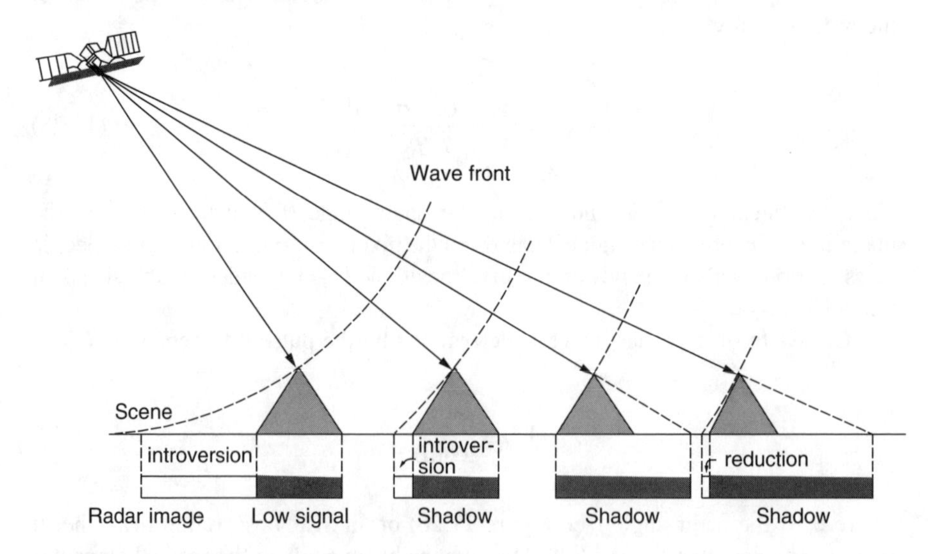

FIGURE 11.3 Artifact of radar image.

with a δX resolution of 2.1 to 2.9 km and a δY resolution of 0.6 to 0.9 km within a swath that is 475 km wide. The wide swath is advantageous for ice patrols. This SLR can prepare a map of the ice situation in the Arctic zone in a period of 3 days. This period is short enough to suggest that the ice situation remains the same throughout. Equation (11.29) is not sufficiently correct in this case, so we must use more exact relations to clarify our geometrical understanding.

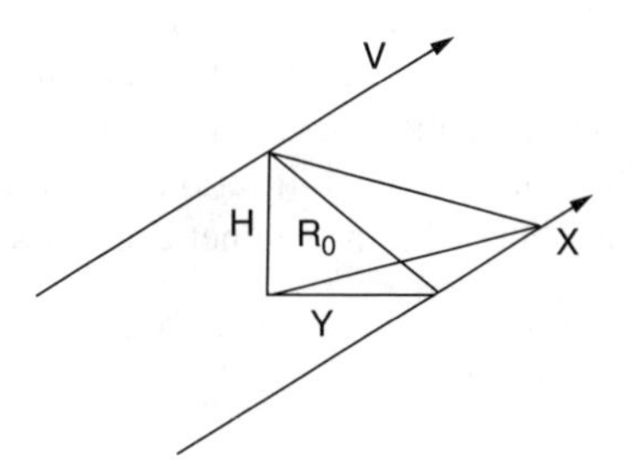

FIGURE 11.4 Geometry of synthetic-aperture radar.

The spatial resolution of the spaceborne SLR *Ocean* is close to limit resolution and it has no difficulty in realizing good range resolution when wide-band signals are applied. For example, a range resolution of 10 m requires pulses with a duration of about $5 \cdot 10^{-8}$ sec or signals with a bandwidth of about 20 MHz; however, to achieve the same azimuth resolution for a spaceborne SLR we must increase the 3-cm antenna to a length of about 3 km! It is not feasible to have such an antenna, but the problem can be overcame with the help of so-called synthetic-aperture radar (SAR).

Let us explain the principles of this kind of radar operation. Figure 11.4 shows the geometry of the problem. The radar platform (P) moves with velocity v in the x-direction. The considered elements of illuminated surface are situated in point (X,Y) and, at time moment t, is removed from the SAR by distance:

$$R = \sqrt{H^2 + Y^2 + (X - vt)^2} \cong R_0 + (X - vt)^2 / 2R_0,$$

$$R_0 = \sqrt{H^2 + Y^2}.$$

Due to the Doppler effect the signals, scattered backward by the surface elements, will have the frequency shift (see Equation (2.98)):

$$f_d = -\frac{2}{\lambda}\frac{dR}{dt} = \frac{2(X - vt)}{\lambda R_0} v. \tag{11.32}$$

The coefficient appears twice because of the doubled Doppler shift for the radio propagation forward and backward. This frequency shift depends on the x-coordinate of the surface element. So, different elements can be resolved by spectral analysis of the received signal. The spatial resolution is related to the spectral resolution by the formula:

$$\delta X = \frac{\lambda R_0}{2v} \delta f_d. \tag{11.33}$$

However, the signal can appear at the filter output after time interval $T_{\text{synth}} = 1/\delta f_{\text{d}}$. This means that signal accumulation takes place during this interval. The platform will pass the distance $d_{synth} = vT_{\text{synth}}$ during this cumulative interval. As a result, the frequency signal filtration is similar to the coherent summation of the signals by the antenna with the synthetic aperture of length d_{synth}. Equation (11.33) can be rewritten in the form:

$$\delta X = \frac{\lambda}{2d_{\text{synth}}} R_0 , \tag{11.34}$$

which is equivalent to Equation (11.24). The difference between physical and synthetic apertures is that the physical aperture provides a parallel summation of the signal, whereas the synthetic aperture relies upon a sequential summation technique.

It might appear that an unlimited increase of the synthetic aperture (or the time of synthesis) would allow a very high resolution to be achieved; in fact, though, this is not true. We have to keep in mind that the process of coherent summation takes place. Only the surface elements lying within the Fresnel zone scatter the signals with the phases. The signals of elements that are outside this zone get an additional phase shift because of the spherical wave diversity. The Fresnel zone size is $\sqrt{\lambda R_0}$ in our case and

$$\delta X = \frac{1}{2}\sqrt{\lambda R_0} . \tag{11.35}$$

This resolution can be improved by correcting the spherical diversity. This operation is called *focusing* and is realized by signal summation with weighting factors of the form:

$$w(x_0) = \exp\left(\frac{2\pi x_0^2}{\lambda R_0}\right) = \exp\left(\frac{2\pi v^2 t^2}{\lambda R_0}\right). \tag{11.36}$$

These weighting factors compensate the summand vt in the bracket of Equation (11.32), which provides the constancy of Doppler shift for the chosen surface element. It is similar to the phase correction in optical lenses by the law $\delta\Phi = kx^2/2F$ (thus the term *focusing*). In the case of the focusing procedure, the maximum synthetic aperture length is equal to the real antenna footprint on the ground; that is, $d_{synth}^{\max} = \lambda R_0/d$, and

$$\delta X = \frac{d}{2} . \tag{11.37}$$

Figure 11.5 provides a comparison of similar radars with and without focusing.

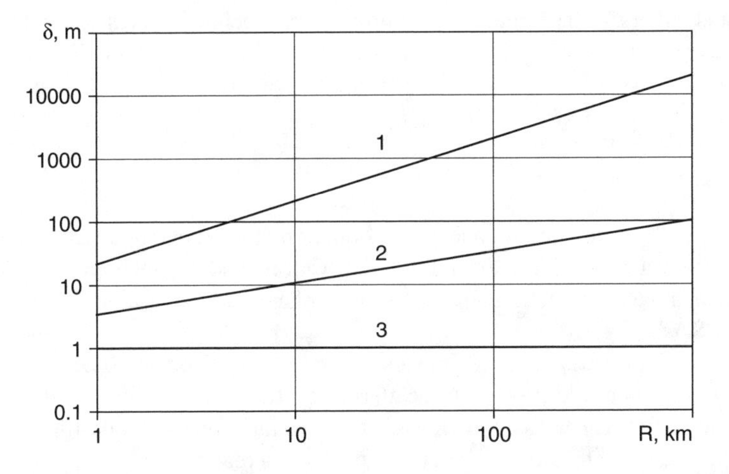

FIGURE 11.5 Linear resolution in a direction of synthesizing for radar with a wavelength of 4 cm and with the antenna size of 2 m. (1) without synthesizing the aperture and without focusing; (2) with synthesizing of the aperture but without focusing; (3) with synthesizing of the aperture and with focusing.

It would appear that an unlimited decrease in the real antenna size would allow us to realize high azimuth resolution, as much as we want; however, it does not. First, a small antenna will not produce an adequate signal-to-noise ratio. Also, the synthesized aperture is similar to the antenna array, with sources separated by distance vT, where T is the pulse repetition time. It is well known[4] that the pattern of such an array has, besides a major maximum in the direction $\varphi = 0$, similar maxima in directions determined by the equality $\sin\varphi_n = n\lambda/2vT$ $(n = 1, 2,\ldots)$. The signals scattered from these directions interfere with the main one and are indistinguishable from it. This means that the angular width of the major lobe of the real antenna has to be sufficiently narrow for suppression of this noise; thus, we can conclude that the inequality $d > 2vT$ must be performed. This last inequality means that the distance passed by the platform at the time interval between two sequential radiations must be less then the best azimuth resolution.

The resolution limit is defined by the time interval $T_{synth}^{\max} = \lambda R_0/vd$ while the point of interest is illuminated by the SAR. This is correct if the SAR antenna pattern is fixed; however, the limit imposed by Equation (11.37) can be overcome if the antenna beam is directed toward the point of interest over a time interval during the platform flight. This mode of operation (i.e., *spotlight*) requires an antenna with a changeable azimuth beam direction. This operation can be realized, for example, by using a phase antenna array. The time of illumination can be greater than $T_{synth}^{\max}$. Correspondingly, the length of the synthetic aperture will be greater than $d_{synth}^{\max}$, and the spatial azimuth resolution improves compared to that given by Equation (11.37).

The range resolution of SAR is realized by short pulse radiation. This method is similar to the one for SLR and was used, for example, in the Russian *Almaz* mission.[146] Today, use of the compressing technique has become more common. This technique has been applied for the European Remote Sensing satellites (ERS-1,2), the Japanese Earth Resources Satellite (JERS-1), the Canadian RADAR

SAT, and others.[146] The range resolution in this technique is determined by the relation:

$$\delta Y \cong \frac{c}{2\Delta f \sin\theta}, \tag{11.38}$$

where Δf is the radar signal bandwidth. Equation (11.29) can be used, as before, to define the swath. In order to extend the observation band, the technique of switching the antenna beam direction in the vertical plane can be used (for example, in RADARSAT).

A typical property of SAR images is the speckle structure. The origin of this effect is the stochastic process of scattering by the surface. Usually, the scattered signal satisfies Gauss's law, which leads to Rayleigh's distribution of its amplitude. In this case, the intensity of the signal power fluctuation equals the mean power itself. Intensive fluctuations are very undesirable for the numerical analysis of image, so to overcome this problem the process of averaging is used. We could see (Equation (6.166)) that the spatial correlation coefficient of the scattered signal is determined by the degree of overlap of the antenna footprints. This means, in our case, that signals of neighboring pixels are statistically independent; therefore, summation of these signal intensities leads to smoothing of the speckle structure. This is a common method for improving the quality of SAR images. The other method involves partly overlapping the neighboring image scenes. Of course, the speckle structure leads to a loss of spatial resolution.

The radar formula for SAR can be obtained from Equation (11.28), in which it is necessary to substitute $2d_{\text{synth}}$ for d. When the pulse compression procedure is used, the factor:

$$G_H = \tau_p \Delta f \frac{d_{\text{synth}}}{vT} \tag{11.39}$$

must be introduced. Here, τ_p is the duration of the uncompressed pulse and Δf is its bandwidth. In the following discussion, it will be useful to take into account the mean radiated power:

$$\bar{P} = \frac{\tau_p}{T} P. \tag{11.40}$$

And we have:

$$W(R) = \frac{\bar{P} A_e^2(0,\theta)\sigma^0 \cdot \delta Y \Delta f}{2\lambda v R^3}. \tag{11.41}$$

Correspondingly, the signal-to-noise ratio is:

$$\left(\frac{S}{N}\right)=\frac{\bar{P}A_{\mathrm{e}}^{2}(0,\theta)\sigma^{0}\cdot\delta Y}{2k_{b}T_{\mathrm{n}}\lambda vR^{3}}, \tag{11.42}$$

where k_b is the Boltzmann constant and T_{n} is the noise temperature of the system. Sometimes, this formula is referred to as the Cutrona equation.

The flow of SAR information is large as a rule but can be easily estimated if we remember that the number of the image elements obtained per second is:

$$M=\frac{2vY_{\mathrm{sw}}}{\delta X\cdot\delta Y}.$$

The coefficient 2 has appeared because two orthogonal components of the signal (sine and cosine) have to be registered. We will suppose that q bits are used to estimate the level of signal power in each element. So, information flow I will be:

$$I=qM=\frac{2qvY_{\mathrm{sw}}}{\delta X\cdot\delta Y}. \tag{11.43}$$

Further, let us take into account the ratios $Y_{\mathrm{sw}}/\delta Y=\Delta f\,T$, in accordance with Equations (11.29) and (11.38), and $vT/\delta X=m^{-1}$, where m is the number of pulses illuminating one image pixel by the SAR. Hence,

$$I=\frac{2q\Delta f}{m}. \tag{11.44}$$

Usually, $q = 4$ to 5 and $m = 2$ to 5. The signal bandwidth (Δf) of modern spaceborne SAR is tens of megahertz. We can easily see that the SAR data flow is several tens of megabits per second, distinguishing values of the order of 100 Mbit/sec in modern space systems. A powerful computer is needed for such a large volume of data processing. In the early days of SAR, when computers did not have sufficient speed and memory, special optical systems were applied for SAR data processing. Films were used to store the radar data in holographic form and lenses played the role of the Fourier processor. Now, due to the fast development of computer technology, all operations are carried out in digital form by comparatively simple computers.

Radio propagation in the atmosphere of Earth can influence spaceborne SAR operation. Summation of waves by synthetic aperture can be effective in the presence of spatial wave coherence, which may be destroyed, even if partially, due to turbulent atmospheric processes. This effect becomes appreciable when the coherence function (see Chapter 7) turns out to have a value of substantially less than unity. This condition is defined, in the case of Kolmogorov turbulence, by Equation (7.39) for plane waves and Equation (7.43) for spherical waves. Thus, we have to keep in mind that when we move the radar platform distance d_{synth}, the radio beam moves in the troposphere at a distance $d_{\text{T}} \cong H_{\text{T}} d_{\text{synth}} / H$, where H_{T} is the tropospheric height. Similarly for the ionosphere, $d_{\text{I}} = H_{\text{I}}\, d_{\text{synth}} / H$, where H_{I} is the ionospheric height. It is easy to see that the beam shift in the troposphere is one-tenth the distance of the synthesis. In the case of the ionosphere, this ratio is about one half or one third.

Let us proceed now to the numerical estimation of possible limitations. The approximation of spherical waves must be used for the troposphere. By a simple transformation, we can use Equation (7.43) to find the minimal azimuth spatial resolution:

$$\delta X_{\min}^{\text{T}} = \frac{5.56\left(\left\langle \mu^2 \right\rangle\right)^{3/5}}{\lambda^{1/5} l^{2/5}} \left(\frac{H_{\text{T}}}{\cos\theta}\right)^{8/5} . \tag{11.45}$$

In this process, we take into account that, due to passing twice through the same inhomogeneities, the intensity of the phase fluctuation decreases four times, not by half, as it would seem at first. The example for $\left\langle \mu^2 \right\rangle = 10^{-12}$, $\lambda = 3$ cm, $l = 1$ km, $H_{\text{T}} = 10$ km, $\theta = 45°$ shows that $\delta X_{\min}^{\text{T}} \cong 20$ cm.

In the case of the ionosphere, the plane wave approximation is more valid. The assumption of Kolmogorov's type of turbulence is more correct for the lower ionosphere. Often, it is necessary to address the presence of turbulence anisotropy and differences of the turbulence spectra compared to that of Kolmogorov; however, we will retain the assumption of Kolmogorov's turbulence to simplify the problem for our required estimations. Note that, for the ionosphere, $\left\langle \mu^2 \right\rangle = \delta N^2 \;\; f_p^4 / f^4$, $\delta N = \sqrt{\langle \Delta N^2 \rangle} / \langle N \rangle$. The beam shift in the ionosphere is comparable to the synthetic aperture; therefore, the assumption that $q_0 s \gg 1$ in Equation (7.31) looks quite reasonable, and the second term can be neglected. So, we shall talk not about the limit for the synthetic aperture size, but about the frequency below which the coherence disturbance becomes sensitive by performing the inequality $2\pi^2 k^2 q_0^2 C_\varepsilon^2 \Delta H_{\text{I}} / (\nu - 1)\cos\theta \geq 1$, where ΔH_{I} is the ionospheric thickness. Considering $\nu = 11/6$ and using the formulae of Chapter 4, Section 4.4, we can show that the last inequality leads to the requirement:

$$f \leq 6.2 \frac{f_{\text{p}}^2 \delta N}{c} \sqrt{\frac{\Delta H_{\text{I}} l}{\cos\theta}} . \tag{11.46}$$

When $f_{\text{p}} = 10$ MHz, $\delta N = 5 \cdot 10^{-3}$, $l = 5$ km, $\Delta H_I = 100$ km, $\theta = 45°$, we get $f = 0.3$ GHz.

There is also a limitation in the range of spatial resolution caused by distortion of the pulse dispersion in the ionosphere. Equation (2.88) may be rewritten as:

$$\Delta F = 13.6 f^{3/2} \sqrt{\frac{\cos\theta}{2N_t}}\,. \tag{11.47}$$

The estimation $N_t = 3 \cdot 10^{13}$ cm^{-2} can be used for the total electron content. If θ = 45°, then we have the approximation:

$$\Delta F \cong 1.5 \cdot !0^{-6} f^{3/2}\,. \tag{11.48}$$

For a frequency of 0.3 GHz, $\Delta F \cong 8$ MHz, which leads to the limitation $\delta Y^{\mathrm{I}}_{\min} \cong 30$ m.

11.5 SCATTEROMETERS

Scatterometers are a type of radar designed to measure the cross section of a surface. The interpretation of these measurements provides information about the physical, biometrical, and other properties of the investigated surfaces. For example, sea wave intensity is associated with wind velocity, so measurement of σ^0 permits us to estimate the height of the sea waves and then draw conclusions about the velocity of the near-surface winds. Knowledge of the forest area σ^0 at the C-band leads to an estimation of the tree's crown biomass.

Any radar can be used as a scatterometer; in particular, calibrated SAR allows us to measure σ^0 with good spatial resolution. This is important, because in many cases it is difficult to depend on the spatial homogeneity of the roughness within boundaries of large areas.

One of the main problems of the measurements discussed here is radar calibration. As $\delta Y \cdot \Delta f = c/2\sin\theta$, Equation (11.41) can be rewritten in the form:

$$W = \frac{\bar{P} A_e^2(0,\theta) f_0}{4 v R^3 \sin\theta}\sigma^0, \tag{11.49}$$

where f_0 is the radar carrier frequency. Here, we have a minimum of unknown parameters, and the radar formula is convenient for future analysis with regard to calibration. The procedure of defining σ^0 is an obvious one. It is necessary to measure the received signal power, and σ^0 can be calculated where all parameters of Equation (11.12) are known (i.e., we have calibration data). Two basic methods of obtaining calibration data are known. One of them is the direct one, when all values that are part of the design formula are measured beforehand or during flight by various methods. The second method can be thought of as a relative one. In this case, the signals scattered by the studied object and by a standard reflector are compared.

The direct method has low accuracy, because such parameters as transmitted and received power are usually measured with gross errors. Also, determining the board antenna parameters is very difficult. In most cases, the accuracy of the direct method is of the order of 5 to 7 dB.[89]

The relative method of measurements is more precise. In this process, it is not necessary to know the absolute value of most of the radar parameters. This is especially true when the standard reflector is situated close to the studied terrain. The corner reflector is commonly used as the reference one. More complicated systems, such as a transponder, are applied in some cases. The standard reflector represents a point target. Its cross section, which is obtained in the calibration procedure, has to be compared with $\sigma^0 \cdot \delta X \cdot \delta Y$ of the radar terrain pixel. The definition of pixel size is one of the main problems of radar calibration. Nevertheless, the mean-square error of the relative method achieves an accuracy of the order of 1 dB.[89] In practice, both of these methods are used.

For other types of scatterometers, the pixels are determined by the antenna beam or by the combination of an antenna beam that is narrow in the vertical plane with time or frequency selection. The spatial resolution in these cases is worse compared to SAR, but these scatterometers allow the detection of weak scattering by small-scale roughness. We can point out at least two cases when the knowledge of low roughness scattering is important. One of them relates to the possibility of soil moisture measurement. We could see in Chapter 6 that, in the case of small-scale roughness, the ratio of backward scattering cross sections for vertical and horizontal polarization is a function of the dielectric permittivity and does not depend on the roughness characteristics (Equations (6.46) and (6.48)). This property of scattering allows soil moisture estimation on the basis of polarization measurements. Another example is concerned with the problem of defining wind velocity by cross-sectional measurement of radiowaves scattered by a rough sea; however, large gravity sea waves, which produce powerful scattering, are directly connected with the wind velocity only at steady-state conditions. The period of the steady-state conditions has to continue for at least several hours; therefore, it is not easy to establish the relationship between gravitational seawave height and wind velocity. The capillary and capillary–gravitational waves (ripples) are more sensitive to the strength of wind. This is the reason why measurement of the scattering cross section of ripples is more useful for wind velocity definition over the ocean; however, the discussed cross section per surface unit is small. So, in order to have a tolerable signal-to-noise ratio it is necessary to have a large pixel size; therefore, in the case of spaceborne systems, special types of scatterometers are applied to measure wind velocity over the ocean. Such a technique was used in the American *SEASAT* mission and is currently being used in the European *ERS-1,2* missions. Incidentally, the pixel size of the European scatterometer is 50 km. Several antennae with differently oriented beams are used to determine wind direction. It is not necessary, in these examples of scatterometer applications, to calibrate with the help of reflectors. In these cases, the radar signal intensity can be directly related to wind velocity and wind direction.

The radar formula is similar to Equation (11.14) and appears in our case as:

$$W(R) = \frac{PA_e(0)\sigma^0}{2R^2 \cos\theta}. \tag{11.50}$$

Here, θ is the inclination angle (relative to the nadir) of the antenna beam.

More complicated is the system intended for analysis of polarization characteristics of scattered waves. The airborne AIRSAR is such a system, and a SAR system was operated during a SIR-C mission by one of the Shuttle flights. The objects of interest are components of the polarization matrix:

$$\left\| \sigma^0 \right\| = \begin{Vmatrix} \sigma^0_{hh} & \sigma^0_{hv} \\ \sigma^0_{vh} & \sigma^0_{vv} \end{Vmatrix}, \tag{11.51}$$

which describes the intensity of the matched and cross-polarized components of the scattered field. More detailed analysis is conducted using the Stokes matrix. Thus, we have two matrices; one of them describes the components of the scattered field when vertically polarized waves are radiated by the radar transmitter, and the second one corresponds to the radiated waves of horizontal polarization. Of course, knowledge of the Stokes matrix essentially expands the possibility to make objective conclusions about the properties of the researched surface. These possibilities give much more information if data about angular, frequency, and other dependencies of the cross section are obtainable; however, such systems are very complicated and are usually applied in the case of ground platforms for experimental investigations (so-called tower measurements).

11.6 RADAR FOR SUBSURFACE SOUNDING

Radar for subsurface sounding is usually referred to as ground-penetrating radar (GPR) and is intended for many applications, including finding buried engineering elements (tubes, cables, mines, etc.). We will restrict ourselves here only to the problems of research into ground layers. It was already pointed out in Chapter 3 that spectral analysis of the reflection coefficient of the layered media gives us the opportunity to reconstruct the fundamental dependence of the dielectric permittivity of the media. The problem can be easily solved in the case of a simple layer with two sharp boundaries (see Chapter 3, Section 3.3). When the depth dependence of the permittivity is more complicated, we encounter the problem of an integral equation solution. One of the variants of such an equation was formulated in Chapter 3, Section 3.9. As was mentioned earlier, this procedure requires spectral analysis of the reflected waves and, correspondingly, a multifrequency radar system. These systems are very complicated; therefore, pulse systems are more common for subsurface sensing. In this case, the pulse traveling time is the main subject of interest.

The simplest GPR is the usual radar with a pencil-beam antenna. Such airborne radar is used to measure the thickness of weakly absorbing layers in, for example, freshwater ice. The time delay (τ) between pulses, reflected from the upper and bottom boundary of the layer, gives information about the layer thickness. The thickness l is determined by the formula:

$$l = \frac{v\tau}{2}, \tag{11.52}$$

where v is the velocity of the pulse propagation in the layer. The permittivity dispersion is negligible in these cases, and the pulse velocity coincides with the wave phase velocity: $v = c/\sqrt{\varepsilon'}$.

Sounding of the ground presents many difficulties. The most significant of them is radiowave absorption and, as a consequence, weak penetration of radiowaves into the ground. Let us suppose that the reflecting layer lies at depth z. The wave attenuation for a two-way pass is represented by the formula:

$$\Gamma_{\text{dB}} = -14.37 \frac{2\pi\sqrt{\varepsilon'}\tan(\Delta/2)}{\sqrt{1-\tan^2(\Delta/2)}} \frac{z}{\lambda} = -90.3 \frac{\sqrt{\varepsilon'}\tan(\Delta/2)}{\sqrt{1-\tan^2(\Delta/2)}} \frac{z}{\lambda}, \tag{11.53}$$

where $\tan\Delta = \varepsilon''/\varepsilon'$ is the loss tangent and λ is the wavelength in air. If we suppose that ε' = 10 and $\tan\Delta$ = 0.1, then $\Gamma_{\text{dB}} \cong -17.3z/\lambda$. We can see that radiowave attenuation due to only the absorption in the ground is very great; therefore, it is difficult to count on the possibility of subsurface sounding appreciably deeper than the radar wavelength in air. This is a reason why only low-flying platforms are used for subsurface remote sensing. More frequently, ground-based platforms are used for these purposes. Thus, the bistatic technique is used, in many cases, when the transmitting and receiving antennae are separated.

The poor penetration of radiowaves into the ground leads to very limited choice of radio frequency for SSR. These frequencies, depending on the problem, lie within the range 50 MHz $< f <$ 1GHz. A bandwidth near 100 MHz is typical for layered soil research. Such frequency choice provides sounding to a depth of several meters. It is necessary to take into account that the penetration depth depends on the soil type, its density, moisture, etc. The requirement of good range resolution reinforces the necessity of wideband signal use. Let us suppose that range resolution Δz has to be 50 cm. To distinguish this, one must use radar with short pulses or signals with the bandwidth $\Delta f \cong c/2\Delta z\sqrt{\varepsilon'} \cong 9.5\cdot 10^7$ Hz, ($\varepsilon' = 10$). So, the bandwidth of the signal is comparable to its median frequency. In other words, we have the ratio:

$$\eta = 2\frac{f_{\max} - f_{\min}}{f_{\max} + f_{\min}} \sim 1, \tag{11.54}$$

where f_{max} and f_{min} are extreme frequencies of the signal bandwidth. As a rule, this ratio is much smaller than unity in traditional radio engineering. They say, in this case, that signals without carrier frequency are in use.

It is very difficult to avoid signal form distortion due to frequency dispersion of the medium. This dispersion at the discussed frequencies is basically determined by the imaginary part of the permittivity. Usually, $\varepsilon'' << \varepsilon'$, and the coefficient of attenuation is defined by the formula (see Equation (2.16)):

$$\gamma = \frac{\pi f \varepsilon''}{c\sqrt{\varepsilon'}} \quad (11.55)$$

If the cause of the radiowave absorption is static conductivity or Debye polarization friction, then $\varepsilon'' \sim 1/f$. In these cases, attenuation is uniform within the bandwidth (α does not depend on the frequency), and the signal does not experience distortion, although its energy is being decreased; however, in reality, distortion does take place, which raises another issue related to signal velocity. The common definition of group velocity (see Chapter 2) is not correct in this case for at least two reasons. First, the notion of group velocity (for example, from Equation (2.72)) is related to narrow-band signals when the ratio $\eta << 1$, and, second, distortion of the signal form does not allow formulation of a universal determination of signal velocity.

The bandwidth of GPR pulses is so wide that spectral characteristics of the antenna become important. One of the antenna variants is the so-called *bow-tie dipole* (Figure 11.6), and another type of wide-band antenna is the *horn*. We will not go into detail here regarding the types of antennae but refer the reader to special publications. These antennae often play the role of a GPR transmitter, and the method of shocking excitation is used for this purpose. The short pulses from the special generator proceed periodically to the antenna and excite oscillations of the antenna (see Figure 11.7). Pulses of this form are radiated by the transmitting antenna and

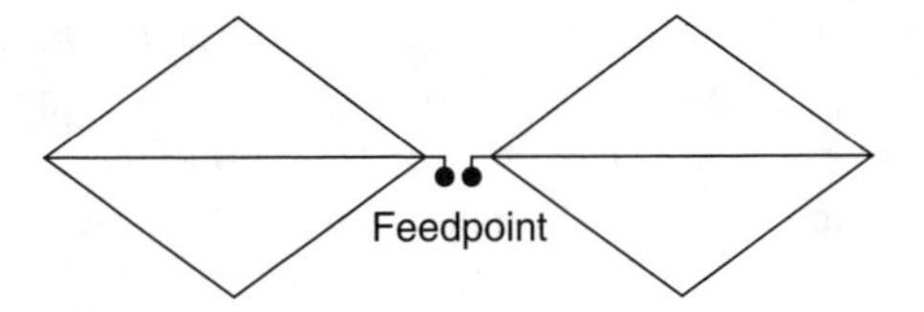

FIGURE 11.6 Bow-tie dipole.

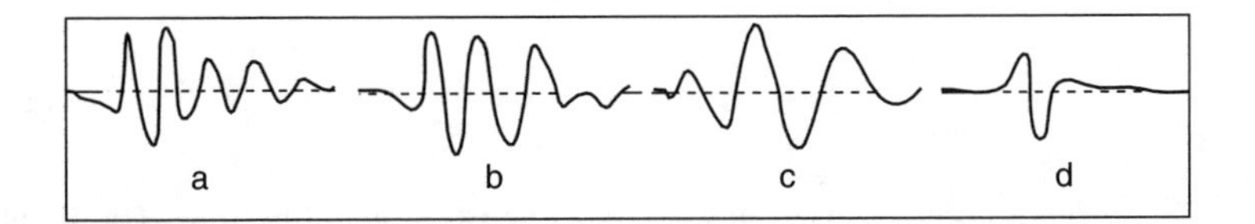

FIGURE 11.7 Pulse responses of antennas: (a) symmetrical vibrator (3.5 nsec/div); (b) monopole (10 nsec/div); (c) logarithmic spiral antenna (5 nsec/div); (d) a horn (1.4 nsec/div).

are received after reflection. The bandwidth of these oscillations is sufficiently wide that they look like short pulses of a complicated form. To record such pulses is not a simple problem; often, this problem is solved by the use of the stroboscopic technique.[147]

Among the various methods used for GPR signal processing are spectral and wavelet analysis, matched filtration, etc. Cepstral analysis is used for improvement of the separation pulses reflected from close boundaries.[147] It is especially effective for measuring the thin-layer thickness.

The logarithm of the energetic spectrum of the received signal is processed, in this case, by the procedure:

$$C(t) = \int_{\omega_{min}}^{\omega_{max}} \ln\left[\left|\tilde{E}(\omega)\right|^2 \left|F(\omega)\right|^2\right] e^{-i\omega t} d\omega \,, \tag{11.56}$$

where $\tilde{E}(\omega)$ is the transmitted signal spectrum, and $F(\omega)$ is the reflection coefficient. The expression for the reflection coefficient is represented by Equations (3.36a) and (3.41). Because $\ln\left[\left|\tilde{E}(\omega)\right|^2 \left|F(\omega)\right|^2\right] = \ln\left|\tilde{E}(\omega)\right|^2 + \ln\left|F(\omega)\right|^2$ and the logarithm of the signal spectrum is a slowly changing function of the frequency comparable to the reflection coefficient, we have an opportunity to define layer thickness. This technique is used as well for separating the sounding and reflecting signals. It allows us to allocate the signal reflected from the bottom border. Besides the often-used pulse technique, other methods also have application in the subsurface sounding of the environment (e.g., interferometric, frequency, holographic).[148]

11.7 MICROWAVE RADIOMETERS

Microwave radiometers (MWRs) are devices for thermal radiation measurement in the radio region. They are used effectively for the measurement of the brightness temperature of the environment. Usually, the 0.5- to 600-GHz band is used for MWR. Industrial sources of interference and space noises influence the microwave measurements in the lower part of this region. The upper part of the region is exposed to electronic noises, thermal radiation of the atmosphere, and so on. At their output, MWRs form voltage U_{out}, which is linearly related to the power at the MWR antenna output:

$$U_{out} = aP_{in}^{r} + b \,, \tag{11.57}$$

where a and b are coefficients usually determined by the calibration. The power at the antenna output flows from all directions (Figure 11.8). Certainly, the antenna perceives these flows differently in accordance with its directivity. The value P_{in}^{r} is characterized by the antenna temperature corresponding to the formula:

$$P_{in}^{r} = k_b T_a \Delta f \ , \tag{11.58}$$

where Δf is the radiometer pass-band. More exactly, it is the noise-equivalent pass-band. In this formula, besides the power at the antenna input, we must add the noise power of the antenna and the transmission line noise. So, the general definition of the antenna temperature becomes:

$$T_a = \int_{4\pi} T_b(\Omega)G(\Omega)d\Omega + T_0(1-\eta) = T_{bm}\eta(1-\beta) + \eta\beta T_{os} + (1-\eta)T_0 \ , \tag{11.59}$$

where $T_b(\Omega)$ is the brightness temperature of the observed scene as a function of the solid body, T_0 is the averaged temperature of the antenna and the feeder, T_{bm} is the averaged brightness temperature within the antenna's major lobe, and T_{os} is the same temperature averaged over the side lobes (see Figure 11.8).

The radiometer output in the temperature scale is realized by the square-law detector. The main problem of microwave radiometry is the fact that the studied oscillations do not differ from those of the measurement system's own noise. Only the noise level increment serves as an indicator of the signal at the antenna input; however, this increment can be detected if its value exceeds the level of the MWR noise fluctuation. The main source of radiometer noise is, as a rule, the input amplifier. If MWR looks at the terrain, thermal radiation is a significant addition. The noise generated by the antenna and transmission line is usually less. The total noise level has a noise temperature of the order of several hundred Kelvin. The

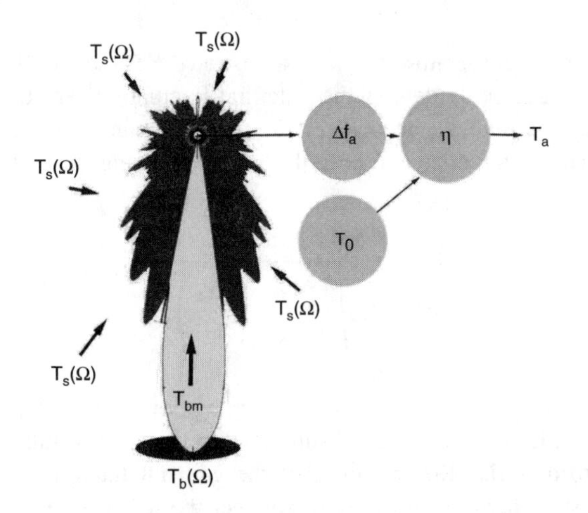

FIGURE 11.8 Components of the antenna temperature: T_o, thermodynamic temperature of antenna; Δf_a, antenna band width; η, antenna efficiency.

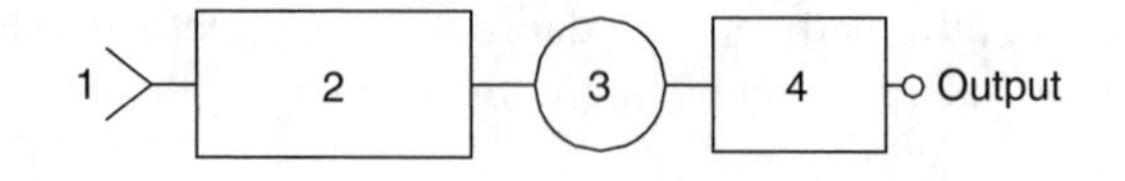

FIGURE 11.9 Block diagram of a total-power radiometer: (1) antenna; (2) receiver-amplifier units, which can provide direct and superheterodyne amplification; (3) square-low detector; (4) integrator.

remote sensing problems must measure brightness temperature with an accuracy of about 1 K. This means that the radiometer must be designed in such manner that guarantees less fluctuation in the noise temperature. The standard deviation of the noise fluctuation intensity is equal to the mean value of the intensity itself due to the Gaussian character of the thermal fluctuations. Hence, the radiometer must have a unit to determine the average of the noise fluctuations in order to decrease their level. This unit can be an integrator or low-pass filter.

Thus, the main elements of MWR are the antenna, the amplifier, the square-law detector, and the integrator (Figure 11.9). The intermediate-frequency (IF) amplifier can also be used to increase the signal power. In this case, the radiometer has to be supplied with a heterodyning system (local oscillator, mixer, etc.). Such a radiometer is called a *total-power* or *compensation* one (if it compensates for the constant component of noise). The time integration or the time of averaging τ_{av} must be much greater than the noise correlation time τ_{cor}. This allows $N = \tau_{av}/\tau_{cor}$ independent samples, which leads to a decrease in the standard deviation of the noise fluctuations at the ratio $\sqrt{N}$. More exactly $\sigma_{av} = \sigma_{in}\sqrt{2\tau_{cor}/\tau_{av}}$. Then, $2\tau_{cor} = \Delta f$, and, as the result, the fluctuation temperature is given by:

$$\delta T = \frac{T_n}{\sqrt{\Delta f \tau_{av}}} = T_n\sqrt{\frac{2\Delta F}{\Delta f}}, \quad \Delta F = \frac{1}{2\tau_{av}}. \tag{11.60}$$

Here, T_n is the noise temperature of the system, and ΔF is the pass-band of the low-pass filter. The last formula defines the potential fluctuation sensitivity of the radiometer. It cannot be realized because of the receiver gain coefficient fluctuations. The gain fluctuation influence can be reflected by reducing the last formula to the form:

$$\delta T = T_n\sqrt{\frac{1}{\Delta f \tau_{av}} + \left(\frac{\delta G}{G}\right)^2}, \tag{11.61}$$

where G is the main value of the gain coefficient by power and $(\delta G)^2$ is the intensity of its fluctuations. Usually, the second summand in the last formula is the dominant one. In order to diminish the influence of the gain fluctuation, it is necessary to calibrate the system periodically (i.e., to connect the reference noise source to the radiometer input), which is done in the modulation radiometer (or Dicke radiometer).

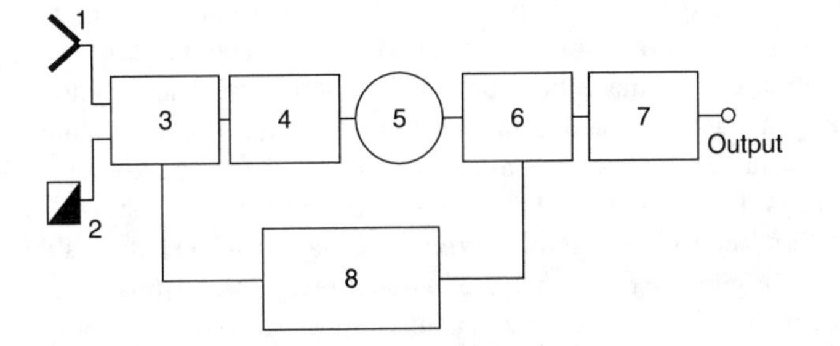

FIGURE 11.10 Block diagram of a modulation radiometer: (1) antenna; (2) reference load; (3) input switch (modulator); (4) receiver–amplifier units, which can provide both direct and superheterodyne amplification; (5) square-low detector; (6) synchronous demodulator (detector); (7) integrator; (8) square-wave generator.

The block diagram of this radiometer is represented by Figure 11.10. Several new elements appear in this scheme. The first one is the switch for the periodic, sequential connection of the receiver to the antenna and to the reference load transmitting the noise of the calibrated level. The switch realizes the modulation of the input signals, and the synchronous detector selects the signal before the detector input. The estimation of the fluctuation sensitivity can be written as:[90]

$$\delta T = \sqrt{\frac{2T_{\text{n}}^2 + 2\left(T_{\text{ref}} + T_{\text{n}} - T_{\text{a}}\right)^2}{\Delta f \tau_{\text{av}}} + \left(\frac{\delta G}{G}\right)^2 \left(T_{\text{a}} - T_{\text{ref}}\right)^2}, \tag{11.62}$$

where T_{ref} is the temperature of the reference load. When $T_{\text{ref}} = T_{\text{a}}$, the influence of the amplifier gain variation is compensated fully, and

$$\delta T = \frac{2T_{\text{n}}}{\sqrt{\Delta f \tau_{\text{av}}}}; \tag{11.63}$$

that is, the fluctuation sensitivity becomes twice as poor compared to the potential sensitivity due to the fact that the signal enters the receiver only half the time.

Some methods allow us to create modulation radiometers for which the equality $T_{\text{ref}} = T_{\text{a}}$ is always true. These radiometers are referred to as *balanced*. Some of them are described in Ulaby et al.[90] Usually, when a balanced radiometer is used, the brightness temperatures of objects are determined on easing temperature of radiation basic noise of generators.

For an unbalanced radiometer, the output voltage is proportional to the difference of antenna and reference temperatures. Equation (11.57) is transformed to the form:

$$U_{\text{out}} = a'\left(T_{\text{a}} - T_{\text{ref}}\right) + b', \quad a' = ak_b \Delta f, \quad b' = b - a'T_{\text{ref}}. \tag{11.64}$$

The constants a' and b' are determined during the procedure of calibration, which relies on the use of two calibrating standards with known brightness temperature. The calibration can be made both by standards connected directly to the receiver input and by standards located in the main antenna lobe. The calibration is referred to as *internal* in the first case; its aim is to provide the calibration and operational stability of MWR without taking into consideration the antenna system. The second type of calibration is classified as *external*; one way to achieve this calibration is to measure the brightness temperature of known test sites. In this case, calibration requires us to take into account the antenna properties and the external conditions of the experiment (in particular, the effects of side lighting). Equation (11.59) is reduced to the form:

$$T_{\text{a}} = a''T_{\text{om}} + b'', \quad a'' = \eta(1-\beta), \quad b'' = \eta\beta T_{\text{os}} + (1-\eta)T_0\,, \tag{11.65}$$

where T_{om} is now the averaged brightness temperature of the test site. It is assumed to be known. Now, it is easy to define the coefficients of the linear dependence:

$$U_{\text{out}} = a'''T_{\text{om}} + b'''\,, \tag{11.66}$$

which is the primary aim of external calibration. Measurement of two test sites with different brightness temperatures solves this problem.

In order to decrease the loss of useful signal time, a two-channel system can be used. Here, the modulation voltages in both channels are delayed at half of the period. The signals are summed after square-law detection, and time losses decrease from 2 to $\sqrt{2}$ (Figure 11.11).

The correlation scheme is a development of the two-channel schemes. In this scheme, constant components of noise are subtracted and the signals are added (Figure 11.12); this approach works due to the fact that the noises of two channels are not correlated but the signals are correlated fully. Sometimes, radiometers with

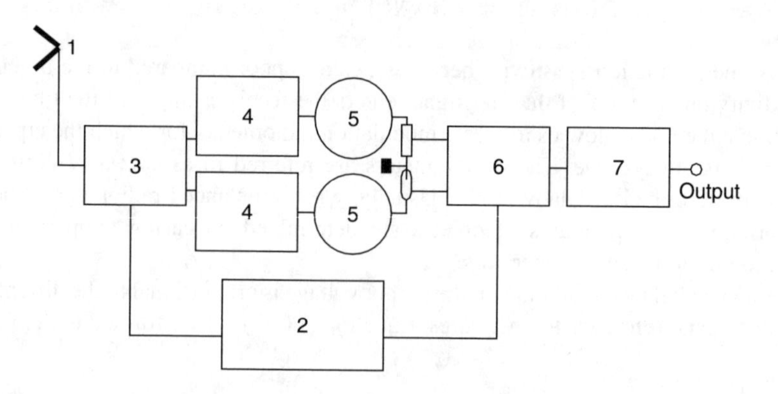

FIGURE 11.11 Two-channel modulation radiometer (see Figure 11.10).

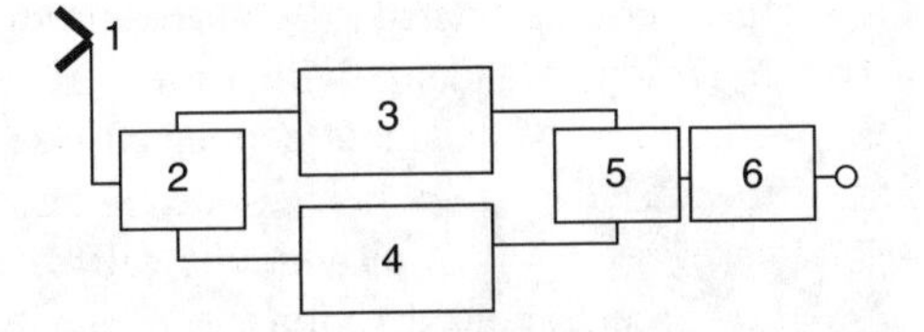

FIGURE 11.12 Scheme of a correlation radiometer: (1) antenna, (2) hybrid joint, (3,4) high-frequency amplifier, (5) correlation unit, (6) integrator.

a reference signal that controls the gain of the system find application in remote sensing.

In conclusion, we can say that the approximate sensitivity of the microwave radiometer is generally determined by the formula:

$$\delta T = \frac{\alpha T_{\mathrm{n}}}{\sqrt{\Delta f \tau_{\mathrm{av}}}}, \tag{11.67}$$

where coefficient α is greater than unity and its value depends on the radiometer design.

Some of the MWR systems are intended for polarization measurements. Such systems have two channels for independent reception of both orthogonal polarization components. These components can be detected independently or the output voltage is made proportional to the polarization coefficient:

$$K_{\mathrm{p}} = \frac{T_{\mathrm{ov}} - T_{\mathrm{oh}}}{T_{\mathrm{ov}} + T_{\mathrm{oh}}}, \tag{11.68}$$

where T_{ov} and T_{oh} are the brightness temperatures of the vertically and horizontally polarized components of radiation.

A view of terrain via airborne MWR is usually accomplished with the aid of scanning by a pencil-beam antenna. Another approach is parallel observation by several antennas. The spatial resolution depends on the angle width of the antenna major lobe. Equation (1.120) can be rewritten as:

$$\rho = \frac{\lambda}{d} H, \tag{11.69}$$

where H is the platform altitude above the investigated surface. For example, $\lambda = 1$ cm, $d = 1$ m, and $H = 1000$ km (typical for spaceborne systems) gives $\rho = 10$ km. Such poor resolution is the main obstacle to the use of spaceborne systems for many applications. Generally, these systems are used to observe objects with large-scale spatial homogeneity (atmosphere, ocean, deserts, etc.). The simplest method of observation is cross-track scanning (Figure 11.13a). Platform movement automatically

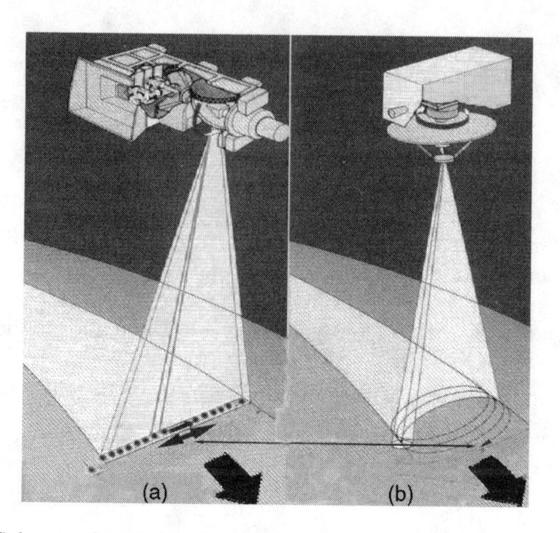

FIGURE 11.13 Schema of scanning observation.

creates the microwave scene formation. Obviously, scan time t_c has to satisfy the inequality:

$$t_c \leq \frac{\rho}{v} = \frac{\lambda}{vd} H, \tag{11.70}$$

where v is the platform flight velocity. The number (n) of elements (pixels) in the scan line depends on the time required to observe one element (i.e., upon integration time τ_{av}). Hence,

$$n = \frac{t_c}{\tau_{av}} \leq \frac{\rho}{v\tau_{av}} = \frac{\lambda H}{v\tau_{av} d} . \tag{11.71}$$

The minimum time of one pixel observation is related to the minimum resolvable contrast, δT_{min}. Equation (11.67) allows us to determine this time. The maximum observation band (swath) is:

$$Y_{sw}^{max} = n_{max}\rho \leq \frac{\rho^2 \left(\delta T_{min}\right)^2 \Delta f}{v\alpha^2 T_n^2} = \left(\frac{\lambda H \cdot \delta T_{min}}{\alpha d T_n}\right)^2 \frac{\Delta f}{v} . \tag{11.72}$$

In accordance with Equation (11.43), the information flow is:

$$I_{max} = q\frac{vY_{sw}^{max}}{\rho^2} \leq q\frac{\left(\delta T_{min}\right)^2 \Delta f}{\alpha^2 T_n^2} = \frac{q}{\tau_{av}} = 2q\Delta F . \tag{11.73}$$

The factor of 2 is omitted here because only one component of the signal takes place.

The angle of observation is changed during the line scanning, which produces some difficulties with regard to thematic processing. These disadvantages may be avoided by the use of conical scanning (Figure 11.13b) when the antenna beam is directed at one angle to the observed surface.

Microwave radar is used for temperature measurements, particularly in medicine. The brightness temperature can be represented as a sum (see Chapters 8 and 9):

$$T_o(f) = \left(1 - |F|^2\right) T_b(f) + |F|^2 T_l(f) = T_b(f) + \left[T_l(f) - T_b(f)\right]|F|^2, \quad (11.74)$$

where:

$$T_b(f) = \int_0^\infty T(z)\Gamma(z,f)\exp\left[-\int_0^z \Gamma(z',f)dz'\right]dz \quad (11.75)$$

is the effective radiation temperature of the studied body, and T_l is the brightness temperature of the outer source (e.g., atmosphere) reflected from the body. We can see that the reflection coefficient and the brightness temperature of the outer source strongly influence the result of measurement. Variation of these values during the process of measurement can lead to considerable errors in the estimation of temperature. Being able to check these troubling factors or lessen their influence on the measurement results allows us to measure the temperature of a body by MWR. Measurements at several frequencies open the way to reconstructing the depth temperature profile $T(z)$ by solving the inverse problem (see Chapter 10).

We can conclude from Equation (11.74) that the reflection coefficient has no influence when the effective radiation temperature and the brightness temperature of the outer source are equal. We usually classify these conditions as thermodynamically balanced (locally). The only way to realize these conditions, or conditions close to them, is by the use of an artificial source of lighting. Usually, partial thermodynamic balance has applications when the temperature of the outer source is chosen on the basis of *a priori* supposed radiation temperature of the body.[150] The simplest way to realize thermodynamic balance is by placing the antenna of the system directly on the surface of the body. These methods are often called application techniques, in contrast to remote methods. In the process, lighting is realized directly through the receiving antenna.

Another method of reflection compensation is connected with direct measurement of the reflection coefficient. In this method, one of the pauses in the signal modulation is used for reflection coefficient measurement with the help of a special well-calibrated noise source. The noise temperature of this source is a little higher than the body temperature. The measurement device also detects the reflected signal.[131] A block diagram of such a radiometer is provided in Figure 11.14.

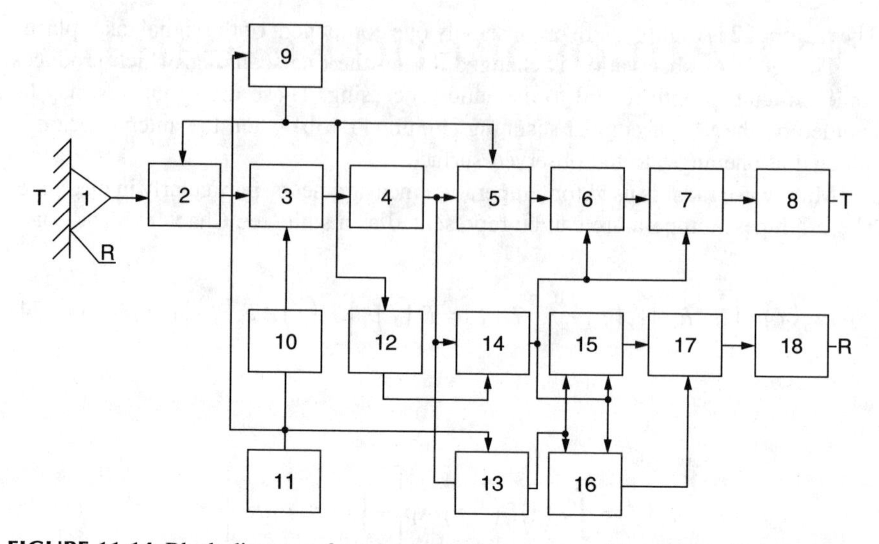

FIGURE 11.14 Block diagram of a microwave thermometer. (1) antenna; (2) input switch (modulator); (3) microwave circulator; (4) HF amplifier and square-low detector; (5), (13), and (14) synchronous demodulators (detectors) and integrators; (6) and (16) adders; (7) and (17) divider; (8) and (18), voltage testers (T and R); (9) and (11) square-wave generators; (10) noise source; (12) phase changer; (15) subtracter.

12 Atmospheric Research by Microwave Radio Methods

Microwave radio methods are finding greater application for research of the atmosphere of Earth. As discussed previously, they are based on the interaction of radiowaves with the atmosphere. This interaction is apparent in the decrease in wave amplitude, change of phase, polarization, and other radiowave parameters. Thermal microwave radiation is also the result of this interaction. The main focus of this chapter is on the neutral part of the atmosphere of Earth — the troposphere. Investigating the ionosphere is addressed in Chapter 13.

The interaction itself depends on the atmospheric components (gases, hydrometeors, etc.) and on general atmospheric parameters such as temperature and pressure. It allows formulation of the two main problems in tropospheric study on the basis of remote sensing technology. One problem is how to obtain information about general atmospheric parameters and their spatial distribution and dynamics. Radiowave interaction with constant atmospheric components provides the basis for solving this problem, where radiowave absorption by atmospheric oxygen is the main feature. The second problem is related to determining changeable atmospheric components, their spatial concentration, and so on. Solving this problem requires consideration of water vapor concentration, liquid water content in clouds, concentrations of minor gaseous constituents, their dynamics, etc. Both problems are in one way or another connected with the inverse problem solution. Solving the second one, as we discussed in Chapter 10, requires *a priori* information. To some extent, parameters of an atmospheric model can play the role of this *a priori* information.

12.1 MAIN *A PRIORI* ATMOSPHERIC INFORMATION

A standard cloudless atmosphere is characterized by such parameters as temperature, density, and pressure. The height temperature profile is described by the broken line function

$$T(h) = \begin{cases} T_0 - ah, & 0 \le h \le 11\text{km}, \\ T(11\text{km}), & 11\text{km} \le h \le 20\text{km}, \\ T(11\text{km}) + h - 20, & 20\text{km} \le h \le 32\text{km}. \end{cases} \quad (12.1)$$

Here, T_0 is the temperature at sea level, and a is the temperature gradient. For the U.S. standard atmosphere, T_0 = 288.15 K, and a = 6.5 K/km. For an approximate calculation, we can use the simplified formula:

$$T(h) = T_0 e^{-h/H_t}, \quad H_t = 44.3 \text{ km}. \tag{12.2}$$

For many preliminary calculations, we can also use the exponential altitude model of atmospheric pressure:

$$P(h) = P_0 e^{-h/H_p}, \quad P_0 = 1013 \text{ mbar}, \quad H_p = 7.7 \text{ km}. \tag{12.3}$$

The exponential model is used to describe the water vapor density:

$$\rho_w(h) = \rho_0 e^{-h/H_w}. \tag{12.4}$$

Here, ρ_0 is the water vapor density at sea level and depends on climate and the locality. On the average, it varies from 10^{-2} g/m^3 for a cold and dry climate up to 30 g/m^3 for a warm and damp climate. For moderate latitudes, the U.S. standard model assumes ρ_0 = 7.72 g/m^3. The height (H_ω) is usually considered to be between 2 and 2.5 km.

The refractive index of air is determined by the semi-empirical formula:

$$(n-1)^6 = \frac{77.6}{T}\left(P + \frac{4810e}{T}\right), \tag{12.5}$$

where e is the water vapor pressure (mbar). The first term of this formula is determined by the induced polarization of air molecules (mainly nitrogen and oxygen), and the second one describes the orientation polarization of water molecules that have a large dipole moment. The procedure for choosing numerical constants can be found in Bean and Dutton.[30] The partial water vapor pressure is associated with density ρ_w by the approximate formula:

$$e = 1.61p\frac{\rho_w}{\rho_{air}}, \tag{12.6}$$

where ρ_{air} is the density of moist air. The standard value of ρ_{air} is $1.225 \cdot 10^3$ g/m^3. The altitude distribution can be approximately represented by the exponential model:

$$n(h) = 1 + (n_0 - 1)e^{-h/H_n}. \tag{12.7}$$

The standard water vapor pressure at sea level for moderate latitudes is 10 mbar. So the standard value of the refractive index is $n_0 - 1 = 3.19 \cdot 10^{-4}$. The standard gradient near the surface of Earth is assumed to be:

$$\left.\frac{dn}{dh}\right|_{h=0} = -\frac{n_0 - 1}{H_\text{n}} = -4 \cdot 10^{-8}\,\text{m}^{-1}. \tag{12.8}$$

It is easy to determine from this that the standard value of height H_n is 8 km. Variation of the air humidity for different climatic zones leads to a variation mean value of n_0 – 1 and height H_n. The surface value of the refractive index can be calculated from meteorological data. With regard to H_n, the following fact can be used. It was established by Bean and Dutton.[30] that the value of the refractive index varies slightly at the tropopause altitude of h = 9 km and is practically independent of the geographical place and period of year. The averaged value $n(9\,\text{km}) - 1 = 1.05 \cdot 10^{-4}$. Hence, the median value of the height can be estimated from the following equation:

$$H_\text{n} = \frac{9}{\ln\left[(n_0 - 1)10^4/1.05\right]}. \tag{12.9}$$

Now, we will turn our attention to radiowave absorption by atmospheric gases. The main absorptive components are water vapor and oxygen. Water vapor has absorptive lines at wavelengths 1.35 cm (f = 22.23515 GHz), 0.16 cm (f = 183.31012 GHz), 0.092 cm (f = 325.1538 GHz), and 0.079 cm (f = 380.1968 GHz), as well as many lines in the submillimetric waves region. The wings of the submillimetric lines influence the absorption at millimetric waves; therefore, they are taken into account for calculation of absorptive coefficients. The resulting computation is comparatively complicated, so we will provide only some examples of the calculation. Figure 12.1 gives attenuation coefficient values of water vapor at sea level vs. frequency. The maximum attenuation of $\gamma_\text{w}/\rho_\text{w} \cong 2.2 \cdot 10^{-2}$ dB/km/g/m³ is at the resonance wavelength λ = 1.35 cm. Thus, $\gamma_\text{w} \cong 0.17$ dB/km near the surface of Earth at moderate latitudes. The altitude dependence can be expressed as a first approach by the exponential model:

$$\gamma_\text{w}(h) = \gamma_\text{w0} \exp\left(-\frac{h}{H_{\gamma\text{w}}}\right). \tag{12.10}$$

$H_{\gamma\text{w}} \cong H_\text{w}$ at transparency windows and rather more at the resonance wavelength, because absorption by water vapor becomes independent of the air pressure. Some data show that this height varies with the season of the year and reaches a value of 5 km.

Oxygen owns the paramagnetic moment and has many lines of absorption in the millimeter-wave region. A separated absorption line occurs at a frequency of

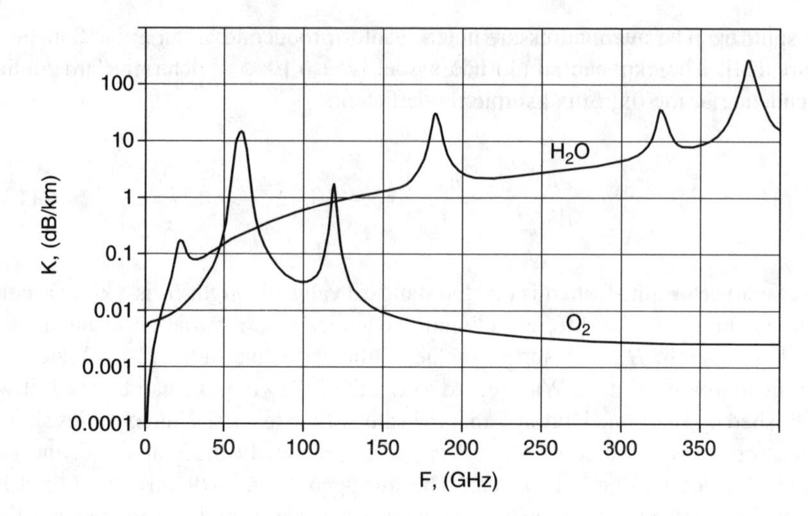

FIGURE 12.1 Computed spectra of attenuation coefficient of oxygen and water vapor at sea level.

TABLE 12.1
Oxygen Absorptive Lines Frequencies

Frequency (f) (GHz)	Wavelength (λ) (mm)	Frequency (f) (GHz)	Wavelength (λ) mm	Frequency (f) (GHz)	Wavelength (λ) (mm)
48.4530	6.19	56.2648	5.33	63.5685	4.72
48.9582	6.13	56.3634	5.32	64.1278	4.68
49.4646	6.06	56.9682	5.27	64.6789	4.64
49.9618	6.00	57.6125	5.21	65.2241	4.60
50.4736	5.94	58.3239	5.14	65.7647	4.56
50.9873	5.88	58.4466	5.13	66.3020	4.52
51.5030	5.82	59.1642	5.07	66.8367	4.49
52.0212	5.77	59.5910	5.03	67.3694	4.45
52.5422	5.71	60.3061	4.97	67.9007	4.42
53.0668	5.65	60.4348	4.96	68.4308	4.38
53.5957	5.60	61.1506	4.91	68.9601	4.35
54.1300	5.54	61.8002	4.85	69.4887	4.32
54.6711	5.49	62.4112	4.81	70.0000	4.29
55.2214	5.43	62.4863	4.80	70.5249	4.25
55.7838	5.38	62.9980	4.76	71.0497	4.22

118.7503 GHz (λ = 2.53 mm) and an absorptive band at the 5- to 6-mm area. The absorptive line frequencies of this band are provided in Table 12.1.

These lines overlap at the lower levels of the atmosphere of Earth, forming a practically continuous band of absorption. Line resolution begins only at altitudes higher than 30 km. Sometimes at this altitude, we have to take into account Zeeman's line splitting. The attenuation coefficient values produced by oxygen are shown in Figure 12.1. The exponential altitude model is also used to determine the altitude dependence of the oxygen absorption coefficient:

$$\gamma_{\mathrm{ox}}(h) = \gamma_{\mathrm{ox0}} \exp\left(-\frac{h}{H_{\mathrm{ox}}}\right). \tag{12.11}$$

In transparent windows, the following empirical relation can be used for the specific height:

$$H_{\mathrm{ox}} = 5.3 + 0.022\left(T_0 - 290\right) \text{ km}. \tag{12.12}$$

This height varies from 8 to 21 kilometers at frequencies coinciding with the absorptive lines. A more detailed discussion of this problem can be found in Deirmendjian.[86]

The total absorption of a cloudless atmosphere is described by the sum:

$$\gamma_{\mathrm{at}} = \gamma_{\mathrm{ox}} + \gamma_{\mathrm{w}} \,. \tag{12.13}$$

The main *a priori* information about hydrometeor formation is based on the statement that they consist of water drops or ice crystals. The drop sizes, their concentration, and the altitude distribution of these parameters are defined by the type of hydrometeor formation. The initial parameters of these formations (temperature, pressure, etc.) depend on their altitude and, initially, can be found using the standard atmospheric model. Very important characteristics of hydrometeors are their geometrical dimensions, motion velocity, and lifetime.

The first point of interest in this discussion is describing the electrophysical properties of fresh water, particularly the water dielectric permittivity and its dependence on wavelength (frequency). This dependence of the real and imaginary parts of permittivity is defined by the Debye formulae, which have the form:

$$\varepsilon' = \varepsilon_{\mathrm{o}} + \frac{\varepsilon_{\mathrm{s}} - \varepsilon_{\mathrm{o}}}{1 + \left(\lambda_r/\lambda\right)^2}, \qquad \varepsilon'' = \frac{\lambda_{\mathrm{r}}}{\lambda} \frac{\varepsilon_s - \varepsilon_o}{1 + \left(\lambda_{\mathrm{r}}/\lambda\right)^2}, \tag{12.14}$$

where ε_{s} is the so-called *static dielectric constant*. This value is reached at $\lambda \to \infty$, from which we derive the label of static dielectric constant. The opposite term (ε_{o}) is often called the *optical permittivity* and is reached at $\lambda \to 0$. The wavelength λ_{r} is related to the relaxation time of water by the equation $\lambda_{\mathrm{r}} = 2\pi c \tau_{\mathrm{r}}$. $\varepsilon_{\mathrm{o}} = 5.5$, and ε_{s} and λ_{r} depend on the water temperature and salinity. The details of these dependencies will be given in the next chapter; here, we shall give the values of these parameters for $T = T_0$. Thus, $\varepsilon_{\mathrm{s}} = 83$ and $\lambda_{\mathrm{r}} = 2.25$ cm.

To estimate hydrometeor reflectivity (see Equation (11.15)), we need to calculate the parameter:

$$K=\left|\frac{\varepsilon-1}{\varepsilon+2}\right|^2. \tag{12.15}$$

For fresh water:

$$K=\left|\frac{(\varepsilon_s-1)^2+(\varepsilon_o-1)^2\chi^2}{(\varepsilon_s+2)^2+(\varepsilon_o+2)^2\chi^2}\right|,\quad \chi=\frac{\lambda_r}{\lambda}. \tag{12.16}$$

It is easy to see that the frequency dependence of this parameter begins developing for the wavelength $\lambda<\lambda_r(\varepsilon_o+2)/(\varepsilon_s+2)\cong 0.1\lambda_r$ (i.e., for wavelengths shorter than 2 mm). For longer wavelengths,

$$K=\left(\frac{\varepsilon_s-1}{\varepsilon_s+2}\right)^2 \tag{12.17}$$

and now depends on temperature only. Usually, K = 0.8 to 0.93 for water. The ice permittivity is about 3.2 at these wavelengths, and $K\cong 0.2$.

The other parameter of our interest is:

$$L=\frac{\varepsilon''}{|\varepsilon+2|^2} \tag{12.18}$$

which is associated with absorption in hydrometeors (see Equation (5.46)). For water,

$$L=\frac{(\varepsilon_s-\varepsilon_o)\chi}{(\varepsilon_s+2)^2+(\varepsilon_o+2)^2\chi^2}. \tag{12.19}$$

The absorption coefficient in clouds is:

$$\gamma=\frac{24\pi^2N\langle a^3\rangle}{\lambda_r}\frac{(\varepsilon_s-\varepsilon_o)\chi^2}{(\varepsilon_s+2)^2+(\varepsilon_o+2)^2\chi^2}=\frac{18\pi w}{\lambda_r\tilde{\rho}_w}\frac{(\varepsilon_s-\varepsilon_o)\chi^2}{(\varepsilon_s+2)^2+(\varepsilon_o+2)^2\chi^2}. \tag{12.20}$$

Here, w is the cloud liquid water concentration, and $\tilde{\rho}_w$ is the water density. For waves longer than 2 mm:

$$\gamma = \frac{18\pi w\lambda_r}{\lambda^2\tilde{\rho}_w}\frac{\varepsilon_s - \varepsilon_o}{(\varepsilon_s + 2)^2} \cong 1.36\cdot 10^{-6}\frac{w}{\lambda^2}\left[\frac{1}{\text{cm}}\right] = 0.59\frac{w}{\lambda^2}\left[\frac{\text{dB}}{\text{km}}\right]. \quad (12.21)$$

The water content here is expressed in g/m^3 and wavelength in cm. For freshwater ice, we can assume $\varepsilon'' \approx 5\cdot 10^{-3}$ to obtain:

$$\gamma \cong 0.01\frac{I}{\lambda\rho_I}, \quad (12.22)$$

where I is the ice concentration and ρ_I is its density.

12.2 ATMOSPHERIC RESEARCH USING RADAR

We briefly examined this problem in Chapter 11 when we discussed weather radar (WR), which will be covered in greater detail here. The main areas of concern include:

- Measurement of the radio echo power from meteorological targets, with the echo being selected against a background of interfering reflection from the top beacons
- Detection and identification of meteorological objects by their reflectivity
- Definition of the horizontal and vertical extent of meteorological formations and determination of their velocity and the displacement direction
- Determination of the upper and lower boundary definitions for clouds
- Detection of hail centers in clouds, determining their coordinates, and defining their physical characteristics

Weather radar operates at centimeter and millimeter wavelengths; some of them are dual frequency. Specific requirements for weather radar depend upon the particular type of meteorological object and include the following:

- Exceptionally wide-scattering cross-section variations of atmospheric formations reaching values of around 100 dB
- Considerable horizontal and vertical sizes of the atmospheric objects relative to the antenna footprint and spatial extent of the radio pulse
- Rather low velocity of moving targets
- Large time–spatial changeability of radio-reflecting and -attenuating characteristics of atmospheric formations

The primary WR value measured is the backward scattering cross section:

$$\sigma_{\mathrm{d}}^{0}(\pi)=\frac{16\pi^{4}}{\lambda^{4}}K\Theta\cong\frac{1.45\cdot10^{3}}{\lambda^{4}}\Theta\,. \tag{12.23}$$

Here, the reflectivity is expressed in cubic centimeters. In some cases, the reflectivity is expressed via the diameter of drops, in which case Equation (12.23) must be increased by a factor of $2^6 = 64$.

As we can see, the drop radius must be known to determine the reflectivity. Various distribution functions are used to calculate $\langle a^6 \rangle$.[39] One of the most commonly used (as noted in Chapter 11), is the distribution proposed by Deirmendjian.[86]

The gamma-distribution (or Pearson's distribution), which has wide application, is a particular case of Deirmendjian's distribution when $\gamma = 1$:

$$f(a)=\frac{a^{\alpha}}{\Gamma(\beta)}\left(\frac{\alpha}{a_{\mathrm{m}}}\right)^{\beta}\exp\left(-\alpha\frac{a}{a_{\mathrm{m}}}\right)=\frac{a^{\alpha}}{\Gamma(\beta)a_{0}^{\beta}}\exp\left(-\frac{a}{a_{0}}\right),\quad \beta=\alpha+1,\ a_{0}=\frac{a_{m}}{\alpha}\,. \tag{12.24}$$

Finally, with regard to $\alpha = 0$, we obtain the following exponential distribution:

$$f(a)=\frac{1}{a_{0}}\exp\left(-\frac{a}{a_{0}}\right). \tag{12.25}$$

In the future, we will most commonly use the gamma-distribution, as this distribution describes well the atomized component of clouds. As shown in Aivazjan,[39] the processing of experimental data to determine coefficient α gives us values of 2 to 6 within the drop radius interval $(1.0 \text{ to } 45.0)\cdot 10^{-4}$ cm, depending on the type of cloud. The radius a_0 value varies in the range $(1.333 \text{ to } 3.500)\cdot 10^{-4}$, and the drop concentration N attains values of 188 to 1987 cm^{-3}. For the mean model (referred to as the *Medi* model by the author), we can assume that $\alpha = 2$, $a_0 = 1.5\cdot 10^{-4}$ cm, $N = 472\ \mathrm{cm}^{-3}$, $a_{\min} = 1.0\cdot 10^{-4}$, and $a_{\max} = 20.0\cdot 10^{-4}$ cm. Calculations based on this model give us values of $W = 0.4\ \mathrm{g/m^3}$, and $\Theta = 9.9\cdot 10^{-17}\ \mathrm{cm}^3$. Aivazjan[39] recommends use of the distribution:

$$f(a)=\frac{\mu-1}{1-\left(a_{\min}/a_{\max}\right)^{\mu-1}}\frac{a_{\min}^{\mu-1}}{a^{\mu}} \tag{12.26}$$

to describe large-drop components of clouds within the radius range $(20 \text{ to } 200)\cdot 10^{-4}$ cm. The value of μ varies from 4 to 10 for different cloud types. Variations in concentration N for this radius interval are $2.0\cdot 10^{-5}\ \mathrm{cm}^{-3}$ to 20 cm^{-3}. The Medi model parameters are $\mu = 6$, $a_{\min} = 2.0\cdot 10^{-3}$ cm, $a_{\max} = 8.5\cdot 10^{-3}$ cm, and

TABLE 12.2
Average Cloud Reflectivity Values (Z, mm^6/m^3)

Type of Clouds								
St	Sc	Cu Cong	Ac	As	Ci	Ns	Cb	Cb with Thunder
0.83	17.61	55.17	1.31	0.78	0.87	350.7	2432.2	19,234

Source: Data from Stepanenko.[87]

$N = 1.54$ cm^{-3}. The results of calculations give us $W = 0.12$ g/m^3 and $\Theta = 1.6 \cdot 10^{-15}$ cm^3. Comparison reveals that cloud water content and radiowave absorption are primarily determined by the atomized component. By contrast, large drops dominate in cloud reflectivity formation. Sometimes super-large drops with radii up to 0.15 cm influence the radar echo of clouds even though their concentration is extremely low. In particular, calculations made on the data given in Aivazjan[39] give us $\Theta = 1.4 \cdot 10^{-14}$ cm^3 for the main conditions, with concentration $N = 2.0 \cdot 10^{-3}$ cm^{-3}. Such drops are often generated in rain clouds. The variety of cloud reflectivity allows us to distinguish the type of clouds by radar data.

In meteorology, the reflectivity is often determined relative to drop diameter (expressed in mm^6/m^3). This value is equal to $Z = (6.4 \cdot 10^{13})\Theta$. The average values of reflectivity for some types of clouds are shown in Table 12.2.

Good spatial resolution, achieved by use of a pencil-beam antenna and wideband signals, allows us to study the inner cloud structure and to detect local motions due to the Doppler effect. Reflectivity changes with altitude and has a maximum at the altitude of the zero-isotherm ($h \cong 1.5$ to 2 km for moderate latitudes in summer). A second maximum at a height of 8 km typically belongs to thunderclouds. For ordinary clouds, the value of Θ decreases smoothly from the lower boundary to the upper one. Cumulus clouds have maximal reflectivity in the middle. The extent of the reflected signal and its change in shape depend on the type of cloud. All of these data are used to identify and classify cloud cover.

The process of radiowave reflection by rain is more complicated compared to the case of clouds. To begin, the theoretical description must include consideration of the velocity of drop fall, which depends on the radius of the drop. The sizes of rain drops are larger then water drops in clouds. For example, the median value of drop radius is:

$$a_{\text{med}} = 0.069 J^{0.182} \tag{12.27}$$

according to the formula proposed by Laws and Parsons.[88] Here, J is the rain intensity (expressed in mm/hr); $a_{\text{med}} \cong 0.1$ cm for rain of strong intensity ($J = 12.5$ mm/hr). When J = 100 mm/hr, the drop radius will be of the order of 2.5 mm. The Rayleigh approximation is not correct for calculation of the cross sections of drops in the millimeter-wave region. The Mie formulae must be used instead because of the necessity to sum the slow convergent series. These calculations for different situations were done (see, for example, Aivazjan[39]). The complexity of the calculations is made greater by the need to know the distribution function. The Marshall–Palmer function, a result of experimental data approximation, is commonly used for first estimations; this function is a variant of the exponential distribution (Equation (12.25)), and the parameters depend on the rain intensity. The Marshall–Palmer distribution describes well the distribution of drop sizes for radii $a > 0.05$ cm. A more accurate picture is given by the Best distribution, which is a gamma-distribution variant. In any case, the dependence of distribution parameters on the rain intensity has to be taken into account. As a result, we cannot express analytically the dependence of the rain reflectivity on its intensity J; therefore, the empirical dependence of the type given by Equation (13.26) is commonly applied.

Another circumstance that must be taken into account is wave attenuation in rain which becomes particularly noticeable in the millimeter-wave region. This means that the radar equation has to be developed by taking into account radiowave extinction inside the rain. So, Equation (11.18) is reduced to:

$$W(L) = \frac{8\pi^4 P A_e(0) l}{\lambda^4 L^2} \left| \frac{\varepsilon - 1}{\varepsilon + 2} \right|^2 \Theta(L) \exp(-2\gamma L) . \qquad (12.28)$$

Here, scattering has to be added to the absorption by the drops. In other words, the total cross section must be used for the extinction coefficient calculation. Equation (12.28) is simplified based on the assumption of spatial homogeneity of the rain. The extinction coefficient also depends on the rain intensity. The empirical formula is similar to Equation (11.22) and has the form:

$$\gamma = \nu J^{\mu} \ \text{dB/km}. \qquad (12.29)$$

The parameters ν and μ depend on the radiowave frequency. Their empirical value can be found in Ulaby et al.[90] and Atlas et al.[91] The data given in Ulaby et al.[90] allow us to determine an approximation in the region of 2.8 to 60 GHz:

$$\nu = 3.97 \cdot 10^{-5} f^{2.377} , \qquad (12.30)$$

We can assume that parameter μ is equal to unity for the first approximation.

Frequently, qualitative assessment for description of the rain is realized by the value of parameter:

$$Y_i = H \lg\Theta_{max} \big/ \left(\lg\Theta_{max} - \lg\Theta_i \right) , \qquad (12.31)$$

where H is the target height (km), and Θ_i is the reflectivity at a level 2 km higher than the maximal reflectivity Θ_{max} zone. A steady downpour takes place at $Y_i < 2$ for

80% of cases. For the remaining 20% of cases, a shower occurs. At $2 < Y_i < 8$, showers occur 75% of the time; a steady downpour, 20% of the time; and thunderstorms, 5% of the time. When $Y_i > 8$, probability of a storm is high (95%), and showers occur only 5% of the time.

We must say a few words about radar detection of hail. Small, as compared with the wavelength, ice particles scatter much more weakly than water drops of similar size because, as we already pointed out, $K \cong 0.18$ for ice; however, ice particles of radius comparable to the wavelength can scatter significantly more than similar water drops because of the transparency of ice particles for radiowaves. Some resonance can take place (see Chapter 5) which leads to growth of the scattering cross section, up to 10 to 15 dB in comparison with water spheres. A simple explanation of this effect is based on the supposition that radio beams are focused on the back wall of an ice particle and reflected backward to the radar.[91,92] In reality, hail particles are often covered by a water film created, for example, during their fall due to an increase in air temperature and decrease of altitude. The thickness of this film forming in the melting regime is of the order of 0.01 cm. The scattering cross section becomes smaller in this case. During the wet growth of hail, this film can have a thickness of the order of 0 to 0.2 cm.[92] In this case, the hail particle behaves as an absorptive sphere, and Equation (5.97) can be used to calculate the cross section in the first approximation, especially in the case of waves in the millimeter range. In light of the relatively large size of hailstones, we can come to some conclusions with regard to the strong reflectivity of hail clouds and hail precipitation. The backward scattering cross section of hail precipitation varies within the range 10^{-8} to 10^{-6} cm^{-1} at the X-band and $(5 \cdot 10^{-10})$ to $(5 \cdot 10^{-6})$ cm^{-1} for the S-band.[92] These cross sections are comparable to those for rain. This similarity of cross sections makes it difficult to distinguish reflection from hail and reflection from rain by one frequency; however, the specific cross-section spectral dependencies for rain and hail are different which allows us to use the two-frequency radar technique. This idea is also used in other areas of weather radar applications. In particular, a high probability of almost all precipitation detection is achieved by the simultaneous use of the X-band and millimeter-wave bands.

Rain drops and ice and snow particles have a non-spherical form which leads to depolarization of the scattered radiowaves. This depolarization can be assessed by the depolarization factor $m = \sigma_c / \sigma_m$, where σ_m is the cross section of the matched polarization (for transmission and reception), and σ_c corresponds to the cross-polarization component. At the X-band, $m \cong -1.8$ dB for dry snow, –4 dB for moist snow, –16.5 dB for a steady downpour, and –19.5 dB for showers. The degree of polarization allows us to identify the type of precipitation.

It is necessary to note that reflection by rain, clouds, etc., has a stochastic character. The scattered field is described by the Gaussian function which leads to the Rayleigh law for amplitude distribution and exponential function of the probability distribution for the scattered signal intensity:

$$P(I) = \frac{1}{\langle I \rangle} \exp\left(-\frac{I}{\langle I \rangle} \right). \tag{12.32}$$

The stochastic character of reflection means that all relations between reflectivity and hydrometeor parameters are statistical. For example, Equation (12.29) reflects a correlative connection but not a deterministic one. Due to the multiplicity of the hydrometeor structures and their dynamics, etc., the accuracy of the calculations compared to experimental data is not very high and has very often the value of tens percents.

Particle movement changes the frequency of backward-scattered waves due to the Doppler effect. This regular motion causes a frequency shift that helps define a particular regular transition. In such a way, we can measure the velocity of the fall of a drop, wind speed, etc. This particle motion results in signal spectrum expansion, which is described by Equation (5.188). Let us assume that the velocity distribution is a Gaussian one:

$$P_z(\mathrm{v}) = \frac{1}{\sqrt{2\pi}\sigma_{\mathrm{v}_z}} \exp\left[-\frac{\left(F - 2\mathrm{v}_{0z} f_0/c\right)^2}{2\sigma_{\mathrm{v}_z}^2 f_0^2/c^2}\right], \quad F = f - f_0 . \tag{12.33}$$

Here, v_{0z} is the velocity of the deterministic motion (wind, for example). The standard deviation (σ_{v_z}) characterizes the velocity pulsation because of, for example, turbulent processes. The maximum of the spectral curve defines the regular motion speed, and its width permits us to obtain the amplitude of the turbulent pulsation. In particular, the spectrum width of the signal reflected by a tornado allows us to estimate the angular velocity of its rotation, which is one of the distinguishing signs of a tornado.

Weather radar is used sometimes to observe lightning. The short lifetime of lightning (0.2 to 1.3 sec) is a problem with regard to their observation. Very often, the reflection from lightning is masked by signals scattered from the rainy zones.

We have stated already that modern radar is able to detect radiowave scattering by weak inhomogeneities of the troposphere. Sometimes these inhomogeneities behave as a point target moving in space and are regarded as the so-called *angels* type.[91] For the first approximation, these "angels" can be assumed to be dielectric spheres for which the dielectric constant differs slightly from the permittivity of the surrounding medium. For these spheres, the cross section of the backward-scattering is expressed in the form:

$$\sigma_{\mathrm{d}}(\pi) = \frac{k^4 a^6 |\varepsilon - 1|^2}{9} G^2(2ka) \cong \frac{a^2 |\varepsilon - 1|^2}{16} \cos^2(2ka) \cong \frac{a^2 |\varepsilon - 1|^2}{32} \tag{12.34}$$

in accordance with Equation (5.51). It was supposed in the process of simplification that $2ka >> 1$, and $\cos^2(2ka)$ was averaged. Assuming that $a = 100$ m = 10^4 cm, and $\varepsilon - 1 = 2 \cdot 10^{-5}$, we obtain $\sigma_{\mathrm{d}}(\pi) \cong 10$ cm^2. Such high values are observed during some summer conditions;[87] however, the assumption about sharp spherical walls appears to be too artificial. More reasonable is an assumption about a smooth change of permittivity. To take this fact into account, let us note Equation (12.34) can be rewritten as:

$$\sigma_{\mathrm{d}}(\pi) = 2F^2 \frac{a^2}{4}, \quad F = \frac{\sqrt{\varepsilon} - 1}{\sqrt{\varepsilon} + 1}. \tag{12.35}$$

In order to consider a smooth permittivity change, we can use Equation (3.108). The cross section will depend on wavelength in this case. It has been reported that two-frequency radar allows us to estimate the values of the temperature or air humidity gradients within a spherical layer bounded by an atmospheric inhomogeneity.[88] Sometimes, the concentration of dielectric inhomogeneities is so high that reflection from them has a diffusive character, and the amplitude of the reflected signal experiences intense fluctuations. This radio echo is referred to as an incoherent one, in contrast to the echo generated by point targets. An incoherent echo is frequently connected with the precloud state.

Other types of "angel"-like reflection are caused by atmospheric formations. In particular, tropospheric layers with high vertical gradients of air permittivity lead to a horizontally stretched echo with a scattering-specific cross section of the order of 10^{-16} to 10^{-15} cm^{-1}.[87]

Zones of high turbulence are also the subject of radar observation. Usually, the Kolmogorov–Obukhov approximation is applied to describe a radar echo. Equation (11.23) is reduced to:

$$\sigma^0_{\mathrm{d}}(\pi) = 0.38 C_n^2 \lambda^{-1/3}, \tag{12.36}$$

where C_n is the so-called structure constant. Its value in the troposphere varies within the interval of $(2 \cdot 10^{-8})$ to $(2 \cdot 10^{-7})$ $\mathrm{cm}^{-1/3}$. We can easily see that, in this case, the specific cross section has a value of 10^{-16} to 10^{-14} cm^{-1} at a wavelength of 10 cm.

12.3 ATMOSPHERIC RESEARCH USING RADIO RAYS

The various atmospheric effects (e.g., radio ray bending, phase delay, frequency shift, intensity attenuation) can provide the basis for determining atmospheric parameters through the processing of radio signals. Airborne platforms (aircraft, balloons, satellites) and the Sun can be sources of the radiowaves. The reception platform, as a rule, is ground based. The refraction angle and relative Doppler shift are frequently independent, and the angular position of the radiation source is the only parameter for which observation data are accumulated. However, the refraction angle value and, correspondingly, the value of the relative tropospheric frequency shift are weakly dependent on the internal details of the air permittivity altitude profile (see Chapter 4) because of the small troposphere thickness relative to the radius of Earth, which leads to the small role of tropospheric sphericity. As a result, the key parameters for these effects are the ground and integral values of the air dielectric constant.

Tropospheric sphericity is important for the occultation technique of observation and gives more reliable results for height profiles based on radio data. This method is rather simple in terms of the basic idea and in interpretation of the data; however, it is comparatively complicated in practice because it requires at least two satellites

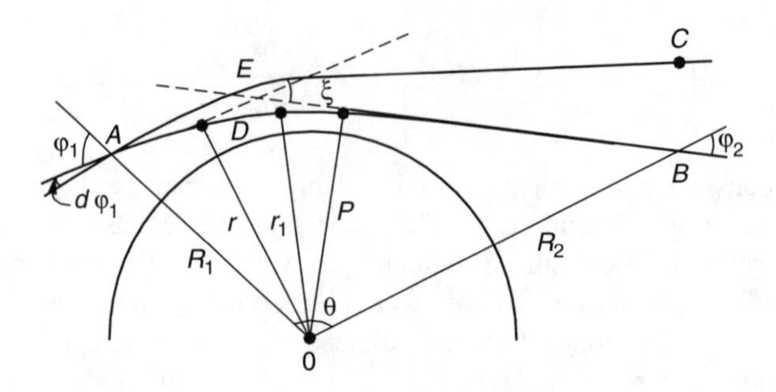

FIGURE 12.2 Researching the atmosphere of Earth by means of radio-occultation.

and communication links for the measurement of data transmission to the ground terminal. This method was shown to be efficient when used for Mars and Venus atmospheric research[32] and is being developed further for use within the atmosphere of the Earth.[35]

The idea behind the method is very simple. Let us imagine that high-orbit satellite A (i.e., a satellite in an orbit above the atmosphere) radiates the radiowaves, and satellite B receives them (Figure 12.2). The radio beam connecting the satellites will sink into the atmosphere during their mutual movement. At least two effects will take place during the sinking of the radio beam that are applicable for interpretation of these measurement data. The first effect relates to the amplitude change due to differential refraction (Equation (4.39)). The second one is the frequency change as a consequence of the Doppler effect. In this case, the corresponding frequency change is described by Equation (4.61), which, strictly speaking, is correct for the case of an infinitely distant transmitter (receiver). In practice, this situation occurs when, for example, the transmitter is onboard a geostationary satellite and the receiver is onboard a satellite rotating around the Earth, which is the simplest case.

Measurement of the frequency shifts of the received waves during the satellite motion allows us to note the refraction angle value as a function of the sighting (tangent) distance $p=\sqrt{\varepsilon\left(R_{\mathrm{m}}\right)}R_{\mathrm{m}}$. We then obtain the integral equation on the basis of Equation (4.44). The variable change by the formula $F=\sqrt{\varepsilon(R)}R$ transforms the equation:

$$\xi(p)=-p\int_{p}^{\infty}\frac{d\ln\varepsilon/dF}{\sqrt{F^{2}-p^{2}}}dF\,, \qquad (12.37)$$

which can be reduced to a known Abel equation, which we will carry out here as it is simpler to state the method of its solution. Let us multiply Equation (12.37) by $(p^{2}-R^{2})^{-1/2}$ and integrate with respect to p within the limit from R to ∞. Later, taking into account the change of integration order,

$$\int_R^\infty \frac{\xi(p)dp}{\sqrt{p^2-R^2}} = -\int_R^\infty \frac{pdp}{\sqrt{p^2-R^2}} \int_p^\infty \frac{d\ln\varepsilon/dF}{\sqrt{F^2-p^2}} dF = -\int_R^F \frac{pdp}{\sqrt{\left(p^2-R^2\right)\left(F^2-p^2\right)}} \int_R^\infty \frac{d\ln\varepsilon}{dF} dF$$

.

By considering the value of the integral:

$$\int_R^F \frac{p\,dp}{\sqrt{\left(p^2-R^2\right)\left(F^2-p^2\right)}} = \frac{\pi}{2}, \tag{12.38}$$

we have:

$$\ln\varepsilon(R) = \frac{2}{\pi}\int_R^\infty \frac{\xi(p)\,dp}{\sqrt{p^2-R^2}} . \tag{12.39}$$

We may assume that the difference $\varepsilon - 1$ is small; therefore,

$$\varepsilon(R) - 1 = \frac{2}{\pi}\int_R^\infty \frac{\xi(p)dp}{\sqrt{p^2-R^2}} . \tag{12.40}$$

The obtained formula is the method of the problem solution for height profile determination of the spherically symmetrical atmosphere on the base of radio occultation data.

Let us now consider how to define atmospheric permittivity based on refraction attenuation. We have described it before, and the effect itself is estimated by simple concepts. Let us assume a parallel radiowave beam and study the change of this beam area inside an interval of sighting distances (p, $p + dp$). The incident beam energy is proportional to the differential dp at this altitude. Due to refraction, the beam turns through an angle $\xi(p)$ at altitude p, and through an angle $\xi(p)+d\xi(p)$ at altitude $p + dp$. As a result, the beam is divergent, and its area is $\left(1+L\,d\xi/dp\right)dp$ at the place of reception. Here, $L \cong -R_r\cos\theta$ is the distance from the turn point to the point of radiowave reception, where θ is the central angle between the satellites ($\theta > \pi/2$) by radio occultation and R_r is the distance from the receiving satellite to the center of the Earth. The refraction attenuation of the radiowave amplitude is described by the formula:

$$V = \frac{1}{\sqrt{1-d\xi/dp\,R_r\cos\theta}} . \tag{12.41}$$

Creating corresponding amplitude measurements, we can determine the derivative value:

$$\frac{d\xi}{dp} = -\frac{1-V^2}{V^2 R_r \cos\theta} . \tag{12.42}$$

It remains now to establish the relation of the permittivity height profile with the derivative we have just found. This can be done easily with the help of the integration operation in Equation (12.40):

$$\varepsilon(R) - 1 = -\frac{2}{\pi}\int_R^\infty \frac{d\xi}{dp} \, arch\frac{p}{R} dp . \tag{12.43}$$

Knowledge of the altitude distribution of air permittivity opens the way for determining the air temperature height profile. The equation for air pressure P has the form:

$$P(R) = P(R_0) - m\int_{R_0}^{R} g(h)N(h)dh , \tag{12.44}$$

where m is the average mass of the air molecules, N is their concentration, and g is the free-fall acceleration. On the other hand, the gas state equation gives us:

$$P = k_b NT \tag{12.45}$$

which allows us to change Equation (12.41), which is for pressure, to a similar equation for temperature. Then, we can use the "dry" part of Equation (12.5) together with the equation of state for the gas molecule concentration to obtain:

$$N = \frac{\varepsilon - 1}{155.2 \cdot k_b} . \tag{12.46}$$

The equation for the vertical temperature profile becomes:

$$T(R) = \frac{T(R_0)\left[\varepsilon(R_0) - 1\right]}{\varepsilon(R) - 1} - \frac{m}{k_b\left[\varepsilon(R) - 1\right]}\int_{R_0}^{R} g(s)\left[\varepsilon(s) - 1\right]ds. \tag{12.47}$$

We need to choose height R_0 and know in advance the value of the temperature at this point. The first argument might be to choose $R = a$ (i.e., assume that the initial condition is at the surface of the Earth because the temperature can be measured there directly); however, it is not the best solution. As a matter of fact, all of our arguments are based on the assumption of a "dry" atmosphere, but this assumption is not valid close to the surface of the Earth, where the influence of water vapor becomes substantial. Also, our definition of the temperature profile is not correct for heights less then several kilometers above the surface of the Earth; therefore, it is better to choose point R_0 in the upper layers of the atmosphere, where the average annual temperature value can be used and inaccuracy in this value has only a small influence on the result. In particular, it could be the tropopause altitude.

The problem of water vapor in the lower layers of the troposphere can be overcome by having a second radio line with a frequency near 22.23 GHz (water vapor absorption line), which allows us to determine the water vapor profile by wave attenuation depending on the sighting parameter. The absorption coefficient can be represented as an integral over the rectified beam trajectory:

$$\Gamma(p) = 2\int_0^\infty \gamma_w\left(\sqrt{p^2+s^2}\right)ds = 2\int_p^\infty \frac{\gamma_w(F)F\,dF}{\sqrt{F^2-p^2}}\,. \tag{12.48}$$

This equation is similar to Equation (12.33), and its solution is:

$$\int_R^\infty F\gamma_w(F)dF = \frac{1}{\pi}\int_R^\infty \frac{p\Gamma(p)dp}{\sqrt{p^2-R^2}}\,,$$

or, after differentiation:

$$\gamma_w(R) = -\frac{1}{\pi R}\frac{d}{dR}\int_R^\infty \frac{p\Gamma(p)dp}{\sqrt{p^2-R^2}} = -\frac{1}{\pi}\int_R^\infty \frac{d\Gamma(p)}{dp}\frac{dp}{\sqrt{p^2-R^2}}\,. \tag{12.49}$$

From this, the water vapor concentration can be defined by formulae for the absorption coefficient. The differentiation procedure of the experimental data is regarded as being an ill-posed problem, which means that the problem as a whole is ill posed. The realization of the method is not simple because of the necessity of taking into account reflection from the surface of the Earth, the use of a pencil-beam spaceborne antenna, etc.

Let us note, particularly, that the radio occultation method is based on the spherical symmetry of the troposphere. This condition does not apply on a global scale; therefore, it would be more correct to talk about a local property of spherical symmetry. Obviously, interpretation of radio occultation data, based, in essence, on the Abel transform, can lead to noticeable errors on board day–night, where monotone change of troposphere parameters occurs due to the change of conditions of solar illumination.

Let us consider the influence of measurement errors on the accuracy of the discussed problem solution. It is necessary to substitute the sum $\xi(p)+\xi_N(p)$ in Equation (12.40) for $\xi(p)$, where $\xi_N(p)$ is the stochastic error of measurements. It follows from this that the stochastic error in the permittivity definition is:

$$\delta\varepsilon(R)=\frac{2}{\pi}\int_R^{R_b}\frac{\xi_N(p)\,dp}{\sqrt{p^2-R^2}}. \tag{12.50}$$

In the upper limit, we have knowingly substituted the height of the upper troposphere border (R_b) for infinity. The error is assumed to be equal to zero at altitudes higher than this border. It is, specifically, the altitude from which (in the case of satellite set) or up to which (in the case of satellite rise) the measurement is carried out. The error is equal to zero on average and its dispersion is:

$$\left\langle\left[\delta\varepsilon(R)\right]^2\right\rangle=\frac{4\left\langle\xi_N^2\right\rangle}{\pi^2}\int_R^{R_b}\int_R^{R_d}\frac{\hat{k}_\xi\left|p'-p''\right|\,dp'dp''}{\sqrt{(p'^2-R^2)(p''^2-R^2)}}.$$

Here, ξ_N^2 is dispersion in determination of the refraction angle value; it is naturally connected with errors in the frequency measurement. The value $\hat{k}_\xi(p)$ is the normalized correlation function of the measurement errors. On the face of it, this coefficient may be assumed to be equal to the delta-function $\delta(p'-p'')$; however, we must not make this assumption because the result will be a diverging integral. It is also necessary to perform narrow bandwidth filtration of the signal for the sake of frequency measurement accuracy. This means that the measurement error has a finite correlation time defined by the filter bandwidth. In accordance with the satellite movement, this correlation in time is evaluated in the correlation via sighting distance. The necessity to regard its finite value means that the accuracy of the permittivity definition is sensitive to the correlation scale quantity.

By a standard change of variables $p'+p''=2Y$ and $p'-p''=p$, the last integral is reduced can be to:

$$\left\langle\left[\delta\varepsilon(R)\right]^2\right\rangle=\frac{8\left\langle\xi_N^2\right\rangle}{\pi^2}\int_0^{R_b-R}\hat{k}_\xi(p)\,dp\int_{R+p/2}^{R_b-p/2}\frac{dY}{\sqrt{\left(Y^2-R^2-p^2/4\right)^2-R^2p^2}}.$$

The transform $Y^2 - R^2 - p^2/4 = Rp\,\mathrm{ch}\tau$ leads us to the compact expression:

$$\left\langle\left[\delta\varepsilon(R)\right]^2\right\rangle = \frac{4\left\langle\xi_N^2\right\rangle}{\pi^2 R}\int_0^{R_b-R}\hat{k}_\xi(p)\,dp\int_0^{\tau_b}\frac{d\tau}{\sqrt{\left(1+p/2Re^{\tau}\right)\left(1+p/2Re^{-\tau}\right)}}. \quad (12.51)$$

The upper limit in the inner integral is determined from the equality:

$$\cosh\tau_b = \frac{R_b^2 - R^2 - R_b p}{Rp}. \quad (12.52)$$

We are justified in assuming that $p << R_b - R$, as the integration over p really takes place within an interval of the order of the correlation scale. In this case, we can assume that the upper limit in the first integral is equal to infinity. The inner integral — let us designate it as $I(p,R)$ — is calculated approximately by:

$$I(p,R) \cong \int_0^{\tau_b}\frac{d\tau}{\sqrt{1+p/2Re^{\tau}}} = \ln\frac{\sqrt{1+p/2R}+1}{\sqrt{1+p/2R}-1}\frac{x_b-1}{x_b+1}, \quad x_b = \sqrt{1+\frac{p}{2R}e^{\tau_b}}.$$

By neglecting the small terms, we now have:

$$\left\langle\left[\delta\varepsilon(R)\right]^2\right\rangle = \frac{4\left\langle\xi_N^2\right\rangle}{\pi^2 R}\int_0^{\infty}\hat{k}_\xi(p)\ln\frac{4(R_b-R)}{p}\,dp. \quad (12.53)$$

The next step depends to a small degree on the specific form of the normalized correlation function (due to slowness of the logarithm change); therefore, using the mean-value theorem, we can take the logarithm away from the integral sign at the argument value $p = p_0/\gamma$, where γ is a value of the order of unity, and p_0 is the correlation scale. It is necessary take into consideration in the future that the integral with respect to the correlation coefficient is equal to the correlation scale. Finally,

$$\left\langle\left[\delta\varepsilon(R)\right]^2\right\rangle = \frac{4\left\langle\xi_N^2\right\rangle}{\pi^2}\frac{p_0}{R}\ln\frac{4\gamma(R_b-R)}{p_0} \cong \frac{4\left\langle\xi_N^2\right\rangle}{\pi^2}\frac{p_0}{a}\ln\frac{4\gamma(z_b-z)}{p_0}. \quad (12.54)$$

The correlation scale p_0 can be expressed through the interference time interval on the basis of the approximate relation $p_0 = v_\perp t_0$. If the question is one of receiver thermal noise, then the correlation time is associated with the filter bandwidth $\left(t_0 \cong 1/\Delta f_0\right)$, and the fluctuations of the refraction angle can be defined via the frequency fluctuations:

$$\left\langle \xi_N^2 \right\rangle = \frac{c^2}{v_\perp^2} \frac{\left\langle \left[\delta f\right]^2 \right\rangle}{f_0^2}. \tag{12.55}$$

Equation (12.55) is determined by many factors, particularly, by the signal-to-noise ratio.

We must say that the given error is not a single one. Apart from tropospheric asymmetry, one has to include into consideration effects of turbulent pulsation, atmospheric fronts, etc.

In conclusion, we can note that generally the problem is not complicated if one of the correspondents is not at infinity; otherwise, it becomes necessary to use more cumbersome formulae that take into account the movement of both satellites.[35]

Now, we will turn our attention to investigating the atmosphere by measuring radiowave absorption when the source of radiation is the Sun or, for example, a spaceborne transmitter. We must first define the integral parameters. Using the radiation at the absorption line of any gas, we can determine the value of the integral coefficient of radiowave attenuation:

$$\Gamma = \int_0^\infty \gamma(h) \frac{dh}{\cos\alpha(h)}. \tag{12.56}$$

The definition of angle α is given in Chapter 4. Usually, the straight-line approximation is sufficiently accurate, and we can set $\alpha = \alpha_0$, where α_0 is the zenith angle of the source. The absorption equations allow us to move from integral attenuation to integral parameters of absorption ingredient. For example, observation at the absorption line of the water vapor allows us to obtain data about the local water vapor content in the troposphere. The ozone content can be measured at 3-mm wavelength. The 8-mm wavelength is appropriate for the cloud water content. Corresponding to Equation (12.21):

$$\Gamma = \frac{18\pi}{\lambda_r \tilde{\rho}_w} \frac{(\varepsilon_s - \varepsilon_o)\chi^2}{(\varepsilon_s + 2)^2 + (\varepsilon_o + 2)^2 \chi^2} \frac{W}{\cos\alpha_0} \cong \frac{2.70 \cdot 10^{-5} \chi^2}{1 + 7.79 \cdot 10^{-3} \chi^2} \frac{W}{\cos\alpha_0}, \tag{12.57}$$

$$W = \int_0^\infty w(h)\,dh.$$

The water content (W) is expressed in g/m^2. The given numerical coefficients have approximate values as they are calculated for the temperature of the atmospheric standard model near the surface of Earth. These coefficients depend on the temperature and have to be calculated according to the cloud height and corresponding air temperature. According to the data given in Aivazjan,[39] the cloud water content W

varies from 15 to 7000 g/m². We can see from this that attenuation will be sufficient only for the millimeter-wave range; thus, the waves chosen must be at the window of atmospheric gas absorption so that attenuation by clouds is not masked by attenuation by gases. This is the reason why the wavelength of 0.8 cm is optimal for measurement of the water content of clouds. In this case, $\Gamma \cong 2.14 \cdot 10^{-4} W/\cos\alpha_0$. So, it is easy to understand that the effect will be noticeable for $W \approx 10^3$ g/m³ and greater. We have to point out that it is necessary to know the temperature of the cloud for correct computing of the coefficients in Equation (12.57). In the first approximation, this temperature can be determined if the cloud altitude is known. In this case, the unknown temperature can be calculated by the standard atmospheric model (see Equation (12.1)).

Another approach is based on the fact that λ_r is most sensitive to temperature change. Therefore, the extremum of the absorption coefficient temperature dependence is reached at the frequency when $\chi \cong (\varepsilon_s + 2)/(\varepsilon_o + 2) \cong 11.3$ (i.e., for wavelength $\lambda \cong 2$ mm). The coefficient of attenuation depends weakly on the temperature, so accurate knowledge in this regard is not as important; however, we must take into account the fact of wave absorption by water vapor, so it is necessary to have a multichannel measurement system that can include the 2-mm line.

In principle, the measurement of wave attenuation should involve a method of plotting the vertical profile of gas distribution or some atmospheric parameters. The method is based on the attenuation measurement at several frequencies. The frequencies are chosen in such a way that they lie on the slope of the absorption line. We will suppose for simplification that we are dealing with the separated line. As a result, we obtain the following series of equations:

$$\Gamma(f_i) = \gamma_0 \int_0^\infty \rho(h) F(f_i, h)\, dh, \quad i = 1, 2, \ldots, n \,. \tag{12.58}$$

Here, γ_0 is a known constant depending on the type of absorbing ingredient, $\rho(h,T)$ is the density of the studied gas, and the weighting function $F(f_i,h)$ is the line form factor. The various types of form factors differ slightly. Most simple is the Lorenz form:

$$F(f,h) = \frac{1}{\pi} \frac{\nu(h)}{\left(f - f_{\text{res}}\right)^2 + \nu^2(h)}, \tag{12.59}$$

where f_{res} is the gas resonant frequency, and $\nu(h)$ is the so-called half-width, which is defined by the collision frequency of the absorbing gas molecules, which, in part, depends on the pressure. The value of the weighting function changes during integration in Equation (12.58), reaching maximum at the height where $\nu^2(h_{\max}) = (f - f_{\text{res}})^2$. In such a way, the height of the maximum is a function of chosen frequency f; that is, $h_{\max} = h_{\max}(f)$. On the other hand, the weight of the integration areas in Equation (12.58) is concentrated close to h_{max}. This means that the gas concentration values at various altitudes are dominant for different frequencies. In other words, the family of form factors for different frequencies plays the role of different altitude filters that select the contribution of different layers of gas in absorption. This make Equation (12.58) more stable relative to the measurement errors, although it does not free this system entirely from being an ill-posed problem.

12.4 DEFINITION OF ATMOSPHERIC PARAMETERS BY THERMAL RADIATION

Microwave radiometry is an important branch of atmospheric research with regard to the interests of science as well as its many applications. Ground-, air-, and spaceborne platforms are used for microwave radiometry. The zenith angle of observation, as a rule, is chosen not to be close to 90°; therefore, the atmosphere can be considered in all calculations as a plane-layered medium[90] (Figure 12.3), and the formulae of Chapter 9 are available for our analyses. We will begin with the case of ground-based platforms. Equations (9.22) and (9.23) describe the intensity of atmospheric thermal radiation received by a microwave radiometer and is acceptable for our purposes here. This intensity is expressed, as we have already explained, in the brightness temperature scale. The first step is to study the so-called integral atmospheric parameters. The integral humidity content of air and the liquid water content are among these parameters. As a rule, it is necessary to determine both parameters simultaneously because the optimal wavelengths chosen for the measurement of the microwave radiation intensity are not usually separated significantly in the considered problems. However, we shall regard the Earth problem separately and then briefly touch on the problem of combined measurements.

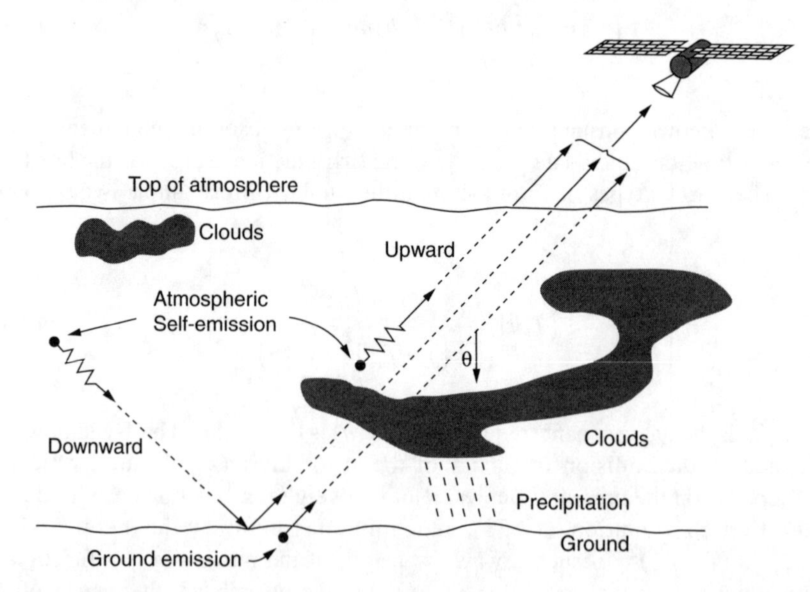

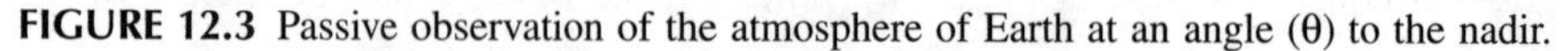

FIGURE 12.3 Passive observation of the atmosphere of Earth at an angle (θ) to the nadir.

Looking upward, we can detect not only atmospheric emission but also sources of extra-atmospheric radiation. Among them is relict radiation with a temperature of 2.7 K and a maximum-intensity wavelength of 1.1 mm, in accordance with Wien's law. The relict radiation is isotropic over space and can be easily taken into account. Then, must also take into account galactic radiation, which depends on the frequency and direction relative to the galactic center. The atmospheric absorption is included in the calculation. Further, we can encounter radiation from the Sun and Moon, if these bodies get into the antenna main beam.

We can rewrite Equation (9.22) in the form:

$$T_o(\mu) = \frac{1}{\mu}\int_0^{\infty} \gamma(z)T(z)\exp\left(-\frac{\tau(0,z)}{\mu}\right)dz, \quad \tau(0,z) = \int_0^{z} \gamma(z')dz' \quad , \qquad (12.60)$$

assuming that all extra-atmosphere sources are taken into account and their radiation is excluded from the general sum of the received power. The value of the discussed brightness temperature depends on the wavelength, as the coefficient of absorption is a function of the latter. Calculation of the atmospheric self-brightness temperature spectra (taking into account relict radiation) is shown in Figure 12.4.

Turning to the problem of water vapor or air humidity content, we note that two absorption lines are candidates for measurement of their radiation. One of them corresponds to the frequency 22.23 GHz (λ = 1.35 cm) and the other one to the frequency 118.31 GHz (λ = 0.16 cm). The center of the second line is not convenient for measurements over the whole range of air humidity variations due to the large value of the absorption coefficient. The brightness temperature in this case differs little from the thermodynamic temperature of air and practically does not react to humidity change; therefore, it is necessary to operate within the line neighborhood,

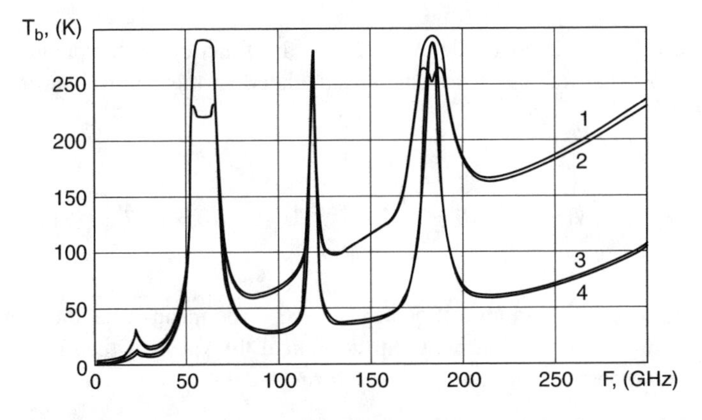

FIGURE 12.4 Brightness temperature of the atmosphere: (1) and (3), downward atmospheric self-emission; (2) and (4), upward atmospheric self-emission; (1) and (2), water vapor density at sea level (7 g m^{-3}); (3) and (4), water vapor density at sea level (2 g m^{-3}).

but the oxygen line absorption wings maintain their role and influence the results of measurement. Thus, the 22.23-GHz line is preferable in many relations. The attenuation at this line is weak, so the optical thickness τ is small and, in the first approximation,

$$T_o = \int_0^\infty \gamma(z) T(z)\, dz, \quad \mu = 1 \tag{12.61}$$

for the zenith observation. On the base of the mean-value theorem:

$$T_o = \bar{T} \int_0^\infty \gamma(z)\, dz \,. \tag{12.62}$$

The last integral is proportional to the integral humidity content. Equation (12.62) allows us to determine how to withdraw data about air humidity on the basis of microwave radiometry; however, the primary problem of how to determine the mean temperature is still unclear. The first step is to use the standard atmosphere or mean climatic data of the chosen area of measurement. The more accurate procedure defines the average temperature by the relation:

$$\bar{T} = \frac{\int_0^\infty T(z)\gamma(z)\exp\left[-\tau(0,z)\right] dz}{\int_0^\infty \gamma(z)\exp\left[-\tau(0,z)\right] dz} \,. \tag{12.63}$$

This relation is rather contradictory because the function of temperature itself is under the integral sign, so it is usually calculated in the approximation of mean climatic data. In this case,

$$T_o = \bar{T} \int_0^\infty \gamma(z) e^{-\tau(0,z)} dz = \bar{T}\left[1 - e^{-\tau(0,\infty)}\right] \cong \bar{T}\tau(0,\infty) = \bar{T} \int_0^\infty \gamma(z)\, dz \,. \tag{12.64}$$

The last expression is similar to Equation (12.62). The mean-value theorem permits us to write down the relation between the optical thickness of the cloudless atmosphere and its integral humidity in linear form:[94]

$$\tau_w(0,\infty) = \int_0^\infty \gamma_w(z)\, dz = \psi(\lambda) Q, \quad Q = \int_0^\infty \rho_w(z)\, dz \,. \tag{12.65}$$

Such a linear dependence is verified by measurements. Mitnik[94] recommends the linear approximation:

$$Q = 0.058T_o - 0.19 = 0.058\left(T_o - 3.28\right) \qquad (12.66)$$

as a result of statistical processing of experimental data. The coefficient of correlation is near 0.9. The integral humidity is expressed in g/cm^2 and the brightness temperature in degrees K. The parameters of such linear regression probably differ slightly according to climate and place of observation.[90]

The line at 1.35 cm is acceptable for integral humidity measurement; however, this line is too weak for humidity vertical profile definition on the basis of measuring the brightness temperature at the neighboring frequencies close to the center of the line. A better line for this purpose is at 0.16 cm (183.31 GHz). Just as for Equation (12.55), the results of measurement give us:

$$T_o\left(f_i\right) = \int_0^\infty K_w\left(f_i, z\right)\rho_w(z)dz, \quad i = 1, 2, \ldots, n\,, \qquad (12.67)$$

where the kernel of the integral equation or the weighting function:

$$K_w\left(f, z\right) = \beta\left[f, T(z)\right]T(z)\exp\left\{-\int_0^z \beta\left[f, T\left(z'\right)\right]\rho_w\left(z'\right)dz'\right\},$$
$$\beta\left[f, T(z)\right] = \frac{\gamma\left(f, z\right)}{\rho_w(z)} \qquad (12.68)$$

itself is a functional of the unknown air humidity. Generally speaking, we deal with the nonlinear equation. As a first approach, the kernel can be calculated using the approximation of mean climatic data. Then, a new function ρ_w, obtained as a result of the integral equation solution, can be substituted into the expression for the kernel. These values of the kernel will be its corrected form. This interactive procedure can be repeated to get more and more accurate values of the humidity. Another approach to solving the problem is to use the Taylor expansion for the kernel relative to the reference function of the humidity. The mean climatic data may be chosen for this reference function. The linearization of the integral equation is determined in this way, then standard procedures of the numerical solution can be carried out. We must point out, however, that the weighting function in this case of looking upward contains other filtering properties as compared to the ones we have talked about above. It does not have a maximum at any altitude and decreases practically exponentially with the height. So, we can talk about a "filter" of upper heights, which somewhat complicates the problem but is not the principal obstacle.

The looking-upward measurement of atmospheric humidity encounters interference from radiation by clouds. It is easy to estimate that clouds' thermal radiation at 22.23 GHz is intensive enough to influence the accuracy of the water vapor determination. In fact, the cloud absorption coefficient is approximately $7.5 \cdot 10^{-5} W$ in accordance with Equation (12.57). Correspondingly, the brightness temperature $T_o \cong 22 \cdot 10^{-2} W$. Assuming that $W = 10^3$ g/cm^2, we have $T_o \cong 22$ K (i.e., a quite noticeable value). This example shows that it is necessary to use a two-channel microwave radiometry system for joint measurement of the atmospheric humidity content and water content in clouds. We can show that the combination of wavelengths 1.35 and 0.8 cm is optimal for this purpose.[11] The brightness temperature at the 1.35-cm wavelength is more sensitive to humidity variation, while the corresponding temperature at the 0.8-cm wavelength reacts better on the liquid water content of clouds. The linear regression connecting the brightness temperature with the cloud water content is acceptable in this case. So, the measurements at two wavelengths give us:

$$T_o(\lambda_i) = a(\lambda_i)Q + b(\lambda_i, T_{cl})W + T_{ox}(\lambda_i), \quad i = 1, 2, \tag{12.69}$$

where $T_{ox}(\lambda_i)$ describes the contribution of the atmospheric oxygen radiation, which can be calculated on the basis of the standard atmosphere or mean climatic data. To achieve a more precise result, a third (oxygen) channel can be added. The main problem encountered when trying to solve Equation (12.69) is knowledge of the cloud effective temperature, T_{cl}. This temperature can be determined if the altitude of the clouds is known. As we have explained already, the model of standard atmosphere or the climatic data can be used in this case for calculation of this temperature; however, determination of the cloud height is itself a problem that can be solved by radar and lidar measurements or with the help of other indirect data. One way to estimate the cloud temperature is based on measurement of cloud radiation at some added frequencies and comparison of the intensity of this radiation with the radiation at other frequencies. The ratio of corresponding brightness temperatures depends on the effective temperature of the cloud.[94] This last statement means it is necessary to increase the number of regressive equations in Equation (12.69).

The last term in Equation (12.69) can be determined more accurately by microwave spectroscopy based on the oxygen absorption band. We are concerned with frequencies in the range of 52 to 59 GHz. The upward-looking measurement of air thermal radiation at these few frequencies allows the following integral equation system to be obtained:

$$T_o(f_i) = \int_0^\infty K_{ox}(f_i, z)T(z)\,dz, \qquad K_{ox}(f, z) = \gamma_{ox}(z)e^{-\tau(0,z)},$$
$$\tau(0, z) = \int_0^z \gamma_{ox}(z')\,dz', \tag{12.70}$$

which is similar to Equation (12.67). The weighting function also depends here on an unknown function and can be defined in the first approximation on the basis of atmospheric models. This function attenuates exponentially with increasing height, so, in a word, we have come to an already known problem. The solution of this inverse problem allows us to retrieve the vertical temperature profile which has independent importance and improves the accuracy of Equation (12.69).

One can see that remote sensing of the atmosphere by the microwave radiometry technique requires the use of a multifrequency system. This system can be improved, for example, by adding radar, lidar, and infrared radiometry data. Infrared data can aid in obtaining more specific information on cloud altitude and temperature. The joint processing of the radiometry and radar data allows us to determine the spatial distributions of the water content of clouds. It is reported in Popova and Shchukin[95] that the water content along line of sight s can be determined by the relation:

$$w(s) = \frac{\sqrt{Z(s)}}{\int_s \sqrt{Z(s)}ds} W \,. \tag{12.71}$$

The microwave radiometer is operated at the frequency of the radar and is connected to the radar antenna at the between pulses transmission.

Now, we will discuss microwave radiometry of the atmosphere from space platforms. The new elements that appear in this case are radiation of the surface of Earth and atmospheric radiation reflected by this surface. The expression for the brightness temperature becomes more complicated compared to Equation (12.60) and is described by Equation (9.27). We assume, in this case, that scattering processes do not play an important role which means the absence of rain. The complication is not very significant if the reflection coefficient and emissivity are known. It is not very easy to obtain this information, but we will assume for now that we can avoid Earth-surface clutter. For example, assume that our aim is to measure the atmospheric water vapor content and the liquid content of clouds. The solution of the problem is simplified by choosing such frequencies for which the atmospheric opacity is so large that the radiation coming from the ground is attenuated practically fully. One can easily estimate the requirements needed for unknown opacity. Let us substitute the mean atmosphere temperature for the integrand in Equation (9.27):

$$T_o = \bar{T}_a\left[1-e^{-\tau(0,\infty)}\right]\left[1+|F|^2 e^{-\tau(0,\infty)}\right]+\kappa_g T_g e^{-\tau(0,\infty)}, \quad \tau(0,\infty)=\int_0^\infty \gamma(z)dz \,. \tag{12.72}$$

It is easy to state that the influence of the Earth can be neglected if the opacity, $\tau(0,\infty)$, is sufficiently large; however, it would not be large enough because only the average atmospheric temperature will be determined in this case, and we will not be able to say anything about the opacity value. On the other hand, opacity is of great interest in our investigations as it defines the unknown integrated parameters;

hence, some compromise has to be found when the opacity is simultaneously large and not very large. So, the first approximation, neglecting the Earth-surface clutter, is:

$$T_0 \cong \bar{T}_a\left[1-e^{-\tau(0,\infty)}\right],\quad \tau(0,\infty)=-\ln\left(1-\frac{T_o}{\bar{T}_a}\right). \tag{12.73}$$

The result obtained permits us to estimate the values of rejected terms. Indeed, the term describing the reflected atmospheric radiation will have the value $T_o|F|^2\left(1-T_o/\bar{T}_a\right)$ in this approximation, and the term of the ground radiation is estimated by the value $\kappa_g T_g\left(1-T_o/\bar{T}_a\right)$. We can find conditions for minimal opacity based, for example, on the requirement $\kappa_g T_g\left(1-T_o/\bar{T}_a\right) << T_o$. Assuming that $T_g \cong \bar{T}_a$ leads to the requirement:

$$\frac{T_o/\bar{T}_a}{1-T_o/\bar{T}_a} >> \kappa_g\,. \tag{12.74}$$

Emissivity κ_g may be estimated as 0.5; hence, the minimal value of the brightness temperature is estimated as the value $T_o/\bar{T}_a = 5/6 = 0.83$ if the equality with the right side equal to 0.5 is substituted for Equation (12.74). It follows from this that the value of opacity $\tau(0,\infty)$ has to be greater than 1.79.

Such large opacity allows the influence of Earth to be avoided, but leads to a loss of opportunity to detect the changes taking place in the lower atmospheric layers. It causes a definite inconvenience, and the practical choice of frequencies leads to an opacity that is not so large. That situation is understandable in view of the fact that traditional frequencies for spaceborne radiometers lie in a band of less than 40 GHz where the opacity of the atmosphere of Earth is small. So, the radiometry system must include frequencies that allow the surface emissions of Earth to be estimated. These principles were used for the first Russian experimental radiometry systems used on the *Cosmos-243*, *Cosmos-384*, and *Cosmos-1151* missions. The chosen frequencies of 3.5, 8.8, 22, 35, and 37 GHz provided information about the thermal radiation of the surface of the Earth, atmospheric water vapor, and clouds. The radiometry information at the frequencies of 3.5 and 8.8 GHz were appropriate for defining emissions from the surface of the Earth, the 22.35-GHz channel was used to measure water vapor content, and the radiometer at 37 GHz permitted the liquid water content of clouds to be determined. The land surface structure is usually small scale, and a few elements with different emissivities are in the antenna footprint or field of view (FOV), such as open soil, forest, or inland reservoir. In practice, it is very difficult to evaluate emissivity from one frequency to another for such a complicated combination of emitters, especially because the combination itself and its structure within the pixel are unknown. This is one of the reasons why interpretation of radiometry system data is more effective over the ocean than over land; to do so, we must take into account the large-scale oceanic structure and knowledge of the water electrophysical properties. The details of this will be discussed in the next chapter.

Several spaceborne radiometry systems have been launched by the United States, Russia, Japan, and the European Space Agency (ESA). Determination of the atmospheric moisture and water content of clouds is one of the main tasks of these systems. Regression analysis is used for data interpretation. As opposed to the cases discussed above, here we use logarithmic regression in accordance with Equation (12.73), combined with elements of linear regression. The typical equations have the form:

$$U_j = a_{j0} + \sum_i a_{ji} \ln\left(1 - \frac{T_{\mathrm{o}}(f_i)}{\overline{T}_{\mathrm{a}}}\right) + \sum_n b_{jn} T_{\mathrm{o}}(f_n). \qquad (12.75)$$

Here, U_j are unknown parameters of the atmosphere (e.g., air humidity, W, and the liquid content of clouds, Q) and of the surface of the Earth, such as the surface temperature of the ocean (T_s) and near-surface wind speed. Linear regression is chosen for the low-frequency channels, which are sensitive to surface radiation, while logarithmic regression is used for the atmospheric load. The reader can find more detailed discussion of the regression algorithm and corresponding reviews of the literature in Ulaby et al.[90] and Stepanenko et al.[93] The recommended value of the average temperature is $\overline{T} = 280$ K. The average temperature and regression coefficients were found as a result of the solution of several thousand direct problems, including practically the entire range of meteorological parameter changes. This algorithm was developed for processing of the Scanning Multichannel Microwave Radiometer (SMMR) data. This radiometer of *Nimbus-7* and SEASAT missions operated at frequencies of 6.6, 10.7, 18, 21, and 37 GHz and had conical scanning and registered microwave radiation at an angle of incidence near 50° from nadir. The radiation of vertical and horizontal polarizations was received separately, which allowed the seawave intensity to be determined, as well as the wind speed. So, the brightness temperatures of both polarizations could take place in the regression equations separately.

The described system continues to develop from mission to mission. There is a Special Sensor Microwave/Imager (SSM/I) of the Defense Meteorological Satellite Program (DMSP), which is a four-frequency microwave radiometry system for measurement at 19.35, 22.235, 37.0 and 85.5 GHz. All channels, except the 22.235-GHz channel, receive the radiation at both linear polarizations: vertical and horizontal. The Delta system, with frequencies of 7.5, 13.3, 22.2, and 37 GHz, was included in the payload of the Russian–Ukrainian *Ocean* mission. Another application is the radiometry system of the Tropical Rainfall Measuring Mission (TRMM). This system, referred to as the TRMM Microwave Imager (TMI), operates at frequencies of 10.7, 19.4, 21.3, 37, and 85.5 GHz and is primarily intended to obtain information about precipitation. This information is processed together with data of Precipitation, which is used for investigation of precipitation.

The presence of rain influences the accuracy of microwave radiometry data. First, the scattering processes within rain have to be included in the consideration; that is, rain drops cannot be regarded as particles with zero albedo. This consideration changes the transfer equation, and the expression for the brightness temperature differs from Equation (12.72). Equation (9.89) shows this difference for the rather simple case of particles with an isotropic scattering indicatrix filling up the half-space. In reality, the analytical result has to be more complicated because drop albedo is a function of the radii of the drops. This leads us to the conclusion that the albedo of the rain drops is a stochastic value, and it is necessary to know of its statistical characteristics based on the distribution function of the radii of the particles. To solve this problem analytically does not seem possible. This is the reason why it is preferable to take a more direct approach to these calculations by using the Mie formulae and some distribution functions (Marshall–Palmer, for example). The result of such calculations shows that the brightness temperature grows with an increase of the rain rate and reaches saturation.[90] The experimental data generally verify these results.

The second problem of rain intensity measurement is the poor spatial resolution of the microwave radiometry system. The ordinary footprint of the spaceborne radiometer antenna is of the order of tens of kilometers. The horizontal size of the rain can be smaller, and the fraction of FOV covered by the rain is unknown beforehand. This is one of the reasons for using spaceborne weather radar for the detection of rain areas and the rain rate definition, as was done in the TRMM mission.

Spaceborne radiometry can be used for the retrieval of a vertical temperature profile. The emission at the oxygen band of 50 to 60 GHz is of primary interest. The weighting function in this downward-looking case has the altitude filtering properties that we talked about earlier. In particular, these properties allow us to neglect clutter from the surface of the Earth. The Microwave Sounding Unit (MSU), which is used on NOAA satellites, is an instrument designed to solve such a problem. Microwave radiometers at frequencies of 50.3, 53.74, 54.96, and 57.05 GHz are included in this unit.

13 Remote Sensing of the Ionosphere

13.1 INCOHERENT SCATTERING

Most information about the ionosphere was and continues to be obtained based on remote sensing data. In the period before artificial satellites, radar (ionosonde), radiated short pulses with carrier frequency change from pulse to pulse, was the main tool. The carrier frequencies of the pulses lie in the decameter band and are chosen in such a manner that the upper frequency is smaller than the plasma frequency of the electron concentration in the ionospheric maximum. Because the maximum electron concentration (at altitude $z_m \cong 300$ km) reaches a value of $2 \cdot 10^6$ cm^{-3} in the daytime, the corresponding upper frequency (see Equation (2.31)) must have a value of the order of 13 MHz. The pulses are reflected by layers whose plasma frequencies coincide with the carrier frequencies of the pulses. The electron concentration of the proper layer is defined by the reflected pulse frequency, and the time of its arrival determines the layer altitude. This is a brief description of the general idea of ionosonde data interpretation, although in reality it is actually much more complicated.[68]

Evidently, it is possible to obtain in this way only knowledge about the lower ionosphere because the pulses are not reflected by layers above the ionospheric maximum (F-layer). Artificial Earth satellites led to the development of onboard ionosonde, which allowed the ionosphere to be sounded from above and for data to be obtained about the height distribution of the electron concentration above the F-layer. The development of satellite communication systems promoted the study of ionospheric propagation processes, experimental research into the various effects (refraction, phase and group delay, polarization plane rotation, etc.), and elaboration of methods for defining ionospheric parameters on the basis of these effects, all of which are considered in this chapter. To begin, we will examine the effect of incoherent scattering. The development of radar (including planetary radar), radio astronomy, and deep space communication promoted use of the incoherent scattering method and led to development of power transmitters of ultrahigh-frequency and microwave bands, as well as large antennae and sensitive receivers.

We have already discussed the incoherent scattering phenomenon by electrons in Chapter 5 and established that the scattering occurs with thermal fluctuation of the electron density. Incoherent scattering differs from radiowave reflection from the ionosphere, which describes the process of backward coherent scattering of radiowaves in the decameter band. We have determined that the intensity of incoherent scattering depends weakly on the frequency (contrary to coherent scattering); therefore, such frequencies may be chosen for investigation of the incoherent scattering for which the ionosphere is transparent. This, in contrast to ionosonde, allows us to obtain information about all of the ionosphere, not only about its lower part (below the F-layer), by means of incoherent scattering radar. Space technology is not needed in this case.

The basic idea of incoherent radar sounding of the ionosphere is extraordinarily simple. As the differential cross section per unit volume is proportional to the electron density (Equation (5.170)), we can measure the electron density inside a layer of given altitude receiving radar signals scattered backward by this layer. The altitude itself is determined by the time it takes the signal to travel along the path of transmitter-scattering layer-receiver. The radar equation is similar in this case to Equation (11.14), which was obtained for radio scattering by particles, except that the backscattering cross section of the electron thermal fluctuation must be substituted in this equation. The absence of radiowave extinction in the ionosphere is assumed, which is reasonable due to the smallness of the total cross section. It is easy to define the theoretical possibility of a backscattering cross section, determined by the use of radar data, and then of the electron concentration value $N(L)$ estimation based on Equation (5.170). This can be done especially easily in the case when the wavelength is chosen to be much smaller than the Debye length, which allows us to avoid the influence of uncertainty in our electron temperature knowledge on the measurement results.

It is necessary to divide the obtained expression by the value $k_b T_n \Delta f$ (where T_n is the receiver noise temperature) to establish the signal-to-noise ratio. The receiver bandwidth is determined by the pulse duration via the known relation $\Delta f_S = 1/\tau$. The pulse duration is associated with the thickness of the sounding layer by the equation $\tau = 2l/c$ (where c is the value of the light speed). We emphasize again that we, of course, assume that the plasma frequency is much smaller than the frequency of the sounding radiowaves. We obtain, as a result:

$$\left(\frac{S}{N}\right) = \frac{P\sigma_d^0(\pi, L)\, l^2 A_e(0)}{c k_b T_n L^2}. \tag{13.1}$$

To estimate the radar parameters, let us assume for simplicity that $\sigma_d^0 = N a_e^2$ and analyze the case of sounding an ionospheric layer with thickness of 100 km and altitude of 1000 km, where $N \cong 10^5$ cm^{-3}. In this case,

$$\left(\frac{S}{N}\right) \cong 2 \cdot 10^{-10} P \frac{A_e(0)}{T_n}.$$

To make simple estimations for a radar whose antenna has an effective area of 1500 m^2 and noise temperature of the system T_n = 100 K, we can see the need to have a transmitter peak power of several megawatts to get a tolerable signal-to-noise ratio. This example shows that we must have tools with difficult to achieve parameters for successful investigation of radiowave scattering processes by ionospheric plasma. These devices are expensive both to manufacture and to operate, which is one of the reasons why such radar is not more widely distributed. On the other hand, the possibility of obtaining greatly expanded intelligence about the ionosphere (regarding not simply electron concentration), has led to the drive of developed countries resolving to create these systems. We will not describe these systems here, as it is more convenient to study the specific literature; instead, we will restrict ourselves to consideration of those ionosphere parameters that are measured by incoherent scattering methods.

The scattering effect depends on the ratio between radio wavelength and Debye length, whose value in the ionosphere varies within the limit fraction of centimeters to several centimeters, dependently on the altitude. So, the question becomes one of determining the different effects for waves a centimeter and smaller compared to the effects for waves several centimeters and greater. At first, it would seem as though we are contradicting the statement made above about the weak frequency dependence of incoherent scattering; however, that discussion was concerned with the scattered wave intensity. Now, we are talking about the spectral density of the scattered radiation power. The chaotic thermal motion of scattering particles leads, due to the Doppler effect, to frequency "smearing" of the scattered signal. This smearing depends strongly on the product value $p = (2kD)^2 = (4\pi D/\lambda)^2$ (where D is the Debye length). When $p >> 1$ (i.e., the wavelength is much less than the Debye length), the electron behavior is like that of free particles, and the frequency broadening of the signals is determined by their thermal velocities. The Doppler broadening of a spectrum is estimated by the value

$$\Delta f = \frac{\Delta\omega}{2\pi} = \frac{2}{\lambda}\sqrt{\frac{2k_b T_e}{m}} \tag{13.2}$$

on the basis of Equation (5.189). Let us compute its value for the case when $\lambda = 1$ cm and $T_e = 1000$ K. We have added the subscript to the temperature sign to emphasize that the question, in this case, concerns the electron temperature $\Delta f \cong 6 \cdot 10^7$ Hz. In the example of energetic computation that we considered earlier, the signal bandwidth can be written as $\Delta f_S = c/2l = 1.5 \cdot 10^3$ Hz (where $l = 100$ km), according to the chosen pulse duration. So, the frequency-broadening smearing of the signal due to chaotic electron motion essentially exceeds the spectral bandwidth of the sounding signal; therefore, when we need to receive the entire signal, it becomes necessary to choose the bandwidth of the scattered signal and not of the transmitted one (matched filtration) for noise bandwidth computing. However, the signal-to-noise ratio worsens by thousands of times, and the reception of the signals scattered by free electrons is not found to be realistic.

The given reasons result in the need to choose a wave bandwidth that substantially exceeds the Debye length ($p << 1$). The electrons are not free in this case (they are connected with heavy ions), and the signal frequency broadening is not so large. It is defined, in the first approximation, by a formula like Equation (13.2), where we must substitute ion mass M for the electron mass. The ion mass is approximately $2.7 \cdot 10^{-23}$ for an ion of atomic oxygen O^+. We obtain $\Delta f \cong 2 \cdot 10^3$ Hz by substituting

$\lambda = 100$ cm and $T_i = 1000$ K. These results are compatible with the radiated signal spectrum.

In fact, the scattered signal spectrum is much more complicated in comparison to the thermal motion spectrum. It is necessary, in the considered wave bandwidth, to pay attention to the presence in plasma of Langmuir (plasma) and ion sound waves. If the length of any wave is Λ, then radiowaves with wavelength $\lambda = 2\Lambda$ will be intensely scattered (compare with discussion toward the end of Section 6.3). We cannot declare the validity of the term *incoherent scattering* in this case, although its use is accepted.

The spectral lines that have shifted relative to the carrier frequency on the values:

$$\Delta f = \pm \frac{2V}{\lambda} \tag{13.3}$$

occur in the scattered signal spectrum. Here, V is the wave speed. The plus/minus (±) sign represents the scattering by waves traveling toward the receiver and away. Langmuir waves are excited when $p \gg 1$, but Landau damping is very strong in this case, and the wave intensity and length are found to be small. So, the scattering by these waves does not play a noticeable role.

Ion sound waves are excited in nonisothermal plasma when the electron temperature is essentially higher than the ion temperature. The theoretical analysis processes are rather complicated;[69] therefore, we will confine ourselves to only their main conclusions.[70–73]

First, let us focus on measurement of the height profile of the ionosphere electron concentration. The total (i.e., integrated over all frequencies) differential cross section of the backscattering is:

$$\sigma_d^0(\pi, L) = \frac{N(L)a_e^2}{1 + T_e/T_i}, \tag{13.4}$$

where the ratio of electron and ion temperatures is in the denominator. Let us point out that, at $p \gg 1$, Equation (13.4) reduces to the asymptotic form of Equation (5.170) for isothermal plasma. We can define the height distribution of the electron concentration by the measured backscattering cross section when we have *a priori* knowledge of the temperature ratio (which, by the way, is also a function of the height). Alternatively, the given measurement data allow determination of the degree of the different atmospheric layers anisothermic when we know the height profile of the electron density.

The second way to determine the electron concentration is measurement of the polarization plane Faraday rotation, which is significant in the decimeter- and meter-wavelength bands used in incoherent scattering radar. The angle of the turn of the polarization plane can be written for the backward scattered radiation as follows:

$$\Psi_F(L)=\frac{4.72\cdot 10^4}{f^2}\int_0^L N(z)H_0(z)\cos\beta(z)\frac{dz}{\cos\gamma(z)}\,. \tag{13.5}$$

Here, γ is the angle between vertical and the radiowave propagation trajectory. This trajectory is not quite vertical because the refraction indexes of ordinary and unordinary waves are functions of all coordinates, not only of z, due to the coordinate dependence on the magnetic field of Earth. The numerical coefficient ($4.72 \cdot 10^4$) in Equation (13.5) is twice as large as the coefficient in Equation (2.44) because of the double pass of the ray. The derivative:

$$\frac{d\Psi_F}{dL}=\frac{4.72\cdot 10^4}{f^2}N(L)\frac{H_0(L)\cos\beta(L)}{\cos\gamma(L)} \tag{13.6}$$

allows us to estimate the electron concentration at altitude L using *a priori* knowledge of all other parameters. In particular, the influence of the magnetic field on the radiowave trajectory is small due to the high frequency used; therefore, we can suppose that $\gamma = 0$. The product $H_0(L)\cos\beta(L)$ is already known for the chosen geographical position.

Finally, it is possible to define the electron concentration by the position of the spectral peaks corresponding to the Langmuir waves. It is known that the velocity of the latter is determined by the relation:[73,74]

$$V_p=\frac{f_p\lambda}{2}\sqrt{1+3p}\,. \tag{13.7}$$

We have taken into consideration the resonant relation between the lengths of the radiowaves and the Langmuir waves. It follows, then, that plasma frequency satellites are separated from a carrier frequency on the intervals:

$$\Delta f_p=\pm f_p\sqrt{1+3p}\cong f_p=9\cdot 10^3\sqrt{N}\,. \tag{13.8}$$

Also, we can define the electron concentration by the frequency position of the plasma peaks. The plasma lines are very weak at night, and it is difficult to reveal them. In the daytime, the plasma wave intensity increases by dozens of times due to the generating ability of photoelectrons occurring under the action of solar ultraviolet radiation. The plasma lines themselves are separated rather widely about the central frequency in comparison to the signal spectral bandwidth, as the plasma frequency values are at least several megahertz. Therefore, we need to have preliminary data about the electron concentration to exhibit them at a given altitude and to tune the receiver appropriately.

Incoherent scattering radar allows us to identify several plasma parameters other than electron density. This is perhaps one of the main advantages of the incoherent scattering method compared to other techniques of ionosphere remote sensing. We should note the possibility of estimating the electron-to-ion temperature ratio on the basis of the backscattering total cross-section measurement using *a priori* knowledge of the electron concentration. If the frequency spectrum analysis of scattering by ion sound waves can be added to this, then we have an opportunity to obtain new information about the ionosphere. The ion sound wave velocity is determined by the expression:[14]

$$V_{\mathrm{S}} = \sqrt{Z\frac{k_{\mathrm{b}}T_{\mathrm{e}}}{M}\left(1+3\frac{T_{\mathrm{i}}}{ZT_{\mathrm{e}}}\right)}, \tag{13.9}$$

where Z is the ionization multiplex, and M is the ion mass. From this, the frequency position of satellites will be:

$$\Delta f_{is} = \pm\frac{2}{\lambda}\sqrt{Z\frac{k_{\mathrm{b}}T_{\mathrm{e}}}{M}\left(1+3\frac{T_{\mathrm{i}}}{ZT_{\mathrm{e}}}\right)}. \tag{13.10}$$

Determining the frequency position together with previous knowledge allows us to define separately the electron and ion temperature differences in layers of the ionosphere when the ion mass is known. The shift of the ion sound line is several kilohertz; therefore, its discovery is realized rather easily. The details of the scattered signal spectrum analyses allow the possibility of determining the concentration ratio of two known ions using the spectral line slope.

Incoherent scattering data analyses permits the parameters of other ionosphere layers to be defined. A more detailed description of the possibilities taking place here, one can find, for example, in Reference 76.

13.2 RESEARCHING IONOSPHERIC TURBULENCE USING RADAR

Radar can be applied to the study of ionospheric turbulence properties. In this case, the scattering takes place on fluctuations generated by dynamic processes in the ionosphere but not on thermal fluctuations. One of the principal differences between the considered types of fluctuations arises because sometimes the local quasi-neutrality is not disturbed, and sometimes it is (i.e., when the local concentrations of electrons and ions are not equal). In the first case, the concentrations of both kinds of particles are changed in a similar way from point to point; therefore, we may refer to the first kind of fluctuations as *macro-pulsation of plasma*, while the term *micro-pulsation* is more acceptable for the second kind. This difference is emphasized by a discrepancy in scales. Turbulence fluctuation scales have values at least greater than 1 meter, which appreciably exceed the Debye length typical for thermal fluctuations. The fluctuation intensity of the electron density in turbulent processes also exceeds that caused by thermal fluctuation, which leads to more intensive radiowave scattering compared to the case of incoherent scattering. For this reason, more spare radar is required than for incoherent scattering.

As we have determined, the differential cross section is in essence measured by radar, and we will return to this value later. We must now take into account Equations (2.31) and (2.38) to relate the permittivity of isotropic plasma ($\varepsilon = n^2$) with the electron concentration. Then, the spatial spectrum of the permittivity fluctuations is defined by the relationship:

$$\tilde{K}_{\varepsilon}(\mathbf{q}) = \left(\frac{4\pi e^2}{m\omega^2}\right)^2 \tilde{K}_{N}(\mathbf{q}) \tag{13.11}$$

with the spectrum of the electron concentration fluctuations. As a result, we obtain:

$$\sigma_{d}^{0}(\mathbf{e}_s) = 8\pi^3 a_e^2 \left[1 - (\mathbf{g}_i \cdot \mathbf{e}_s)^2\right] \tilde{K}_N\left(k(\mathbf{e}_s - \mathbf{e}_i)\right), \tag{13.12}$$

which must be regarded in light of turbulence anisotropy, which is especially pronounced at altitudes higher than 100 km and high latitudes, where the elongation of ionospheric inhomogeneities along the magnetic field of Earth is typical. The spatial spectrum of the electron density fluctuations can be represented in the first approach in a form similar to that of Equation (4.80):

$$\tilde{K}_N(\mathbf{q}) = \frac{C_N^2 \exp\left(-q_\perp^2/q_{m\perp}^2 - q_=^2/q_{m=}^2\right)}{\left(1 + q_\perp^2/q_{0\perp}^2 + q_=^2/q_{0=}^2\right)^{\nu}}, \tag{13.13}$$

where the wave numbers longitudinally and across the magnetic field direction are designated by the subscripts = and ⊥.

In the lower ionosphere, where the collision number exceeds the cyclotron frequency, turbulence is defined by the properties of the neutral gas component and is characterized by Kolmogorov's spectrum. Anisotropy is missing at these altitudes ($z < 90$ km), $\nu \cong 11/6$, and the internal scale is of the order of 10 to 30 m. The outer scale naturally substantially exceeds this value. The relative amplitude of the electron density fluctuations, determined as:

$$\delta N = \left.\frac{\sqrt{\langle \Delta N^2 \rangle}}{\langle N \rangle}\right|_{l_\perp \cong 1\text{km}},$$

is estimated here by a value of the order 10^{-3}.

It is necessary to take into consideration the anisotropy in the F-layer of ionosphere. We may assume here, in the first approximation, that $l_{m\perp} \cong 1/q_{m\perp}$ is equal to a value of several meters, while $l_{m=} \cong 1/q_{m=} \cong 5 \text{ to } 10 \text{ km}$ in conditions of high latitudes. Under the same conditions, $\nu \cong 1.25$. In the middle latitudes, $\delta N =$ (1 to 3) $\cdot 10^{-3}$. The last value reaches 10 percent under some conditions.

The anisotropy of ionospheric inhomogeneities leads to foreshortened scattering of radiowaves along the cone surface with an axis lengthwise on the magnetic field of Earth. In this case, multiposition radar technology, when receivers of the scattered radiation are distributed on the curve of the foreshortened cone crossing the surface of Earth, is applicable. Determination of the frequency shift of the scattered waves (due to Doppler effect) and the velocities of the inhomogeneous movement allows us to define and study the dynamic processes in the ionosphere at various altitudes. A more detailed description of radar sounding of the turbulent ionosphere can be found in, for example, Gershman et al.[13] and Roettger.[76]

13.3 RADIO OCCULTATION METHOD

Soon after the launch of the first artificial Earth satellite, ionosphere remote sensing methods started to be developed based on analysis of the effects of radiowaves propagation in the ionosphere. We have pointed out in Chapter 4 that a series of effects (phase shift, polarization plane rotation, etc.) are strong and can be applied successfully for the study of the environment. We will begin our discussion here with the radio occultation method, the principles of which were briefly described in Chapter 11. All of the formulae of this chapter can be applied to the interpretation of ionospheric effects. It is only necessary to remember that dielectric permittivity is expressed in this case via electron concentration, which means, for example, that we can obtain a formula for the electron concentration height profile:

$$N(R) = -\frac{m\omega^2}{2\pi^2 e^2} \int_R^\infty \frac{\xi(p)\,dp}{\sqrt{p^2 - R^2}} = -7.58 \cdot 10^{-9} f^2 \int_R^\infty \frac{\xi(p)\,dp}{\sqrt{p^2 - R^2}} . \qquad (13.14)$$

This method is simple in its basic idea and in the possibility of its data interpretation. This is determined by many factors, particularly by the signal-to-noise ratio. The problem is straightforward if the receiver or the transmitter is not at infinity. For the troposphere, it becomes necessary to use more cumbersome formulae that take into account the positions and movements of both satellites. In reality, both tropospheric and ionospheric effects impact on the radio signal, and some problems are encountered when trying to distinguish the influences of these media. A system of two coherent frequencies allows us to solve this problem and to determine the Doppler shift caused by the ionosphere.

13.4 POLARIZATION PLANE ROTATION METHOD

Extensive developments have been made in ionosphere research based on analysis of the properties of radiowaves radiated by artificial satellites and received by ground terminals. These methods began to be developed just after the launch of the first satellite in 1957. Among them, the method based on measurement of radiowave polarization plane rotation due to the Faraday effect is most actively used. A formula similar to Equation (4.74) is the main one used here, although it is often represented in a somewhat different form. The main objective of this measurement is determining the integral electron concentration.

Two principal obstacles are encountered when trying to apply this method. The first one is related to the fact that, at a sufficiently low frequency (hundreds of MHz), the polarization plane experiences many turns, but the effect itself cannot be measured as the rotation angle can be defined within 2π. The second difficulty is connected with the need to know the direction of the radiated wave polarization vector to determine the turn angle. In order to do this, it is necessary to have a satellite with a three-axis orientation, which is not always possible.

One of the simplest ways of overcoming these difficulties is based on determination of the difference of the plane polarization angle turn by the change in the zenith angle of the satellite during its movement within some interval $\Delta\alpha_0$, such that the difference would be less than π. Asymptotically, this means that the question is concerned with measurement of the derivative $d\Psi_F/d\alpha_0$ (i.e., of the rotation speed of the Faraday angle). It is assumed in the process that the satellite orientation is not practically changed within the time interval Δt between two sequential samples.

The second technique uses the analysis of radiation of a transmitter at two close frequencies (f_1 and f_2). As a result, we can determine the difference:

$$\Delta\Psi_F = \Psi_F(f_1) - \Psi(f_2) = \\ = -\frac{e^3 M}{2\pi m^2 c^2}\left(\frac{1}{f_1^2} - \frac{1}{f_2^2}\right)\frac{1}{\sqrt{1-0.91\sin^2\alpha_0}}\frac{\partial}{\partial\tau'}\left(\frac{\cos\mu}{R'^2}\right)_{R'=R_m} N_t. \tag{13.15}$$

We assume that the satellite orbit is significantly higher than the ionospheric maximum altitude, which is the usual case. It is important that the frequency chosen conforms with the condition $\Delta\Psi_F < \pi$. This means, using the example given in Section 4.3, that the frequency difference $\Delta f = f_2 - f_1$ has to be a dozen times smaller than the central frequency.

13.5 PHASE AND GROUP DELAY METHODS OF MEASUREMENT

Initially, it would seem that the most sensitive method of determining ionospheric parameters would be a method based on phase measurement of the radiowaves transmitted by a satellite and received on the ground. In fact, the question becomes one of defining the eikonal (Equation (4.68)). However, use of such a technology has been hampered because the addition of the ionosphere to the eikonal substantially exceeds the wavelength, and the insuperable component $2\pi n$ (where n is an integer) appears in the phase definition. The situation is similar to measurement of the rotation angle due to the Faraday effect. We can overcome this difficulty using a two-frequency system with the frequencies f_1 and $f_2 = pf_1$. The phases of the corresponding signals will be:

$$\Phi(f_1) = \frac{2\pi L}{c} f_1 - \frac{I}{f_1}, \qquad \Phi(f_2) = \frac{2\pi L}{c} pf_1 - \frac{I}{pf_1}, \tag{13.16}$$

where L is the distance between a ground receiving station and a satellite, and I is the function (see Equation (4.68)) depending on the ionosphere parameters and independent of the frequency. The so-called equivalent phase difference is determined during the processing:

$$\widehat{\Phi} = \Phi(f_2) - p\Phi(f_1) = \frac{(p^2-1)I}{pf_1}. \tag{13.17}$$

The troposphere influence is excluded by the operation of processing, which, due to the frequency independence, is automatically reflected in the L value. The choice of frequencies has to be made according to the validity of the condition $\widehat{\Phi} < 2\pi$. The example given in Section 4.3 using Equation (4.69) leads us to the conclusion that the chosen frequencies have to be sufficiently close ($p - 1 << 1$). The oscillations at these frequencies have to be coherent for phase measurement to occur. Although, the technology of frequency synthesis is able to satisfy these needed requirements, the resulting system is rather complicated.

A simpler system is based on measurement of the traveling time of the pulses. In this system, the radiation of two transmitters at frequencies f_1 and f_2 (optionally coherent) are modulated by similar pulses. The difference in arriving times of the pulses can be written as:

$$\Delta t_g = \frac{1}{c}\left[\Delta L_g(f_1) - \Delta L_g(f_2)\right] \cong$$
$$\cong \frac{e^2}{2\pi mc}\left(\frac{1}{f_1^2} - \frac{1}{f_2^2}\right)\int_0^z \frac{N(\zeta)d\zeta}{\sqrt{\cos^2\alpha_0 + 2\zeta/a\sin^2\alpha_0}}. \tag{13.18}$$

This expression refines Equation (4.68) and can be obtained by simplification of Equation (4.38). Naturally, the frequency choice has to be made in such a way that the relative delay value is less than the pulse time repetition; otherwise, the phase measurements are likely to be indefinite.

Inserting the radicand expression $\zeta = z_m$, we obtain a formula for the electron content (compare with Equation (4.68)); however, this can be considered as a first approach to solving the problem. More exactly, we are dealing with an integral equation, where $\Delta t_g(\alpha_0)$ is the known function, $N(\zeta)$ is the function to be determined, and

$$K(\alpha_0, \varsigma) = \frac{1}{\sqrt{\cos^2\alpha_0 + 2\zeta/a\sin^2\alpha_0}}$$

is the kernel. The formulated equation is a first-order Fredholm equation and can be solved numerically, which opens the way to defining the height profile $N(\zeta)$ of the electron concentration based on the experimental data regarding the pulse group delays. However, the numerical conversion of a Fredholm equation of the first order, as we have stated earlier, is an ill-posed problem that requires *a priori* information (see Chapter 10).

By dividing all of the measurement intervals on subintervals of the observation zenith angle, α_0, we may regard the ionosphere as being locally spherically symmetric within the subintervals of such division. Formulating the problem in this way allows us, in principle, to determine the height profiles of the electron concentration for each subinterval of angles or distances from the observation point and to discover the horizontal gradients of the electron density in the orbit plane. The problem solution is essentially developed by simultaneous reception of signals at several spaced points and subsequent analysis of all the data, in which case we arrive at a tomography problem.[75,76]

13.6 FREQUENCY METHOD OF MEASUREMENT

The inconvenience of studying ionosphere parameters on the basis of phase measurements can be overcome by frequency measurements (i.e., by measurements of the phase derivative with respect to time). They are sometimes referred to as *phase-differential measurements*, and the problem is reduced to determining the angle δ (Equation (4.65)). It is more convenient to perform the measurements at two frequencies, which automatically lets us exclude tropospheric effects and frequency changes due to the Doppler shift. Therefore, in the case considered here, our discussion will be about the equivalent angle $\hat{\delta} = \delta(f_1) - \delta(f_2)$. We will assume that the satellite orbit is so high that at points of the satellite position the permittivity is assumed to be equal to unity due to low electron concentration. Then, we can establish that:

$$\hat{\delta}(\alpha_0) = \frac{e^2}{2\pi m}\left(\frac{1}{f_2^2} - \frac{1}{f_1^2}\right)\frac{a\sin\alpha_0\sqrt{R^2 - a^2\sin^2\alpha_0}}{\sqrt{R^2 - a^2\sin^2\alpha_0} - a\cos\alpha_0} \times \\ \times \int_a^\infty \frac{dN}{dR'}\frac{dR'}{\sqrt{R'^2 - a^2\sin^2\alpha_0}} \quad . \tag{13.19}$$

We have arrived at an equation similar to Equation (12.35), but we can speak only about formal resemblance, not about identity. The essential difference is the fact that the integrand is not reduced to infinity at the lower limit. This circumstance means that we cannot obtain an explicit analytical expression to solve the integral equation, which forces us to apply numerical methods. This problem is also ill posed, and it is necessary to use specific methods (for example, the regularization method) to solve it. Let us point out that, in frequency technology, the problem of indeterminacy does not appear, in contrast to the phase method; therefore, the frequency separation can be made sufficient large to make application of this method easier from a technology point of view.

13.7 IONOSPHERE TOMOGRAPHY

The combined processing of satellite signals received by receivers distributed in space opens the possibility of ionospheric tomography, which was mentioned in passing earlier. This gives us the opportunity to obtain more detailed information about the irregular structure of the ionosphere than would be provided by the approaches we have talked about previously. Three directions of ionospheric tomography are recognized.[75,76] One of them is related to the study of large-scale irregularities; the geometrical optics approximation (see Chapter 4) is sufficient to describe the interaction of electromagnetic waves with these irregularities. This is why we use the term *beam radio tomography*. In fact, the methods described earlier are related to this approach. Beam interpretation is inadequate for the research of localized inhomogeneities with sizes of the order of dozens of kilometers, so it is necessary to be guided by radiowave diffraction; thus, it becomes a problem of *diffraction tomography*. An integral equation can be formulated in this case to describe the diffraction in which a small-angle approximation is applied for the Green function (see Section 1.6). The problem again becomes an ill-posed one, but several methods have been developed for its solution.[75] The distribution of receivers along a line perpendicular to the plane of the satellite orbit is effective for improving this method. The reception of radiation from different directions allows tomographic data transformation at planes perpendicular to the orbit. Such data, together with analysis of the signal time evolution in the satellite motion process, allow us to reconstruct a two-dimensional electron concentration distribution within localized inhomogeneities. The described antenna array is convenient for researching the scattered signal statistical properties in the presence of a great number of random irregularities. For the ionospheric *statistical tomography* problem, the second moments, introduced in Chapter 7, are studied.

14 Water Surface Research by Radio Methods

14.1 GENERAL PROBLEMS OF WATER SURFACE REMOTE SENSING AND BASIC *A PRIORI* INFORMATION

The main characteristics of water surface determined by means of remote sensing techniques are surface temperature, salinity, roughness, ice cover properties, properties of thin films on the water surface, geoid form, and topography (Figure 14.1). The ocean surface temperature changes within a range of –1.9 to +30°C. The maximal value of temperature is observed between latitude 5° and 10° north. The minimal temperature is typical for waters of the Arctic basin. The variation in daytime temperature seldom exceeds 0.5°C.

As a rule, the properties of water surfaces, topographies, and physico-chemical characteristics are statistically homogeneous, which makes them very convenient for the application of remote sensing technology. As we have established, the interaction of radiowaves with surfaces is determined by the dielectric permittivity of the medium bounded by the surface and depends on the topography, particularly the roughness, of the studied surface. As for water permittivity, the Debye formula (see Equation (12.14)) is augmented by a term representing the conductivity:

$$\varepsilon'_{w} = \varepsilon_{o} + \frac{\varepsilon_{s} - \varepsilon_{o}}{1 + (\lambda_{r}/\lambda)^{2}}, \quad \varepsilon''_{w} = \frac{\lambda_{r}}{\lambda}\frac{\varepsilon_{s} - \varepsilon_{o}}{1 + (\lambda_{r}/\lambda)^{2}} + \frac{2\sigma}{c}\lambda . \tag{14.1}$$

The terms ε_o, ε_s, and λ_r are explained in Chapter 12, and σ is the water conductivity. All parameters depend on the water temperature and its salinity. Various approximation formulae can be used. We will take advantage of the formulae given by Rabinovitch and Melentjev[97] to which we add salinity terms. These formulae have the form:

$$\begin{aligned}
&\varepsilon_{o} = 5.5, \\
&\varepsilon_{s} = 88.2 - 0.4088t + 8.1\cdot 10^{-4}t^{2} - 0.303S, \\
&\lambda_{r} = 1.87 + 1.47e^{-0.063t} - 0.027t + 1.36\cdot 10^{-4}t^{2} - 3.45\cdot 10^{-3}S, \\
&\sigma = 6.995\cdot 10^{8}S(1 + 0.034t).
\end{aligned} \tag{14.2}$$

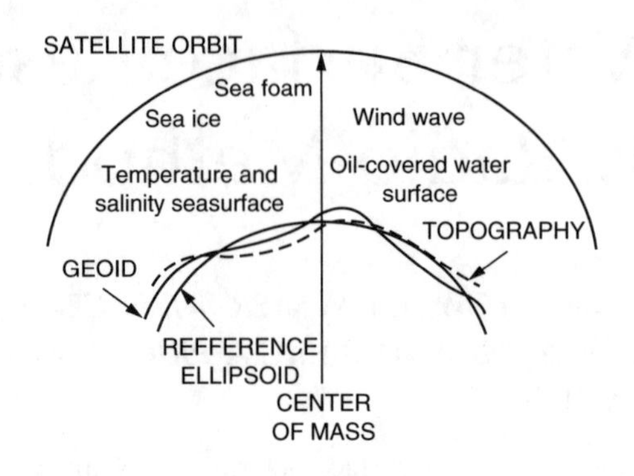

FIGURE 14.1 Main characteristics of a water surface as determined by means of remote sensing.

Here, the temperature ($t = T - 273.16$) is expressed on the Celsius scale, and salinity S is expressed per milligram (grams per liter). The formula for salinity is obtained by interpolating the data given in Rosenberg.[96] More detailed formulae considering the normality of the solution and its mixture can be found in Stogryn.[98] It is assumed that the average value of salinity in the oceans is 35‰. The Red Sea has the highest salinity: 42‰. The polar seas and relatively closed basins (e.g., Baltic, Black Sea) have salinities ranging from 2 to 18‰. The salinity of fresh basins is usually 1 to 2‰. For the estimations carried out in this chapter, we will assume that t = 20°C and S = 35‰ for seawater, and t = 20°C and S = 1‰ for freshwater; thus, $\varepsilon_s = 70$, $\lambda = 1.7$ cm, and $\sigma = 4 \cdot 10^{10}$ CGSE for seawater, and $\varepsilon_s = 80$, $\lambda = 1.8$ cm, and $\sigma = 1.2 \cdot 10^9$ CGSE for freshwater.

We now introduce the small parameter:

$$\mu = \frac{2\sigma\lambda_r}{c\left(\varepsilon_s - \varepsilon_o\right)}, \tag{14.3}$$

where $\mu \cong 0.07$ for seawater and 0.002 for freshwater. The real part of the water permittivity is the monotonous function of the radiofrequency (or freespace wavelength), which changes from static permittivity ε_s to optical permittivity ε_o. The imaginary part of the water permittivity has a maximum at the wavelength:

$$\lambda_{max} \cong \lambda_r \tag{14.4}$$

and a minimum at the wavelength:

$$\lambda_{min} \cong \frac{\lambda_r}{\sqrt{\mu}}. \tag{14.5}$$

Thus,

$$\varepsilon''_{max} \cong \frac{\varepsilon_s - \varepsilon_o}{2}, \quad \varepsilon''_{min} \cong 2\left(\varepsilon_s - \varepsilon_o\right)\sqrt{\mu} \,. \tag{14.6}$$

So, we have $\lambda_{max} \cong 1.7$ cm, $\varepsilon''_{max} \cong 32$, $\lambda_{min} \cong 6.4$ cm, and $\varepsilon''_{min} \cong 34$ for seawater; $\lambda_{max} \cong 1.8$ cm, $\varepsilon''_{max} \cong 37$, $\lambda_{min} \cong 40$ cm, and $\varepsilon''_{min} \cong 6.5$ for freshwater. It is easy to find that the skin depth (see Equations (2.8) and (2.13)) is about 0.5 cm for seawater at the minimum wavelength. The corresponding value for freshwater is about 18 cm. These examples show that microwave penetration into water is very weak, which means that radiowave interaction with water basically takes place in the surface layer and only its characteristics are determined in the process of remote sensing.

The expression for many water parameters can be simplified as $|\varepsilon_w| >> 1$; in particular, the expressions for Fresnel coefficients have the form:

$$F_h \cong \frac{\cos\theta_i - \sqrt{\varepsilon}}{\cos\theta_i + \sqrt{\varepsilon}}, \quad F_v \cong \frac{\sqrt{\varepsilon}\cos\theta_i - 1}{\sqrt{\varepsilon}\cos\theta_i + 1} \,. \tag{14.7}$$

We have already used a similar approach to analyze the thermal radiation of bodies bounded by rough surfaces (see Equation (8.60)). The assumed property of water permittivity allows us to consider the radiowave scattering to be scattering by an ideally conducting surface. So, Equation (6.50) becomes useful for estimation of the scattering by small-scale waves. With regard to large-scale waves, Equation (6.79) is reduced to:

$$\sigma^0 = \frac{\left(1 + \cos\nu\right)^2}{\left(\cos\theta_i + \cos\theta_s\right)^4} P_1\left(\nabla\zeta^\circ\right), \tag{14.8}$$

and in Equation (6.80) unity can be substituted for the reflection coefficient.

The dielectric permittivity of ice formed in the process of freshwater freezing is described by formulae of the Debye type, where $\varepsilon_s = 74.6$, $\varepsilon_o = 3.2$, $\sigma = 10^{-1}$ to 10^{-2} CGSE, and[99,100]

$$\lambda_s = \lambda_s^0 e^{-\beta t}, \quad \lambda_s^0 = 3.48 \cdot 10^6 \text{cm}, \quad \beta = 0.101\,(^\circ\text{C})^{-1} \,. \tag{14.9}$$

This formula allows us to obtain the correct permittivity values of the real and imaginary parts for frequencies less than 100 MHz. For higher frequencies, the calculations lead to very understated results for the imaginary part of the dielectric constant. The experimental data show that the real part of permittivity does not depend on the frequency and the ice temperature at this spectral region; $\varepsilon' = 3.2$ is

TABLE 14.1
Temperature Dependence of the Imaginary Part of Pure Ice Permittivity

	Ice Temperature (°C)		
	0	**−10**	**−20**
ε''	$8.59 \cdot 10^{-3}$	$2.83 \cdot 10^{-3}$	$1.97 \cdot 10^{-3}$

regarded as a reliable value. The imaginary part also does not depend on frequency but is a function of the ice temperature; this dependence is reflected in Table 14.1.

The characteristics of the sea ice dielectric permittivity have not been studied definitively. Usually, the sea ice is considered to be a mixture of a dry ice and brine. The cells of the brine are formed during the process of the seawater freezing. The salted solution gradually flows out, although the brine pockets migrate below the ice thickness due to the permanent presence of a temperature gradient. This process reduces the salt concentration in the brine cells as the ice ages. Radiowave volumetric scattering by pockets of brine can be described in the first approximation by representation of the effective dielectric permittivity. The change of this permittivity as the ice ages influences its reflectivity and can be detected by radar.

The brine volumetric content (ν_b) is defined by the relation:

$$\nu_b = \frac{S_i \rho_i}{S_b \rho_b}, \tag{14.10}$$

where S_i and S_b are the salinity of the sea ice and the brine, respectively, and ρ_i and ρ_b are their densities. The brine volumetric content depends on temperature and is expressed by the simple formula:

$$\nu_b = 10^{-3} S_i \left(0.532 - \frac{49.185}{T} \right) \tag{14.11}$$

within a temperature interval of –0.5 to 22.9°C. The Maxwell–Garnet formula (Equation (5.157)) approximately calculates the permittivity of sea ice; however, in our case, it is necessary to choose a permittivity relative to the dielectric constant of pure ice, which reduces the Maxwell–Garnet formula to one developed by Odelevsky:[101]

$$\frac{\varepsilon_e - \varepsilon_i}{\varepsilon_e + 2\varepsilon_i} = \frac{\varepsilon_b - \varepsilon_i}{\varepsilon_b + 2\varepsilon_i} \nu_b \, . \tag{14.12}$$

Separating the real and imaginary parts of the permittivity, we obtain:

$$\varepsilon_e' \cong \varepsilon_i \frac{1-\nu_b}{1+2\nu_b}, \quad \varepsilon_e'' \cong \frac{9\varepsilon_i^2 \varepsilon_b''}{\left(\varepsilon_b'\right)^2 + \left(\varepsilon_b''\right)^2} \frac{\nu_b}{\left(1+2\nu_b\right)^2}, \tag{14.13}$$

where we have assumed that the imaginary part of the pure ice permittivity is negligible. The following simplifications are based on the assumptions that $\nu_b < 1$ and $\varepsilon_b' >> \varepsilon_i$. Both parts of the brine permittivity can be computed by means of the Debye formula, Equation (14.1).

As mentioned earlier, the brine cells are replaced by air bubbles due to the flow of the salted solution. In this case, the bubble permittivity is equal to unity, and revised calculations give us the following for the effective dielectric constant:

$$\varepsilon_e \cong \varepsilon_i \left(1 - 3\rho \frac{\varepsilon_i - 1}{\varepsilon_i + 2}\right). \tag{14.14}$$

A typical property of the sea surface is its rough character. The roughness is caused by sea surface waves generated by forces due to interactions of the sea with the atmospheric boundary layers. The source of energy for sea waves is the wind, so many parameters for sea waves are functions of the wind speed. The surface waves of low amplitude can be described within a framework of linear theory,[102,103] which defines the major roles of gravity forces and surface tension. These waves are frequently referred to as *gravity* and *capillary waves*. The dispersion equation that establishes the relationship between the frequency and wave number for these waves has the form:[103,104]

$$\omega(q) = \sqrt{\left(gq + \alpha q^3\right) \tan(qH)}\,. \tag{14.15}$$

Here, g is the gravitational acceleration, $\alpha = 74.45$ cm³/sec² is the coefficient of the surface tension on the water-air boundary, and H is the depth of the bottom. In the future, we will primarily consider deep water and will assume that $qH >> 1$. At $q << (g/\alpha)^{1/2} \cong 3.63$ cm^{-1}, sea waves are referred to as *gravity waves*; in the opposite case, when $q >> 3.63$ cm^{-1}, they are referred to as *capillary waves*. The specific length $\Lambda_s = 2\pi\sqrt{\alpha/g} \cong 1.7$ cm divides both types of waves.

Sea-wave damping in the absence of wind depends strongly on the wave number. The formula given in Reference 102 (Equation (2.30)) allows us to estimate the damping time for gravitational waves by the formula:

$$t_d = \frac{3.08 \cdot 10^2}{q^{5/4}\left(1 + 6.16 q^{3/4}\right)}. \tag{14.16}$$

The wave number here has to be expressed per centimeter and the time in seconds. This formula is valid for capillary waves because the second term in the denominator of Equation (14.16) is much closer to unity for short waves. It is easy to see that a wave of length 1 cm ($q = 6.28$ cm^{-1}) damps at a time of 1.3 sec, while the time required is about 30 days for a wave of length 100 m ($q = 6.28 \cdot 10^{-4}$ cm^{-1}). This means that long sea waves are also excited very slowly and is the reason why it is necessary to discriminate between a developing and steady-state (fully developed) rough sea.

In reality, nonlinear phenomena take place. They are caused by the nonlinearity of the hydrodynamic equation and atmospheric turbulence. As a result, a statistical description of sea waves is more suited to the phenomenon. Experimental data show that the probability distribution of water surface elevations is close to a Gaussian one.[103] The distribution function of the slopes can also be assumed to be Gaussian in the first approximation, with some parameters differing along and transverse to the wind direction.

The spatial spectrum of sea waves is a very complicated function.[149] There are only the first steps in the theoretical description of this spectrum in all region of frequencies; therefore, semi-empirical formulae obtained for various spectral bands are usually applied.[104] We will not describe all parts of this spectrum within all intervals of wave numbers but will discuss only some of them. The Pierson–Moskowitz formula is often used to describe the frequency spectrum for gravity sea waves. This formula is a generalization of the experimental data.

$$\frac{S(\omega)g^3}{\hat{U}^5} = a\left(\frac{\hat{U}\omega}{g}\right)^{-5} \exp\left[-0.74\left(\frac{\hat{U}\omega}{g}\right)^{-4}\right], \quad a = 4.05 \cdot 10^{-3}. \tag{14.17}$$

Here, $\hat{U}$ is the wind velocity at a height 19.5 m above sea level. Usually, determination of the wind velocity at height z is based on the logarithmic dependence

$$U(z) = \frac{u_*}{\kappa}\ln\left(\frac{gz}{0.035u_*^2}\right) = \frac{u_*}{\kappa}\ln\left(\frac{2.8 \cdot 10^2 z}{u_*^2}\right), \tag{14.18}$$

where u_* is the velocity of friction, and $\kappa = 0.4$ is Karman's constant. The height is expressed in meters and the velocity in meters per second. The formula for $\hat{U}$ can be represented as:

$$\hat{U} = \frac{u_*}{0.2}\ln\left(\frac{73.9}{u_*}\right). \tag{14.19}$$

The relation between u_* and $\hat{U}$ is nearly linear with coefficients of proportionality ranging from 0.03 to 0.05, depending on the interval of values. The maximum spectral density is reached at a frequency of $\omega_m = 0.88\,g/\hat{U}$ when the sea-wave phase velocity is approximately equal to the wind speed. We can see that this frequency has the order of part of a Hertz. The corresponding wavelength in meters is $\Lambda_m = 2\pi\hat{U}^2/0.88g = 0.73\hat{U}^2$, where the velocity is expressed in meters per second. The spectrum is rather narrow: $\Delta\omega/\omega_m \cong 0.6$. The integral over the frequency gives us the value of the mean squared height of the waves:

$$\langle\zeta^2\rangle = \int_0^\infty S(\omega)\,d\omega = \frac{a}{4\cdot 0.74}\frac{\hat{U}^4}{g^2} = 1.36\cdot 10^{-3}\frac{\hat{U}^4}{g^2}. \tag{14.20}$$

Equation (14.20) allows us to estimate the wave height with regard to $\sqrt{\langle\zeta^2\rangle} = 0.04\hat{U}^2/g$.

Being able to define $\langle\zeta^2\rangle$ by integration of the spatial spectrum $\tilde{K}_\zeta(q)$ (see Chapter 6) over the wave numbers allows us to obtain an expression for this spectrum using the dispersion equation for gravitational waves. In the first approximation, we assume statistical isotropy of the roughness. The corresponding relation can be written as:

$$\tilde{K}_\zeta(q) = \frac{\sqrt{g}\,S\left(\sqrt{gq}\right)}{4\pi q^{3/2}}. \tag{14.21}$$

Thus,

$$\tilde{K}_\zeta(q) = \frac{a}{4\pi}q^{-4}\exp\left(-0.74\frac{g^2}{\hat{U}^4 q^2}\right). \tag{14.22}$$

Equation (14.22) is not valid for wave numbers greater than q_m, which is understandable because large wave numbers correspond to gravity and capillary waves, which have other spatial spectra.

Equation (14.22) allows us to estimate the mean squared slopes by assuming statistical isotropy. In this case,

$$\left\langle(\nabla\zeta)^2\right\rangle = 2\pi\int_0^{q_m}\tilde{K}_\zeta(q)q^3\,dq = -\frac{a}{4}E_i\left(-t_m\right),\quad t_m = \frac{0.74g^2}{\hat{U}^4 q_m^2}, \tag{14.23}$$

where $E_i(x)$ is the exponential integral.[44] The maximal wave number q_m is a function of the wind speed and is estimated by the relation:[90]

$$q_{\mathrm{m}}(u_*) = 0.359\left(\frac{u'_*}{u_*}\right)^2, \quad u'_* = 12\ \mathrm{cm/sec}. \tag{14.24}$$

It is apparent that, after some algebra, that the following value is much less than unity:

$$t_{\mathrm{m}}(u_*) = \frac{0.427}{\ln^4(73.9/u_*)}. \tag{14.25}$$

The asymptotic approximation of the exponential integral is convenient in the given case, and we have:

$$\left\langle (\nabla\zeta)^2 \right\rangle = -\frac{a}{4}\ln\left[1.78 t_{\mathrm{m}}(u_*)\right]. \tag{14.26}$$

The numerical calculations show that, in this approach, the slopes are small and practically independent of the wind velocity. This occurs because low-frequency components of the spectrum are suppressed during the process of slope formation. Note that the spectrum is multiplied by the factor q^2 during the slope calculation; therefore, it is necessary to take into account the height-frequency part of the spectrum which plays a more effective role in this case.

According to Kitaigorodskii,[90] the spectrum form within the wave number interval $q_{\mathrm{m}} < q < q_2 = 0.359\ \mathrm{cm}^{-1}$ has the form:

$$\tilde{K}_\zeta(q) = \frac{a}{2\pi\sqrt{q_{\mathrm{m}} q^7}}. \tag{14.27}$$

Adding the result of integration over this interval reduces Equation (14.20) to:

$$\begin{aligned}\left\langle (\nabla\zeta)^2 \right\rangle &= -\frac{a}{4}\ln\left[5.31\cdot 10^3 t_{\mathrm{m}}(u_*)\right] + 2a\frac{u_*}{u'_*} = \\ &= -1.01\cdot 10^{-3}\ln\left[5.31\cdot 10^3 t_{\mathrm{m}}(u_*)\right] + 6.75\cdot 10^{-2} u_*,\end{aligned} \tag{14.28}$$

where the friction speed (u_*) is expressed in meters per second. The obtained result is consistent with the approximation of experimental data by the linear regression $\langle(\nabla\zeta)^2\rangle = \beta_1 + \beta_2\hat{U}$.

We have discussed the sea-wave spatial spectrum for fully developed roughness; therefore, this spectrum depends only on the dimensionless parameter $\tilde{\omega} = \omega u_*/g$ (compare with the Pierson–Moskowitz formula, Equation (14.17), where the wind velocity is substituted for friction). In the general case of nonstationary and spatially inhomogeneous wind roughness, the frequency spectrum is a function of three dimensionless parameters: $\tilde{\omega}, \tilde{T} = gt/u_*$ and $\tilde{X} = gX/u_*^2$, where t is time, and X is the distance from the lee shore, commonly referred to as the *fetch*. One example is the spectrum obtained by the Joint North Sea Wane Project (JONSWAP) regarding distance X. We will not describe the properties of this spectrum, but instead refer readers to a more detailed discussion found in Hasselman et al.[105] It is assumed that the Pierson–Moskowitz formula applies when $\tilde{X} > 10^4$.

Isotropy of the sea roughness spectrum is, certainly, an idealization of the phenomenon. In reality, this spectrum is anisotropic and is a function not only of the wave number but also the angle ϕ relative to the wind speed direction. So, the general view of this spectrum is usually represented as:

$$S(\mathbf{q}) = \tilde{K}_\zeta(q)\, \Phi(q,\phi), \quad \int_{-\pi}^{\pi} \Phi(q,\phi)\, d\phi = 1 . \tag{14.29}$$

A conventional view of the angular function is not available; therefore, we will not discuss various particular approximations. Instead, we will discuss differences in approximations of the variances in sea-wave slopes in the upwind and crosswind directions developed by Cox and Munk[106] as a result of experimental data processing. Let the x-axis be aligned with the wind direction. Then,

$$\begin{aligned} \langle \gamma_x^2 \rangle &= \left\langle \left(\frac{\partial \zeta}{\partial x} \right)^2 \right\rangle = 3.16 \cdot 10^{-3} U_{12.5}, \\ \langle \gamma_y^2 \rangle &= \left\langle \left(\frac{\partial \zeta}{\partial y} \right)^2 \right\rangle = 0.003 + 1.92 \cdot 10^{-3} U_{12.5}, \end{aligned} \tag{14.30}$$

where wind speed $U_{12.5}$ is expressed in meters per second and is measured at an altitude of 12.5 m above sea level.

The ripple spectrum that accompanies the capillary wave spectrum has the form:

$$\begin{aligned} \tilde{K}_\zeta(q) &= \frac{0.875(2\pi)^{p-2}}{q} \frac{g^{(1-p)/2}(1+3q^2/q_s^2)}{[q(1+q^2/q_s^2)]^{(1+p)/2}}, \\ p &= \log_{10}\left(\frac{10^3}{u_*} \right), \qquad q_s = \sqrt{\frac{g}{\alpha}} = 3.63\,\text{cm}^{-1} \end{aligned} \tag{14.31}$$

in the wave number interval $0.942\ \text{cm}^{-1} < q < q_s$, as recommended by Ulaby et al;[90] the friction speed is expressed here in meters per second. The form of the last spectrum can be radically simplified and represented as an approximation:

$$\tilde{K}(q) = \mu(u_*)\left(\frac{q_s}{q}\right)^{16/5}, \quad \mu(u_*) = 1.173 \cdot 10^{-4} u_*^{1.128}. \tag{14.32}$$

Continuation of this part of the spatial spectrum belongs to capillary waves, although we have not defined this spectrum; for example, Ulaby et al.[90] discuss the spectrum in the form:

$$\tilde{K}_\zeta(q) = \frac{1.473 \cdot 10^{-4}}{2\pi q} u_*^3 q_s^6 / q^9, \tag{14.33}$$

where u_* is expressed in centimeters per second. Remember that capillary wave generation takes place on the relief of long gravity waves. The turbulent structure of the wind near the water surface is partly due to the interaction between the wind and this type of sea wave and leads to the formation of an inhomogeneous field of sea strength followed by inhomogeneous capillary waves. As a rule, the maximum amplitude of capillary waves is found on gravity wave crests.

The light excitability of capillary waves leads to their burnishing and to the appearance of slicks. These slicks are usually observed with rather weak winds of velocities less than 5 m/sec. The slicks are formed by different factors, among which are the already-mentioned wind field inhomogeneities, currents, inner waves, and films of surface-active substances and oils.[103]

To this picture of ocean roughness must be added some details related to nonlinear phenomena typical for sea waves of large amplitude. Wave breaking is a function of some value of the wind speed. A whirlwind generated at the wave crest entrains air into the water. This dynamic mixture of water with air rises up and becomes apparent in the form of whitecaps, which are followed by foaming, splashing clouds, and other dispersed water-air mixtures. We will not go into detail regarding the formation of these structures, but reviews of this problem can be found, for example, in Raizer and Cherny.[107] Let us note here, though, that the foaming appears under conditions when the wind speed exceeds 4 m/sec, and fraction w of the foam covering depends on the wind speed and water temperature. An approximate formula can be written as:[108]

$$w = a\hat{U}^b \exp\left[0.0861 \cdot (T_a - T_w)\right], \quad a \cong 2 \cdot 10^{-5}, \quad b \cong 2.2. \tag{14.34}$$

Here, T_a and T_w are the temperature differences at the air-water boundary. The foam thickness varies within this interval from several millimeters to several centimeters. The density is estimated by the value 0.01 g/cm^3. Using the Maxwell–Garnet formula (Equation (5.157)), we can calculate the foam permittivity and show that $\varepsilon_f \cong 1.03$.

With regard to the sea surface temperature (SST), the processes of heat exchange at the water-air boundary lead to formation of a water layer with a large temperature gradient up to 10 K/cm. The thickness of this layer is of the order of 0.1 cm,[109] and this layer is referred to as the *cold ocean film*. The temperature drop in the cold ocean film can reach a value of 1.5 to 2 K. The cold ocean film exists at any period of the day, is very stable, and is restored 10 to 12 sec after a storm.[109]

14.2 RADAR RESEARCH OF THE WATER SURFACE STATE

Radar research of the water surface state is based on radiowave scattering processes due to water roughness. We will distinguish between scattering of radiowaves caused by gravity sea waves and ripples. In the first case, the sea surface inhomogeneities can be considered as large compared to radiowaves, and Equation (6.81) is applicable for estimating computations. For ripples, which can be considered as small inhomogeneities, Equation (6.50) provides the basis for the necessary calculations.

Let us choose a radio wavelength of $\lambda = 9$ cm. The resonance scattering is defined by the gravity and capillary waves for which the spectra are described by Equation (14.30). Let us assume that wind speed $\hat{U}$ is equal to 10 m/sec, which corresponds to a friction velocity value near 0.4 m/sec. Equation (14.28) gives us the opportunity to estimate the slope $\sqrt{\langle(\nabla\zeta)^2\rangle} = 0.161$.

Figure 14.2 shows the result of all these calculations. Here, $\sigma(\theta)$ is the specular backscattering coefficient vs. incident angle θ, and $\sigma_h(\theta)$ and $\sigma_v(\theta)$ are the backscattering coefficients of the gravity and capillary waves for the horizontally and vertically polarized radio waves. One can see that at incident angles greater than 40° the diffuse scattering by ripples begins to exceed the specular scattering by large gravity waves; we discussed this phenomenon in Section 6.5. We have to keep in mind that, in reality, the picture changes for large incident angles where the shadowing effects and breaking of sea waves play essential roles. Thus, the backscattering coefficient must be determined other than simply by the Bragg resonance; however, the Bragg scattering contribution in the general mechanism of backward scattering is significant at intermediate values of the incident angle, and this fact must be taken into account for radar systems that are used to monitor water areas.

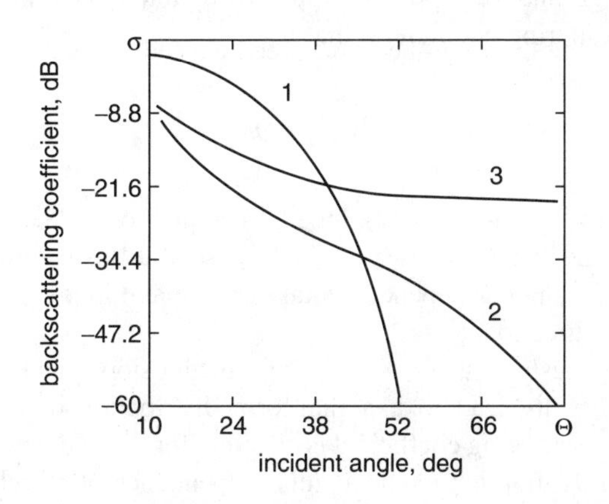

FIGURE 14.2 Backscattering coefficient vs. incident angle: (1) $\sigma(\theta)$; (2) $\sigma_h(\theta)$; (3) $\sigma_v(\theta)$.

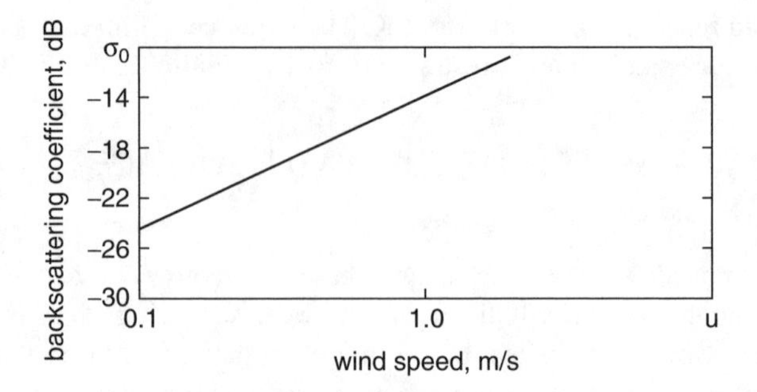

FIGURE 14.3 Total backscattering coefficient at a 45° incident angle vs. the friction velocity value.

Certainly, the intensity of the radio-wave backscattering depends on the wind speed. Calculation for X-band radio waves is based on the assumption that well-developed roughness takes place as shown in Figure 14.3. Here, the total backscattering coefficient at incident angle $\theta_i = 45°$ is shown vs. the friction velocity value. The total coefficient is the sum of the specular reflection described by Equation (6.81) and diffuse scattering by ripples described by Equation (6.50) for vertically polarized radiowaves. The specific scattering cross section of small roughness variations due to large gravity waves was neglected for simplicity. In other words, it was supposed that, in the first approximation, the intensity of scattering by ripples does not depend on slopes of sea waves. In this case, Equation (6.93) is reduced to the primitive sum of backscattering reflectivity for specular and diffuse components. A spatial spectrum of the form shown in Equation (14.30) was used for the computation.

The result of computing generally agrees with the experimental data. In particular, the wind linear dependence becomes apparent on the log-linear plot, which is typical for experimental data.[90] Naturally, it is difficult to talk about full numerical agreement due to the simplicity of the roughness model. The log-linear dependence provides the background for exponential approximation of the relation between wind speed and backscattering coefficient; that is,

$$\sigma^0\left(u_*\right) = D u_*^{\beta} \;. \tag{14.35}$$

In accordance with Equation (14.32), index β is equal to 1.128; however, this value is independent of the incident angle only in the case of diffuse scattering domination. In the inverse case, angular dependence takes place and index β becomes negative close to the nadir direction.

As noted previously, statistical anisotropy of the slopes and near-surface wind strength inhomogeneity lead to a nonuniform distribution of ripple intensity. In particular, the backscattering coefficient is different for up- and downwind directions. This supports our assumption of the angular dependence of σ^0 relative to the wind direction. One such dependence, based on experimental data, has the form of the first terms of the Fourier-series expansion:

$$\sigma^0 = A(u_*) + B(u_*)\cos\phi + C(u_*)\cos 2\phi\ . \tag{14.36}$$

A comparison of Equations (14.35) and (14.36) shows that coefficients D and β have to be functions of the angle ϕ.

The small inertia of ripples relative to the wind action allows, as we mentioned in Chapter 11, the use of radar scatterometers to measure wind speed when the ocean roughness is not fully developed. This technique has been successfully applied in various space programs. The gigahertz frequencies are used for this purpose in order to realize the effectiveness of resonant scattering. The multibeam antenna system allows measurement of specific cross sections under different directions of the wind speed vector. For example, the first spaceborne radar scatterometer of the SEASAT mission (S-193) had four beams. A three-beam system has been used for the scatterometers of the ERS-1,2 missions. The data processing procedure is rather complicated in this case but does allow us to achieve an accuracy of wind direction definition of ±20°.

Altimeters are becoming typical spaceborne radar instruments. They find application for geoid shape refinement, detection of gravitational anomalies and water level variations connected with sea currents, determination of the seawave height, measuring wind speed, etc. The short pulse technique measures height with an accuracy of about ±10 cm.

Let us consider in more detail the ALT altimeter of the TOPEX/POSEIDON mission. Equation (6.149) will be the object of our analysis. As we explained earlier, this is a two-frequency altimeter of C-bands (5.3 GHz) and Ku-bands (13.6 GHz). The frequency bandwidth of the pulses is 330 MHz, which corresponds to a pulse duration of $\tau_0 \cong 3 \cdot 10^{-9}$ sec = 3 nsec. The corresponding angular beamwidths are $\theta_0 = 2.7° \cong 4.7 \cdot 10^{-2}$ rad for the C-band, and $\theta_0 = 1.1° \cong 1.9 \cdot 10^{-2}$ rad for the Ku-band. Orbit height $L_0 \cong 1300$ km, and let us suppose that wind speed $\hat{U} = 10$ m/sec. Then, $\sqrt{\langle\zeta^2\rangle} \cong 0.37$ m and $\sqrt{\langle(\nabla\zeta)^2\rangle} \cong 0.16$. After some simple algebra we find that $\vartheta \cong \theta_0$ (Equation (6.150)) in this case, $\tilde{\tau}_0 \cong 5.8 \cdot 10^{-9}$ sec (Equation (6.113)) at $\theta_i = 0$, and for the Ku-band.

Figure 14.4 shows the averaged form of backscattered pulse as a function of the relative delay time $x = \sqrt{2}\tau/\tilde{\tau}_0$. One can see the rise-time region before the pulse maximum and lengthy trailing edge region after the maximum. The time delay of the pulse maximum relative to zero time is significant. Equation (6.153) gives the estimation $\tau_m \cong 9.6$ nsec for this time for the Ku-band. The determination of this time allows the value $\tilde{\tau}_0$ to be computed and then the sea-wave intensity $\langle\zeta^2\rangle$. For fully developed choppiness, Equation (14.20) allows us to estimate the wind speed value. Another way is based on the definition of the scattered pulse energy and calculation of the backscattering coefficient σ^0 with the help of Equation (6.158). The assumption of a normal law for seawave distribution permits us to determine the slope value by using Equation (6.81) at $\theta_i = 0$; however, the described technology of wind speed measurement breaks down in the case of the swell when, despite a lack of wind, large-scale waves occur due to their weak damping.

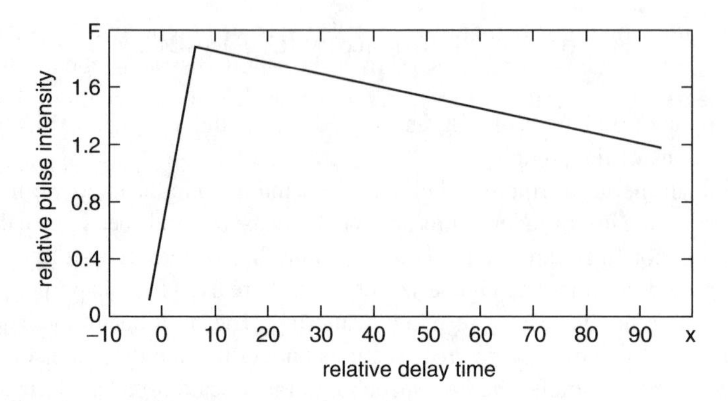

FIGURE 14.4 Backscattered pulse.

The time of the scattered pulse maximum depends on the frequency only via beam width θ_0 as it follows from Equation (6.153). So, by measuring this time for C- and Ku-band pulses, we can determine the value:

$$\tilde{\tau}_0 = 2\sqrt{2}\,\frac{\tau_{mc} - \tau_{mKu}}{\ln\left(\theta_{0c}/\theta_{0Ku}\right)}\,. \tag{14.37}$$

Knowledge of this time constant opens the way for determining seawave heights using Equation (6.113). A similar operation can be carried out by comparing the decreases in velocity of pulses at the trailing edges for both bands.

These examples show that analysis of the scattering pulses gives us information about the sea roughness. Certainly, our formulae have an approximate character that can be refined by the experimental data analysis. In general, this analysis is comparatively complicated, especially in cases when we need to achieve a high level of accuracy; an example would be the precise measurement of the distance between a spacecraft and the mean level of the sea surface.

We can see that several sources cause altitude bias for altimetric measurements over the ocean. One of them is the discussed surface roughness effect. The correction of this effect is provided by careful analysis of the scattered pulse shape. Another source is antenna beam deflection from the nadir (see, for example, Brown[110]) which can be corrected by precise control of the antenna beam position and to some extent by the processing of the received signal. In particular, the trailing edge is more expanded in the case of angular deviation of the antenna beam.[111] The third source of the bias is ionospheric delay, which changes from point to point and with time. The two-frequency system allows this effect to be corrected by detecting the arrival time differences of similar pulses radiated at different carrier frequencies. In the case of the ALT system, this difference, in accordance with Equation (13.18) is:

$$\Delta\tau = 2\Delta t_g = 8.1\cdot 10^{-21} N_t\,.$$

Letting the total electron content $N_t = 3 \cdot 10^{13}$ cm^{-2}, we obtain $\Delta\tau = 2.4 \cdot 10^{-9}$. Such a small time difference can be detected by a short-pulsed system, such as the ALT.

The next source of bias is found in the troposphere (see Equation (4.57)). We know from the data given in Chapter 4 that this additional two-way is of the order of 5 m. To correct such a large value, we must know the integral value of the tropospheric index of refraction. The "dry" part of the refractive index can be calculated by the tropospheric model. The "moist" part cannot be computed in advance due to changeability of the water vapor content. This part of the refractive index, on the other hand, makes a significant contribution to the value of radiowave velocity. By using Equations (12.4), (12.5), and (12.6), it is easy to establish that the corresponding contribution into two-way for the standard atmosphere is in the neighborhood of 20 cm. Naturally, this effect at real tropospheric situations can be significantly greater and must be corrected. This correction is achieved on the TOPEX/POSEIDON mission by microwave radiometry measurements at frequencies of 18.21 and 37 GHz.

Synthetic-aperture radar (SAR) technology is widely applied for oceanic research. Both airborne and spaceborne systems are used. The angle of incidence is chosen for these systems such that diffuse scattering caused by gravity and capillary waves dominates. The intensity of scattering depends here, as we know, on the local angle of incidence, which depends, in turn, on the large-scale gravity wave slopes. So, modulation of the diffuse scattering intensity occurs due to the large energy-carrying waves. If the SAR resolution is less than the spatial period of these waves, then the SAR image indicates the presence of these waves and can be the basis for their structural analyses. A peculiarity of oceanic wave images obtained by SAR is the motion of the scatterers due to choppiness. Because the process of SAR image formation is based on frequency analysis, image distortion has to be caused by the Doppler effect of moving targets. We will not enter into the details of these effects here, but some discussion of this topic can be found in Ulaby et al.[90]

The height spatial resolution of SAR allows us to identify slicks and various surface films, such as polluting oil slicks. An interesting effect is the revelation of underwater features at a depth greater than the skin-layer thickness. Bottom inhomogeneities can cause disturbances in currents that lead to changes in the ripple amplitude which are detected as brightness variations in SAR images.

The radar technique is widely applied for the investigation of ice properties. In the case of freshwater ice, our primary interest lies in measurement of the ice thickness. Determining the thickness is based (as noted earlier in Section 11.6) on measurement of the delay time between pulses reflected from the upper and bottom boundaries of ice (see Equation (11.52)). The data in Table 14.1 allow us to estimate the coefficient of attenuation of microwaves in this ice; $\gamma \cong 5 \cdot 10^{-3}/\lambda$ for a 10°C temperature, or 0.43 dB/m for f = 3 GHz. Obviously, radar systems operating at gigahertz frequencies are effective for determining ice thickness of the order of tens of meters. This benefit was discovered as a result of analyses of unusual indications of airborne altimeters while the aircraft were parked at an ice airfield.

Strong radio-wave absorption does not permit the use of radar of microwave frequencies for determination of sea ice thickness. For example, the experimental data represented in Odelevsky[101] show that the coefficient of attenuation for radio

TABLE 14.2
Selection of Conversion Factor *K*

Characteristics of the Temperature and Salinity of First-Year Ice	Observed Value of Ratio r	K (cm/nsec) Thin Ice (τ< 15 nsec)	K (cm/nsec) Media and Thick Ice (τ> 15 nsec)
	Cold		
Salted	0.3–0.8	6.5	7.2
Desalted	0.8–1.2	7.0	7.5
Fresh	>1	7.5	7.5
	Warm		
Salted	<0.8	6.5	≅7.0
Desalted	>0.8	7.2	7.3

waves of 7.9-cm wavelength varies within the interval of 4 to 30 dB/m, corresponding to a change in ice salinity from 2 to 8‰ at an ice temperature of 14°C; therefore, frequencies close to 100 MHz are more suitable for this purpose. This problem was discussed in Section 11.6, where we concluded that the requirement to use wideband pulses leads to application of signals without the carrier frequency, such as video pulses. Certainly, the reflected signal experiences some distortion of its shape. This distortion depends on the ice characteristics (salinity, age, temperature). Some relation does exist between amplitude ratio r of the signals reflected from the bottom and upper ice edges with the temperature and salinity.[112] For freshwater or strongly desalinated sea ice, $r > 1$; for ice with salinity S 2 to 4‰, r 1, and at salinity $S > 4‰$, $r < 1$.

Thickness value l is associated with the delay time by the relation:

$$l\,(\text{cm}) = K\tau\,(\text{nsec})\ . \tag{14.38}$$

Conversion factor K is defined by the ice permittivity, which is unknown in advance; therefore, Table 14.2 is useful for determining this factor on the basis of experimental data.

The shape of a received signal gives us some information about the ice surface state. The fluctuation of the signal amplitude and its duration is detected for multiyear ice. A flattening of the bottom signal occurs above a hummocked ice ridge. Dilatation at all periods and its confluence with the upper signal can occur above ice with confused hummocking.

The sea ice thickness is estimated by means of indirect data obtained by side-looking radar (SLR) and SAR. This technique is based on the fact that the scattering properties of sea ice depend on its age. The thickness of the sea ice also depends on its age, which follows from Table 14.3, which characterizes ice types according to the recommendations of the World Meteorological Organization (WMO).[114] We can conclude on the basis of the data in this table that some correlation exists between the ice thickness and its properties to scatter microwaves. This correlation has a

TABLE 14.3
Sea Ice Classification

Ice Type	Ice Thickness (m)	Comment
Ice free	0	Open water
Sea ice crust	0–0.10	Dark and light sea with ice crust
Gray ice	0.10–0.15	Pressed ice
Gray-white ice	0.15–0.30	Hummocking under pressure ice
Thin first-year ice	0.30–0.70	Developing from young ice
Medium/thick ice	> 0.70	Developing from thin first-year ice
Old multiyear ice	3.0 and more	Ice endured one melting season

rather clear physical explanation. As with many simplifications, the surface scattering dominates in the case of first-year (FY) ice. The presence of volumetric scattering is typical for multiyear (MY) ice, which leads to the difference in backscattering coefficient values for both types of ice. FY ice contains a distribution of salt in the form of brine and crystals. The FY ice salinity is especially high at the upper layers, reaching a density near 20‰. The approximate value of FY ice permittivity is $\varepsilon = 3.3 + i0.25$.[90] This type of ice intensively absorbs microwaves, which at the X-band penetrates into the ice to a depth of only several centimeters (see Equation (2.16)); therefore, volumetric scattering is weak in this case and surface scattering dominates.

The roughness of ice is estimated by the value $\sqrt{\langle\zeta^2\rangle} \sim 1$ cm. The correlation length is assumed to be 10 to 30 cm.[115] Other models give similar values.[90] The following correlation function suggests that we are dealing with a fractal surface of Brownian dimension (see Chapter 6):

$$\hat{K}(\rho) = \langle\zeta^2\rangle \exp\left(-\frac{\rho}{l}\right). \tag{14.39}$$

The corresponding spatial spectrum is:

$$\tilde{K}(q) = \frac{l^2\langle\zeta^2\rangle}{2\pi\left(1+q^2l^2\right)^{3/2}}. \tag{14.40}$$

Figure 14.5 shows the results of calculating the backscattering coefficients for both matched polarizations at the X-band (f = 10 GHz). The chosen ice parameters are $\sqrt{\zeta^2} = 2.5$ cm and $l = 30$ cm. The choice of parameters is based on the idea of similarity between the computed and some of the experimental data.[91] The roughness is assumed to be small for this calculation, which allows us to base our computations on Equations (6.48) and (6.49). This assumption is not very appropriate for the case

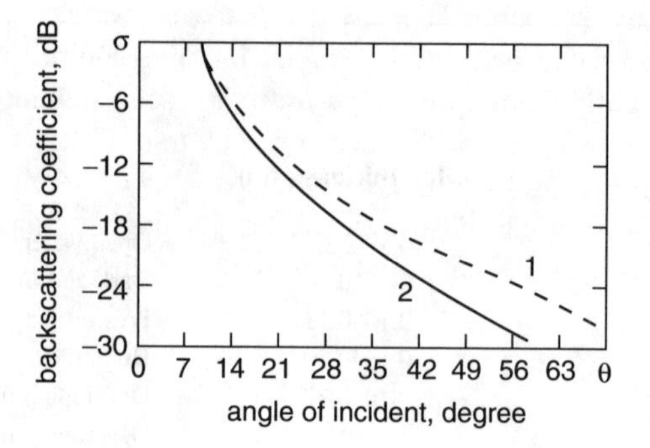

FIGURE 14.5 Backscattering coefficient vs. incident angle: (1) σ_v; (2) σ_h.

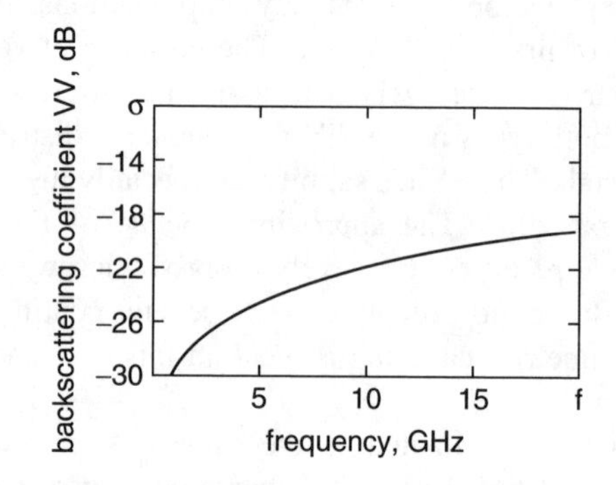

FIGURE 14.6 Backscattering coefficient vs. electromagnetic frequency.

of the chosen radiofrequency band; nevertheless, the results of the computation appear to be acceptable.

We can consider the product $(2kl)^2 >> 1$ and, therefore, $\sigma^0 \sim f$ when Equation (14.40) is used because they reflect the weak frequency dependence of the ice surface scattering; however, this conclusion is correct for frequencies above the S-band when the mentioned inequality is valid. At the L-band, the inverse inequality is more correct, and $\sigma^0 \sim f^4$. Figure 14.6, computed for the backscattering coefficient of vertical polarization under the conditions formulated above and at an incident angle of 40°, illustrates this dependency. In reality, the scattering conditions are varied, which leads to dispersion of the backscattering values within the interval 5 to 10 dB. This is easy to understand if we note that the backscattering coefficient increases by 10 dB with an increase in roughness height of approximately three times.

The salinity of MY ice is much lower than for FY ice. The ice permittivity can be represented as $\varepsilon = 3.0 + i0.03$.[90] It is easy to see that the penetration depth is several decimeters for the X-band and, naturally, the volumetric scattering is included in the process. The main scattering elements are air bubbles formed as a result of brine drainage. The scattering produces two effects. One of them is the change of the effective dielectric constant (see Equation (5.143)), and the second one is backscattering from the bubble density fluctuations, which is equivalent to the scattering by the effective permittivity spatial inhomogeneities. The effective permittivity differs slightly from pure ice permittivity (see Equation (14.14)), but this change can be neglected in many cases. For calculation of the volumetric backscattering coefficient, it is useful to remember the necessity of taking into consideration the radiowave extinction caused by the absorption and scattering; therefore, the volumetric backscattering coefficient is represented in the form:

$$\sigma_v^0(\theta_i) = \left[1 - \left|F(\theta_i)\right|^2\right]^2 \sigma_d\left(-\mathbf{e}_i'\right) \int_0^d N(z)\exp\left(-\frac{2\tau(z)}{\cos\theta_i'}\right) dz,$$

$$\tau(z) = 2\int_0^z \gamma\left(\dot{z}'\right) dz', \ \sin\theta_i' = \frac{\sin\theta_i}{\varepsilon_e}. \qquad (14.41)$$

It is assumed, for simplicity, that we are dealing with similar particles and the differential scattering cross section does not depend on the depth in the ice. Only bubble concentration N and extinction coefficient γ are supposed to be functions of depth z. The differential cross section of the bubbles is:

$$\sigma_d\left(-\mathbf{e}_i'\right) = k^4 a^6 \left|\frac{\varepsilon_i - 1}{2\varepsilon_i + 1}\right|^2 \qquad (14.42)$$

in the first approach, where a is the bubble radius.

If the ice thickness is large enough for the radio waves to be absorbed fully, then infinity can be substituted for thickness d at the upper limit of the integral. The scattering by the bottom ice interface has to be added in the opposite case. This scattering will be important when the bottom roughness is large compared to the wavelength. Equation (6.197) at a moderate incident angle can be available for this situation.

The processes described here are a very simplified picture of reality. First, the scattering properties of ice vary according to season; for example, in the summer, the melting processes and the appearance of pools of water change the scattering process, and the effects of snow must be added to the scattering picture. Second, the classification of ice used here is rather primitive, as it is necessary to take into account intermediate ice states, some of which are mentioned in Table 14.1; therefore, the real situation is often not as simple as described here. The experimental data statistics indicate the vital importance of these conditions, especially for ice patrols. A great volume of data has been accumulated by the Russian–Ukrainian spaceborne SLR of the *Okean* satellite family[114] and during the ICEWATCH[115] project with the use of European spaceborne SAR during the ERS-1 mission.

14.3 MICROWAVE RADIOMETRY TECHNOLOGY AND OCEANOGRAPHY

Microwave radiometry is being applied more frequently for research into oceanic phenomena. Several space missions have used microwave radiometers to carry out various functions. Many seawater parameters affect thermal radiation, including the sea surface temperature, salinity, water surface roughness, foam, and oil films. The radiation dependence of these parameters can be used to define them. On the other hand, some sources of radiation interfere with the process of measurement and reduce the accuracy of data used to define unknown parameters. The thermal radiation of the atmosphere, sun glitter, space relict, and galactic radiation are examples of such sources. We can also add industrial noise to our list of sources of interference.

We will consider the effects that complicate the general picture one at a time. First of all, we will discuss the rather idealized case of a smooth water surface, which in reality can occur during calm weather, and will disregard the influence of external factors. In this case, the emissivity is defined by the reflection coefficients (see Equation (8.18)) and, for both polarizations, the coefficients of emission are functions of the angle of incidence θ, radiofrequency f, water temperature t (°C), and salinity S; that is, $\kappa_h = \kappa_h(\theta, f, t, S)$ and $\kappa_v = \kappa_v(\theta, f, t, S)$. At zero incident angle (nadir observation), we will use the simpler designation $\kappa(f, t, S)$. Figure 14.7 shows the frequency dependence of emissivity for the nadir observation and at an incident angle of 45° for typical values of seawater parameters.

The difference between seawater and freshwater emissivity becomes apparent only at the L-band which is reflected by the sharp fall of the emissivity curves. The curves change slowly at others parts of the frequency region, which suggests a lack of preferred bands (except the L-band) for temperature measurement in the considered case of calm water. The main difficulty is now determining the temperature dependence of the emissivity itself; that is, the lack of temperature knowledge leads to vagueness in the emissivity value which puts an obstacle in the way of correct

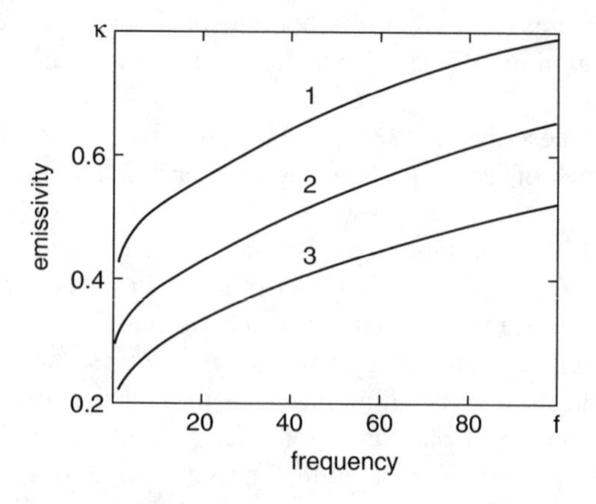

FIGURE 14.7 Frequency dependence of emissivity: (1) at a 45° incident angle (κ_v); (2) for the nadir observation; (3) at a 45° incident angle (κ_h).

determination of the SST on the basis of measuring the brightness temperature. The relative error of the SST determination is:

$$\frac{\delta T}{T} = \sqrt{\left(\frac{\delta T_o}{T_o}\right)^2 + \left(\frac{\delta \kappa}{\kappa}\right)^2}. \tag{14.43}$$

Errors in determining the brightness temperature represent a measurement data analysis problem, as discussed in Chapter 11.

We will now address errors in defining emissivity. The SST value is known due to preliminary measurements of some accuracy. Our task is to understand how uncertainty in our surface temperature knowledge influences uncertainty in the emissivity definition. The following derivatives can help answer this question:

$$\eta_{hT}(\theta, f, t, S) = \frac{d\kappa_h(\theta, f, t, S)}{dT}, \quad \eta_{vT}(\theta, f, t, S) = \frac{d\kappa_v(\theta, f, t, S)}{dT}, \tag{14.44}$$
$$\eta_T(f, t, S) = \eta_{hT}(0, f, t, S).$$

Their values are represented in Figure 14.8 for nadir observations in freshwater ($S = 1‰$) and seawater ($S = 35‰$). It is easy to see that the discussed derivatives are small and have an absolute value of the order of 0.002 K^{-2}. The corresponding error of SST determination has a relative order 0.004 for 1 degree of uncertainty regarding the preliminary temperature knowledge. That cannot be estimated as a small one, keeping in mind the requirement of the temperature accuracy measurement near 0.5 K. Therefore, it is necessary to search the frequency bands where the module of the analyzed derivative has minimal values. The curves of Figure 14.8 allow us to find those bands, one of which corresponds to a frequency near 2.7 GHz and the other to a frequency near 11.8 GHz. The values of the emissivity derivative tend to zero at both bands for seawater and only at the 11.8-GHz band for freshwater. We will restrict our discussion here to primarily the 2.7-GHz band, taking into account that a longer wavelength is actually preferable.

We must point out that the derivative zero value at the 2.7-GHz frequency corresponds to 20°C of the SST. The curve of Figure 14.9 demonstrates the temperature dependence of this derivative and permits us to regard its maximal absolute value as $2 \cdot 10^{-4}$ K^{-2}. In this case, the relative error of the SST measurement will be of the order of $4 \cdot 10^{-4}$ at 1 degree of uncertainty, which can be considered satisfactory. The next step of our discussion is the angular dependence of the emissivity, which is shown in Figure 14.10 for 2.7 GHz. Note the emissivity maximum for the vertically polarized waves at the quasi-Brewster angle. In this case, the value of this angle is near 83.5° and the emissivity derivative has the small value of –1.7 K^{-2}. Initially, the observation at this angle is very effective for the SST measurement by vertically polarized microwave radiometer; however, let us emphasize again that we are assuming a calm water surface. Most promising is making

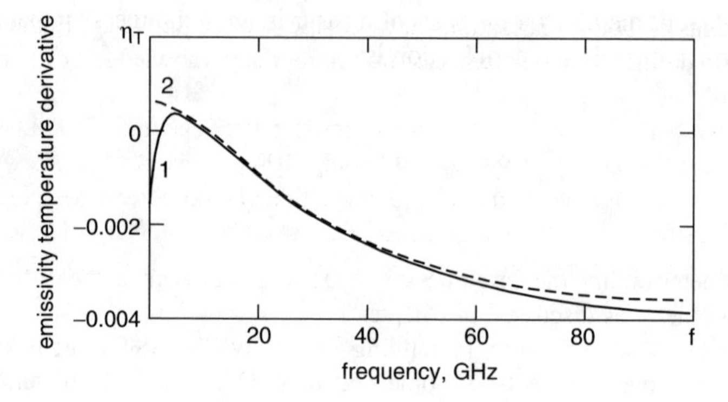

FIGURE 14.8 Dependence of the emissivity temperature derivative on frequency at temperature $t = 20°C$: (1) $S = 35‰$; (2) $S = 1‰$.

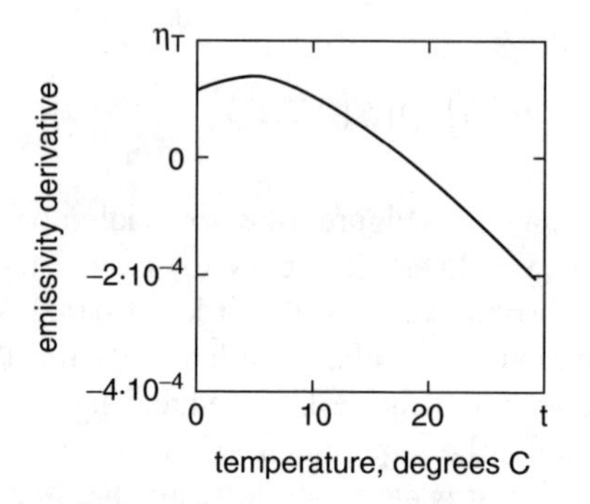

FIGURE 14.9 Temperature dependence of emissivity derivative of seawater for the frequency 2.7 GHz.

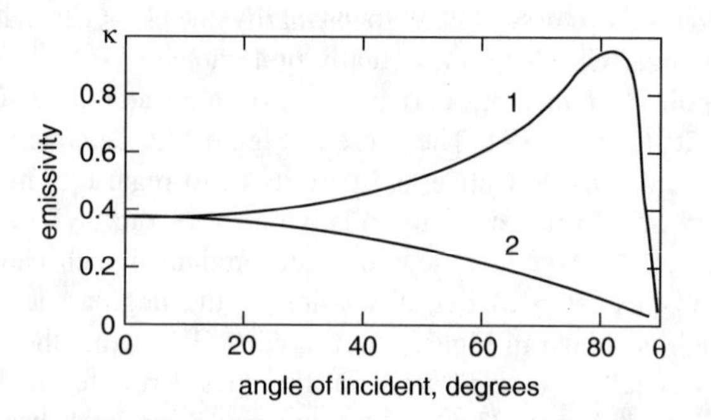

FIGURE 14.10 Angular dependence of emissivity: (1) η_v; (2) η_h.

measurements of both polarizations at a 45° incident angle. Then, Equation (8.20) allows us to define the water temperature without any knowledge of the emissivity coefficients.

Now, we will discuss the effects that interfere with accurate SST measurement by microwave technology. First, we will study the ocean cold film effect. Equation (8.24) can provide the basis for such analysis. Specific numerical computing shows that the first factor of the integrand in Equation (8.24) changes little within a temperature interval of 1°; therefore, Equation (8.25) will be acceptable for future calculations, where the absorption coefficient (Γ) must be assumed to be constant for the same reason as the first factor in Equation (8.24). Assuming the linear depth dependence of the temperature inside of a thin film, we can obtain

$$T_o = \left(1-\left|F\right|^2\right)\left[T(0)+\frac{T(d)-T(0)}{\Gamma d}\left(1-e^{-\Gamma d}\right)\right], \tag{14.45}$$

where d is the film thickness and $T(d)$ is the water temperature at the film lower interface. Obviously, if $\Gamma d \ll 1$ then $T_o = (1-|F|^2)T(d)$; that is, the brightness temperature is defined by the depth. In the opposite case, when $\Gamma d \gg 1$, T_o is expressed by the corrected surface temperature:

$$T_{cor} = T(0)+\frac{T(0)-T(d)}{\Gamma d}. \tag{14.46}$$

Figure 14.11 shows the frequency dependence of attenuation coefficient Γd. It is easy to see that only measurements at the L-band are not influenced by the ocean thin film. In all other cases, the existence of this film must be taken into account.

Now, we need to research the role of water roughness in the formation of microwave thermal emission. The small-scale sea waves and ripples produce a small change in seawater emissivity which leads to a brightness temperature change of the order of 0.5 K.[107] The so-called critical phenomena are discussed in several publications, including Raizer and Cherny.[107] The waves of the second-order approx-

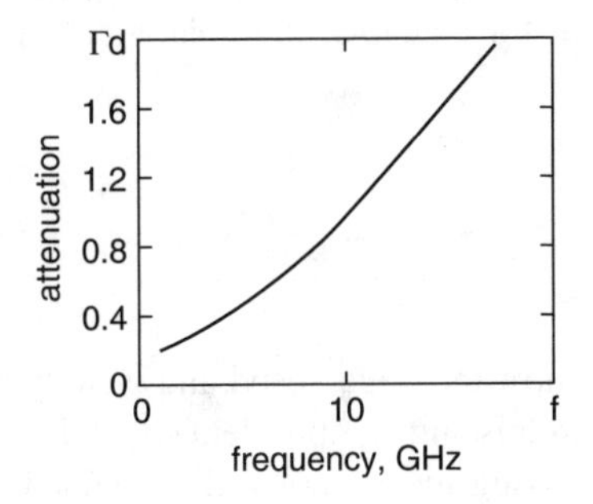

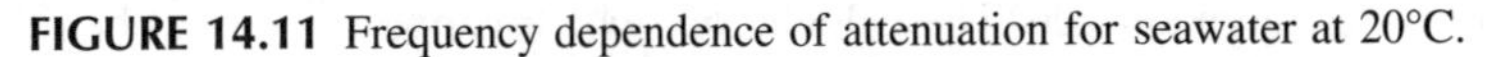
FIGURE 14.11 Frequency dependence of attenuation for seawater at 20°C.

imation in the method of disturbance are under study and are responsible for this unusual emission contrast. We briefly described second-order scattering in Chapter 6 and showed that the effect is especially strong in the case of sinusoidal scattering waves; in that case, it leads to a brightness temperature contrast of the order of 10°.[107] The case of sea roughness is not as clear, however. Our calculations[52] show that the effect does not occur for stochastic surfaces. It is assumed that gravity and capillary waves are not fully stochastic and some periodicity takes place.[108] The analysis given in Reference 107 demonstrates the dependence of brightness temperature contrast on the spectral index of gravity and capillary waves (see Figure 3.15 of Raizer and Cherny's book). The spectral index chosen by us is $n = (16/5) - 1 = 2.2$ (see Equation (14.32)), where n is the spectral index defined in Reference 107, which corresponds to the brightness temperature contrast within an interval of 1 to 3 K from the L-band to the Ku-band for a nadir observation.

More influential is the role of the slopes of the gravity waves; these effects are described by Equations (8.63) and (8.64). Because slopes increase with wind speed, water emissivity and its brightness temperature depend on the wind strength; however, it is interesting to point out that, in the case of vertical polarization, the emissivity does not depend on wind velocity when $\gamma_x = \gamma_y$; that is, the water roughness is isotropic.

This difference between horizontally and vertically polarized emissions agrees with calculations based on the Cox and Munk approximation, Equation (14.30). Figure 14.12 shows the emissivity change vs. wind speed for the nadir direction. The radiowave is polarized along the x-axis. For the y-axis polarization, the contrast will be negative at the chosen parameters of wave slopes. Figure 14.13 demonstrates the varying wind speed dependence for emissivity increments of horizontally and vertically polarized waves. The observation is assumed to be in the wind direction. The numbers in parentheses show the emissivity change depending on wind speed at (1) vertical and (2) horizontal polarization at the angle of observation (45° relative to the nadir), frequency (1 GHz), water temperature (20°C), salinity (35‰). It is easy to see that the emissivity of vertically polarized waves depends weakly on the wind speed. This phenomenon has been verified by experimental data.[90] This almost neglected dependence provides the background for the preferred application of vertically polarized radiation to water temperature measurement. Polarization analysis of microwave sea emissions allows us to estimate the intensity of the sea waves together with wind speed; for example, the polarization coefficient (2.63) reduces in our case to the ratio:

$$m = \frac{\kappa_v - \kappa_h}{\kappa_v + \kappa_h}, \tag{14.47}$$

which is a simple function of the wind speed and weakly depends on the SST.

Let us turn now to addressing the problem of SST measurement by applying Equation (8.20). This tempting idea is not well grounded because Equation (8.20) performs badly for a rough surface. The calculation, which is not carried out here, shows that definition of SST by this relation can lead to an error rate reaching 100%.

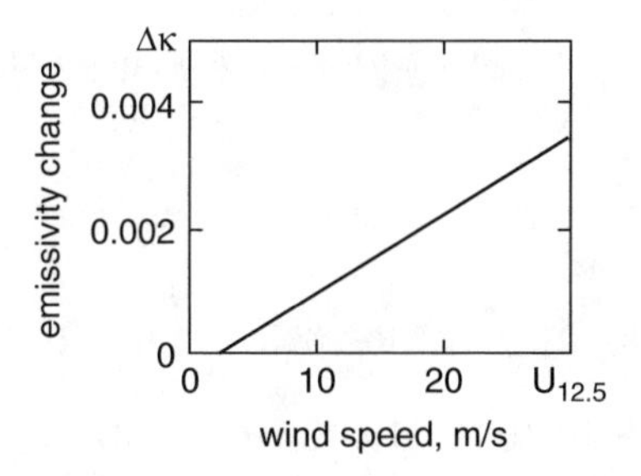

FIGURE 14.12 Emissivity change vs. wind speed for the nadir direction.

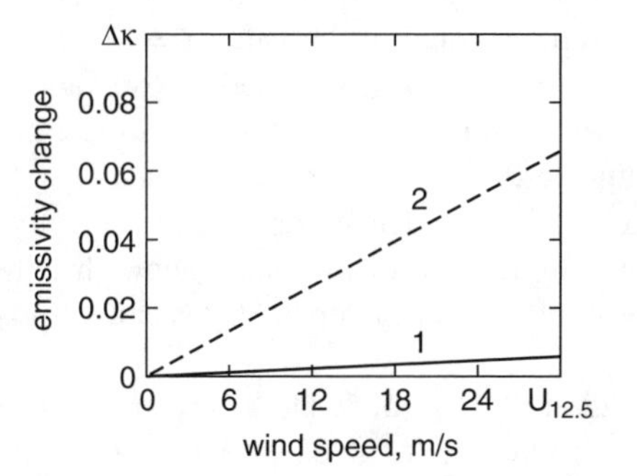

FIGURE 14.13 Wind speed dependence on changes in emissivity: (1) vertically polarized waves; (2) horizontally polarized waves.

Another phenomenon to be addressed is the effect of foam on the seawater emissivity properties. As noted earlier, foam appears when the wind velocity exceeds a value near 4 m/sec. The foam thickness is approximately 1 cm, and its permittivity is usually calculated using formulae for the mixture of air and water.[90,107] From the electrodynamics point of view, foam plays the role of an intermediate layer that partially matches the air and the seawater. This means that the reflection coefficient from the water decreases, and the emissivity of the water and foam system increases compared to foam-free water. The first approximation for the nadir reflection coefficient can be written as:

$$\bar{F}^2 = \left|F^{23}\right|^2 \left(1 - 2\left|F^{12}\right|^2\right) e^{-2\tau} \tag{14.48}$$

in accordance with Equation (3.58), where F^{23} is the coefficient of reflection at the interface of foam and water, F^{12} is the coefficient of reflection at the interface of air and foam, and τ is the attenuation coefficient of the foam.[108] Rather than going into the full details here, we will limit our discussion to approximating the water–foam system emissivity, as proposed by Stogryn:[117]

$$\hat{\kappa}_{h,v}(f,\theta) = \frac{1}{T}(208 + 1.29f)\,\Phi_{h,v}(\theta)\,, \tag{14.49}$$

where

$$\Phi_h(\theta) = 1.0 - 1.748\cdot 10^{-3}\theta - 7.336\cdot 10^{-5}\theta^2 + 1.044\cdot 10^{-7}\theta^3,$$

$$\Phi_v(\theta) = 1.0 - 9.946\cdot 10^{-4}\theta + 3.218\cdot 10^{-5}\theta^2 - 1.187\cdot 10^{-6}\theta^3 + 7.0\cdot 10^{-20}\theta^{10}.$$

The frequency in these formulae is expressed in gigahertz, and the angle of observation is expressed in degrees. The ratio of the water–foam system emissivity to the emissivity of foam-free water is shown in Figure 14.14 for the two frequencies of 10 and 30 GHz. This ratio is equal to the value $\zeta = 1 + (\hat{\kappa}/\kappa - 1)w$, where w is the fraction of the surface covered by foam. Nadir observation is assumed, and the function w is determined by Equation (14.34), where temperature equality is assumed for simplicity. We can see that a foam covering leads to a brightness temperature contrast of several Kelvin degrees depending on wind velocity. This contrast weakly depends on frequency. Some experimental data show that the L-band emission is not as exposed to foam, and frequency dependence is most apparent in this frequency region.

Atmospheric radiation has to be added to the various surface effects. Expressions such as Equation (9.27) must be included in our consideration, which means that atmospheric lighting has to be taken into account to interpret microwave radiometric data. Radio-wave absorption by atmospheric gases and hydrometeors must also be included in the analysis of experimental data. This is especially important when interpreting satellite data. Thus, we must apply multifrequency and polarizing microwave tools to monitor the ocean–atmosphere system.

Sea surface temperature measurement by microwave technology with any height accuracy is a difficult problem waiting for a solution. So, now we will turn to the problem of determining sea salinity. Figure 14.15 shows the frequency dependence of the emissivity derivative with respect to salinity. When multiplied by the sea

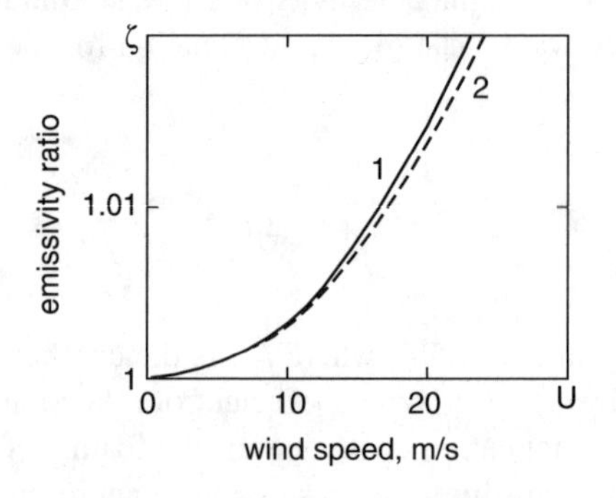

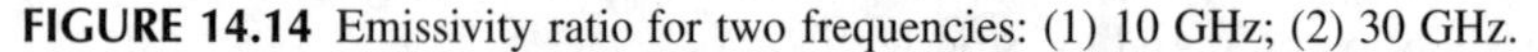
FIGURE 14.14 Emissivity ratio for two frequencies: (1) 10 GHz; (2) 30 GHz.

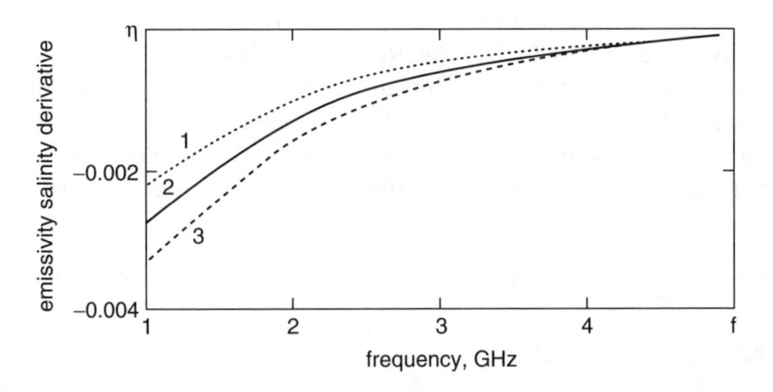

FIGURE 14.15 Frequency dependence of the emissivity derivative with respect to salinity: (1) $\eta_h(\theta = 45°)$; (2) $\eta(\theta = 0°)$; (3) $\eta_v(\theta = 45°)$.

temperature, this derivative determines the sensitivity relative to the salinity measurement by means of microwave radiometry. The median curve η corresponds to the nadir observation, and the other two curves correspond to an incident angle of 45° for both polarizations. It is easy to show that the sensitivity to the salinity change is weak, and only salinity variations of 10°‰ can be detected.

Pollution changes the emissivity of open water. This effect is most apparent in the case of an oil spill. In this case, the oil film takes on the role of a matching layer which increases emissivity. The temperature contrast can reach several tens of degrees, dependently on frequency.

We will now move on to a discussion of sea ice. Flying platforms (aircraft, spacecraft) are becoming more available for sea ice monitoring. A spatial resolution problem appears in this case, so microwave radiometers operating at frequencies above 5 to 10 GHz are suitable for this purpose; however, the extent of radiowave penetration into any type of ice is, as a rule, much less than the ice thickness. This allows us to consider the emissivity of an ice-water system as emissivity of a semi-infinite ice medium. In this case, the emissivities of the FY and MY ice differ due to changes in the ice structure over time (decrease of the ice salinity, drainage of the brine, etc.) and development of the snow cover. So, a two-layer system (ice and snow) must be taken into account. The role of the snow layer is different for the cases of dry and melting snow. Dry snow plays the role of matching layer and increases the ice emissivity by several percentage points. On the other hand, melted snow masks the ice microwave emissions, and the emissivity of the system is defined only by the snow. Experimental data and computer modeling show that emissivity of the FY ice does not depend on frequency, and this dependence is definitely overlooked for MY ice.[90]

This table in Reference 90 shows the emissivity values obtained by the Scanning Multichannel Microwave Radiometer (SMMR) system together with a temperature/humidity infrared radiometer (THIR) mounted on the *Nimbus*-7 satellite. The angle of incidence was 50°. It can be seen that the ice emissivity is sufficiently higher relative to the water emissivity that we can differentiate between open water and ice-covered water by means of microwave measurements. The frequency dependence for MY ice emissivity and the difference between horizontally and vertically polarized waves allow us to distinguish FY from MY ice. Several cluster-type algorithms can be applied. A difficult problem arises when areas of different types of ice and open water appear inside of the radiometer antenna footprint. Such a problem makes some of these algorithms inaccurate. On the other hand, analysis of such mixed radiation allows us to estimate the sea-ice concentration.

15 Researching Land Cover by Radio Methods

15.1 GENERAL STATUS

Land covers are various — open soil, open water, vegetation, forests — and are characterized by both geometrical structure and electro-physical properties that produce a variety of electromagnetic wave interactions with natural objects, thus providing the background for research by microwave remote sensing. Generally, the infrastructures of land objects are comparatively small in geometric size, perhaps a few tens of meters (farming fields, for example); therefore, one of the main problems here, especially in the case of observation from space, is the problem of spatial resolution, which we have mentioned repeatedly. In the case of airborne instruments, the problem of space resolution is not as important for such observations. For observation from space platforms, the synthetic-aperture radar (SAR) systems are more effective for obtaining images with high spatial resolution. Other microwave instruments (for example, radiometers) are widely used, as a rule, for the sounding of statistically homogeneous areas, such as forest tracts, steppes and deserts, and some tundra regions. The situation can be improved by carrying out joint processing of data provided by instruments with different degrees of resolution. Some study could verify the effectiveness of several processing procedures, but this technology has still not found wide application. We will use the following classification of the land covers when we discuss radio methods for remote sensing:

- Bare terrain and geological structures
- Hydrology structures
- Vegetation canopy
- Internal basins
- Snow cover and ice

15.2 ACTIVE RADIO METHODS

Active radio methods are synonymous with radar technology, which is widely employed now due to the development of spaceborne radar techniques. SAR systems have become instrumental for systematic monitoring of the surface of Earth. Radar images reflect many peculiarities of the researched area, such as landscape elements, hydrology network, vegetable canopy, and artificial construction. Gathering this complicated information allows us to assemble special thematic maps: topographic, geologic, hydrologic, forestry, etc. One of the main requirements for radar images is an accurate tie-in to the terrain which is based partly on navigation data and partly on position data of the fixed points. Certainly, knowledge of the appropriate antenna orientation is included on a list of information required for correct radar data interpretation.

In fact, the radar image is a map of the backscattering coefficient, which depends not only on surface scattering of the radiowaves by soil but also on volumetric scattering by elements of a vegetable canopy. To begin, we will address scattering by bare soil and primarily consider the inclined incidence of radiowaves typical for SAR and scatterometer systems. In this case, the separation of large- and small-scale roughness is not as clear as for the ocean; therefore, it is difficult to distinguish between specular and resonant scattering. This is one reason why empirical or semi-empirical models have found wide application for the interpretation of experimental data. Theoretical models and experimental data show the distinct dependence of soil scattering intensity on surface roughness parameters and on soil permittivity, which, in turn, is strongly dependent on soil moisture. The simplest model of soil permittivity can be described by the so-called *refractive formula*:[43,116]

$$\sqrt{\varepsilon} = \xi\sqrt{\varepsilon_w} + (1-\xi)\sqrt{\varepsilon_g} \ . \tag{15.1}$$

Here, ε_w is the water permittivity, described by Equations (14.1) and (14.2); ε_g is the dry ground dielectric constant; and ξ is the water volumetric content (i.e., the volume part occupied by water in mixture). Numerically, this value coincides with volumetric soil moisture m_v (g/cm^3). The absence of a numerical difference indicates that we should not separate these terms. We must be careful when comparing our moisture definition to moisture determined by the gravimetric method with oven drying. The full water content determined by the gravimetric method includes both free water and bound moisture, while electromagnetic waves react only to free moisture. The quantity of bound moisture depends on soil type; it is about 2 to 3% in sandy soils and can reach values of 30 to 40% of dry soil mass in clay and loess grounds. The soil permittivity depends weakly on the soil moisture at small concentration. Equation (15.1) has no theoretical background and is a suitable approximation only in the microwave region;[116] other approximations can be found in Ulaby et al.[90] The permittivity of dry soil depends only on its density in the first approach. This dependence can be expressed in the form:[118]

$$\sqrt{\varepsilon_g} = 1 + 0.5\rho_g \ , \tag{15.2}$$

where ρ_g is the ground density (g/cm^3).

For our discussion here, we will use the following values for the various numerical calculations: $\rho_g = 2$ g/cm^3; $t = 20$°C; $S = 2$‰; $\varepsilon_g = 4$, $\varepsilon_w \cong 80$, and $\sigma \cong 2.4 \cdot 10^9$ CGSE. For many practical applications, however, $\rho = 1.5$ g/cm^3 is more realistic. These values indicate that the behavior of various types of ground is our primary focus. The soil permittivity dependence on moisture is plotted in Figure 15.1. The practical independence of soil permittivity at the C- and L-bands is apparent. This is understandable, as the water permittivity real part is practically constant at these

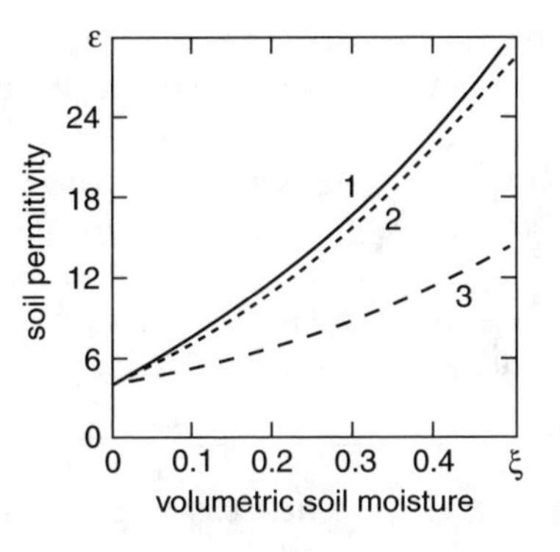

FIGURE 15.1 Average values of soil permittivity dependence on moisture: (1) 1.3 GHz; (2) 5.3 GHz; (3) 37.5 GHz.

frequencies and the imaginary part is small. The frequency dependence occurs closer to millimeter-wavelength bands which is reflected by the 37.5-GHz curve. More detailed analysis can be found, for example, in Ulaby et al.[90] and Shutko.[116]

The predominance of specular or diffuse scattering mechanisms is primarily determined by the radio-wave frequency. The analysis of experimental data provided by Shi et al.[120] gives the values $\sqrt{\langle \zeta^2 \rangle} = 2 \text{ to } 3$ cm for roughness amplitude and $l = 20$ to 30 cm for correlation length. The data of Dierking[121] provided values of 1 to 7 cm for field roughness amplitude and 2 to 37 cm for correlation length. Profilometer measurements allow us to conclude that the exponential autocorrelation function $\rho(x) = \exp\left(-x/l\right)$ is the best approximation of experimental data. Shi et al.[120] tested the autocorrelation function in the form $\rho(x) = \exp\left[-\left(x/l\right)^{\nu}\right]$, where the most probable value of index ν is again unity. More exactly, about 76% of the measured profiles of roughness could be described by the correlation function with ν 1.4. The index difference produces a difference in the spatial spectrum which is proportional to the function:

$$H(\nu,q) = \frac{1}{\nu}\int_0^{\infty} e^{-s} J_0\left(qs^{1/\nu}\right) s^{2/\nu-1} ds, \quad q = 2kl\sin\theta_i \; . \tag{15.3}$$

The curves of Figure 15.2 demonstrate the weak dependence of the spatial spectrum on the index ν value; therefore, it would be enough to be confined by the case ν = 1. This means that we are dealing with fractal surfaces of Brownian type, and the spectrum given by Equation (14.40) is suitable for our purposes.

The data reported by Dierking[121] suggest that many natural surfaces have stationary random processes with a power-law spatial spectrum of the form $\tilde{K}_{\zeta}(q) \approx 1/q^{\alpha}$, where 3 α 3.7, which indicates the fractal character of surfaces

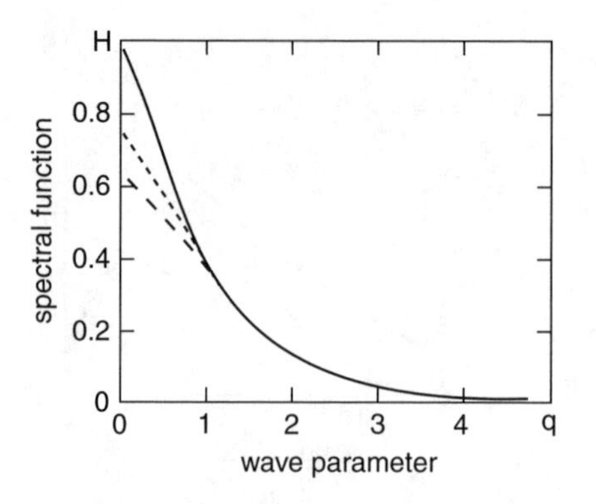

FIGURE 15.2 Graphics demonstrate the weak dependence of spatial spectrum on the index value: ——, $\nu = 1$; ······, $\nu = 1.2$; – – – –, $\nu = 1.4$.

bounded by natural media. The available analytical approximation of these spectra with conservation of roughness magnitude and correlation length can be written as:

$$\tilde{K}_{\zeta}(q) = \frac{(\nu - 1)\langle\zeta^2\rangle l^2}{\left(1 + q^2 l^2\right)^{\nu}}, \qquad \alpha = 2\nu, \tag{15.4}$$

which is similar to the spectra used to describe turbulence (see Chapter 7). The Brownian type of spectrum is from this family.

Now, we are ready to analyze the processes of scattering by the terrain. First, we will consider the P-band waves scattered by bare soil. The parameters specified above allow us to employ the perturbation method approximation; that is, we will base all of our computations on Equations (6.48) and (6.49). Figure 15.3 shows the angular dependence of backscattering coefficients for horizontally and vertically polarized waves of the P-band. It was assumed for our computations that $\xi = 0.2$, $\sqrt{\langle\zeta^2\rangle} = 2\text{ cm}$, and $l = 20$ cm.

Calculation of the backscattering coefficient for higher frequencies cannot be based on an approximation of the perturbation method, because, for example, product $kl > 1$ for the C-band. Unfortunately, no theoretical models have been developed that produce good numerical agreement of computed and experimental data. All models, regardless of soil type, describe the wave scattering processes in forms that allow us only to explain qualitatively some aspects of the phenomenon, primarily because the radio wavelength does not essentially differ from the correlation length of soil surfaces. For this reason, we cannot apply any modern asymptotic approaches of scattering theory. One of the best alternatives is to use semi-empirical models that approximate the experimental data. One of these models is the Oh, Sarabandi, and Ulaby (OSU) model,[122] which provides an expression for the cross-polarized ratio $q = \sigma^0_{hv}/\sigma^0_{vv}$:

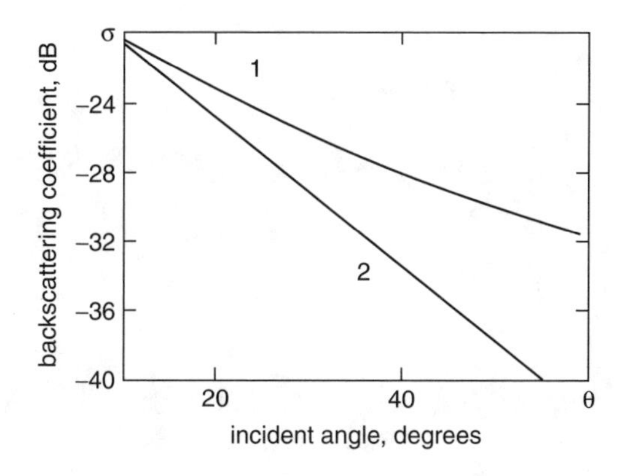

FIGURE 15.3 Angular dependence of backscattering: (1) σ_{vv}; (2) σ_{hh}.

$$q = \frac{\sigma^0_{hv}}{\sigma^0_{vv}} = 0.23\left|F(0)\right|\left[1 - \exp\left(-k\sqrt{\langle\zeta^2\rangle}\right)\right]. \tag{15.5}$$

We must use a two-term subscript now in order to emphasize the fact we are dealing with matched and cross-polarized components of the scattered signal. An equation that applies to the copolarized ratio $p = \sigma^0_{hh}/\sigma^0_{vv}$ is:

$$p = \frac{\sigma^0_{hh}}{\sigma^0_{vv}} = \left[1 - \left(\frac{2\theta_i}{\pi}\right)^{0.33/\left|F(0)^2\right|} \exp(-k\sqrt{\langle\zeta^2\rangle}\right], \tag{15.6}$$

where the incident angle θ_i is expressed in radians. The backscattering coefficient of vertical polarization is approximated by the formula:

$$\sigma^0_{vv} = \frac{g\cos^3\theta_i}{\sqrt{p}}\left[\left|F_h(\theta_i)\right|^2 + \left|F_v(\theta_i)\right|\right], \quad g = 0.7\left\{1 - \exp\left[-0.65\left(k\sqrt{\langle\zeta^2\rangle}\right)^{1.8}\right]\right\}. \tag{15.7}$$

The plots of Figure 15.4 show the dependence of backscattering coefficients on the incident angle at the L- and C-bands. The volumetric moisture value is chosen to be equal to 0.2. The "m" added to the subscript reflects the fact that this model was developed at the University of Michigan. We can see that the backscattering coefficients of horizontal and vertical polarizations differ slightly at the C-band. This weak difference takes place at the chosen parameter of roughness $\sqrt{\langle\zeta^2\rangle} = 2\ \text{cm}$.

Obviously, this difference will be bigger for a smoother surface. This small difference is emphasized by the Kirchhoff (or geometrical optics) approximation (see Chapter 6), which reflects its qualitative correctness. Figure 15.5 shows the

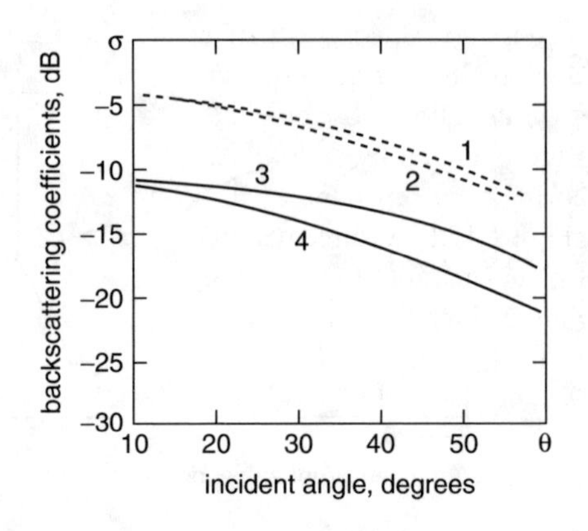

FIGURE 15.4 Dependence m of backscattering coefficients at the L-band (——) and C-band (– – –) on the incident angle: (1) and (3) σ_{vv}; (2) and (4) σ_{hh}.

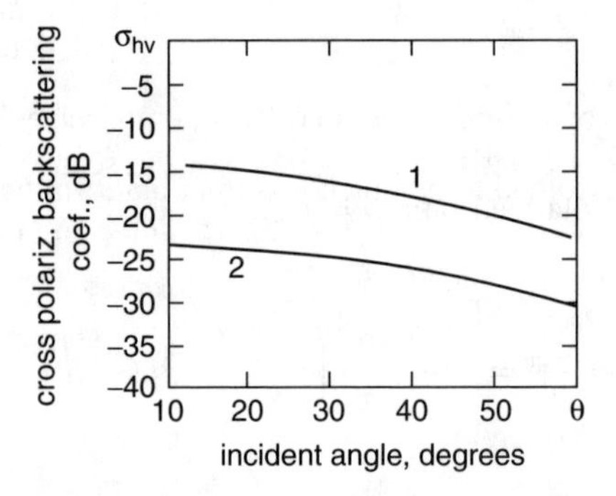

FIGURE 15.5 Cross-polarization coefficient dependence on the incident angle for the (1) L-band and (2) C-band.

dependence of values of the cross-polarization coefficient on the incident angles for the L- and C-bands. The fact that backscattering coefficients are determined relative to only one parameter is an advantage of the OSU model. Recall that, in geometrical optics approximations, the backscattering coefficient depends on only one parameter: the slope.

Large elements of the terrain caused by changes in slope and variations in the roughness parameters are distinguished on radar images by varying brightness. Radar images provide a good representation of the peculiarities of a landscape. This is one of the reasons why radar mapping has found application in geology. The specific method of observation at normal viewing angles from the Earth's surface allows us to detect faintly marked relief elements of slightly rugged terrain such as hills, valleys, etc. Radar maps often have more contrast compared to aerial photographs due to their employment of polarization methods. It is important to note that the use of radar mapping overcomes the screening effect of vegetation to reveal various features of geological structures, including lineaments and circular structures.

Interferometry technology is very effective for mapping landscape details. A brief explanation of this technology is as follows. Imagine that similar radar scenes are obtained by subsequent flights over a neighboring orbit separated by distance d, as shown in Figure 15.6. Assume that the satellite orbits lie at planes parallel to the x-axis and that base d is oriented along the y-axis. Then, let us also assume summation of the signals reflected by any pixel of the surface which is possible due to the high coherency of the radar system itself. The intensity of the summarized signals will depend on their phase difference, which occurs because of the different radar positions. Each reflected signal is noise-like, but, in the case of small distances between the two satellite passes, a high level of coherency between the signals reflected by the same pixel is maintained, and the phase difference has a definite value. More correctly, the discussed phase difference is stochastic but its mean value is not zero.[9] The latter depends on the pixel position and the base size. If the investigated surface is flat, on average, then the lines of constant phase difference will be straight along the flight direction. The values of the x-coordinate are assumed to be much less than platform height H and horizontal distance y from the flight trace and observation point (point O in Figure 15.6). When observing some hill elements above a flat terrain, the equiphase will differ from a straight line and its curvature will depend on the hill topography. It is possible to show (neglecting small values) that equiphase lines are described by the equation:

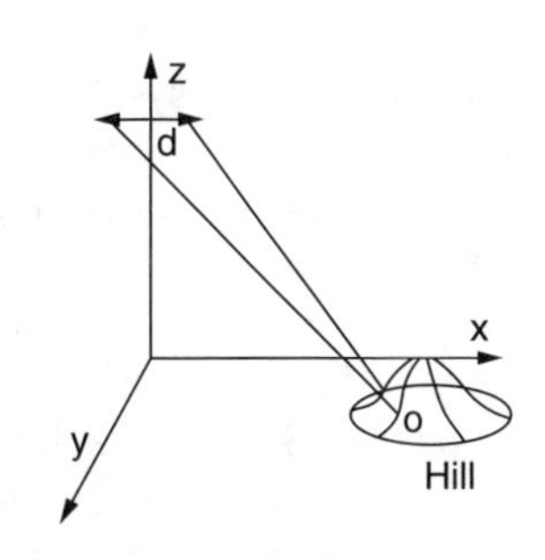

FIGURE 15.6 Interferometry technology for mapping landscape details.

$$\frac{\Delta\varphi}{kd} = \Lambda = \frac{y}{\sqrt{y^2+H^2}} - \frac{2H}{d}\frac{hf(x,y)}{\sqrt{y^2+H^2}}. \tag{15.8}$$

Here, h is the hill height, and $f(x,y)$ is the function describing the hill shape.

The case $h = 0$ corresponds to a flat surface. Having subtracted the first term from the experimental data, we can obtain the equiphase lines related directly to the hill topography. An example of equiphase counters is given by Figure 15.6. The two-pass positions of the radar antenna can be compared to the two-antenna interferometer system, and we can talk about a synthetic interferometer. The real one was realized during the shuttle radar topography mission (SRTM) when the second antennae of the C- and X-band radars were situated at the end of a 60-m boom. This mission provided interferograms within Earth latitudes ±60°.

The processing of interferogram data permits retrieval of a terrain topography with high accuracy. This accuracy is due to the high interferometry sensitivity, particularly when the interferometer lobe angular width is much greater than the angular size of the investigated pixel; that is, $\lambda L/d\cos\theta >> \rho$, where θ is the incident angle. This inequality can be rearranged into $d\cos\theta << 2D$, where D is the synthesis length of the SAR.

Now, we will turn our attention to the problem of surface parameter measurement by means of radar systems. As backscattering coefficients depend only on surface roughness and on the permittivity of the sounded medium, our discussion will focus on measurement of the roughness parameters and the permittivity value. Defining surface roughness parameters is a very important problem for many reasons. For example, pedology is an area where information about the roughness properties is important for our understanding of many processes, such as flooding, infiltration, erosion, etc. Surface roughness is connected with the properties of some materials and this information is of value in geology.

The bare soil backscattering coefficient is determined by both roughness and moisture. One problem with the radar data processing procedure is separation of these effects, but this can be accomplished by polarization measurements. One way to select the roughness effect is to define the correlation coefficient between the signals of orthogonal polarizations.[120] If U_p is the complex amplitude of the received signal of matched p polarization, then the unknown correlation coefficient is defined as:

$$\rho_{pq} = \frac{\left\langle U_p U_q^* \right\rangle}{\sqrt{\left|U_p\right|^2 \left|U_q\right|^2}}, \tag{15.9}$$

where U_q is the signal of the other orthogonal matched polarization. Shi et al.[120] used ρ_{pq} for their calculations, although the value $1/2\mathrm{Re}\rho_{pq}$ is more logical for such applications. Their reason for using ρ_{pq} was that the magnitude of the copolarized correlation coefficient is weakly affected by calibration errors. Analysis of experimental data and theory[120] suggest that the coefficient of correlation between right and left circularly polarized signals depends on the roughness amplitude and weakly reacts to changes in soil moisture. This assumption allows us to consider the correlation coefficient mentioned as being representative of bare soil roughness.

Another parameter frequently under discussion is soil moisture. It is a very important value that plays an essential role in the various phenomena of hydrology, meteorology, climatology, agronomy, etc. Such areas as meteorology and climatology require moisture data on a large spatial scale (even global). Moreover, the soil moisture content is a changeable parameter that must be monitored periodically. These data cannot be obtained by onsite measurements; therefore, remote sensing technology becomes particularly significant. The multiplicity of different terrain areas in the landscape requires relatively high spatial resolution for the sounding tools. Among microwave devices, only SAR, as we noted earlier, satisfies this condition, especially in the case of observation from space. This explains the great interest in soil moisture estimation on the basis of SAR data. The complexity of the radar signal returning from a rough surface makes this a difficult estimation problem.

Figure 15.7 demonstrates the soil moisture sensitivity:

$$\eta_{\mathrm{h,v}}(f,\xi,\theta) = \frac{d\sigma^0_{\mathrm{hh,vv}}}{d\xi}, \tag{15.10}$$

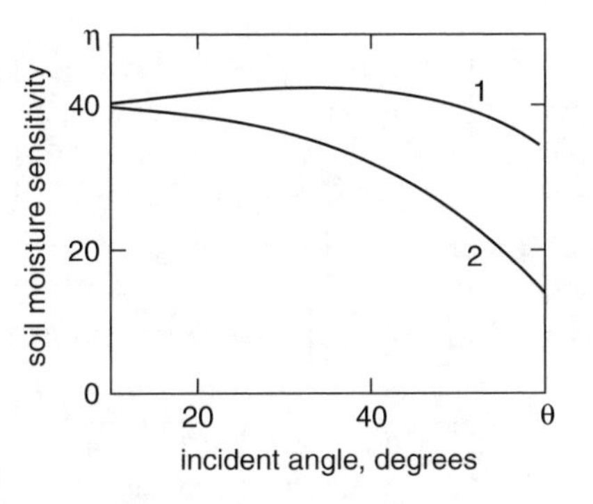

FIGURE 15.7 Angular soil moisture sensitivity: (1) η_v; (2) η_h.

at the L-band for both polarizations as a function of the incident angle. The mentioned sensitivity is expressed in decibels and corresponds to the point $\xi = 0$. This sensitivity is sufficient, especially in the case of vertical polarization. At vertical polarization, a weak maximum of the sensitivity occurs at an incident angle of 40°. Such dependence is the basis for development of soil moisture measurement by radar technology; however, it cannot be the basis for an algorithm to solve the inverse problem (i.e., soil moisture retrieval) because the backscattering coefficient depends on both soil electrophysical properties and the roughness parameters. It is difficult to state the cause of changes in the backscattering coefficient, as they can be the result of a change in roughness or variations in moisture.

It is necessary to know in advance the terrain roughness characteristics in order to evaluate the radar data against the soil moisture value. It is impossible to have such preliminary information on a large scale, so such methods can hardly be considered successful. An investigation conducted at test areas to determine roughness parameters is more likely to help determine the correctness of various scattering models than develop a retrieval algorithm. It is important, then, to have an algorithm of radar data processing that does not take into account the roughness parameters. This is a reason why algorithms based on polarimetric analysis of radar data are more effective.

It is easy to see that, within the framework of the perturbation method, the ratio of the backscattering coefficient does not depend on the roughness spectrum (see Equations (6.48) and (6.49)); that is, the ratio depends only on soil permittivity and angle of incidence:

$$\gamma(f,\xi,\theta) = \frac{\sigma^0_{vv}}{\sigma^0_{hh}} = \left|1 + 2C(\theta)\right|^2 \tag{15.11}$$

This means that P-band radar, for which the perturbation method can be valid, can be used for this kind of measurement. This ratio is shown in Figure 15.8 as a function of volumetric soil moisture at the P-band for incident angles of 30° and 50°.

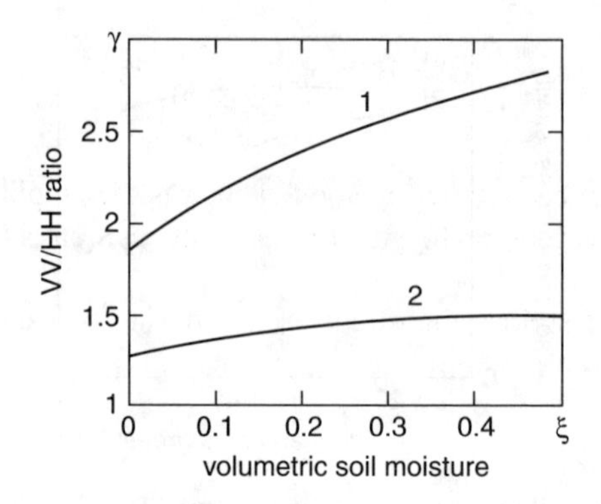

FIGURE 15.8 Soil moisture sensitivity at the L-band for both polarizations as a function of incident angle: (1) 30°; (2) 50°.

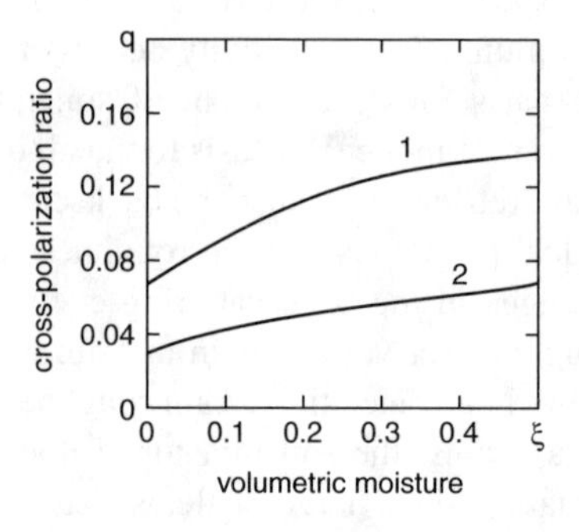

FIGURE 15.9 Volumetric soil moisture ratio for the (1) L-band and (2) C-band.

Obviously, the ratio under discussion is more sensitive to moisture change at large incident angles.

Another way to estimate soil moisture is to determine the cross-polarization ratio (see Equation (15.5)). This ratio vs. volumetric soil moisture is represented in Figure 15.9 for the C- and L-bands. It does not depend on the incident angle in the approximation given by the OSU model. At the C-band, this ratio is more sensitive to moisture content change compared to the L-band. However, this advantage is the seeming one, in the general case, taking into account the scattering and screen effects of vegetation. Before investigating this problem further, we should note that the procedures of polarization data processing presented here reflect only basic approaches to the problem solution; other procedures can be found in the literature.[90,120,124]

For our discussion of soil covered by vegetation, we will first consider grassland. The backscattered signal, in this case, consists of at least five components. The first of them is directly scattered by the soil roughness component (the ground-bounce term) and is attenuated by extinction due to vegetation elements (absorption and spatial scattering). This component can be represented in a very simplified form:

$$\sigma_1^0 = \sigma_{soil}^0 \exp\left(-\frac{4\gamma h}{\cos\theta_i}\right). \tag{15.12}$$

Here, σ_{soil}^0 is the soil backscattering coefficient, γ is the amplitude extinction coefficient, and h is the vegetation height. Two-pass attenuation is taken into account in Equation (15.12).

The next term describes the direct backscattering by vegetation elements. In order to simplify the problem, let us assume similarity of the cross sections of all elements. Based on this assumption:

$$\sigma_2^0 = \frac{\cos\theta_i \sigma_{vg}(-\mathbf{e}_i)}{4\gamma}\left[1 - \exp\left(-\frac{4\gamma h}{\cos\theta_i}\right)\right], \tag{15.13}$$

where $\sigma_{vg}(-\mathbf{e}_i)$ is the backscattering cross section of vegetation per volume unit.

The third term represents ground/grass scattering when the wave reflected by the soil is scattered by vegetation canopy elements (the ground-bounce term). The formula for this term is:

$$\sigma_3^0 = h\left|F(\theta_i)\right|^2 \vec{\sigma}_{vg} \exp\left(-\frac{4\gamma h}{\cos\theta_i}\right), \tag{15.14}$$

where $\vec{\sigma}_{vg}$ is the differential cross section of scattering outward by unit volume of vegetation.

A similar fourth summand corresponds to grass/ground scattering when the waves, initially scattered by the canopy elements, are then reflected by the soil. In this case:

$$\sigma_4^0 = h\left|F(\theta_i)\right|^2 \overleftarrow{\sigma}_{vg} \exp\left(-\frac{4\gamma h}{\cos\theta_i}\right), \tag{15.15}$$

where $\overleftarrow{\sigma}_{vg}$ is the cross section of scattering inward; generally, $\overleftarrow{\sigma}_{vg} \neq \vec{\sigma}_{vg}$.

Finally, the fifth component describes the ground/grass/ground process of scattering (the double-bounce term). The corresponding formula can be represented in the form:

$$\sigma_5^0 = \frac{\cos\theta_i \left|F(\theta_i)\right|^4 \overleftrightarrow{\sigma}_{vg}(-\mathbf{e}_i)}{4\gamma}\left[1 - \exp\left(-\frac{4\gamma h}{\cos\theta_i}\right)\right]\exp\left(-\frac{4\gamma h}{\cos\theta_i}\right). \tag{15.16}$$

Here, the cross section $\overleftrightarrow{\sigma}_{vg}(-\mathbf{e}_i)$ reflects inward vegetation backscattering. In each case, only single scattering by the vegetation elements was considered, and the contribution of multiscattering effects was omitted. It is practically assumed that a lack of soil surface roughness leads to a coherent process of scattering, as reflected by the use of the Fresnel reflection coefficient to describe soil effects.

Forestry areas are characterized by similar terms of radio-wave scattering; however, it is necessary to distinguish more clearly the scattering cross sections of the elements of crown, brushwood, and trunks. Sometimes this is not done, and generalized parameters have to be introduced (e.g., for the cloud model).

The dielectric properties of canopy constituents depend on the water content in the vegetation elements: leaves, stalks, trunks, and branches. It is difficult to justify, theoretically, the development of a procedure to calculate the electromagnetic parameters of complicated elements such as canopy constituents; therefore, we have to resort to experimental data interpolation. Ulaby et al.[125] proposed such an interpolation formula:

$$\varepsilon_{\Phi} = A + B\varepsilon_{w} + C\left[2.9 + \frac{55}{1+\sqrt{-5.556\cdot 10^{-9} if}}\right]. \tag{15.17}$$

Here, ε_w is the water complex permittivity defined by Equations (14.1) and (14.2), where it is necessary to let $S = 0$ and assume that $\sigma = 1.137 \cdot 10^{10}$ CGSE. The coefficients A, B, and C are governed by volumetric moisture content m_v:

$$\begin{aligned} A(m_v) &= 1.7 + 3.2m_v + 6.5m_v^2, \\ B(m_v) &= m_v(0.82m_v + 0.166), \\ C(m_v) &= \frac{31.4m_v^2}{1+59.5m_v^2}. \end{aligned} \tag{15.18}$$

In this considered case, both free water and bound water assume a role. The volumetric moisture content is related to gravimetric moisture content m_g by the equation:

$$m_v = \frac{m_g\rho}{1-(1-\rho)m_g}, \tag{15.19}$$

where ρ is the bulk density of the vegetation material. The volumetric moisture content varies from 0.5 to 0.8.[116]

In principle, knowledge of the canopy element permittivities allows computation of their cross sections of scattering and absorption, as well as determination of radio-wave attenuation in a vegetation canopy and the intensity of the backscattered waves. Several publications have reported attempts to do so. The canopy elements are considered as bodies of simplified shapes (e.g., cylinders of bonded length, thin plates), which allows us to calculate their cross sections in analytical or very simplified numerical forms. It is necessary to include in these calculations the effects of coherency that can occur, especially in the case of a grass canopy,[127,128] when these effects arise during synchronized movement of leaves on the same stalk and on neighboring stalks. This coherency amplifies the effect of scattering. Another coherency effect takes place in a cultivated grass canopy when tillage forms relatively straight, periodic rows that act as a diffraction grating.

Taking into account all of these circumstances leads to a very complex vegetation canopy scattering model. This model often contains many unknown parameters, the determination of which is realized only when the calculated results fit the experimental data. It is difficult to avoid the impression that each such model would have an individual character according to the vegetation species. This is the reason why we do not provide all of the details of such models here, restricting ourselves instead to a rough estimation of radiowave extinction in the vegetation canopy and backscattering processes.

With regard to extinction, we can model a vegetation canopy as a layer of a medium that is a mixture of green mass and air. Then, we can use Equation (15.1) to calculate an approximation of the complex refractive index of this mixture:

$$n_{vg} = 1 - \xi_g\left(\sqrt{\varepsilon_\Phi} - 1\right). \tag{15.20}$$

The imaginary part of this index defines the coefficient of radiowave attenuation (see Equation (2.8)). The results of our computations are plotted in Figure 15.10 for a green mass concentration of $\xi_g = 0.02$ (relatively dense canopy) and two values of volumetric moisture m_v: 0.6 and 0.8. It is easy to see that attenuation in a grass canopy can be considerable and can mask radiowave scattering by the soil, especially in the cases of C- and X-bands. Thus, we can conclude that P- and L-band radar is preferable for soil moisture measurement; with regard to monitoring a grass canopy, the C- and X-band radar is more effective. As a rule, the grass backscattering coefficient is correlated with canopy biomass, and the biomass value is the main product of radar data interpretation. Polarimetric analysis provides an opportunity to distinguish various grass species. Reflections by both ground and vegetation are detected by radar, so, ideally, a multifrequency and fully polarized system that is sensitive to all Stokes components would be best for remote sensing of soil. Such a system can be realized relatively easily on an aircraft platform (for example, the AIRSAR system of JPL) but requires a large satellite platform for monitoring from space. Currently, this has occurred only for a SAR–C/X system during several Shuttle missions.

The application of radar technology for monitoring the state of forests and their dynamic parameters is acquiring greater significance. The complexity of the forest canopy architecture has led, as noted earlier, to attempts to construct very complicated models to describe the radiowave scattering. One of these is the Michigan Microwave Canopy Scattering (MIMICS) model,[125] which divides the canopy into three regions: the crown region, the trunk region, and the underlying ground region. Each region is characterized by its own scattering and absorbing properties and

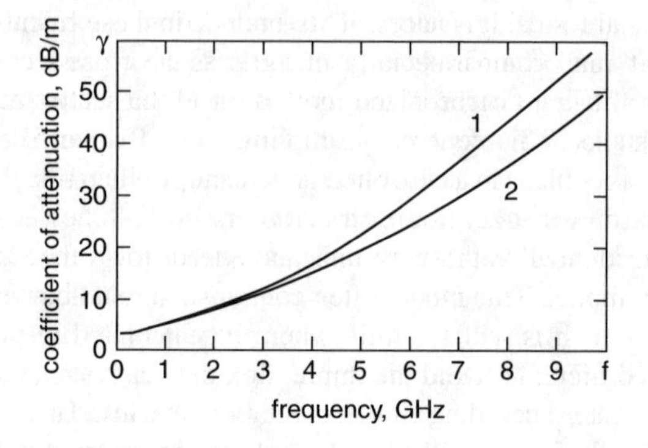

FIGURE 15.10 The coefficient of attenuation for green mass vs. two values of volumetric moisture: (1) $m_v = 0.8$; (2) $m_v = 0.6$.

parameters. The result of radiowave scattering is described in terms of a 4×4 Stokes-like transformation matrix that provides corresponding expressions for backscattering coefficients of matched and cross-polarizations. The effect of forest structure and the biomass on remote sensing is analyzed, for example, in Imhoff[128] for tropical forests.

Kurvonen et al.[129] studied a simpler model for boreal forests, based essentially on the cloud model together with experimental data. Let us introduce the parameters:

$$C_1 = \frac{h\sigma_{\text{vg}}(-\mathbf{e}_{\text{i}})}{V},$$

$$C_2 = -\frac{2\gamma h}{V\cos\theta_{\text{i}}},$$

$$C_3 = \sigma^0_{\text{soil}} + h\left|F(\theta_{\text{i}})\right|^2\left(\vec{\sigma}_{\text{vg}} + \vec{\sigma}_{\text{vg}}\right) + \frac{\left|F(\theta_{\text{i}})\right|^4}{-2C_2V}h\ddot{\sigma}_{\text{vg}}\left(1 - e^{2C_2V}\right),$$

where V is the stem value (biomass; m³/ha). Then, all of the scattering summands are concentrated into two groups:

$$\sigma^0 = \frac{C_1}{-2C_2}\left(1 - e^{2C_2V}\right) + C_3 e^{2C_2V}. \tag{15.21}$$

The first component here represents the direct forest canopy backscattering and the second one is related mostly to ground effects. We can connect the coefficients C_1 and C_2 to the vegetation moisture by the relations:

$$C_1 = \alpha_1 m_v^2, \quad C_2 = \alpha_2 m_v, \tag{15.22}$$

where α_1 and α_2 are scaling factors. This connection is obvious in the case of scatterers that are small compared to wavelength, as the cross section of absorption is proportional to the first degree of their volume and the scattering cross section is proportional to the second degree of this volume (see Chapter 5).

It is easy to see that a saturation effect occurs when $-2C_2V >> 1$, and the backscattering coefficient does not react on the steam volume change. Because the value of coefficient C_2 increases with radiowave frequency, the saturation effect is more remarkable for the C-band than, for example, the L- or P-bands. For this reason, Mougin et al.[130] suggest that the P-band is most promising for biomass measurement. This band is also preferable for soil moisture determination. When $-2C_2V << 1$, the moisture effect will appear to happen twice: the first time due to direct backscattering by the soil itself and the second time due to double-bounce scattering between tree trunks and the ground.

The parameters of Equation (15.22) were determined by Kurvonen et al.[129] by comparing calculated and experimental data with real measurements taken on a terrestrial surface. The SAR experimental data were obtained from multitemporal ERS-1 and JERS-1 images. It was possible to compare the data obtained in different frequency bands, and this data processing has produced values of C_1 that vary from $(8.26 \cdot 10^{-4})$ to $(4.09 \cdot 10^{-3})$ ha/m^3 for the C-band and from $(1.56 \cdot 10^{-3})$ to $(1.86 \cdot 10^{-3})$ ha/m^3 for the L-band, depending on the season. Correspondingly, values of C_2 range from $(-2.51 \cdot 10^{-3})$ to $(-5.99 \cdot 10^{-3})$ ha/m^3 for the C-band and from $(-1.06 \cdot 10^{-3})$ to $(-4.54 \cdot 10^{-3})$ ha/m^3 for the L-band. Finally, values of C_3 range from 0.092 to 0.238 for the C-band and from $(5.73 \cdot 10^{-2})$ to $(8.53 \cdot 10^{-2})$ for the L-band. The saturation levels of the steam value were 125 to 175 m^3/ha for the C-band and 225 m^3/ha for the L-band. The universality of these parameters and their dependence on the incident angle is the subject of future investigation. It would be useful to have special training areas to provide definitions for such parameters for various territories.

The experimental data analysis of Kurvonen et al.[129] could reveal a possible approach to data processing that would achieve a high level of correlation between retrieval and ground truth steam values. These authors have shown the importance of processing data for a block of forest larger than 30 ha and applying the weighted averages for various SAR images. The correlation coefficient at the L-band gives a value of 0.85; for the C-band, this value is about 0.65. This result verifies once more the advantage of the L-band for the measurements discussed here. A more progressive way to measure biomass is to use a multifrequency, multipolarization system. Mougin et al.[130] reported that application of the AIRSAR system significantly improves biomass measurement. The P-band HV and L-band HV have the largest sensitivity to total biomass of mangrove forests.

Another problem is the classification of vegetation type and species. It is not easy to come up with such physical descriptions based on a theory of radiowave scattering for various vegetation canopies. We have already pointed out some examples of such an approach and explained the complexity of these models, but a simpler approach is based on the statistical analysis of scattered signal parameters. An example of recognizing trees types has been provided by Nazarov.[137,138] Their cluster analysis, based on the statistics of backscattering coefficients and their standard deviations, has allowed identification of coniferous, deciduous, and mixed woods. Another type of selection, of arctic forests, is based on different gradations of the backscattering coefficient and is discussed in Proisy et al.[132]

The application of multifrequency, multipolarization radar systems allows us to realize more effective identification. An investigation conducted by the NASA program in 1965 in western Kansas demonstrated the effectiveness of polarization measurement. Generally speaking, the identification procedures are based, to a noticeable degree, on information about the peculiarities of a studied territory and on the experience and qualification of the interpreter. A source of this preliminary information can be data obtained by optical sensors. Certainly, the joint processing of information provides an opportunity for more valid interpretation of remote sensing data. Nevertheless, no reliable algorithms are currently available for vegetation canopy identification based on radar data.

15.3 PASSIVE RADIO METHODS

Passive radio methods do not have as many applications for the investigation of land compared to remote sensing of the atmosphere and ocean. The main reason, as we have said before, is the poor spatial resolution, which makes it difficult, for example, to employ space platforms and high-altitude aircraft. The situation becomes more complicated by the fact that P- and L-bands are preferable for soil research, which magnifies the spatial resolution problem. In general, low-altitude platforms (airplanes and helicopters) are useful primarily for terrain monitoring by microwave radiometry methods.

An important application of the passive technique is soil moisture measurement. In the case of a bare soil, the emissivity is mostly defined by the reflection coefficient, i.e.,

$$\kappa(\varepsilon,\theta) = 1 - \left|F(\varepsilon,\theta)\right|^2, \tag{15.23}$$

where θ is the zenith angle of observation. Because soil permittivity ε depends on moisture (see Equation (15.1)), the soil emissivity and soil brightness temperature are functions of moisture.

More detailed analyses try to consider bounded water, a more accurate mixture formula for soil permittivity, soil roughness, etc.[116,133] The emissivity dependence of soil moisture content computed on the basis of Equation (15.23) is shown in Figure 15.11 for both polarizations for the L-band and an observation angle of 30°. We can see the significant variation of emissivity that leads to a change of soil brightness temperature within an interval of 100 K.

The moisture sensitivity is determined by the derivative:

$$e = \frac{d\kappa}{d\xi}. \tag{15.24}$$

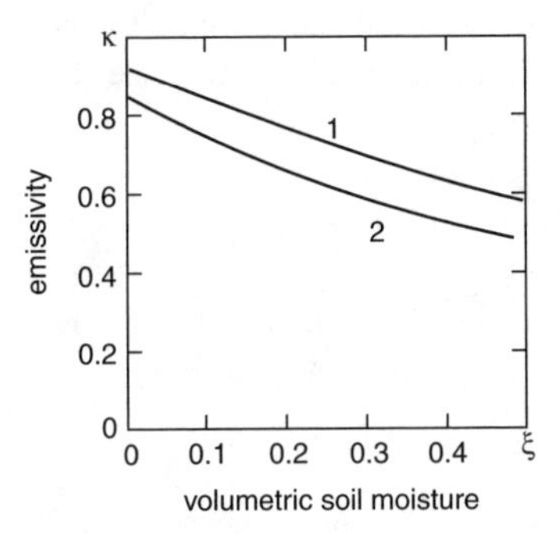

FIGURE 15.11 Emissivity dependence of soil moisture content: (1) $\hat{e}_v$; (2) $\hat{\hat{e}}_h$.

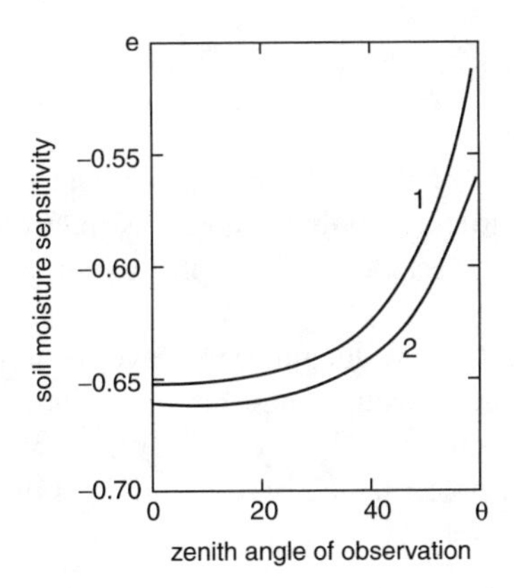

FIGURE 15.12 Moisture sensitivity for horizontal and vertical polarizations: (1) e_v; (2) e_h .

The values of moisture sensitivity for horizontal and vertical polarizations vs. angle of observation are represented in Figure 15.12. The L-band and 0.3 moisture volumetric content are assumed. It is noticeable that the soil moisture sensitivity does not depend on the observation angle within the interval of 30°.

The soil roughness leads, as a rule, to a variation in the brightness temperature of the order of not more than 10 K. This causes an emissivity variation of the order of 0.03 which leads to an error in moisture determination of the order of 0.02 to 0.03 g/cm^3. These last quantities can be regarded as limiting values in the accuracy of bare soil moisture definition.

It is possible to detect subsoil water by microwave radiometry based on the fact that the surface moisture correlates with the depth of the saturated water layer.[116] This correlation is due to the capillary phenomenon which causes flow from the water table to the soil surface. In this case, we are dealing with depth-distributed moisture, which changes the reflection coefficient value compared to the simple Fresnel coefficient. The details of this distribution for various soil types can be find, for example, in Shutko.[116] For some types of soil, the moisture grows smoothly with depth. In this case, Equation (3.101) can be used to define the reflection coefficient with vertical observation. It is necessary to assume that $F_{23}(d) = 0$. Without going into detail, we have:

$$F = F_{12} - \frac{\left(1 - F_{12}^2\right)}{4ik} \frac{d}{dz}\left(\frac{1}{\sqrt{\varepsilon}}\right)_{z=0}$$

and, correspondingly, the emissivity is:

$$\kappa = \kappa_0 + \frac{1}{4ik}\left[\Phi \frac{d}{dz}\left(\frac{1}{\sqrt{\varepsilon}}\right)_{z=0} - \Phi^* \frac{d}{dz}\left(\frac{1}{\sqrt{\varepsilon^*}}\right)_{z=0}\right], \qquad (15.25)$$

$$\Phi = F_{12}^*\left(1 - F_{12}^2\right)$$

where κ_0 is the emissivity of the soil upper layer. The second term of the second formula is the correction to the emissivity caused by the soil moisture depth gradient. This correction can be detected, for example, by employing a two-wavelength radiometry system.

In other cases, a thin (in wavelength scale) layer of moisture is located close to the water table depth. The moisture concentration changes within this layer from the low moisture of the upper horizon of the soil to water-saturated layers. The reflection coefficient can be estimated on the basis of Equation (3.36b), which can be rewritten in our case as:

$$F = F_{12} + F_{23}\left(1 - F_{12}^2\right) e^{2i\varphi(d)} .$$

Here, d is the water table depth, and it is assumed that radiowave absorption in the upper layers of soil makes the term $F_{23}e^{2i\varphi(d)}$ small. The reflection coefficient (F_{23}) can be calculated using Equation (3.117), and the emissivity becomes:

$$\kappa = \kappa_0 - \Phi F_{23} e^{2i\varphi(d)} - \Phi^* F_{23}^* e^{-2i\varphi(d)} . \qquad (15.26)$$

Frequency dependency occurs in this case, a fact that is more likely to be used to estimate water table depth than to determine the moisture profile.[116] In any case, the surface moisture correlates with the water table depth for small depths (Figure 15.13).

The presence of a vegetation canopy changes the general picture of radiation. The radio-wave attenuation in vegetation screens the soil radiation, sometimes partly and at other times fully. The emissions of the vegetation are added to the emissions of the soil. Some secondary effects (as described for calculation of the backscattering coefficient) contribute to the general picture of the microwave radiation of a

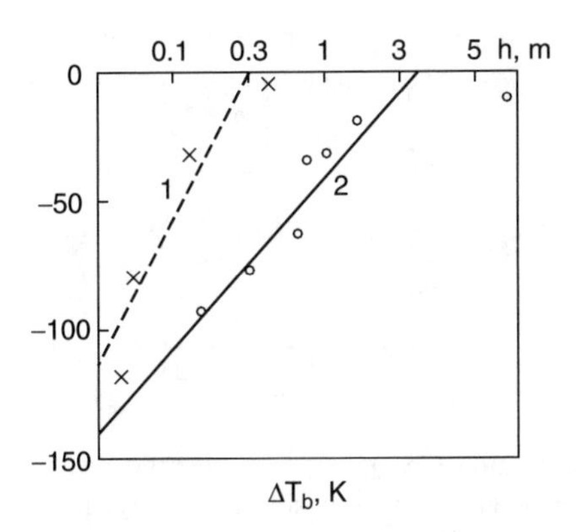

FIGURE 15.13 Correlation of increment of brightness temperature and stratification depth of groundwater: (1) clay loam; (2) sandy loam.

soil-vegetation system. The emissions of the vegetation depend on the albedo of the scatterers. Equation (9.89) represents a simplified description of the emission of a scattering layer of infinite thickness. We can see that in the case of small albedo the thick vegetation cover behaves like a black body; that is, its emissivity is close to unity. This property is typical for forests and small-leaved crops.

The brightness temperature of the soil-vegetation system can be estimated in the first approximation by:[134,151]

$$T_{\mathrm{osv}} = \left[1-\eta(1-e^{-\tau})\right]T_{\mathrm{os}} + \eta(1-\hat{A})\,(1-e^{-\tau})\left[T_{\mathrm{v}} + \kappa_{\mathrm{s}}\left(T_{\mathrm{v}} - T_{\mathrm{s}}\right)e^{-\tau}\right]. \qquad (15.27)$$

Here, η is the degree to which vegetation covers the soil; k_s, T_s, and T_{os} are the emissivity, the temperature, and the brightness temperature of the bare soil, respectively; T_v and $\hat{A}$ are the temperature and the albedo of the vegetation, respectively; and τ is the radiation attenuation in the vegetation layer. The vegetation empty spaces are supposed to be much greater than the radiowavelength. The experimental spectral dependence of the emissivity and straight lines of the regression for various vegetation types are represented in Figure 15.14. The given dependence is in satisfactory agreement with the computed data.

Snow cover is also a subject of investigation by microwave radiometry. Most effective in this case is spaceborne remote sensing, which provides useful information that can be applied to a number of different areas. The scattering processes caused by a grain of snow play significant roles in the formation of snowpack emission; therefore, a theoretical description of snow emission over soil is rather complicated, as shown by the examples provided in Chapter 9. The problem is complicated by the poor spatial resolution; thus, interpreting the behavior of the microwave radiometry pixels can become very complex when they react to a snow area partly covered by forests. This explains why empirical models are frequently employed for retrieving snow characteristics on the basis of microwave radiometry data.

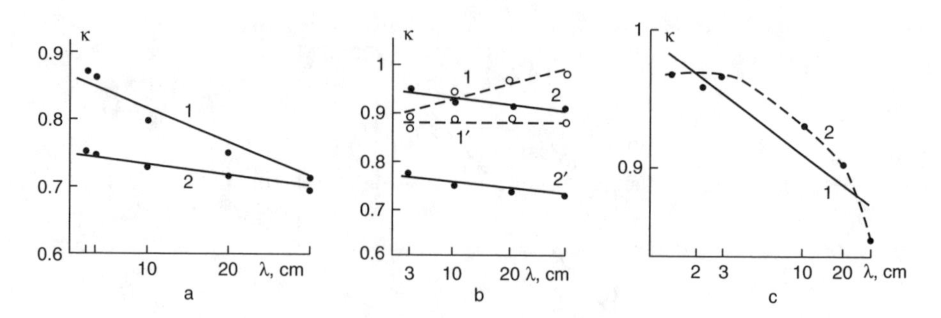

FIGURE 15.14 Spectral relationships of the radiant emittance of various types of vegetative covers: (a) field in winter — (1) annual erypsipelas and (2) tillage; (b) corn on (1) dry and wet (1-inch) soils and (2) dry and wet (2-inch) tillage; (c) underflooding forest — (1) straight line of regression and (2) approximation.

The extent of snow cover and snow water equivalent (SWE) and snow depth are important parameters for various types of application. The extent of snow cover is usually detected by radiometry contrast. Multifrequency and multipolarization observations redouble the validity of such detection. It has been found difficult to distinguish terrain covered by snow from moist soil when using microwave radiometry measurements at one frequency only. For determination of SWE, a regression algorithm is normally used. For example, Special Sensor Microwave/Imager (SSM/I) data have allowed construction of an algorithm for SWE retrieval based on the spectral and polarization difference (SPD):

$$\text{SWE (mm)} = \alpha + \beta\left[2T_{\text{ov}}(19) - T_{\text{oh}}(19) - T_{\text{ov}}(37)\right]. \qquad (15.28)$$

The numbers in parentheses signify that the microwave radiometry channel frequency is expressed in gigahertz. The main problem with this approach is that regression coefficients are not universal and have a regional and sometimes temporal character.

The internal structure of old snow differs from that of fresh snow due to processes of metamorphism; therefore, the emissivity of snow toward the end of winter can change because of changes in the snow albedo (not SWE). This situation has been observed during analysis of SSM/I data.[135] In particular, at 85 GHz, the snow brightness temperature of some territories of the former Soviet Union begins to increase significantly from the middle of February without a change of SWE, which means that not only regional but also temporal changes of regression coefficients in SPD algorithms must be addressed.

Another area of microwave radiometry application is determination of the thickness of freshwater ice in covered lakes. In accordance with the data given in Table 14.1, the attenuation of microwaves in freshwater ice is small, and the penetration depth is estimated by values of the order of 10 m at the C-band. It means that lake ice does not fully absorb radiowaves longer than the C-band waves, and the thermal microwave radiation at these waves depends on the ice thickness.

Ice temperature changes with depth. The temperature of the upper ice interface is close to the temperature of the air, while the temperature of the lower ice boundary does not differ noticeably from that of the water. In general, the ice-water system is a nonequilibrium one; however, this nonequilibrium state is not considered to be a strong one and, for the first approximation, can be reduced to the equilibrium state with some mean temperature. In this quasi-equilibrium approximation, the emissivity of the ice-water system can be expressed via the reflection coefficient of the system. Equation (3.58) can be used in this case, as we are dealing with wide-band radiation. The Fresnel coefficient (F^{12}) is applied, in this case, to the air-ice interface, and F^{23} is applied to the ice-water interface; hence, the emissivity of the system is:

$$\kappa = 1 - F_0^2 = \left(1 - \left|F^{12}\right|^2\right) \frac{1 - \left|F^{23}\right|^2 e^{-2\tau}}{1 - \left|F^{12}F^{23}\right|^2 e^{-2\tau}}, \tag{15.29}$$

where τ is, in the given case, the optical thickness of the ice sheath.

Ulaby et al.[90] calculated the lake ice brightness temperature for various frequencies vs. the ice sheath thickness. They found that the saturation effect appears rather quickly for short waves, which explains the preference of waves longer than those of the C-band for determining ice thickness. Sometimes, it is necessary to also consider the salinity of the lake water.

References

1. Stratton, J.A., *Electromagnetic Theory*, McGraw-Hill, New York, 1941.
2. Jackson, J.D., *Classical Electrodynamics*, John Wiley & Sons, New York, 1962.
3. De Broglie, L., *Problems de Propagations Guidees des Ondes Electromagnitiques*, Paris, 1941.
4. Silver, S., Ed., *Microwave Antenna Theory and Design*, Vol. 12, MIT Radiation Laboratory Series, McGraw–Hill, New York, 1949.
5. Vainshtein, L.A., *Electromagnetic Waves* [in Russian], Sovetskoe Radio, Moscow, 1957.
6. Feld, J.N., *Theoretical Principles of Slot Antennas* [in Russian], Sovetskoe Radio, Moscow, 1948.
7. Ufimtsev, P.J., *Method of Edge Waves in Physical Theory of Diffraction* [in Russian], Sovetskoe Radio, Moscow, 1962.
8. Fock, V.A., *Problems of Diffraction and Propagation of Electromagnetic Waves* [in Russian], Sovetskoe Radio, Moscow, 1970.
9. Levin B.R., *Theoretical Principles of Statistical Radiophysics* [in Russian], Sovetskoe Radio, Moscow, 1969.
10. Born, M. and Wolf, E., *Principles of Optics*, Pergamon Press, London, 1965.
11. Goodman, J.W., *Statistical Optics*, John Wiley & Sons, New York, 1985.
12. Ginsburg, V.L., *Propagation of Electromagnetic Waves in Plasma* [in Russian], Nauka, Moscow, 1967.
13. Gershman, B.N., Eruchimov, L.M., and Yashin, Y.Y., *Wave Phenomena in Ionospheric Plasma* [in Russian], Nauka, Moscow, 1984.
14. Vinogradova, M.B., Rudenko, O.V., and Suchorukov A.P., *Theory of Waves*, Nauka, Moscow, 1979.
15. Fock, V.A., *Theory of Space, Time, and Gravity* [in Russian], GIT-TL, Moscow, 1955.
16. Tonnelat, M.-A., *Les Principes de la Theorie Electromagnetique et de la Relativite*, Masson et Cie, Paris, 1959.
17. Titchmarsh, E.C., *Introduction to the Theory of Fourier Integrals*, 2nd ed., Oxford University Press, London, 1948.
18. Heading J., *An Introduction to Phase–Integral Methods*, John Wiley & Sons, New York, 1962.
19. Froeman, N. and Froeman P.O., *JWKB Approximation*, North-Holland, Amsterdam, 1965.
20. Kamke, E., *Differentialgleihungen, Loesungmethoden und Loesungen*, Leipzig, Germany, 1959.
21. Brekhovskikh, L.M., *Waves in Layered Media* [in Russian], Soviet Academy of Science, Moscow, 1957.
22. Morse, P.M. and Feshbach, H., *Methods of Theoretical Physics*, McGraw-Hill, New York, 1953.
23. Tichonov, A.N. and Arsenin, V.Y., *Methods of Ill–Posed Problems Solution* [in Russian], Nauka, Moscow, 1979.
24. Landau, L.D. and Lifshits, E.M., *Statistical Physics* [in Russian], Nauka, Moscow, 1976, p. 1.

25. Nussenzveig, H.M., *Causality and Dispersion Relations*, Academic Press, New York, 1972.
26. Kratzer, W.F., *Transzendente Funktionen*, Academische Verlagsgesellschaft, Leipzig, Germany, 1960.
27. Kravtsov, Y.A. and Orlov, Y.I., *Geometrical Optics of Inhomogeneous Media* [in Russian], Nauka, Moscow, 1980.
28. Babitch, V. and Buldyrev, V.S., *Asymptotic Methods in Diffraction Problems of Short Waves* [in Russian], Nauka, Moscow, 1972.
29. Norden, A.P., *Differential Geometry* [in Russian], Utchpedgiz, Moscow, 1971.
30. Bean, B.R. and Dutton E. J., *Radio Meteorology*, Dover Publications, New York, 1966.
31. Kolosov, M.A., Armand, N.A., and Yakovlev, O.I., *Radio Propagation in Space Communication*, Svjaz, Moscow, 1969.
32. Yakovlev, O.I., *Radio Propagation in Solar System* [in Russian], Sovetskoe Radio, Moscow, 1974.
33. Zharkov, V.N., *Inner Structure of Earth and the Planets* [in Russian], Nauka, Moscow, 1978.
34. Tatarskiy, V.I., *Propagation of Waves in Turbulent Atmosphere* [in Russian], Nauka, Moscow, 1967.
35. Yakovlev, O.I., *Space Radio Physics* [in Russian], Nauchnaya Kniga, Moscow, 1998.
36. Ginzburg, V.L., *Theoretical Physics and Astrophysics* [in Russian], Nauka, Moscow, 1975.
37. Van de Hulst, H.C., *Light Scattering by Small Particles*, Dover Publications, New York, 1981.
38. Shifrin, K.S., *Light Scattering in Cloudy Medium* [in Russian], GIT-TL, Moscow, 1956.
39. Aivazjan, G.M., *Propagation of Millimetric and Submillimetric Waves in Clouds* [in Russian], Gidrometeoizdat, Leningrad, 1991.
40. Janke, E., Emde, F., and Loesch F., *Tafeln Hoeherer Funktionen*, B.G. Teubner Verlagsgesellschaft, Stuttgart, 1960.
41. Landau, L.D. and Lifshits, E.M., *Electrodynamics of Continuum*, Nauka, Moscow, 1982.
42. Froelich, H., *Theory of Dielectrics*, Clarendon Press, London, 1958.
43. Brown, W.F., *Dielectrics*, Springer-Verlag, Berlin, 1955.
44. Gradstein, I.C. and Ryghik, I.M., *Tables of Integrals, Sums, and Products* [in Russian], GIFML, Moscow, 1962.
45. Crosignani, B., di Porto, P., and Bertolotti, M., *Statistical Properties of Scattered Light*, Academic Press, New York, 1975.
46. Ryasanov, M.I., *Electrodynamics of Condensed State* [in Russian], Nauka, Moscow, 1984.
47. Klimantovitch, Y.L., *Statistical Physics* [in Russian], Nauka, Moscow, 1982.
48. Bass, F.G. and Fucks, I.M., *Wave Scattering on Statistically Inhomogeneous Surface* [in Russian], Nauka, Moscow, 1972.
49. Ishimaru, A., *Wave Propagation and Scattering in Random Media*, Academic Press, New York, 1978.
50. Beckman, P. and Spizzichino, A., *The Scattering of Electromagnetic Waves from Rough Surfaces*, Pergamon Press, London, 1963.
51. Smirnov, V.I., *Course of Higher Mathematics* [in Russian], Vol. 4, GIT-TL, Moscow, 1953.

52. Armand, N.A., Statistical approach to radiowave scattering by sea surface, in *Methods, Procedures, and Facilities for Air-Space Computing Radio Tomography of Near-Surface Earth Areas* [in Russian], Nesterov, S.V., Shamaev, A.S., and Shamaev, S.I., Eds., Nautchnii Mir, Moscow, 1996.
53. Klyatskin, V.I., *Statistical Description of Dynamic Systems with Fluctuating Parameters* [in Russian], Nauka, Moscow, 1975.
54. Shpolsky, E.V., *Atomic Physics* [in Russian], Vol. 2, Nauka, Moscow, 1974.
55. Rytov, S.M., *Theory of Electrical Fluctuations and Thermal Emission* [in Russian], Soviet Academy, Moscow, 1953.
56. Levin, M.L. and Rytov, S.M., *Theory of Equilibrium Fluctuations in Electrodynamics* [in Russian], Nauka, Moscow, 1967.
57. Bogorodsky, V.V. and Kozlov, A.I., *Microwave Radiometry of Earth's Covers* [in Russian], Gidrometeoizdat, Leningrad, 1985.
58. Apresyan, L.A. and Kravtsov, Y.A., *Theory of Radiation Transfer* [in Russian], Nauka, Moscow, 1983.
59. Sobolev, V.V., *Transfer of Radiant Energy in Stars' and Planets' Atmospheres* [in Russian], GIT-TL, Moscow, 1956.
60. Case, K.M. and Zweifel, P.F., *Linear Transport Theory*, Addison-Wesley, Reading, MA, 1967.
61. Cercighnani, C., *Mathematical Methods in Kinetic Theory*, Macmillan, New York, 1969.
62. Gelfand, I.M. and Shilov, G.E., *Generalized Functions and Operation with Respect to Them* [in Russian], GIFML, Moscow, 1954.
63. Musheleshvily, N.I., *Singular Integral Equations*, GIFML, Moscow, 1962.
64. Tricomi, F.G., *Integral Equations*, Wiley-Interscience, New York, 1957.
65. O'Neill, E.L., *Introduction to Statistical Optics*, Addison-Wesley, Reading, MA, 1963.
66. Madelung, E., *Die Mathematischen Hilfsmittel des Physikers*, Springer-Verlag, Berlin, 1957.
67. Leontovich, M.A., On approximated boundary conditions on the surface of well conducting bodies, in *Radio Propagation Research* [in Russian], Vol. II, Wvedensky, B.A., Ed., Soviet Academy, Moscow, 1948.
68. Alpert, Y.L., *Propagation of Electromagnetic Waves and Ionosphere* [in Russian], Nauka, Moscow, 1972.
69. Ahiezer, A.I., Ahiezer, I.A., Polovin, R.V., Sitenko, A.G., and Stepanov, K.N., *Electrodynamics of Plasma* [in Russian], Nauka, Moscow, 1974.
70. Brunelly, B.E. and Namgladse, A.A., *Physics of Ionosphere* [in Russian], Nauka, Moscow,1988.
71. Brunelly, B.E., Kotchkin, M.I., Presnyakov, I.N., Tereshchenko, E.D., and Tereshenko, V.D., *Method of Radio Wave Incoherent Scattering* [in Russian], Nauka, Leningrad, 1979.
72. Evans, J.V., Theory and practice of ionosphere study by Thomson scatter radar, *Proc. IEEE*, 57, 496, 1969.
73. Beynon, W.J.G. and Williams, P.J.S., Incoherent scatter of radio waves from the ionosphere, *Rep. Prog. Phys.*, 41(6), 909, 1978.
74. Kunitsyn, V.E. and Tereshchenko, E.D., *Tomography of the Ionosphere* [in Russian], Nauka, Moscow, 1991.
75. Kunitsyn, V.E. and Tereshchenko, E.D., Radio tomography of the ionosphere, *IEEE Antennas Propagation Mag.*, 34(5), 22, 1992.

76. Roettger, J., Radar systems in ionospheric research, in *Modern Radio Science*, Oxford University Press, London, 1999, p. 213.
77. King, R.W. and Wu, T.T., *The Scattering and Diffraction of Waves*, Harvard University Press, Cambridge, MA, 1959.
78. Elder, J., *Fractals*, Plenum Press, New York, 1988.
79. Fung, A.K., *Microwave Scattering and Emission Models and Their Applications*, Artech House, Norwood, MA, 1994.
80. Chen, K.S., Wu, T.D., Tsay, M.K., and Fung, A.K., A note on multiple scattering in an IEM model, *IEEE Trans. Geosci. Remote Sensing,* 38(1), 249, 2000.
81. Evans, J.V. and Hagfors, T., Eds., *Radar Astronomy,* McGraw-Hill, New York, 1968.
82. Grane, R.K., Fundamental limitations caused by RF propagation, *Proc. IEEE*, 59, 196, 1981.
83. Himmelblau, D.M., *Process Analysis by Statistical Methods*, John Wiley & Sons, New York, 1970.
84. Ryzin, V., Ed., Classification and clustering, in *Proc. of Advanced Seminar Conducted by the Mathematics Research Center of the University of Wisconsin at Madison,* Academic Press, New York, 1977.
85. Sokolov, A.V. and Sukchonin E.V., Millimetric wave attenuation in the atmosphere [in Russian], *Adv. Sci. Technol.*, 20, 107, 1980.
86. Deirmendjian, D., *Electromagnetic Scattering on Spherical Polydispersions*, Elsevier, New York, 1969.
87. Stepanenko, V.D., *Radar Meteorology* [in Russian], Hydrometeoizdat, Leningrad, 1966.
88. Laws, J.O. and Parsons, D.A., The relationship of raindrop size to intensity, in *Transactions of the American Geophysical Union 24th Annual Meeting*, April, 1943, p. 452.
89. Melnic, Y. et al., *Radar Technique for Earth Research* [in Russian], Sovetskoe Radio, Moscow, 1980.
90. Ulaby, F.T., Moore, R.K., and Fung, A.K., *Microwave Remote Sensing*, Addison-Wesley, Reading, MA, 1981.
91. Atlas, D., Landsberg, H.E., and Mieghem, V., Advances in radar meteorology, *Adv. Geophys.*, 10, 318, 1964.
92. Abshaev, M.T., Burtsev, I.I., Vaxenberg, S.I., and Shevela, G.F., *Instructions for Application of MRL-4, MRL-5, and MRL-6 Radar in Hail Prediction Systems* [in Russian], Hydrometeoizdat, Leningrad, 1980.
93. Stepanenko, V.D., Shchukin, G.G., Bobilev, L.P., and Matrosov, S.Y., *Radio Radiometry in Meteorology* [in Russian], Hydrometeoizdat, Leningrad, 1987.
94. Mitnik, L.M., Determination of effective temperature of liquid clouds by atmospheric emission at MW region: satellite meteorology [in Russian], *Proc. Hydrometeocenter*, 148, 115–125, 1974.
95. Popova, N.D. and Shchukin G.G., Study of the opportunity for joint application of passive–active radars for precipitation rate definition, in *Proc. of VII All-Union Conference on Radio Meteorology* [in Russian], Hydrometeoizdat, Leningrad, 1989, p. 20.
96. Rosenberg, V.I., *The Scattering and Attenuation of Electromagnetic Emission by Atmospheric Particles* [in Russian], Goshydromet, Leningrad, 1972.
97. Rabinovitch., J.A. and Melentjev, V.V., Influence of temperature and salinity on the emission of plane water surface at cantimetric wave band [in Russian], *Proc. GGO*, 235, 78, 1970.
98. Stogryn, A., Equations for calculating the dielectric constant for saline water, *JEEE Trans. Micr. Theor. Technol.*, 19(8), 733, 1971.

99. Pounder, E.R., *The Physics of Ice*, Pergamon Press, London, 1956.
100. Bogorodsky, V.V. and Gavrilo, V.P., *Ice* [in Russian], Hydrometeoizdat, Leningrad, 1980.
101. Odelevsky, V.I., The calculation of conductivity generalized of heterosystems, *JTPh*, 11(6), 1951.
102. Zaslavsky, M.M. and Monin, A.S., Wind-waves, *Oceanology*, 2, 146, 1978.
103. Monin, A.S. and Krasitsky, V.P., *The Phenomena on the Ocean Surface* [in Russian], Hydrometeoizdat, Leningrad, 1985.
104. Karaev, V.Ju. and Balandina, G.N., Modified wave spectrum and remote sensing of ocean, *Issledovanie Zemli iz Kosmosa*, 5, 45, 2000.
105. Hasselman, K., Barnet, N.R., Bouws E., et al., Measurement of wind-wave growth and swell during the Joint North Sea Wave Project (JONSWAP), *Dt. Hydrodr. Z. Reihe A*, 8(12), 95, 1984.
106. Cox, C. and Munk, W., Slopes of the sea surface deduced from photographs of sun glitter, *Bull. Scripps Inst. Oceanogr.*, 6, 401, 1956.
107. Raizer, V. Ju. and Cherny, I.V., *Microwave Diagnostics of the Ocean Surface* [in Russian], Hydrometeoizdat, St. Petersburg, 1994.
108. Monahan., E.C. and O'Muircheartaigh, I.G., Optimal power-law description of oceanic whitecap coverage dependence on wind speed, *J. Phys. Oceanogr.*, 10(12), 2094, 1980.
109. Aksenov, V.N. and Hundgua, G.G., Cold film and temperature of the ocean surface[in Russian], *Space Physical Geography*, Mguizdat, 1992.
110. Brown, G.S., The average impulse response of a rough surface and its application, *IEEE J. Oceanogr. Eng.*, OE-2(1), 67, 1977.
111. Vasljeva, A.B. and Tichonov, N.A., *Integral Equations* [in Russian], Moscow University, Moscow, 1989.
112. Steffen, K. and Heinricks, J., Feasibility of sea typing with synthetic aperture radar (SAR): merging of Landsat thematic mapper and ERS-SAR satellite imagery, *J. Geophys. Res.*, 99(C11), 22, 413, 1994.
113. Wagapov, R.H., Gavrilov, V.P., Kozlov A.I., et al., *Remote Research of Sea Ice* [in Russian], Hydrometeoizdat, St. Petersburg, 1993.
114. Mitnic, L.M. and Viktorov, S.V., Eds., *Radar Investigation of the Earth's Surface from Space* [in Russian], Hydrometeoizdat, Leningrad, 1990.
115. ICEWatch, *Real-Time Sea Ice Monitoring of the Northern Sea Route Using Satellite Radar Technology: Final Report*, Technical Report No. 113, Nansen Environmental and Remote Sensing Centre (NERSC), Bergen, Norway, 1997.
116. Shutko, A.M., *MW Radiometry of Water Surface and Soil* [in Russian], NAUKA, Moscow, 1986.
117. Stogryn, A., The emissivity of sea foam at microwave frequencies, *J. Geophys. Res.*, 77(9), 1658, 1988.
118. Kroticov, V.D., Some electrical characteristics of Earth's rocks and comparison with characteristics of the upper layer of the moon [in Russian], *Izvestia Vuzov, Radiofizika*, 5, 1057, 1962.
119. Mattia, F., LeToan, T., Souyris, J.C., De Garolis, G., Floury, N., and Posa, F., The effect of surface roughness on multifrequency polarimetric SAR data, *IEEE Trans. Geosci. Remote Sensing*, 35(4), 954, 1997.
120. Shi, J., Wang, J., Hsu, A.Y., O'Neil, P., and Engman, E.T., Estimation of bare surface soil moisture and surface roughness parameter using L-band SAR image data, *IEEE Trans. Geosci. Remote Sensing*, 35(5), 1254, 1997.

121. Dierking, W., Quantitative roughness characterization of geological surfaces and implications for radar signature analysis, *IEEE Trans. Geosci. Remote Sensing*, 37(5), 2397, 1999.
122. Oh, Y., Sarabandi, K., and Ulaby, F.T., An empirical model and an inversion technique for radar scattering from bare soil surfaces, *IEEE Trans. Geosci. Remote sensing*, 30(2), 370, 1992.
123. Tough, R.A., Blacknell, D., and Quegan, S.A., A statistical description of polarimetric and interferometric synthetic aperture radar data, *Proc. R. Soc. London Ser. A*, 449, 567, 1995.
124. Oevelen, P.J. and Hoekman, D.H., Radar backscatter inversion techniques for estimation of surface soil moisture: EFEDA-Spain and HAPEX-Sahel case studies, *IEEE Trans. Geosci. Remote Sensing*, 37(1), 113, 1999.
125. Ulaby, F.T., Sarabandi, K., Mc Donald, K., Whitt, M., and Dobson, M.C., Michigan microwave canopy scattering model, *Int. J. Remote Sensing*, 11(7), 1223, 1990.
126. Stiles, J.M. and Sarabandi, K., Electromagnetic scattering from grassland. Part I. A fully phase-coherent scattering model, *IEEE Trans. Geosci. Remote Sensing*, 38(1), 339, 2000.
127. Stiles, J.M., Sarabandi, K., and Ulaby, F.T., Electromagnetic scattering from grassland. Part II. Measurement and modelling results, *IEEE Trans. Geosci. Remote Sensing*, 38(1), 349, 2000.
128. Imhoff, M.L., A theoretical analysis of the effect of forest structure on synthetic aperture radar backscatter and the remote sensing of biomass, *IEEE Trans. Geosci. Remote Sensing*, 33(2), 341, 1995.
129. Kurvonen, L., Pullianen, J., and Hallikainen, M., Retrieval of biomass in boreal forests from multitemporal ERS-1 and JERS-1 SAR images, *IEEE Trans. Geosci. Remote Sensing,* 37(1), 198, 1999.
130. Mougin, E., Proisy, C., Marty, G., Fromard, F., Puig, H., Betoulle, J.L., and Rudant, J.P., Multifrequency and multipolarization radar backscattering from mangrove forests, *IEEE Trans. Geosci. Remote Sensing*, 37(1), 94, 1999.
131. Polyakov, V.M., Measurements of absolute thermodynamic temperatures of biological objects by a radiophysical method [in Russian], *Radioteknica*, 8, 88, 1998.
132. Proisy, Ch., Mougin, E., and Le Dantec, V., Monitoring seasonal changes of a mixed temperate forest using SAR observations, *IEEE Trans. Geosci. Remote Sensing*, 38(1), 540, 2000.
133. Komarov, S.A. and Mironov, V.L., *Microwave Sounding of Soils* [in Russian], Scientific Publishing Center of Siberian Branch of R.A.S., Novosibirsk, 2000.
134. Basharinov, A.E., Gurvich, A.S., and Yegorov, S.T., *Radiation of the Earth as a Planet* [in Russian], Scientific Publishing House, Moscow, 1974.
135. Rosenfeld, N.S. and Grody, C., Metamorphic signature of snow revealed in SSM/I measurement, *IEEE Trans. Geosci. Remote Sensing*, 38(1), 53, 2000.
136. Livingstone, C.E., Sinch, K.P., and Gray, A.L., Seasonal and regional variations of active/passive microwave signatures of sea ice, *IEEE Trans. Geosci. Remote Sensing*, GE-25(9), 159, 1987.
137. Nazarov, L.E., Application of multilayer artificial neural networks for type classification using SAR images [in Russian], *Issledovanie Zemli iz Kosmosa*, 3, 63, 2000.
138. Nazarov, L.E., Forest type classification algorithms based on analysis of SAR images [in Russian], *Issledovanie Zemli iz Kosmosa*, 4, 56, 1999.
139. Uberla, K., *Faktoranalyse*, Springer-Verlag, Berlin, 1977.
140. Mudrov, V.I. and Kushko, V.L., *Methods of Measurements Processing* [in Russian]. Sovetskoe Radio, Moscow, 1974.

141. Faddeev, D.K. and Faddeeva, V.N., *Computing Methods of Linear Algebra* [in Russian], Physmatgis, Moscow, 1963.
142. Fedorov, V.V., *The Theory of Optimum Experiment* [in Russian], Nauka, Moscow, 1971.
143. Turchin, V.F., Kozlov, V.P., and Malkevich, M.C., Use of statistics methods for the solution of ill-posed problems [in Russian], *Successes Phys. Sci.,* 102(3), 345, 1970.
144. Skolnik, M.I., Ed., *Radar Handbook*, McGraw-Hill, New York, 1970.
145. Chelton, D.B., Walsh, E.J., and MacArthur, J.L., Pulse compression and sea level tracking in satellite altimetry, *J. Atm. Oceanic Technol.,* 6, 407, 1989.
146. Kramer, H.J., *Observation of the Earth and Its Environment: Survey of Mission and Sensors*, German Remote Sensing Data Center DLD–DFD, Oberpfaffenhofen, 1993.
147. Finkelshtein, M.I., Mendelson, V.L., and Kutev, V.A., *Radiolocation of Plane-Layered Earth Covers* [in Russian], Sovetskoe Radio, Moscow, 1977.
148. Finkelshtein, M.I., Kutev, V.A., and Zolotorev, V.P., *Application of Radar Subsurface Sounding in Engineering Geology* [in Russian], Nedra, Leningrad, 1986.
149. Bjerkaas, A.W. and Riedel, F.W., *Proposed Model for the Elevation Spectrum of a Wind-Roughened Sea Surface*, Applied Physics Lab. Report TG 1328, The Johns Hopkins University Press, Baltimore, MD, 1979.
150. Ludeke, K.M., Schick, B., and Kohler, J., Method and Arrangement for Measuring the Physical Temperature of an Object by Means of Microwaves, U.S. Patent 4,235,107, Cl.3, G01J5/52, 1980.
151. Kirdiashev, K.P., Chukhlantsev, A.A., and Shutko, A.M., Microwave radiation of the Earth's surface in the presence of vegetation cover, *Radio Eng. Electron. Phys. Transl.*, 24, 256, 1979.

Index

A

B

C

D

E

L

M

N

O

P

Q

R

S

T

V

W